THE SHORTER
BERGEY'S MANUAL OF

DETERMINATIVE BACTERIOLOGY

THE SHORTER BERGEY'S MANUAL OF DETERMINATIVE BACTERIOLOGY

Eighth Edition

JOHN G. HOLT
EDITOR

SANS TACHE
WILLIAMS & WILKINS
Baltimore • Hong Kong • London • Sydney

Accurate indications, adverse reactions, and dosage schedules for drugs are provided in this book, but it is possible that they may change. The reader is urged to review the package information data of the manufacturers of the medications mentioned.

Made in United States of America

Library of Congress Cataloging in Publication Data Main entry under title:

The Shorter Bergey's manual of determinative bacteriology, eighth edition.
Bibliography:
Includes index.
1. Bacteriology—Classification. 2. Bacteriology—Terminology. 1. Bergey, David Hendricks, 1860–1937. II. Holt, John G. III. Bergey's manual of determinative bacteriology. [DNLM: 1. Bacteria—Classification. 2. Bacteria—Terminology. QW4 S559]
QR81.A5 1977 589.9′001′2 76-30898
ISBN 0-683-04105-3

The illustration on the cover is one of the Myxobacteria, *Stigmatella aurantiaca (Stigmatella media)* (× 55), kindly supplied by Dr. Howard McCurdy, and reproduced by permission from the *Canadian Journal of Microbiology 15:* 1453–1461, Figure 20, 1969.

89 10 9 8

Preface to *The Shorter Bergey's Manual of Determinative Bacteriology*, Eighth Edition

This is not the first time that *Bergey's Manual of Determinative Bacteriology* has been produced in an abridged form. In 1948 the editors compiled and published an abridged edition of the Sixth Edition of the *Manual,* especially "... for the use of students in identifying the cultures of bacteria used in beginning courses. . . ." The Board of Trustees of Bergey's Manual Trust has decided once again to publish an abridged version of the *Manual* for students and laboratory workers who desire a more portable, less monetarily valuable compendium of information useful for the identification of bacteria.

Briefly, *The Shorter Bergey's Manual* contains the outline classification of the bacteria; the descriptions of all genera and higher taxa; all of the keys and tables for the diagnosis of species; the glossary; all of the illustrations; and two of the introductory chapters. The *Shorter* version does not contain the descriptions of species; most of the taxonomic comments; the etymology of names; and references to authors. This latter omission brings up a special point about reference to the *Manual* in published works. All the material in the complete *Manual* is the work of many contributors who are identified by obvious bylines at the beginning of each section. The *Shorter* version is derivative of the larger book and often has material added and much original material deleted in the abridgment process. We believe that it is not appropriate to refer to the *Shorter* version in a scientific publication; rather, reference should always be to the complete *Manual*. Additionally, reference to the *Manual* should be to the *author* of the section being noted and not to the editors (Buchanan and Gibbons in the case of the 8th edition). A reminder and example of the proper form of citation is given throughout the text of the *Shorter* version.

The reader is well advised to read the introductory chapter on "The Mechanism of Identification" before starting out to use either form of the book as a guide for identification. Use of the criteria and suggestions for methodology outlined in that chapter will often lead the reader to the proper Part. There is also a short key to the 19 Parts which follows the chapter to provide an alternative access to the proper Part.

Once the proper placement in a Part has been made, the reader should study the keys and tables in that Part to determine which additional tests will need to be performed in order to use the identification devices. Key characteristics are listed for the reader's convenience for the longer, more involved keys. Most keys are short and the characteristics used are quite obvious. A word about methods is needed here. There are few standardized methods in use in bacteriology, and each taxon or major group may be classified on characteristics which are based on very specific methods. Usually, in the *Shorter* version, reference is made to methodology if it differs from the usual laboratory practice. The reader should consult the text before planning further tests.

Sometimes synopses are used instead of dichotomous keys. A synopsis is not a key; i.e., it does not offer the user alternative statements that logically lead to a determination of identity of an unknown organism. Rather, a synopsis is a device which gives abbreviated descriptions for a group of related taxa. Some, fortunately not many, of the original keys in the *Manual* are synopses; this is a reflection of a paucity of taxonomic data, which does not allow a logical classification. One must remember that an identification device is only as good as the classification upon which it is constructed and that in many groups of bacteria there still is not a good, logical classification possible. The goal of the *Manual,* of course, is to continue to present the latest thinking on the classification of all of the bacteria and the best keys and tables which can be constructed from these classifications.

All species in a genus are mentioned in some way (with a few exceptions): by key, table, or editorial note. If there is no mention of constituent species, then the genus is monotypic and only the type species will be given. Once a determination has been made, say to the level of species, it may be necessary to consult the *Manual* for further information. This *Shorter* version does not contain detailed descriptions of species, except in the case of some tables, and the reader will have to return to the *Manual,* which is, it is hoped, close by in a safe place, to check the more complete description for confirmation. Or, one may need to consult the *Manual* for more information on where one went wrong or for a lengthy discussion of the taxonomic problems of the group. The *Manual* is liberally salted with "further comments" by the authors on classification or identification problems.

There has been an addition of some new material to this *Shorter* version of the *Manual*. Occasionally there were no keys to the constituent sections or families in a Part or there was no way to gain access to the genera of uncertain affiliation. To correct these situations new keys or synopses have been added that should help to provide better determinations of some of the groups. Also, there have been a number of new genera described since the completion of the *Manual,* and many of these have been noted in the proper place.

JOHN G. HOLT

Acknowledgments

The *Bergey's Manual* Trustees wish to acknowledge the following individuals who contributed to *Bergey's Manual of Determinative Bacteriology*, 8th edition, and whose material was synopsized, providing much of the basis for this book.

O. N. Allen†
T. V. Aristovskaya
H.-D. Babenzien
A. C. Baird-Parker
Elio Baldacci
J. H. Becking
E. L. Biberstein
C. E. Bland
W. J. Brinley-Morgan
Thomas D. Brock
Marion A. Brooks
M. P. Bryant
R. E. Buchanan†
Jeffrey C. Burnham
L. Leon Campbell
Ercole Canale-Parola
Patricia Carpenter†
J. Geoffrey Carr
Sylvia G. Cary
J. N. Couch
S. T. Cowan†
Koby T. Crabtree
Tom Cross
C. S. Cummins
R. H. Deibel
J. De Ley
Robert B. Dienst
M. Doudoroff†
Douglas W. Dye
James B. Evans
Oscar Felsenfeld
Joseph Frateur†
E. A. Freundt
Ellen I. Garvie
Lucille K. Georg
N. E. Gibbons
Thomas Gibson†
Marion Gilmour
Morris A. Gordon
Ruth E. Gordon
David Gottlieb
Wilma Kane Hanton
Margaret S. Hendrie
Aino Henssen
Peter Hirsch
Geoffrey Hobbs
Lillian V. Holdeman
A. J. Holding
John G. Holt
D. B. Johnstone
D. C. Jordan
R. M. Keddie
K. Kitahara
Miloslav Kocur
Julius P. Kreier
Noel R. Krieg
George P. Kubica
Daisy A. Kuhn
E. Küster
S. P. Lapage
J. W. M. la Rivière
H. Lautrop
Edward R. Leadbetter
R. A. Lelliott
L. Le Minor
Ralph A. Lewin
John Liston
R. Locci
George M. Luedemann
S. Maier
Karl Maramorosch
Norvel M. McClung
Elizabeth McCoy
N. B. McCullough
H. D. McCurdy
H. H. Mollaret
W. E. C. Moore
J. W. Moulder
E. G. Mulder
R. G. E. Murray
C. F. Niven
Frits Ørskov
Ida Ørskov
Cora R. Owen

† Deceased.

Leslie A. Page
N. J. Palleroni
John E. Peterson
Norbert Pfennig
C. B. Philip
J. E. Phillips
Leo S. Pine
Margaret Pittman
Jeanne S. Poindexter
J. R. Postgate
John R. Preer, Jr.
T. G. Pridham
Alice Reyn
Miodrag Ristic
John Robinson
Morrison Rogosa
R. Rohde
Ernest H. Runyon
Riichi Sakazaki
R. H. W. Schubert
Jiri Sedlák†
H. W. Seeley, Jr.
H. P. R. Seeliger
James Shewan
V. B. D. Skerman
H. L. Skuja†
J. M. Slack
R. M. Smibert
J. E. Smith
L. DS. Smith
P. H. A. Sneath
James T. Staley
R. Y. Stanier
Pamela D. Steed-Glaister
E. Thal
Josef E. Thiemann
H. D. Tresner
Hans G. Trüper
L. H. Turner
W. L. van Veen
M. Véron
W. V. Vishniac†
S. W. Watson
Lawrence G. Wayne
Owen B. Weeks
David Weinman
Emilio Weiss
H. J. Welshimer
Ruth G. Wittler
S. A. Zahler
G. A. Zavarzin
K. S. Zinnemann

† Deceased.

Contents

Proper Form of Citation

Published reference to any or all material from *Bergey's Manual* must always be made to the complete eighth edition, *Bergey's Manual of Determinative Bacteriology,* with credit being given to the author(s) of the section being cited. For example:

Krieg, N. R. 1974. Genus *Spirillum* Ehrenberg. *In* Buchanan and Gibbons (Editors), *Bergey's Manual of Determinative Bacteriology,* 8th Ed., The Williams & Wilkins Co., Baltimore, pp. 196–207.

INTRODUCTION

A Place for Bacteria in the Living World

R. G. E. MURRAY

"What's in a name? that which we call a rose
By any other name would smell as sweet . . . ."
Romeo and Juliet, Act II. W. Shakespeare

One of the most difficult tasks for the Editor-Trustees of all previous editions of *Bergey's Manual* was the provision of an intelligible and recognizable description of the bacteria as a taxon and a scientifically acceptable placing among the realms of living things. Without doubt most thoughtful bacteriologist-taxonomists since the time of Ferdinand Cohn have been convinced on an intuitive base and slender morphological evidence, that there was something unique about bacteria and that a degree of morphological similarity existed between the bacteria (*Schizomycetes*) and the blue-green algae (*Cyanophyceae* or *Schizophyceae*) whatever the arrangements made at the highest hierarchical levels. This was recognized in the many and various versions of the familiar classifications of the living world by placing these two taxa together as equivalent classes in the Division *Protophyta* of the Plant Kingdom, as was done in the 7th edition. A consequence of the morphological alliance, together with the evident physiological similarity of photosynthesis by the blue-green algae to that of plants, was a continuing adherence to descriptions in a botanical context. The definition of the *Schizomycetes* in the 7th edition was framed in terms of this long established prejudice despite misgivings (see 7th edition, p. 9). The essential operative statement is that the bacteria are "typically unicellular plants," which was not then and certainly is not now defensible in terms of the description of a typical or idealized plant cell. In fact, it is surely true that the intuitive feeling of the bacteriologist rejected this view, but without solid evidence to buttress his faltering steps, as shown by the acceptance and continual refinement of the Code of Bacteriological Nomenclature independent of the Botanical Code. Resolution of the dilemma became possible as soon as the microbes could be described in terms of cellular organization.

The change in our view of the nature of bacteria was slow in coming to full expression. It derives from the application of three sets of experimental-observational approaches to the microbial world and bacteria in particular: (1) comparative cytology using the light microscope and classical staining methods to describe the form and behavior of the DNA-containing portions of nuclei; (2) the development of appropriate techniques of electron microscopy for the extension of comparative cytology to the ultrastructural level; and (3) the extension of biochemical and biophysical observations to the definition of unique features of cellular organization.

Cytologists who studied the protists became aware that the chromatin of bacteria, which they assumed correctly to be the equivalents of nuclei, was markedly different in appearance and behavior during segregation in comparison to the nuclei and component chromosomes of other protists, plant and animal cells. This was sufficient for some to make fresh attempts to circumscribe a taxon at the highest level to include the bacteria. Among these were the proposals of Copeland (1938), Stanier and van Niel (1941) and Whittaker (1959). In essence, these proposals (derived in part from attempts to rationalize Haeckel's concept of the Protists) placed the bacteria in a kingdom of anucleate organisms in contrast to those with true nuclei. Stanier and van Niel (1941) invoked two other negative features—the seeming absence of sexual reproduction and of plastids; however, it was soon to be established that photosynthetic bacteria possessed chromatophores derived from the plasma membrane and that the bacterial genome was capable of recombinative processes. Most significant of all, early electron microscopic observations of sections of cells showed that the genophore of the bacteria and the blue-green algae consisted of a nucleoplasm that was not separated from the cytoplasm by a nuclear membrane as found in fungi, protozoa, plants and animal cells. Further structural and biochemical studies established new constellations of features unique to bacterial cells in the form of heteropolymers and constituent molecules of the intimate make-up of cell walls (see Salton, 1964). In the face of this level of understanding, the kingdom designations of *Mychota* (Enderlein, 1925) or *Monera* (Copeland, 1938) were not appropriate in their definitions. However, Chatton (1937) had proposed a most appropriate conceptual basis for taxa at the highest level by recognizing two general patterns of *cellular organization*—the procaryotes and the eucaryotes. The truth of this prescient generalization was recognized by Stanier (1961) and is now amply supported by a wealth of data derived from comparative cytology involving microscopical, biochemical and physiological approaches. Stanier and van Niel (1962) put the matter in clear perspective, with an accompanying statement of the evidence: "The distinctive property of bacteria and blue-green algae is the procaryotic nature of their cells." Furthermore, it was possible to justify the use of the inclusive term, the cell, for the unit of structure of both procaryotic and eucaryotic organisms as an expression of equivalence of function; the differences between the broad groupings concern the detailed organization of the cellular machinery. The broad canvas was now sketched in and the bacteria could be confidently placed in perspective with other cellular organisms. The viruses, alone among organisms, remain without a clear position in the scheme of things.

There is no real need to repeat here all the arguments that have been presented to embroider and support the cytological recognition of procaryotic organization as the cornerstone for a coherent view of the nature of bacteria. The case has been made several times (Stanier, 1961; Stanier and van Niel, 1962; Murray, 1962; Allsopp, 1969; Stanier, 1970) and the student should consult these essays for details and references.

The essential features are:

I. The nature of the genophore (a term used by Ris (1961) to avoid the connotations of "chromosome" and "nucleus") constituting the morphologically distinct nucleoplasm of the procaryotic cell. This consists of a skein of double-stranded DNA fibrils that is not separated from cytoplasm by any membranous boundary. Structural and genetic evidence, based on bacterial examples, indicate that the genophore is in the form of a closed loop (often described as a "circular chromosome") and the constituent genes form a single linkage group. The fibrils are not associated in any regular way with a protein in contrast to the histones of the eucaryotic chromosome. For the moment it must be assumed, because of fragmentary genetic or cytogenetic evidence, that the morphologically similar nucleoplasm of blue-green algae has parallel properties.

II. The lack of unit membrane-bounded cytoplasmic organelles is the second most important distinction of the procaryotes. This belies their metabolic diversity and the simplest form is provided by some bacteria whose "membrane system" consists of the plasma membrane alone, smoothly enclosing the protoplasm. Simple intrusions of membrane are common and complex systems of unit membranes do exist (e.g. for photosynthesis, nitrification, etc.), but in all cases there is good support for the view that they are derived by invagination of the plasma membrane into the cytoplasm and in most cases the connection is maintained. There is a possible exception in the blue-green algae whose photosynthetic apparatus is located in an extensive system of thylakoids, which do not appear to be in direct continuity with the plasma membrane.

III. A further distinction attributable to all procaryotes is that the ribosomes are of the small 70 S type as opposed to the consistently larger 80 S ribosomes of the eucaryotes (Taylor and Storck, 1964). They are distributed in the cytoplasm and are not arrayed on membranes as in the endoplasmic reticulum of eucaryotes.

There are a number of positive characters that can be considered as features of many but not all procaryotes, which further strengthen the distinction. For example (1) *cell walls* are not confined to procaryotes but the components are unusual to a degree that is taxonomically significant. Without doubt the now well known peptidoglycan (murein or mucopeptide) component of the cell walls, shared by the majority of bacteria and the blue-green algae, is a uniquely constituted heteropolymer with distinctive subunits in both the amino sugar backbone (muramic acid) and the peptide (D-amino acids and unusual diamino acids) portions. It cannot be used to characterize the procaryotes because the wall-less *Mycoplasma* and exceptional bacteria (e.g. *Halobacterium*) do not possess it. Other constituents may yet prove to have taxonomic value but the peptidoglycan is particularly important as a bridging character with the blue-green algae. (2) *Flagella* are distinctive organs of swimming motility confined to certain genera of bacteria. This complex tubular assembly of protein subunits with an elaborate anchorage in both wall and plasma membrane cannot be confused with the cilia of eucaryotes. (3) *Gas vacuoles* are common in blue-green algae and are found in a few widely separated genera of bacteria. *Chlorobium* vesicles are unique containers for the chlorophylls and carotenoids of the green bacteria. What they have in common is that they are organellar inclusions that are not bounded by unit membranes but rather by special single layers. They are without counterpart among eucaryotes.

It is no longer possible to support two negative features that have been proposed as characterizing procaryotes: lack of cytoplasmic microtubules and an inability to synthesize sterols. It is now apparent that some *Treponema* may possess microtubules and sterols have been isolated from some *Mycoplasma* and blue-green algae.

Bacteria are remarkable for an extraordinary variety of metabolic mechanisms and, particularly, for a wide range of anaerobic energy-yielding reactions; these are in marked contrast overall to the glycolysis utilized by the eucaryotic cell. The result is that many bacteria are obligate anaerobes and share this condition of existence with only a few protozoa among the eucaryotes. Apart from this generalization and the synthesis of unique cell wall polymers, already mentioned, it should be noted that the ability to fix nitrogen and to accumulate poly-β-hydroxybutyrate as a reserve material are metabolic attributes widely distributed among procaryotes but completely absent from eucaryotes.

The fossil record, although indicative of microbial life long ages before recognizable complex forms of life appeared, is not able to tell us anything of the order of appearance and thus contribute to phylogeny. Those groupings that we can observe today seem likely to represent a coherent segment of the terminal branches of an evolutionary tree. Photosynthesis probably originated when the procaryotic stem was already well developed. The precise forms of photosynthesis represent the derivatives of, most likely, a single evolutionary event (Stanier, 1970) since many elements of the machinery are common to all existing phototrophs.

It has been pointed out by Stanier (1970) that the plasma membrane of procaryotic organisms is not adapted to the transfer of particulates or large molecules either in or out of the living cell; indeed, transforming DNA fragments are about the only example. Furthermore, there is no evidence for endocytosis (viz. phagocytosis and pinocytosis) and its directional counterpart, exocytosis, as mechanisms of import or export of soluble material or particulates. Yet this is a very generalized characteristic of eucaryotic cells. Many protozoa ingest procaryotic organisms as food into food vacuoles. It is not surprising, then, that a good number of stable associations have been set up so that distinctly recognizable endosymbionts are characteristic of the cytoplasm of certain cells. These are also enclosed in a sac of host cell membrane. The most notable studies are on the endosymbionts of *Paramecium aurelia*. Although they are generally not cultivable, many of them are structurally recognizable as procaryotes and some even exhibit bacterial flagella and bacteriophages of the tailed type (Beale *et al.*, 1969; Preer *et al.*, 1972). Some endosymbionts have the characteristics of a eucaryotic green alga (as in *Paramecium bursaria*) and other protozoa have blue-green inclusions (cyanelles) that have been interpreted, with insufficient rigor, as blue-green algae. It is clear that endosymbiotic associations are widespread in a great variety of metazoa (Buchner, 1965), as well as in the vascular plants, such as the well known *Rhizobium* association in the root nodules of legumes. Identification, description and classification rests, in the absence of verifiable cultivation, on morphological and biochemical features. It appears that the capacity to take up external cells brings with it the possibility of endosymbiosis. No stable endosymbiosis within procaryotes has yet been identified. (Bacterial parasitism by *Bdellovibrio* is best described as a nonsymbiotic multiplication between cell wall and protoplast causing death of the host.)

One of the excitements of recent years is the realization that the mitochondria and the plastids of eucaryotes could represent the most extreme form of endosymbiotic parasitism by procaryotes. These unit membrane-bounded structures have physical, genetic and biochemical features that suggest the possibility of their procaryotic nature; furthermore, the DNA from the very small genophore can be isolated in circular form and has a very different base composition in comparison with that of the host cell. At this stage the observations provide the base for fascinating and possibly unverifiable speculation concerning an event (or set of events) of the greatest significance in eucaryotic evolution (see Stanier, 1970; Allsopp, 1969).

There is little doubt, at this time and in the face of the arguments that have been presented, that biologists can accept the division of cellular life (as opposed to virus "life") into two groupings at the highest level expressing the encompassing characters of procaryotic and eucaryotic cellular organization. Those who have written concerning the topic have generally avoided committing themselves to a formal nomenclature. So far as we know there are only two formal proposals, both based on recognizable descriptions. Murray (1968) proposed *Procaryotae* as a taxon "at the highest level" and described it as "a kingdom of microbes . . . characterized by the possession of nucleoplasm devoid of basic protein and not bounded from cytoplasm by a nuclear membrane." *Eucaryotae* was suggested as a possible taxon at the same level to include other protists, plants and animals. Allsopp (1969) declared that these groups merit the status of "kingdom or even superkingdom" and he proposed with an extended assessment of characters that the Kingdoms should be *Procaryota* and *Eucaryota* (Allsopp, 1969, 607). The Bergey's Manual Trust has discussed these names and the various alternatives on many occasions and agreed that *Procaryotae* was the most appropriate, as a plural feminine noun, for such a taxon.

The assumption of a new kingdom is both appropriate and helpful to the bacterial taxonomist, but a kingdom including all the eucaryotes would be disturbing to botanists and zoologists causing a realignment of their respective hierarchies. It is probably best to leave matters as they have been expressed above and only recognize, at the moment, the Kingdom *Procaryotae*.

How are the bacteria to be considered in relation to the blue-green algae within the Kingdom? There is no doubt of their relationship at that level and of the many parallels in the forms and arrangements of their cells. As proposed by Stanier *et al.* (1971), "the blue-green algae must now be recognized as a major group of bacteria, distinguished from other photosynthetic bacteria by the nature of their pigment system and by their performance of aerobic photosynthesis." It is a departure from tradition but it seems just, and it respects the fact that there are parallel groups of photosynthetic bacteria equally distinguished by the nature of their photopigments and their ability to perform anaerobic photosynthesis without oxygen production. For the present, the Bergey's Manual Trustees have no intention of attempting a reassessment of the blue-green algae. It is suggested, by the present designation (p. 22), that they are truly allied to bacteria and at a level of equal importance without hierarchical prejudice to the remainder of the Kingdom, the Bacteria.

The future will have to bring a regrouping of higher taxa to express a more coordinate view. This example is constructed without hierarchical prejudice (e.g. no attempt is made to indicate that non-photosynthetic procaryotes must have preceded the photosynthetic groups on the evolutionary scale) and using common names or tentative names in parentheses, as a suggestion of a possible arrangement.

Kingdom: *Procaryotae*

Division I: Phototrophic procaryotes ("Photobacteria")
- Class I: Blue-green photobacteria
- Class II: Red photobacteria } Present Part 1
- Class III: Green photobacteria } Present Part 1

Division II: Procaryotes indifferent to light ("Scotobacteria")
- Class I: The bacteria—Present Parts 2 to 17
- Class II: Obligate intracellular Scotobacteria in eucaryotic cells—Rickettsias—Present Part 18
- Class III: Scotobacteria without cell walls—*Mollicutes*—Present Part 19

No doubt time will settle the problems involved in making a reasonable arrangement within the *Procaryotae*. Haste is unwise; all previous classifications seem to have suffered infinite rearrangement due to insufficient information. The groupings suggested above, even if the rank designations may need adjustment, might appear to be stable until a new level of understanding overtakes us. The new insights are likely to come from a clear understanding, on a comparative level, of the components of the genome of procaryotic cells. This may allow us to understand why and how vastly different physiological groupings of phototrophic, lithotrophic and organotrophic organisms have remarkably similar and complex structural components (as exemplified by the wide range of spirilla). Surprises may well be in store for us; hence, our adherence to simplicity.

The Mechanism of Identification

S. T. COWAN
J. LISTON

Bacteria isolated directly from animals, plants or the general environment are rarely obtained in pure culture. The first step in identification is to obtain each organism in pure culture, since the reactions of mixed cultures in characterization tests cannot be used for identification. This step is so obvious that it is seldom mentioned, but in fact it cannot be stressed too much if we are to obtain clearcut characterization of our unknown organism. Techniques for obtaining pure cultures vary with the organism being studied; they range from repeated plating on non-inhibitory media to single cell isolations. The commonest single source of impure cultures is the picking of a colony from a selective or inhibitory medium; consequently, these media are to be avoided in the purification or checking process.

Characterizing tests range from the descriptive (morphology and tinctorial reactions) through simple biochemical tests for the detection of metabolic products or enzyme action, to highly specialized techniques required for the estimation of the GC percentage of the bacterial DNA; these advanced techniques are not likely to be within the competence of the worker responsible for making an identification, which must normally be based on characterizations made by simpler methods.

From a practical point of view, it is important that the number of tests applied to identify a bacterium be kept to a minimum. Hundreds of characterization tests have been proposed, and it is necessary to select a test pattern which will quickly and simply identify the organism under study. Knowledge of the source of the strain and the conditions under which it was originally isolated is often helpful in delimiting the genera to which the organism may belong. Indeed this information, together with details of morphology and Gram's staining reaction, which can easily be obtained during the purification procedure, will frequently enable the investigator to narrow his search to a small group of genera. Thus a Gram positive motile spore-forming rod isolated from soil by aerobic incubation of a non-selective medium is almost certainly a *Bacillus* and there is no need to apply tests to differentiate this organism from Gram negative rods, obligate anaerobes, Gram positive cocci, etc.

The desirability of careful morphological examination of strains as a first step in identification cannot be overemphasized. The phase contrast microscope is particularly useful, since it can be used to view living organisms without the distortion imposed by staining procedures or by drying and embedding, as in electron microscopy. It is simple to use and provides a rapid method for detecting motility, as well as shape and cell arrangements, in most cases. The Gram reaction is still the most generally useful of the staining methods and should be applied routinely to all fresh isolates. This may be done conveniently using the phase contrast preparations. Flagella staining and tests of acid-fastness are the only other two staining procedures which are likely to be useful with any great frequency, and they need only be applied when seen to be necessary.

Electron microscopy is not yet an everyday tool of the routine bacteriologist, but where facilities are available it can provide most useful information on the procaryotic nature of an unknown organism, wall structure, the exact mode or site of flagellar insertion and other taxonomically useful details.

Typically, the detailed classification of bacteria is based on results of physiological and biochemical tests, as well as the more specialized techniques often used for strain identification, such as serology, bacteriophage typing and genetic analysis, including measurement of percentage of GC and DNA-DNA homology analysis.

The major physiological divisions of bacteria into chemolithotrophic autotrophs, photosynthetic bacteria and chemoheterotrophic bacteria will normally present little difficulty. These are obvious separations, based on characteristics which should be apparent during the initial isolation procedures. Similarly, the striking morphological attributes of the larger and more complex organisms (formerly called "Higher Bacteria"), including holdfasts, slime-encased chains, tubules, etc., will distinguish these heterotrophic forms from the smaller, morphologically simpler types (formerly called "True Bacteria") routinely isolated in most bacteriological laboratories. It is these latter forms which consistently present most difficulty in identification. In practice the simple, or true, bacteria are classified primarily on a combination of morphology and physiology, with emphasis on oxygen requirement and the manner of utilization of carbohydrates (particularly glucose). Thus, we have the aerobic Gram negative rods, which are obligate aerobes utilizing sugars oxidatively, and the fermentative Gram negative rods which ferment sugars and are, in fact, facultative anaerobes. Each of these groups contains a number of genera further subdivided by other tests, and this is representative of the general pattern of classification within which the diagnostic bacteriologist must work. In practice, therefore, it is commonly useful to test whether an organism is obligately aerobic or anaerobic, or a facultative anaerobe, and to find out the mode of dissimilation of sugar (usually glucose).

The simple tests described lead, as noted, to groups of genera, and further classification depends on tests which vary with each group but usually involve more physiology and biochemistry. It is a good idea, wherever possible, to grow the pure strain in a relatively simple liquid medium such as peptone water to provide a non-interfering inoculum for these tests (this is also useful earlier for phase contrast examination). Again, however, it is important to apply common sense at this point. A polarly flagellate oxidative Gram negative rod which produces a green fluorescent pigment is obviously a *Pseudomonas* species and requires little further testing to identify to genus. A Gram positive facultatively anaerobic fermentative rod could belong to a number of genera, but a simple catalase test will establish with a high degree of probability whether or not it is a *Lactobacillus*. It is well to scan the general characteristics of the targeted group and select the tests necessary for identification. The tests which best distinguish between one group of organisms and another are shown in **boldface** in the *Manual*.

In the case of bacteria suspected to be pathogenic, experimental inoculation of plants or animals may be a necessary step in identification. Here (as in most other tests) only a positive result is significant, since a negative result may merely indicate that the wrong animal or plant species was chosen or that conditions of inoculation were inappropriate.

Antigenic analysis is often complex and highly specialized, being frequently used for strain identification in epidemiological or clinical studies. However, the serological method can provide a rapid, unambiguous identification of bacterial isolates in the middle of an epidemic or where an early clue to the genus is available. A good example of this is in identification of *Salmonella* serotypes, for which there is a readily available supply of polyvalent and monospecific antisera.

Simpler biochemical and physiological tests which are frequently useful in identification of bacteria at the species level include oxidase, catalase, urease, nitrate reduction, H_2S production, acid or gas production from sugars, tests of amino acid metabolism, temperature range of growth, response to NaCl and antibiotic sensitivity. In some genera, the nutritional requirements, ability to utilize specific substrates and production of polyhydroxybutyrate are useful.

In summary, the rules of the game are:

1. Use all the information available to you.
2. Apply common sense at each step.
3. Use the minimum number of tests to make the identification.

The practical steps to follow are:

1. Make sure you have a pure culture.
2. On the basis of the isolation procedure, establish whether you have a chemolithotrophic autotroph, a photosynthetic organism or a chemoheterotrophic organism.
3. Examine living cells by phase contrast and Gram-stained cells by light microscopy. Apply other stains if this seems appropriate. If some outstanding morphological property, such as endospore production or possession of

holdfasts, is obvious, confine your further efforts in identification to groups having these properties.

4. Examine gross growth appearances for pigments or other unique characteristics.
5. Test for oxygen requirements.
6. Test the dissimilation of glucose (or other simple sugar)—oxidative or fermentative.
7. Complete additional tests selected by scanning the characteristics for the cluster of genera to which you have assigned your organism on the basis of the tests listed above.

If, as may well happen, you cannot identify your isolate from the information contained in the *Manual*, neither despair nor immediately assume that you have isolated a new species; many of the problems of microbial classification and nomenclature are the results of people jumping to this conclusion prematurely. Remember that descriptions in the *Manual* are based on majorities, and give the reactions regarded, by those familiar with them, as characteristic of the species; the Editors have generally, but alas not always, deleted such words as generally, usually and sometimes, which so often befog descriptions of species. When you fail to identify your culture check (1) its purity, (2) that you have carried out the appropriate tests, (3) that your methods are satisfactory (in several places in the *Manual* the need for specified methods is emphasized; one way to make sure is to include a known control culture in your tests) and (4) that you have used correctly the various keys and tables of the *Manual*. It has been said that the most frequent cause of mistaken identity of bacteria is errors in the determination of shape, Gram reaction and motility. In most cases you should have little difficulty in placing your isolate into a genus; allocation to a species or subspecies may need the help of a specialized reference laboratory.

Key to the 19 Parts

aa. Cells not filamentous and ensheathed
- b. Products of binary fission not equivalent (have appendages other than flagella and pili or reproduce by budding) .... Part 4, p. 53
- bb. Not as above
  - c. Cells not rigidly bound
    - d. Have cell wall, spiral-shaped .... Part 5, p. 63
    - dd. Have no cell wall .... Part 19, p. 320
  - cc. Cells rigidly bound
    - d. Gram negative
      - e. Obligate intracellular parasites .. Part 18, p. 308
      - ee. Not as above
        - f. Curved rods .... Part 6, p. 70
        - ff. Not curved rods Part 9, p. 142
          - g. Rods
            - h. Aerobic .... Part 7, p. 77
            - hh. Facultatively anaerobic .... Part 8, p. 99
            - hhh. Anaerobic .... Part 9, p. 142
          - gg. Cocci or coccobacilli
            - h. Aerobic .... Part 10, p. 166; Part 7, p. 77
            - hh. Anaerobic .... Part 11, p. 170
    - dd. Gram positive
      - e. Cocci
        - f. Endospores produced .... Part 15, p. 201
        - ff. Endospores not produced .... Part 14, p. 184
      - ee. Rods or filaments
        - f. Endospores produced .... Part 15, p. 201
        - ff. Endospores not produced
          - g. Straight rods .... Part 16, p. 218; Part 17, p. 224
          - gg. Irregular rods (coryneform) or tend to form filaments or filamentous ..... Part 17, p. 224

KINGDOM PROCARYOTAE

Murray 1968, 252

Single cells or simple associations of similar cells (0.2–10 μm in smallest dimension) forming a Kingdom defined by cellular, not organismal, properties. The nucleoplasm (genophore) is never separated from the cytoplasm by a unit-membrane system (nuclear membrane) and is not associated with a basic protein. Cell division is not accompanied by cyclical changes in the texture or staining properties of either nucleoplasm or cytoplasm; a microtubular (spindle) system is not formed. The plasma membrane is frequently complex in topology and forms vesicular, lamellar or tubular intrusions into the cytoplasm; vacuoles and replicating cytoplasmic organelles enclosed by unit membranes are absent. Cytoplasmic organelles independent of the plasma membrane system (chlorobium vesicles, gas vacuoles) are relatively rare, and are enclosed by non-unit membranes. Respiratory and photosynthetic functions are associated with the plasma membrane system in those members possessing these physiological attributes. Ribosomes of the 70 S type are dispersed in the cytoplasm; an endoplasmic reticulum with attached ribosomes is not present. The cytoplasm is immobile; cytoplasmic streaming, pseudopodial movement, endocytosis and exocytosis are not observed. Nutrients are acquired in molecular form. Enclosure of the cell by a rigid wall is common but not universal.

In organismal terms, these ubiquitous inhabitants of moist environments are predominantly unicellular microorganisms, but filamentous, mycelial or colonial forms also occur. Differentiation is limited in scope (holdfast structures, resting forms and modifications in cell shape). Mechanisms of gene transfer and recombination occur but these processes never involve gametogenesis and zygote formation.

Editorial Note. Acceptance of the Kingdom *Procaryotae* logically calls for an integrated treatment of the blue-green algae and the bacteria. One possibility is outlined in "A Place for Bacteria in the Living World." However, in this edition of the *Manual*, the Kingdom is divided into two divisions, one for the blue-green algae or cyanobacteria and one for "procaryotic organisms that are not blue-green algae." The Board of Editors is not prepared to subdivide the *Procaryotae* further at this time.

DIVISION I. THE CYANOBACTERIA

Phototrophic procaryotic organisms that use water as an electron donor and hence produce oxygen in the light.

Cells are always enclosed by a rigid, multilayered wall with an inner peptidoglycan layer. The wall may in turn be surrounded by a gelatinous or fibrous sheath. Most cyanobacteria are motile at some stage of development; motility is always of the gliding type, dependent on surface contact. The cytoplasmic region is traversed by an extensive system of paired photosynthetic lamellae (thylakoids), the outer surfaces bear characteristic granules (phycobilisomes) composed of aggregates of the phycobiliprotein pigments.

Some cyanobacteria are unicellular, others consist of chains of cells (filaments), either simple or branched. Reproduction of unicellular forms may occur by binary fission, by multiple fission or by the serial release of apical cells (exospores) from a sessile individual. Forms that consist of filaments grow by repeated intercalary cell divisions and reproduce either by random fragmentation of the filament or by terminal release of short motile chains of cells (hormogonia). Certain filament-formers can produce specialized cells, known as akinetes and heterocysts. Akinetes are larger than the vegetative cells in the filament and represent a resting stage, which germinates with release of a hormogonium. Heterocysts are non-reproductive cells, distinguishable from the adjoining vegetative cells by the presence of refractile polar granules and of a thick outer wall; they are believed to be physiologically specialized cells that serve as sites of nitrogen fixation.

The characteristic photopigments include chlorophyll *a* as the only chlorophyllous pigment and phycobiliproteins (allophycocyanin, phycocyanin and sometimes phycoerythrin). The cellular absorption spectrum has a peak at approximately 680 nm attributable to chlorophyll *a* and a broad absorption band with one or more peaks, attributable to phycobiliproteins, between 560 and 630 nm.

The Cyanobacteria are not considered further in this edition.

DIVISION II. THE BACTERIA

Unicellular procaryotic organisms or simple associations of similar cells based upon growth habit, planes of division and cell separation. Cell multiplication involves growth and division, usually binary but occasionally unequal and by budding. Cells may remain attached after division thus causing recognizable arrangements; true branching may occur in certain species. In fluid environments many exhibit swimming motility and possess flagella of bacterial type; others show gliding, twitching, snapping and darting movements on the surface of solid media. Endospores are formed by some species, arthrospores and cysts by some others; heterocysts are not formed.

With few exceptions (*Mollicutes*) the cells are contained within a rigid or semirigid cell wall conferring a constancy of form; with few exceptions the cell wall polymers include peptidoglycans.

Those that perform photosynthesis do so under anaerobic conditions, require an electron donor other than water, do not produce oxygen in the process, utilize one or more of the bacteriochlorophylls (*a*, *b*, *c* or *d*) and never contain chlorophyll *a* or phycobiliproteins.

Those that perform chemosynthesis may require either aerobic or anaerobic conditions or be facultative.

Important Notes
for
Users of this Edition

1. Always read both generic and species descriptions because characters listed in the generic description are not usually listed in the species descriptions.
2. In tables, characters common to all taxa are not shown but may be listed in footnotes.
3. Generally in tables (exceptions are clearly indicated in footnotes, q.v.) the meanings of symbols are as follows:

+	more than 90% strains positive
−	more than 90% strains negative
d	11–89% strains positive
( )	delayed reaction
w	weak reaction
D	Different reactions in different taxa (species of a genus or genera of a family)
v	strain instability (NOT differences between strains)

PART 1

THE PHOTOTROPHIC BACTERIA

ORDER I. RHODOSPIRILLALES PFENNIG AND TRÜPER 1971, 17

The phototrophic (photosynthetic) bacteria represent a physiological community of different, predominantly aquatic bacteria. Cells spherical, rod-, vibrio- or spiral-shaped. Diameter of individual cells from 0.3 to over 6 μm. In most cases multiplication by binary fission, some species of the *Rhodospirillaceae* have a polar type of cell growth and multiply by budding. Gram-negative. Colors of cell suspensions from purple-violet, purple, red, orange-brown to brown or green. May contain sulfur globules.

Common to all is the presence of bacteriochlorophylls (see Table 1.1) and of carotenoid pigments (see Table 1.2). The photosynthetic metabolism differs from that of the Cyanobacteria and green plants in that it occurs under anaerobic conditions and oxygen is not produced; it depends on the presence of oxidizable external electron donors, such as reduced sulfur compounds, molecular hydrogen or organic compounds. Carbon dioxide is photoassimilated through the reductive pentose phosphate cycle and further carbon dioxide-incorporating reactions. As far as studied all strains contain cytochromes, ubiquinones and non-heme iron proteins of the ferredoxin type. Fixation of molecular nitrogen has been demonstrated in representatives of all groups.

The photopigments are located in internal membrane systems continuous with the cytoplasmic membrane or in vesicles characteristic of the genus *Chlorobium* which underlie and are attached to the cytoplasmic membrane. DNA base ratios from 45–73 moles % guanine + cytosine (buoyant density). The type and neotype strains were proposed by Pfennig and Trüper (1971).

The present classification is—with the exception of a few species—based on pure culture studies. The phototrophic bacteria comprise the two essentially different groups: purple bacteria and green sulfur bacteria. Molisch's (1907) differentiation of two physiological groups among the purple bacteria, the *Chromatiaceae* (formerly *Thiorhodaceae*) and *Rhodospirillaceae* (formerly *Athiorhodaceae*) is maintained. For each of the three families an addendum has been added of species which are incompletely described, or are hardly recognizable, or are not available in enrichment or pure culture.

TABLE 1.1

Nomenclature of bacterial chlorophylls

Designation of Jensen *et al.* (1964)	Former designations	Characteristic absorption maxima of living cells
		nm
Bacteriochlorophyll *a*	Bacteriochlorophyll	375, 590, 805, 830–890
Bacteriochlorophyll *b*	Bacteriochlorophyll *b*	400, 605, 835–850, 1020–1040
Bacteriochlorophyll *c*	Chlorobium chlorophyll 660	Long wave length abs. max. 745–755
Bacteriochlorophyll *d*	Chlorobium chlorophyll 650	Long wave length abs. max. 705–740

TABLE 1.2

Carotenoid groups of phototrophic bacteria[a]

Group	Name	Major components
1	Normal spirilloxanthin series	Lycopene, rhodopin, spirilloxanthin
2	Alternative spirilloxanthin series and ketocarotenoids of spheroidenone type	Spheroidene, hydroxyspheroidene; spheroidenone, hydroxyspheroidenone, spirilloxanthin
3	Okenone series	Okenone
4	Rhodopinal series (former warmingone series)	Lycopenal, lycopenol, rhodopin, rhodopinal, rhodopinol
5	Chlorobactene series	Chlorobactene, hydroxychlorobactene, β-isorenieratene, isorenieratene

[a] Liaaen-Jensen, 1963, 1965; Schmidt *et al.*, 1965; Pfennig, 1967.

Key to the suborders and families of order **Rhodospirillales**

Key characteristics

- Photosynthesis carried out under anaerobic conditions
- Color of culture
- Presence of bacteriochlorophyll *a*, *b*, *c* or *d* (see Table 1.1)
- Nature of electron donor (organic or sulfur-containing)
- Oxidation of accumulated sulfur to sulfate
- Elemental sulfur as electron donor
- Accumulation of elemental sulfur

Cells capable of carrying out a photolithotrophic and/or a photoorganotrophic metabolism under anaerobic conditions.

I. Cells contain bacteriochlorophyll *a* or *b* and various carotenoids; photopigments located on internal membrane systems of different morphology, which are continuous with the cytoplasmic membrane.

Purple bacteria: Suborder *Rhodospirillineae*

A. Cells photoassimilate simple organic substances; most species unable to grow with sulfide as the sole photosynthetic electron donor. When sulfide or thiosulfate are used as an electron donor, elemental sulfur is not an intermediate oxidation product. When elemental sulfur is formed from sulfide, it is not further oxidized to sulfate. None of the species able to utilize elemental sulfur as an electron donor.

Purple nonsulfur bacteria:

Family I. *Rhodospirillaceae*, p. 26*

B. Cells able to grow with sulfide and sulfur as the sole photosynthetic electron donor. In the presence of sulfide, globules of elemental sulfur are formed inside or outside the cells and further oxidized to sulfate.

Purple sulfur bacteria:

* All page numbers cited refer to *Bergey's Manual of Determinative Bacteriology*, 8th edition, The Williams & Wilkins Company, Baltimore, 1974.

Family II. *Chromatiaceae*, p. 34

II. Cells contain bacteriochlorophyll *c* or *d* as the major bacteriochlorophyll components and various carotenoids; photopigments located in "chlorobium vesicles" which underlie and are attached to the cytoplasmic membrane.

A. Green sulfur bacteria: Suborder *Chlorobiineae*

Family III. *Chlorobiaceae*, p. 51

Note: Trüper (*International Journal of Systematic Bacteriology 26:* 74–75, 1976) has proposed a revision of the classification of the higher taxa of phototrophic bacteria necessitated by the discovery of a genus of gliding, green bacteria named *Chloroflexus*. He proposes a second sub-order *Chlorobiineae* containing the family *Chlorobiaceae* and a new family *Chloroflexaceae*, with the genus *Chloroflexus* as type genus. This genus is distinguished by the possession of gliding motility, flexible cells, chlorobium vesicles, chlorophyll *c* and *a*, a preference for photoorganoheterotrophic growth, ability to grow aerobically in the dark and the presence of β-carotene as the main carotenoid.

FAMILY I. **RHODOSPIRILLACEAE** PFENNIG AND TRÜPER 1971, 17

Cells spherical, short or long rod-shaped, vibrio- or spiral-shaped; **multiplication by binary fission or budding.** Cells of two genera **motile** by means of polar flagella; in the third genus, *Rhodomicrobium*, cells peritrichously flagellated. **Internal photosynthetic membrane system continuous with the cytoplasmic membrane** and of vesicular, lamellar or tubular type (Plate 1.2). The known genera **do not contain gas vacuoles.**

Generally microaerophilic, although **many representatives may grow at full atmospheric oxygen tension in the light or dark.** In strains able to grow under microaerophilic or aerobic conditions, the photopigment content and the internal membrane system decrease as the concentration of dissolved oxygen increases; the formation of photopigments and internal membranes becomes derepressed below certain oxygen concentrations. Under strictly anaerobic conditions normal growth is possible only in illuminated cultures; however, very slow multiplication has been observed in the dark (Uffen and Wolfe, 1970).

Phototrophic development is dependent on simple organic substrates which are photoassimilated or serve as electron donors for carbon dioxide assimilation. Molecular hydrogen can serve as an electron donor in many strains. When sulfide or thiosulfate is used as an electron donor, elemental sulfur is not an intermediate oxidation product. When elemental sulfur is formed from sulfide, it is not further oxidized to sulfate. **None of the species able to utilize elemental sulfur as an electron donor. Most strains require one or more vitamins** as growth factors; some do not require growth factors. The fixation of molecular nitrogen has been demonstrated in a number of strains.

Bacteriochlorophylls *a* or *b* and carotenoids of groups 1, 2 and 4, as well as some carotenoids not yet included in one of the existing series, **occur.** In general, cultures of strains with carotenoids of group 1 appear orange-brown to brownish red or pink to purple-red, of group 2 dirty greenish yellow under anaerobic conditions, brownish red to purple under aerobic conditions.

Storage materials: polysaccharides, poly-β-hydroxybutyrate and polyphosphate. Range of DNA base ratios: 61.2–72.4 moles % guanine + cytosine. Widely distributed in nature.

In the 7th edition of the *Manual* the genus *Rhodomicrobium* Duchow and Douglas (1949) was included with the nonphototrophic Hyphomicrobiaceae. Following the definitions used here the genus *Rhodomicrobium* is now included with the phototrophic bacteria in *Rhodospirillaceae*.

The type genus: *Rhodospirillum* Molisch 1907, 24, designated by Pfennig and Trüper 1970, 33.

Key to the genera of family **Rhodospirillaceae**

I. Cells spiral-shaped

Genus I. *Rhodospirillum*

II. Cells rod-shaped, ovoid or spherical, not forming filaments

Genus II. *Rhodopseudomonas*, p. 29

III. Cells ovoid to elongated ovoid, forming filaments

Genus III. *Rhodomicrobium*, p. 33

Genus I. **Rhodospirillum** *Molisch 1907, 24*

Spiral-shaped bacteria, multiplying by binary fission, motile by means of polar flagella. Gram-negative. **Contain bacteriochlorophyll *a* and group 1 carotenoids,** both pigments being located at internal membrane systems (vesicles, tubes, stacks of lamellae, thylakoids) which originate from and may be continuous with the cytoplasmic membrane (Plate 1.2, Figs. 1 and 2). Do not contain gas vacuoles.

Anaerobic phototrophs; in addition, **most species** able to carry out an **oxidative metabolism in the dark under microaerophilic to aerobic conditions.**

Capable of photosynthesis in the presence of simple organic carbon compounds which serve two functions: direct photoassimilation and source of reducing power for carbon dioxide fixation. Molecular hydrogen may be used as electron donor. Molecular oxygen is not evolved during photosynthesis. Cell suspensions appear in various shades of red to brown, due to the photosynthetic pigments.

Range of DNA base ratios: 61.7–65.8 moles % guanine + cytosine (buoyant density).

Type species of genus: *Rhodospirillum rubrum* (Esmarch) Molisch 1907, 25.

Key to the species of genus **Rhodospirillum**

Key characteristics

Color of culture
Cell size
Relationship to oxygen

I. Cultures deep red to purple red without brownish tinge; characteristic absorption band around 550 nm. Cells 0.7–1.0 μm wide, facultatively microaerophilic to aerobic:
1. *Rhodospirillum rubrum*

II. Cultures orange, reddish brown or purple-violet; characteristic absorption maximum around 530 not 550 nm.
A. Cells 0.3–0.5 μm wide, facultatively aerobic; cultures reddish brown or purple-violet.
2. *Rhodospirillum tenue*
B. Cells 0.5–0.7 μm wide, no growth under aerobic conditions.
3. *Rhodospirillum fulvum*
C. Cells more than 0.7 μm wide, no growth under aerobic conditions.
a. Cells 0.7–1.0 μm wide, 5–10 μm long.
4. *Rhodospirillum molischianum*
b. Cells 1.2–1.5 μm wide, 14–30 μm long.
5. *Rhodospirillum photometricum*

Genus II. **Rhodopseudomonas** *Kluyver and van Niel in Czurda and Maresch 1937, 119*

Nom. cons. proposed by Pfennig and Trüper 1969, 153.

Rod-shaped and ovoid to spherical bacteria, multiplying either by binary fission or asymmetrical division (budding without stalk formation). Motile by means of polar flagella. Gram-negative. **Contain bacteriochlorophyll *a* or *b* and carotenoids,** both types of pigments being located at internal membrane systems of either vesicular, tubular (in the species multiplying by binary fission) or lamellar type (in the species multiplying by budding) (Plate 1.2, Fig. 3). Do not contain gas vacuoles.

Anaerobic phototrophs; in addition, **some species** able to carry out an **oxidative metabolism in the dark under microaerophilic or aerobic conditions.**

Capable of photosynthesis in the presence of simple organic carbon compounds which serve two functions: direct photoassimilation and source of reducing power for carbon dioxide fixation. Molecular hydrogen may be used as electron donor. Not able to utilize elemental sulfur as a photosynthetic electron donor. Molecular oxygen is not evolved during photosynthesis. Cell suspensions appear in various shades of yellow-green to brown and red due to the photosynthetic pigments.

Range of DNA base ratios: 62.2–72.4 moles % guanine + cytosine (buoyant density).

Type species of genus: ***Rhodopseudomonas palustris*** (Molisch) van Niel 1944, 89.

Key to the species of genus **Rhodopseudomonas**

Key characteristics

Cell shape (tested in media above and below pH 7.0)	Cell arrangement
Mode of multiplication	Color of culture
Formation of stalk if budding occurs	Utilization of 0.2% propionate, mannitol and sorbitol, and 0.05% higher fatty acids
Cell size	

I. Cells clearly rod-shaped in all media; do not become spherical in media below pH 7.
  A. Multiplication by budding: daughter cells originate as buds at the pole opposite the flagellum-bearing end and are either sessile or grow at the tip of a slender opaque tube.
    1. Cells rod-shaped to elongate-ovoid, somewhat curved; size of young and short cells 0.6–0.8 by 1.2–2.0 μm. Daughter cells (buds) grow at the tip of a slender tube, giving a dumbbell appearance before division.
      a. Cultures red to dark brown-red. Thiosulfate is used as an electron donor:
        1. *Rhodopseudomonas palustris*
      b. Cultures yellowish green, green or dirty green. Contain bacteriochlorophyll *b*; long wave length absorption maximum of living cells around 1020 nm.
        2. *Rhodopseudomonas viridis*
    2. Cells rod-shaped to elongate-ovoid, somewhat curved, 1.0–1.3 μm wide and 2.0–6.0 μm long. Daughter cells (buds) sessile. Grows well in media of pH 5.0–5.5. Cultures orange-brown to purple-red:
        3. *Rhodopseudomonas acidophila*
  B. Multiplication by binary fission; cells slender rods, 0.5 by 1.2 μm, usually clumped together in extensive slime masses. Cultures pale brown to peach. Gelatin liquefied; propionate, mannitol and sorbitol not used:
        4. *Rhodopseudomonas gelatinosa*

II. Cells more or less spherical in media at pH below 7.
  A. In media above pH 7, cells clearly rod-shaped, 1.0 by 1–2.5 μm, frequently in chains with characteristic zigzag arrangement. 0.2% propionate and higher fatty acids (0.05%) used. Molecular hydrogen is usually a good electron donor:
        5. *Rhodopseudomonas capsulata*
  B. In media above pH 7, cells predominantly spherical, 0.7–4 μm in diameter. Mostly singles, little tendency to chain formation. Does not develop in media with 0.2% propionate; mannitol and sorbitol (0.2%) used:
        6. *Rhodopseudomonas sphaeroides*

Genus III. **Rhodomicrobium** *Duchow and Douglas 1949, 415*

Ovoid to elongate-ovoid **bacteria, multiplying by budding.** Daughter cells originate as spherical **buds at the end of filaments** from one to several times the length of the mother cell. Mature buds may separate from the filament; they are **motile** by means of **peritrichous flagella.** Gram-negative. **Contain bacteriochlorophyll *a* and carotenoids,** both pigments being located at **internal membranes of lamellar type.** Do not contain gas vacuoles.

Anaerobic phototrophs; in addition, **most strains** able to grow and to carry out an **oxidative metabolism in the dark under microaerophilic to aerobic conditions.**

Capable of photosynthesis in the presence of simple organic carbon compounds which serve two functions: direct photoassimilation and source of reducing power for carbon dioxide fixation. Molecular hydrogen may be used as electron donor. Molecular oxygen is not evolved during photosynthesis. Cell suspensions appear in various shades of pink to reddish brown due to the photosynthetic pigments.

Range of DNA base ratios: 61.8–63.8 moles % guanine + cytosine (buoyant density).

Type species of genus: *Rhodomicrobium vannielii* Duchow and Douglas 1949, 415.

FAMILY II. **CHROMATIACEAE** BAVENDAMM 1924, 125

Nom. cons. proposed by Pfennig and Trüper 1971, 16.

Individual cells spherical, ovoid, rod-, vibrio- or spiral-shaped; motile or non-motile; with or without gas vacuoles.

Motile forms have polar flagella; in the largest forms, the tuft of polar flagella is visible under the light microscope. The very slow movements of groups of cells and jerky movements of individual cells seen in gas vacuole-containing forms and described by Winogradsky (1888) for *Amoebobacter* arise as artefacts of the microscopic observation technique and have no connection with the movements of gliding bacteria. **Cells contain bacteriochlorophylls *a* or *b*, carotenoids of groups 1, 3 and 4 or tetrahydrospirilloxanthin.** In general, cultures of strains with carotenoids of group 1 appear orange-brown to brownish red or pink, of group 3 purple-red and of group 4 purple-violet. **Internal photosynthetic membrane system continuous with the cytoplasmic membrane** and vesicular (in most species), tubular or lamellar (Plate 1.3, Figs. 1 and 2).

Most species are strictly anaerobic (only one species, *Thiocapsa roseopersicina*, is able to grow in the dark under microaerophilic conditions) and capable of **photolithotrophic carbon dioxide assimilation in the presence of elemental sulfur and sulfide. Elemental sulfur accumulates intermediarily** as globules **inside the cells or in one genus** (*Ectothiorhodospira*) **outside the cells.** Sulfate is the ultimate oxidation product of sulfur compounds. Many strains are able to use molecular hydrogen as an electron donor. **All strains photoassimilate** a number of **simple organic substrates** of which **acetate** and **pyruvate** are most widely used. Strains lacking assimilatory sulfate reduction utilize the organic substrates only in the presence of sulfide or other reduced sulfur compounds as a source of cell sulfur; the majority of strains is able to utilize organic substrates in the absence of hydrogen sulfide or sulfur; thus **all forms are potentially mixotrophic.** The fixation of molecular nitrogen has been demonstrated in a number of strains. Storage materials are polysaccharides, poly-β-hydroxybutyrate and polyphosphate.

Vitamin B_{12} is required by the larger forms.

Range of DNA base ratios: 45–70.4 moles % guanine + cytosine (buoyant density).

In nature, the *Chromatiaceae* occur in the anaerobic and sulfide-containing parts of all kinds of aquatic environments from moist and muddy soils to ditches, ponds, lakes, rivers, sulfur springs, salt lakes, estuaries and marine habitats.

Two physiological-ecological subgroups can be differentiated on the basis of the presence or absence of gas vacuoles and the associated selective advantage in enrichment cultures. Forms without gas vacuoles have a selective advantage in cultures with medium to high sulfide concentrations and light intensities. Forms containing gas vacuoles require very low sulfide concentrations, low light intensities and low incubation temperatures (10–20 C). The cells of the gas vacuole-containing forms rise to the top of the culture bottles at lower temperatures (4–10 C).

Depending on the culture conditions or environmental conditions in nature, all forms are able to develop either in the form of single cells or in non-motile cell aggregates or families of variable size and shape embedded in slime. The different modes of growth described by Winogradsky (1888) may be obtained experimentally by varying sulfide concentration, light intensity, pH, salinity, temperature and oxygen tension. At high sulfide concentration and high light intensity, all flagellated forms become non-motile as the result of slime formation embedding groups of cells. For a given light intensity the various genera and species differ from each other with respect to the sulfide concentration at which the cells develop functional flagella and become motile; for a given sulfide concentration they differ with respect to the light intensity and mode of illumination which allow the cells to become motile. The lower the light intensity the higher will be the sulfide concentration at which a given organism can persist in the motile stage. These characteristics must be considered in the classification of a particular organism.

For these reasons the classification of the *Chromaticaeae* (*Thiorhodaceae*) used in previous editions of the *Manual* has not been followed. The present classification is based primarily on the characteristics of single cells (Pfennig, 1967).

The type genus: *Chromatium* Perty 1852, 174.

Key to the genera of family **Chromatiaceae**

Key characteristics

Position of stored sulfur
Presence of gas vacuoles
Motility
Type of flagellation
Cell shape and arrangement

I. Sulfur globules stored inside the cells when grown with sulfide as electron donor.
  A. Cells do not contain gas vacuoles
    1. Cells motile by means of polar flagella
      a. Cells ovoid to rod-shaped:
        Genus I. *Chromatium*, p. 36
      b. Cells spherical; typically diplococcus-shaped before cell division:
        Genus II. *Thiocystis*, p. 39
      c. Cells spherical to ovoid, grouped as regular sarcina packets:
        Genus III. *Thiosarcina*, p. 40
      d. Cells clearly spiral-shaped:
        Genus IV. *Thiospirillum*, p. 41
    2. Cells non-motile
      a. Cells spherical; typically diplococcus-shaped before cell division; individual cells often surrounded by a more or less thick slime layer:
        Genus V. *Thiocapsa*, p. 42
  B. Cells contain gas vacuoles
    1. Cells motile by means of polar flagella
      a. Cells spherical; typically diplococcus-shaped before cell division:
        Genus VI. *Lamprocystis*, p. 43
    2. Cells non-motile
      a. Cells rod-shaped:
        Genus VII. *Thiodictyon*, p. 44
      b. Cells spherical to ovoid, characteristically arranged in regular platelets (flat sheets):
        Genus VIII. *Thiopedia*, p. 45
      c. Cells spherical:
        Genus IX. *Amoebobacter*, p. 46
II. Sulfur globules appearing outside the cells when grown with sulfide as electron donor.
        Genus X. *Ectothiorhodospira*, p. 47

Genus I. **Chromatium** *Perty 1852, 174*

Ovoid, bean- or rod-shaped bacteria, multiplying by **binary fission, motile by means of polar flagella.** Under certain unfavorable conditions and in many natural environments cells **may be immotile and grow in** more or less regular **aggregates surrounded by slime.** Unfavorable mineral salt concentrations cause pronounced swellings, club- and spindle-shaped forms formerly recognized as a separate genus, *Rhabdomonas* Cohn 1875, 167 (synonym *Rhabdochromatium* Winogradsky 1888, 100). **Gram-negative. Contain bacteriochlorophyll *a* and carotenoids,** both pigments being located at **internal membranes of vesicular type. Do not contain gas vacuoles.**

Anaerobic.

Capable of photosynthesis in the presence of hydrogen sulfide, during which they produce and **store,** as an intermediate oxidation product, **elemental sulfur** in the form of **globules inside the cells.** Molecular hydrogen may be used as electron donor. Molecular oxygen is not evolved during photosynthesis. Cell suspensions appear in various shades of orange-brown and pink to violet and purple-red, due to the photosynthetic pigments.

Storage materials: polysaccharides, poly-β-hydroxybutyrate, polyphosphates.

Range of DNA base ratios: 48.0–70.4 moles % guanine + cytosine (buoyant density).

Type species of genus: *Chromatium okenii* (Ehrenberg) Perty 1852, 174.

Key to the species of genus **Chromatium**

Key characteristics

Cell shape and size
Requirement for vitamin B_{12}
Color of culture
Type of carotenoid (see Table 1.2)
DNA base ratio

I. Cells large, more than 2.5 μm wide; vitamin B_{12} usually required as a growth factor.
  A. Culture purple-red, major carotenoid okenone
    1. Cells 4.5–6 μm wide:
      1. *Chromatium okenii*
    2. Cells 3.5–4.5 μm wide:
      2. *Chromatium weissei*
  B. Culture purple-violet, major carotenoid rhodopinal
    1. Cells 3.5–4.0 μm wide; sulfur globules predominantly located at the two poles of the cell:
      3. *Chromatium warmingii*
    2. Cells 3.0–4.5 μm wide; sulfur globules evenly distributed within the cell; 1–3% NaCl required, otherwise cells pleomorphic:
      4. *Chromatium buderi*

II. Cells small, less than 2.5 μm wide. Vitamin B_{12} usually not required as a growth factor.
  A. Culture purple-red, major carotenoid okenone
      5. *Chromatium minus*
  B. Culture purple-violet, major carotenoid rhodopinal
      6. *Chromatium violascens*
  C. Culture orange-brown to brownish red; carotenoids of the normal spirilloxanthin series
    1. Cells ovoid, about 2 μm wide; DNA base ratio 61–66 moles % guanine + cytosine:
      7. *Chromatium vinosum*
    2. Cells slender, rod-shaped, 1.2–1.5 μm wide; DNA base ratio around 70 moles % guanine + cytosine:
      8. *Chromatium gracile*
    3. Cells almost spherical to ovoid, about 1.2 μm wide; DNA base ratio around 64 moles % guanine + cytosine:
      9. *Chromatium minutissimum*

Genus II. **Thiocystis** *Winogradsky 1888, 60*

Spherical to ovoid bacteria, before cell division typically diplococcus-shaped, multiplying by **binary fission, motile** by means of **polar flagella.** Under certain unfavorable conditions and in many natural environments, **cells may be immotile and grow in more or less regular aggregates surrounded by slime. Gram-negative. Contain bacteriochlorophyll *a* and carotenoids,** both pigments being located at **internal membranes of vesicular type. Do not contain gas vacuoles.**

Anaerobic.

Capable of photosynthesis in the presence of hydrogen sulfide, during which they produce and **store,** as an intermediate oxidation product, **elemental sulfur** in the form of **globules inside the cells.** Molecular hydrogen may be used as electron donor. Molecular oxygen is not evolved during photosynthesis. Cell suspensions appear in various shades of brown to purple-red and violet due to the photosynthetic pigments.

Storage materials: polysaccharides, poly-β-hydroxybutyrate, polyphosphates.

Catalase and hydrogenase present.

Range of DNA base ratios: 61.3–67.9 moles % guanine + cytosine (buoyant density).

Type species of genus: *Thiocystis violacea* Winogradsky 1888, 65.

Key to the species of genus **Thiocystis**

I. Sulfur globules appear evenly distributed throughout the cell; cells contain carotenoids of the rhodopinal series; culture purple-violet:

1. *Thiocystis violacea*

II. Sulfur globules occur only in the peripheral part of the cell; cells contain carotenoids of the okenone series; culture purple-red:

2. *Thiocystis gelatinosa*

Genus III. **Thiosarcina** *Winogradsky 1888, 104*

Individual bacteria spherical, forming regular cubical packets, resulting from consecutive divisions in three perpendicular planes. Packets commonly containing 8–64 cells. **Motile by means of polar flagella.** Not spore-forming. **Gram-negative. Contain bacteriochlorophyll *a* and carotenoids. Do not contain gas vacuoles.**

Anaerobic.

Capable of photosynthesis in the presence of hydrogen sulfide, during which they produce and **store,** as an intermediate oxidation product, **elemental sulfur** in the form of **globules inside the cells.** Molecular hydrogen may be used as electron donor. Molecular oxygen is not evolved during photosynthesis. Cell suspensions appear in various shades of purplish to red due to the photosynthetic pigments.

Type species of genus: *Thiosarcina rosea* (Schroeter) Winogradsky 1888, 104.

Genus IV. **Thiospirillum** *Winogradsky 1888, 104*

Spiral-shaped bacteria, dividing by **binary fission, motile by means of polar flagella.** Under certain unfavorable conditions cells may be immotile and grow in more or less regular aggregates surrounded by slime. **Gram-negative. Contain bacteriochlorophyll *a* and carotenoids,** both pigments being located at **internal membranes of vesicular type. Do not contain gas vacuoles.**

Anaerobic.

Capable of photosynthesis in the presence of hydrogen sulfide, during which they produce and **store,** as an intermediate oxidation product, **elemental sulfur** in the form of globules **inside the cells.** Molecular hydrogen may be used as electron donor. Molecular oxygen is not evolved during photosynthesis. Cell suspensions appear in various shades of orange-brown to purple-red due to the photosynthetic pigments.

DNA base ratio: 45.5 (one species) moles % guanine + cytosine (buoyant density).

Type species of genus: *Thiospirillum sanguineum* (Ehrenberg) Winogradsky 1888, 104. (Designated by Enlows 1920, 94.)

Key to the species of genus **Thiospirillum**

I. Cells 2.5–4.0 μm wide, generally 30–40 μm long.

A. Individual cells and cell masses purple-red:

1. *Thiospirillum sanguineum*

B. Cultures orange-brown:

2. *Thiospirillum jenense*

II. Cells 1.5–2.5 μm wide, 4–12 μm long.

3. *Thiospirillum rosenbergii*

Genus V. **Thiocapsa** *Winogradsky 1888, 84*

Spherical to slightly ovoid, multiplying by **binary fission, non-motile. Tetrads may be formed** as a result of consecutive divisions in two perpendicular planes. Under certain unfavorable conditions and in many natural environments cells **may grow in more or less regular aggregates surrounded by slime. Gram-negative. Contain bacteriochlorophyll *a* or *b* and carotenoids,** both types of pigments being located on **internal membranes of vesicular or tubular type. Do not contain gas vacuoles.**

Anaerobic phototrophs; in addition, one species, *T. roseopersicina,* able to carry out an oxidative metabolism in the dark under microaerophilic to aerobic conditions.

Capable of photosynthesis in the presence of hydrogen sulfide, during which they produce and **store,** as an intermediate oxidation product, **elemental sulfur** in the form of **globules inside the cells.** Molecular hydrogen may be used as electron donor. Molecular oxygen is not evolved during photosynthesis. Cell suspensions appear in various shades of orange-brown to pink and purple-red due to the photosynthetic pigments.

Range of DNA base ratios: 63.3–69.9 moles % guanine + cytosine (buoyant density).

Type species of genus: *Thiopcasa roseopersicina* Winogradsky 1888, 84.

Key to the species of genus **Thiocapsa**

I. Cells contain bacteriochlorophyll *a* and spirilloxanthin as the major carotenoid; culture definitely pink. Photosynthetic membrane system of vesicular type; all strains tested able to grow under microaerophilic to aerobic conditions in the dark:

1. *Thiocapsa roseopersicina*

II. Cells contain bacteriochlorophyll *b* and tetrahydrospirilloxanthin as the major carotenoid; culture light orange to orange-brown. Photosynthetic membrane system consisting of bundles of ribbon-like branched tubes; strictly anaerobic:

2. *Thiocapsa pfennigii*

Genus VI. **Lamprocystis** *Schroeter 1886, 151*

Spherical to ovoid bacteria, before cell division typically diplococcus-shaped, multiplying by **binary fission, motile by** means of **polar flagella.** Depending on conditions, cells may remain as individuals or attached in tetrads which **aggregate into areas of considerable size, the whole being embedded in slime.** The elongated colonies may be branched or even somewhat netlike. The aggregates may break up into smaller clusters and **more or less spherical colonies which become motile** by the flagella of the composing individual cells. **Gram-negative. Contain bacteriochlorophyll *a* and carotenoids,** both pigments being located at **internal membranes of vesicular type. Contain gas vacuoles** in the central part of the cell.

Anaerobic.

Capable of photosynthesis in the presence of hydrogen sulfide, during which they produce and **store,** as an intermediate oxidation product, **elemental sulfur in the form of globules in the gas vacuole-free peripheral part of the cell.** Molecular hydrogen may be used as electron donor. Molecular oxygen is not evolved during photosynthesis. Cell suspensions appear in various shades of purple-red, due to the photosynthetic pigments.

DNA base ratio: 63.8 (one species) moles % guanine + cytosine (buoyant density).

Type species of genus: *Lamprocystis roseopersicina* (Kützing) Schroeter 1886, 151.

Genus VII. **Thiodictyon** *Winogradsky 1888, 80*

Individual bacteria **rod-shaped with rounded ends, sometimes** appearing **spindle-shaped,** multiplying by **binary fission, non-motile under all conditions. May form aggregates** in which the cells become arranged end to end in an **irregular netlike structure,** the shape of which is not constant; cells may also form more compact masses or break up into individual cells. **Gram-negative. Contain bacteriochlorophyll *a* and carotenoids,** both pigments being located at

internal membranes of vesicular type. Contain gas vacuoles in the central part of the cell.

Anaerobic.

Capable of photosynthesis in the presence of hydrogen sulfide, during which they produce and **store,** as an intermediate oxidation product, **elemental sulfur** in the form of **globules in the gas vacuole-free peripheral part of the cells.** Molecular hydrogen may be used as electron donor. Molecular oxygen is not evolved during photosynthesis. Cell suspensions appear in various shades of purple-violet due to the photosynthetic pigments.

Range of DNA base ratios: 65.3–66.3 moles % guanine + cytosine (buoyant density).

Type species of genus: *Thiodictyon elegans* Winogradsky 1888, 82.

Key to the species of genus Thiodictyon

I. Cells able to form typical netlike aggregates:

1. *Thiodictyon elegans*

II. Cells not able to form netlike aggregates, usually present in the form of free individual cells:

2. *Thiodictyon bacillosum*

Genus VIII. Thiopedia *Winogradsky 1888, 85*

The description is based in part on: Hirsch, 1969.

Individual bacteria **spherical to ovoid,** multiplying by **binary fission. Cells arranged in flat sheets with typical tetrads as the structural units;** due to consecutive divisions in two perpendicular planes, platelets with 4, 8, 16, 32 and 64 cells are formed. **Non-motile. Gram-negative. Contain bacteriochlorophyll *a* and carotenoids,** both pigments being located at **internal membranes of vesicular or stack-lamellar type. Contain gas vacuoles in the central part of the cell.**

Anaerobic.

Capable of photosynthesis in the presence of hydrogen sulfide, during which they produce and **store,** as an intermediate oxidation product, **elemental sulfur** in the form of **globules in the gas vacuole-free peripheral part of the cells.** Molecular hydrogen may be used as electron donor. Molecular oxygen is not evolved during photosynthesis. Cell suspensions appear in various shades of pink to purple due to the photosynthetic pigments.

Type species of genus: *Thiopedia rosea* Winogradsky 1888, 85.

Genus IX. Amoebobacter *Winogradsky 1888, 71*

The genus *Amoebobacter* Winogradsky (1888) has been redefined in accordance with the characteristics of the type species *A. roseus* as representing only the spherical, non-motile, gas vacuole-containing *Chromatiaceae*. Therefore, the rod-shaped species of the genus *Amoebobacter*, *A. bacillosus* Winogradsky (1888), was transferred to the genus *Thiodictyon* Winogradsky (1888) and became *Thiodictyon bacillosum.*

The genus *Rhodothece* Molisch (1907), comprising spherical, non-motile, gas vacuole-containing forms, is treated as a later synonym of the genus *Amoebobacter*.

Spherical to ovoid bacteria, multiplying by **binary fission, non-motile. Gram-negative. Contain bacteriochlorophyll *a* and group 1 carotenoids,** both pigments being located in **internal membranes of vesicular type. Contain gas vacuoles in the central part of the cell.**

Anaerobic.

Capable of photosynthesis in the presence of hydrogen sulfide, during which they produce and **store,** as an intermediate oxidation product, **elemental sulfur** in the form of **globules in the gas vacuole-free peripheral part of the cells.** Molecular hydrogen may be used as electron donor. Molecular oxygen is not evolved during photosynthesis. Cell suspensions appear in various shades of pink to purple due to the photosynthetic pigments.

Hydrogenase and catalase present.

Range of DNA base ratios: 64.3–65.3 moles % guanine + cytosine (buoyant density).

Type species of genus: *Amoebobacter roseus* Winogradsky 1888, 77.

Key to the species of genus Amoebobacter

I. Cells 2–3 μm in diameter:

1. *Amoebobacter roseus*

II. Cells 1.5–2.5 μm in diameter, pronounced slime formation under certain conditions resulting in a highly viscous culture medium:

2. *Amoebobacter pendens*

Genus X. **Ectothiorhodospira** *Pelsh 1936, 63*

Spiral to slightly bent rod-shaped bacteria, multiplying by **binary fission, motile** by means of **polar flagella.** Under certain unfavorable conditions and in many natural environments cells may be immotile and grow in more or less regular aggregates surrounded by slime. Gram-negative. **Contain bacteriochlorophyll *a* and carotenoids,** both pigments being located at **lamellar membrane stacks** which are continuous with the cytoplasmic membrane (Plate 1.3, Fig. 1). **Do not contain gas vacuoles.**

Anaerobic.

Capable of photosynthesis in the presence of hydrogen sulfide, during which they produce and deposit, as an intermediate oxidation product, **elemental sulfur in the form of globules outside the cells** in the medium. Molecular hydrogen may be used as electron donor. Molecular oxygen is not evolved during photosynthesis. Cell suspensions appear in various shades of brown to red due to the photosynthetic pigments.

Range of DNA base ratios: 62.2–69.9 moles % guanine + cytosine (buoyant density).

Type species of genus: *Ectothiorhodospira mobilis* Pelsh 1936, 63.

Key to the species of genus **Ectothiorhodospira**

Key characteristics

Requirement for NaCl
Flagellation
Cell shape
DNA base ratio

I. Little or no NaCl required for growth. Polar tuft of flagella.
   A. Cells curved; 2–3% NaCl required for growth; DNA base ratio around 67 moles % guanine + cytosine:

      1. *Ectothiorhodospira mobilis*

   B. Cells rather straight; no added NaCl required; DNA base ratio around 63 moles % guanine + cytosine:

      2. *Ectothiorhodospira shaposhnikovii*

II. Extreme halophiles, 14–23% NaCl required for growth. One sheathed polar flagellum:

   3. *Ectothiorhodospira halophila*

FAMILY III. **CHLOROBIACEAE** COPELAND 1956, 31

Cells spherical, ovoid- or rod-shaped; **multiplication by binary fission or binary plus ternary fission. In one genus (*Chloropseudomonas*) cells motile by means of polar flagella; all other genera non-motile.** Genera with or without gas vacuoles.

Depending on culture conditions or environmental conditions in nature, most forms are able to develop either in the form of single cells or in cell masses of various shapes and sizes with more or less slime. **In the presence of sulfide, globules of elemental sulfur are deposited outside the cells, never inside the cells. Single cells and cultures green or brown. The photopigments are characteristically located in the chlorobium vesicles which underlie and are attached to the cytoplasmic membrane** (Plate 1.4).

Bacteriochlorophylls *c* or *d* occur as major components in addition to small amounts of bacteriochlorophyll *a*; carotenoids of group 5 present. Major carotenoid of the green forms is chlorobactene; major components of the brown forms are the isorenieratenes. Many strains require vitamin B_{12} for growth.

All forms are **strictly anaerobic** and **obligately phototrophic**; capable of **photolithotrophic assimilation of carbon dioxide in the presence of sulfide or sulfur** which are photooxidized to sulfate. Many strains are able to use molecular hydrogen as an electron donor for growth provided hydrogen sulfide is present in addition as a source of cell sulfur. Most forms lack an assimilatory sulfate reduction. A number of simple organic substrates are photoassimilated in the presence of both sulfide and carbon dioxide; acetate is the most widely used compound; the *Chlorobiaceae* are therefore potentially mixotrophic. The fixation of molecular nitrogen has been demonstrated in a number of strains. Storage material: usually polyphosphates; poly-β-hydroxybutyrate has not been found in the strains studied. Polysaccharides were identified only in *Chloropseudomonas ethylica*. Range of DNA base ratios: 48.5–58.1 moles % guanine + cytosine (buoyant density).

Two physiological-ecological subgroups can be differentiated on the basis of the presence or absence of gas vacuoles and the resulting selective advantage in enrichment cultures. The first group comprises the forms without gas vacuoles which have a selective advantage in cultures with medium to high sulfide concentrations and light intensities. The second subgroup includes the gas vacuole-containing forms which require very low sulfide concentrations, low light intensities and growth temperatures (10–20 C). The cells of the gas vacuole-containing forms rise to the top of the culture bottles at lower temperatures (4–10 C).

In nature the green sulfur bacteria occur in the anaerobic and sulfide-containing parts of all kinds of aquatic environments, e.g. ditches, ponds, lakes, rivers, sulfur springs, estuaries and other marine habitats. The brown-colored forms occur only in deeper layers of ponds and lakes and in the hypolimnion of meromictic lakes.

The type genus: *Chlorobium* Nadson 1906, 190, designated by Trüper and Pfennig 1971, 9.

Key to the genera of family **Chlorobiaceae**

Key characteristics

Association with other microbes
Presence of gas vacuoles
Cell shape and arrangement
Motility

I. Cells free living not intimately associated with other microbes.
  A. Cells do not contain gas vacuoles.
    1. Cells non-motile.
      a. Cells ovoid to rod-shaped or vibrioid.
        Genus I. *Chlorobium*, p. 52*
      b. Cells irregular spherical to starlike, covered with extrusions.
        Genus II. *Prosthecochloris*, p. 55
    2. Cells motile by means of polar flagella; rod-shaped.
        Genus III. *Chloropseudomonas*, p. 55
  B. Cells contain gas vacuoles; non-motile.
    1. Cells rod-shaped to ovoid; may unite into characteristic aggregates.
        Genus IV. *Pelodictyon*, p. 56
    2. Cells spherical to ovoid; united into characteristic aggregates.
        Genus V. *Clathrochloris*, p. 57
II. Cells live in symbiotic aggregates with other microorganisms (consortia).
        See addenda, p. 58

Note: Gorlenko and Lebedeva (*Mikrobiologiya 40:* 1035–1039, 1971) have described a new genus and species of prosthecate green bacteria, *Ancalochloris perfilievii*, which produce irregular cells 0.5–1.0 μm in diameter with up to six pointed appendages 2 μm long. The organisms produce gas vacuoles, have chlorobium vesicles, and do not deposit sulfur inside the cells or become closely associated with other microbes.

Genus I. **Chlorobium** *Nadson 1906, 190*

Rod-shaped, ovoid or vibrio-shaped bacteria, multiplying by **binary fission, non-motile.** Gram-negative. Contain either **bacteriochlorophyll *c* or *d*** as the major bacteriochlorophyll component as well as **carotenoids of group 5,** all pigments being located in elongated,

* All page numbers cited refer to *Bergey's Manual of Determinative Bacteriology*, 8th edition, The Williams & Wilkens Company, Baltimore, 1974.

ovoid vesicles ("chlorobium vesicles") underlying and attached to the cytoplasmic membrane. **Do not contain gas vacuoles.**

Anaerobic.

Capable of photosynthesis in the presence of hydrogen sulfide, during which they produce and **deposit** as an intermediate oxidation product, **elemental sulfur** in the form of globules **outside the cells in the medium.** Molecular hydrogen may be used as electron donor. Molecular oxygen is not evolved during photosynthesis. **Cell suspensions** appear in various shades of **yellowish green and chocolate brown** due to the photosynthetic pigments.

Storage material: polyphosphate, not polysaccharides or poly-β-hydroxybutyrate.

Range of DNA base ratios: 49.0–58.1 moles % guanine + cytosine (buoyant density).

Type species of genus: *Chlorobium limicola* Nadson 1906, 190.

Key to the species of genus **Chlorobium**

Key characteristics

Color of culture
Type of carotenoid (see Table 1.2)
Cell shape and size
Utilization of thiosulfate

I. Culture green; major carotenoid chlorobactene.
  A. Cells rod-shaped, 0.7–1.1 μm wide.
    1. Thiosulfate not utilized:
      1. *Chlorobium limicola*
    2. Thiosulfate utilized:
      1a. *Chlorobium limicola* forma sp. *thiosulfatophilum*
  B. Cells short vibrio-shaped, 0.5–0.7 μm wide.
    1. Thiosulfate not utilized:
      2. *Chlorobium vibrioforme*
    2. Thiosulfate utilized:
      2a. *Chlorobium vibrioforme* forma sp. *thiosulfatophilum*

II. Culture brown; major carotenoids isorenieratene and β-isorenieratene.
  A. Cells rod-shaped, 0.6–0.8 μm wide.
      3. *Chlorobium phaeobacteroides*
  B. Cells short vibrio-shaped, 0.3–0.4 μm wide.
      4. *Chlorobium phaeovibrioides*

Genus II. **Prosthecochloris** *Gorlenko 1970, 148*

Spherical to ovoid bacteria, forming prosthecae, dividing by binary fissions in many directions. When separation is incomplete, cells form groups and branched chains, the configuration of which depends on the direction of fissions. **Non-motile. Gram-negative. Contain bacteriochlorophyll *c*** as the major bacteriochlorophyll component as well as **carotenoids,** all pigments being located in elongated, ovoid vesicles ("chlorobium vesicles") underlying and attached to the cytoplasmic membrane. **Do not contain gas vacuoles.**

Anaerobic.

Capable of photosynthesis in the presence of hydrogen sulfide, during which they produce and **deposit** as an intermediate oxidation product, elemental **sulfur** in the form of globules **outside the cells.** Molecular oxygen is not evolved during photosynthesis. **Cell suspensions** appear **green** due to the photosynthetic pigments.

DNA base ratio: 50.0–56.1 moles % guanine + cytosine (buoyant density).

Type species of genus: *Prosthecochloris aestuarii* Gorlenko 1970, 148.

Genus III. **Chloropseudomonas** *Czurda and Maresch 1937, 123*

Nom. cons. proposed by Pfennig and Trüper 1969, 153

Rod-shaped bacteria, multiplying by **binary fission, motile by means of polar flagella. Gram-negative. Contain bacteriochlorophyll *c*** as the major bacteriochlorophyll component and

group 5 carotenoids, all pigments being located in elongate-ovoid vesicles ("chlorobium vesicles") underlying and attached to the cytoplasmic membrane. **Do not contain gas vacuoles.**

Anaerobic.

Capable of photosynthesis in the presence of hydrogen sulfide, during which they produce and **deposit** as an intermediate oxidation product, **elemental sulfur** in the form of globules **outside the cells** in the medium. Molecular hydrogen may be used as electron donor. Molecular oxygen is not evolved during photosynthesis. **Cell suspensions** appear in various shades of **yellowish green,** due to the photosynthetic pigments.

Range of DNA base ratios: 57.2–57.6 (one species) moles % guanine + cytosine (chemical analysis), 55.1 (buoyant density).

Type species of genus: *Chloropseudomonas ethylica* Shaposhnikov *et al.* 1960, 168.

Note: The validity of this genus has been questioned (Pfennig, personal communication). It is stated to be a syntrophic mixture of a phototroph and a sulfur-reducing bacterium.

Genus IV. **Pelodictyon** *Lauterborn 1913, 98*

Rod-shaped to ovoid bacteria, multiplying by **binary fission, non-motile. Branching may occur as a result of ternary fission and netlike three dimensional aggregates** or more or less **spherical colonies may be formed.** Gram-negative. Contain either **bacteriochlorophyll *c* or *d*** as the major bacteriochlorophyll components and **carotenoids of group 5,** both pigments being located in elongated-ovoid vesicles ("chlorobium vesicles") underlying and attached to the cell membrane (Plate 1.4). **Contain gas vacuoles.**

Anaerobic.

Capable of photosynthesis in the presence of hydrogen sulfide, during which they produce and **deposit** as an intermediate oxidation product, **elemental sulfur** in the form of globules **outside the cells** in the medium. Molecular hydrogen may be used as electron donor. Molecular oxygen is not evolved during photosynthesis. **Cell suspensions** appear in various shades of **yellowish green** due to the photosynthetic pigments.

Range of DNA base ratios: 48.5–58.1 moles % guanine + cytosine (buoyant density).

Type species of genus: *Pelodictyon clathratiforme* (Szafer) Lauterborn 1913, 98.

Key to the species of genus **Pelodictyon**

I. Cells rod-shaped, forming typical netlike aggregates due to their ability to undergo both binary and ternary fissions:

1. *Pelodictyon clathratiforme*

II. Cells ovoid to short vibrio-shaped, occurring either in the form of free individual cells or united into hollow spherical or irregular round colonies:

2. *Pelodictyon luteolum*

Genus V. **Clathrochloris** *Geitler in Geitler and Pascher 1925, 457*

Spherical to ovoid bacteria, multiplying by **binary fission, non-motile. Cells in chains which are usually arranged in loose trellis-like aggregates.** Color is yellowish green. **Contain gas vacuoles** which have been mistaken for sulfur globules.

Anaerobic.

Capable of photosynthesis in the presence of hydrogen sulfide, during which they produce and **deposit** as an intermediate oxidation product, **elemental sulfur** in the form of globules **outside the cells. Cell suspensions** appear in various shades of **yellowish green.**

Type species of genus: *Clathrochloris sulfurica* (Szafer) Geitler in Geitler and Pascher 1925, 457.

PART 2

THE GLIDING BACTERIA

Key to the orders of gliding bacteria

I. Cells aggregate to form fruiting bodies (under suitable environmental conditions).

Order I. *Myxobacterales*, p. 76.*

II. Cells do not form fruiting bodies.

Order II. *Cytophagales*, p. 99, see also families and genera of uncertain affiliation, p. 120.

ORDER I. MYXOBACTERALES THAXTER *emend.* STANIER 1957, 854

The Fruiting Myxobacteria

Unicellular rods, <1.5 μm in diameter, typically embedded in a more or less tough layer of slime. The rods may be uniformly cylindrical with blunt rounded ends or somewhat tapered at the ends. Multiplication is by binary transverse fission. Capable of slow gliding movement in contact with a solid surface or air-water interface but lacking detectable locomotor organelles. Gram-negative.

Under appropriate environmental conditions the cells aggregate to form fruiting bodies constructed of slime and cells and which are often brightly colored and of macroscopic dimensions. The cells within the fruiting bodies become resting cells, termed myxospores. In some genera the myxospores are not readily distinguishable from vegetative cells; in others they are optically dense or refractile, encapsulated, shortened rods or spheres known as microcysts. The fruiting body may consist simply of a mass of slime and cells or the myxospores may be enclosed in sporangia of characteristic shape and dimensions which may be raised above the substrate on simple or branched stalks.

Chemoorganotrophs: strictly aerobic: energy-yielding metabolism respiratory, never fermentative. Typically produce enzymes capable of hydrolyzing such macromolecules as proteins, nucleic acids, fatty acid esters and various polysaccharides including, in some species, cellulose. Many are capable of lysing other eucaryotic and procaryotic microorganisms. Photosynthetic pigments are lacking but carotenoid pigments characteristically, and melanin pigments often, are produced.

The G + C content of the DNA ranges from 67–71 moles % for all species examined.

Further Comments

Most known myxobacteria occur in soil and frequently develop on decomposing plant material, the bark of living trees or animal dung. Both in nature and in the laboratory their presence may be detected through the appearance of fruiting bodies. Although, one species has been described as aquatic (Geitler, 1924) and a number of myxobacteria have been isolated from fresh water environments (Jeffers, 1964; Shilo, 1970), further work will be required to determine whether these occurrences are incidental or reflect true aquatic habitats.

Most species of myxobacteria fail to grow or grow poorly on conventional bacteriological media (e.g. nutrient broth, etc.). They may conveniently be divided into two physiological groups: the bacteriolytic group, which includes all *Myxococcaceae*, *Archangiaceae*, *Cystobacteraceae* and

* All page numbers cited refer to *Bergey's Manual of Determinative Bacteriology*, 8th edition, The Williams & Wilkins Company, Baltimore, 1974.

most *Polyangiaceae*; and the cellulolytic group. The bacteriolytic myxobacters may usually be cultivated on agar media containing living or killed bacterial or yeast cells, which they lyse. Their frequent observation on animal dungs probably reflects the abundance therein of bacteria as a source of food. Most will also grow well on agar media containing enzymatically hydrolyzed protein and salts. However cultivation in liquid media may present special problems and a solid medium which supported good growth, may fail to do so if agar is omitted.

The minimal nutritional requirements of the bacteriolytic myxobacters are still very poorly known. When cultivated on chemically defined media, the few strains studied (Dworkin, 1962; Mayer, 1967; Hemphill and Zahler, 1968; McCurdy and Khouw, 1969) required complex amino acid mixtures and relatively high concentrations of Mg^{++} or Ca^{++} for growth. No definite, obligate vitamin requirements have been demonstrated for most species but some of the *Polyangiaceae* seem to require vitamin B_{12} and perhaps other factors as well (McCurdy, 1964, 1969). None appears capable of utilizing carbohydrates, for carbon or energy. In spite of this the addition of a polysaccharide (as well as other soluble polymers) such as starch, glycogen, carboxymethylcellulose, etc. (Schurmann, 1967; Dworkin, 1962; McCurdy, 1969), will often stimulate growth in liquid media which would otherwise fail to support growth.

The cellulolytic myxobacters examined so far (Krzemieniewska and Krzemieniewski, 1937; Coucke, 1969) have simple requirements and can develop in a medium with an inorganic base supplemented with single carbohydrates (cellulose or a simple sugar).

Although distinguished from other bacteria by their gliding motility and the active bending and flexing which they frequently exhibit, the vegetative cells of myxobacters have not been found, either chemically or in ultrastructure, to differ, in any significant way, from other Gram-negative bacteria (Voelz and Dworkin, 1962; McCurdy, 1968; White *et al.*, 1968; Reichenbach and Voelz, 1969). There is a tendency to autolyse or to form spheroplasts under a variety of conditions (anaerobiosis, in old cultures, high temperature, in the presence of some cations). In general they stain poorly by conventional methods and are distorted by heat fixation; it is therefore best to study vegetative cells with phase-contrast microscopy. Motility is slight or absent in cells from liquid culture and is most conveniently observed in cells growing on the surfaces of (especially nutritionally dilute) agar media.

The vegetative colony of myxobacters, often termed a swarm or pseudoplasmodium, has a characteristic appearance as a result of slime production and gliding movement of the cells. It is usually flat and thin with many concentric folds and/or radiating lines and spreads extensively, sometimes rapidly, over the surface as a result of the outward movement of cells at the periphery. Frequently, the latter advance as groups forming tongue-like extensions or isolated clumps and streams. Especially in the *Polyangiaceae*, the underlying agar is etched, eroded and penetrated, thereby conferring an irregular topology to the colony.

Continued cultivation in nutrient-rich media frequently yields variants which produce mucoid colonies with little tendency to spread, which may no longer fruit and which therefore resemble non-gliding bacteria.

Media supporting good vegetative development are seldom suitable for fruiting body formation. In general fructification is favored by dilute media* (e.g. water agar, *E. coli* medium (ECM), dung pellet agar, etc.). In some cases, the suppression of fruiting body formation on nutrient-rich media furnished only with soluble nutrients, appears due to the inhibiting effect of certain amino acids (Dworkin, 1963; McCurdy, 1964; Hemphill and Zahler, 1968).

The fruiting bodies of the myxobacters are the resistant or resting stage. In those organisms producing microcysts, the hard slime capsule surrounding the cell confers considerably more resistance to heat, desiccation, physical disruption and ultraviolet radiation than is possessed by vegetative cells. Microcysts may in general be induced to form in liquid media without fruiting body formation by the addition of 0.5 M glycerol (Dworkin and Gibson, 1964; McCurdy, 1969; Reichenbach and Dworkin, 1970; McCurdy, unpublished results). In the *Polyangiaceae*, which do not produce microcysts, resistance appears to be a

* Media for Myxobacters: DPA, dung pellet agar. Three or four sterile antibiotic-free rabbit dung pellets are placed in a sterile Petri dish and partially covered with water agar or Ca^{++} agar (McCurdy, 1968).

ECM, *Escherichia coli* medium. Dried *E. coli* cells, 100 mg; agar, 1.5 g; water, 100 ml. Autoclave.

Cas.-Mg^{++}. Casitone (Difco), 2.5 g; soluble starch, 5.0 g; $MgSO_4 \cdot 7H_2O$, 0.5 g; K_2HPO_4, 0.25 g; agar, 15 g; water, 1000 ml. Autoclave.

W. A., water agar. K_2HPO_4, 0.25 g; $MgSO_4 \cdot 7H_2O$, 0.5 g; agar, 15 g; water 1000 ml. Autoclave. Living or killed bacterial cells (*E. coli*, *Micrococcus* (*Sarcina*) *luteus*, etc.) heavily streaked over the surface are the source of organic nutrients.

function of the modified slime which provides the wall material of the intact sporangium.

The taxonomy of the *Myxobacterales* is still largely based on the characteristics of the fruiting bodies, even though these structures are known to vary greatly in response to environmental influences and either may not be formed or are arrested in their development as a result of mutation or unfavorable conditions. Imperfectly formed fruiting bodies of certain species simulate the mature fruiting bodies of other members of the group. Accordingly it is essential, in identifying these organisms, to study fruiting body formation, preferably with pure cultures, under a wide range of defined nutritional and environmental conditions and over a considerable period of time. This principle has not always been followed in the taxonomy of the myxobacters; the biochemical data conventionally used in bacterial taxonomy are often lacking.

Key to the families of order **Myxobacterales**

I. Vegetative cells tapered, microcysts produced.
  A. Microcysts spherical or oval.
    Family I. *Myxococcaceae*, p. 79
  B. Microcysts rod-shaped.
    1. Microcysts not in sporangia.
      Family II. *Archangiaceae*, p. 83
    2. Microcysts in sporangia.
      Family III. *Cystobacteraceae*, p. 86

II. Vegetative cells of uniform diameter with blunt, rounded ends. Myxospores resemble vegetative cells.
    Family IV. *Polyangiaceae*, p. 92

FAMILY I. **MYXOCOCCACEAE** JAHN 1924, 84

Vegetative cells slender, straight to slightly tapered, flexible rods with rounded ends. **Refractile myxospores** (microcysts) **spherical or ellipsoidal.**

The type and only genus: *Myxococcus* Thaxter 1892, 403.

Genus I. **Myxococcus** Thaxter 1892, 403

Vegetative cells slender rods with tapering or rounded ends, 0.4–0.7 by 2.0–10.0 μm. Motile by gliding motility. Fruiting bodies contain **refractile spherical or ellipsoidal microcysts** up to 2.3 μm in the largest diameter.

The **myxospores** are **not** enclosed **in a sporangium.**

Chemoorganotrophs: Metabolism respiratory.

TABLE 2.1

Differential characteristics of species of **Myxococcus**

Characteristics determined on Casitone-Mg^{++}. See McCurdy (1969) for methods.

	1. *M. fulvus*	2. *M. virescens*	3. *M. xanthus*	4. *M. stipitatus*	5. *M. coralloides*	6. *M. macrosporus*
Fruiting bodies deliquescent	+	+	+	+	−	−
Fruiting bodies raised on a well defined, persistent stalk	±	−	−	+	−	−
Myxospores 1.5 μm or more in diameter	−	+	+	−	−	+
Color of vegetative cell masses						
Yellow	−	+	+	+	−	−
Buff to tan	−	−	−	−	d	d
Flesh-colored to reddish orange	+	−	−	−	d	d
Greenish diffusible pigment on agar	−	+	−	−	−	−
Hydrolyzes starch in 3 days or less	−	−	−	−	+	+*
Hydrolyzes esculin	−	−	−	−	±	−
Oxidase	−	−	−	+	−	−

* Only one strain tested.

Those which have been studied require several amino acids, and are capable of hydrolyzing protein, starch, nucleic acids and various fatty acid esters. **Bacteriolytic; non-cellulolytic.**

Strict aerobes.

Fail to grow at 40 C or above.

Colonies adsorb congo red.

Resistant to 10-unit discs of penicillin. Sensitive to 10-μg discs of neomycin and tetracycline and to 5-μg discs of erythromycin.

Methods suitable for studying the physiology of members of the genus are described by McCurdy (1969).

The molar G + C content of the DNA of those species examined ranges from 68–71% (Mandel and Leadbetter, 1965).

Type species: *Myxococcus fulvus* (Cohn) Jahn 1911, 198.

FAMILY II. **ARCHANGIACEAE** JAHN 1924, 66

The **vegetative rods** are slender **with tapered ends.** The **microcysts** are very **short rods, ellipsoids or spheres** which are **never enclosed in sporangia;** the **fruiting bodies** consist of **irregular masses** or projections of variable dimensions.

The type and only genus is *Archangium* Jahn 1924, 66.

Genus I. **Archangium** *Jahn 1924, 66*

Vegetative cells slender, **tapered,** flexible **rods.**

Sporangia lacking; fruiting bodies irregular in form, swollen or brainlike consisting internally of intestine-like twisted or intertwined masses. There is **no definite slime wall.**

Microcysts are very **short rods, ellipsoids or spheres,** refractile or optically dense.

Vegetative colonies do not etch or erode agar media. Congo red is adsorbed.

Type species: *Archangium gephyra* Jahn 1924, 67.

FAMILY III. **CYSTOBACTERACEAE** McCURDY 1970, 286

Vegetative cells tapered flexible **rods** which are converted to refractile or phase-dense, **rod-shaped microcysts** enclosed **in sporangia** of definite shape. The sporangia may be sessile, occurring singly or in groups and enclosed in a slime membrane or envelope; or borne on stalks (sporangiophores) which may be simple or branched. The sporangia may be solitary or in clusters at the tips of the stalks.

The vegetative colonies **do not etch or erode agar. Congo red is adsorbed** by the vegetative slime. All of the species examined grow well on media containing enzymatically hydrolyzed protein, starch and salts. **Bacteriolytic. Non-cellulolytic.**

Key to the genera of family **Cystobacteraceae**

I. Sporangia sessile.

Genus I. *Cystobacter*, p. 87

II. Sporangia stalked.

A. Sporangia borne singly on an unbranched stalk.

Genus II. *Melittangium*, p. 89

B. Sporangia in clusters on branched or unbranched stalks.

Genus III. *Stigmatella*, p. 90

Genus I. **Cystobacter** *Schroeter 1886, 170*

Vegetative cells slender, **tapered,** flexible rods.

Sporangia sessile, occurring singly or in groups; rounded, elongate or coiled and surrounded by a definite slime envelope or wall; either free or embedded in a second slimy layer.

Microcysts rod-shaped, phase-dense or **refractile,** rigid.

Vegetative colonies do not etch or erode agar media, **congo red is adsorbed.**

The minimum nutritional requirements are not known but all species are easily cultivated on media containing enzymatically hydrolyzed protein, salts and starch or glycogen. The latter utilized. **Cellulose is not digested.**

Type species: *Cystobacter fuscus* Schroeter.

Key to the species of genus **Cystobacter**

I. Sporangia rounded to spherical.
  A. Mature sporangia dark brown, >60 μm diameter.
    1. *Cystobacter fuscus*
  B. Mature sporangia pink to light brown, <60 μm
    2. *Cystobacter minus*

II. Sporangia elongate, coiled.
    3. *Cystobacter ferrugineus*

Genus II. **Melittangium** *Jahn 1924, 78*

Vegetative cells tapered rods.

Sporangia borne singly on a stalk.

Microcysts rod-shaped, optically dense or refractile.

Vegetative colonies do not etch or erode agar media; congo red is adsorbed.

Minimum nutritional requirements unknown but cultivatable on media containing enzymatically hydrolyzed protein.

Type species: *Melittangium boletus* Jahn 1924, 78.

Key to the species of genus **Melittangium**

I. Sporangia flattened like a mushroom cap.
    1. *Melittangium boletus*

II. Sporangia spherical or ellipsoidal.
  A. Stalk short, <40 μm
    2. *Melittangium lichenicola*
  B. Stalk long, >40 μm, bent.
    3. *Melittangium alboraceum*

Genus III. **Stigmatella** *Berkeley and Curtis in Berkeley 1875, 97*

Vegetative cells are rods with tapered ends.

Sporangia borne singly or in clusters on stalked fruiting bodies (the stalks often occurring in groups arising from a common hypothallus).

Myxospores short, rigid, phase-dense or refractile rods surrounded by a definite slime capsule.

Vegetative colonies do not etch, erode or penetrate agar media. Congo red is adsorbed.

The minimum nutritional requirements are not known but easily cultivated on media containing enzymatically hydrolyzed protein.

Urea is usually hydrolyzed.

Aerobic.

Temperature range 18–37 C; optimum 30 C.

G + C content of species examined 68.5–68.7 by T_m (McCurdy and Wolf, 1967).

Type species: *Stigmatella aurantiaca* Berkeley and Curtis in Berkeley 1875, 97.

Note: The two species in this genus, *S. aurantiaca* and *S. erecta*, are very similar and appear to be separated on the basis of fine differences in the fruiting bodies. Refer to Plate 2.2, Figs. 11–14 for a visual comparison.

FAMILY IV. **POLYANGIACEAE** JAHN 1924, 75

Vegetative cells cylindrical, of uniform diameter **with blunt, rounded ends. Myxospores resemble vegetative cells,** not refractile or phase-dense, **lacking a slime capsule** in species examined (McCurdy, 1968; Peterson, 1969). **Sporangia sessile, solitary or in groups** (sori) bound by a common membrane or they may be borne singly or in clusters on branched or unbranched stalks.

The **colonies** of all species examined **etch,** **erode** and **penetrate agar media. Congo red** is **not adsorbed** by the vegetative slime of any of the species examined.

Cellulose may or may not be hydrolyzed.

The type genus is *Polyangium* Link 1809, 42.

TABLE 2.1A

Characteristics of the species of genus **Polyangium**

Species	Vegetative rods (μm)	Sporangia					Sorus envelope color	Size of myxospore (μm)	Vegetative colonies	Cellulose digestion	Saprophytic on algae
		Color	Shape	Size (μm)	Wall	Arrangement					
P. vitellinum	0.9–1.2 by 4–10	Golden yellow, reddish yellow, brownish orange	Oval, spherical or cushion-like	75–400	Definite	Single or groups of 20	White	0.9 by 3–5			−
P. luteum	0.8 by 6–7	Yellow	Oval, spherical or shapeless	37–115 by 39–90	Thin, colorless		Yellow	0.7–0.8 by 3.8–5.8	Colorless to yellow, penetrate agar	−	−
P. aureum		Light to reddish brown	Oval or spherical	20–60 by 15–50	Orange-yellow, 3.5 μm thick		Colorless	0.7–0.9 by 2.8–5.3	Colorless, penetrate agar		−
P. parasiticum	0.7 by 4–7	Reddish brown	Spherical or elongate	15–50	Double-contoured	Groups of 2–8	Colorless	Slightly shorter than vegetative rods			+
P. fumosum	0.7–1.0 by 5–7	Colorless	Spherical or elongate	17–72 by 13–60	Definite, often double-contoured	Groups of 2–50	Pink to smoky gray	0.7–0.8 by 2.5–5	Light pink, divide into migrating masses		−
P. spumosum		Transparent to brownish	Spherical or ellipsoid	8–26 by 7–20	Colorless to brownish	Numerous in double or single rows		0.4–0.5 by 2–3		+	−
P. minor		Orange-brown	Spherical, oval or bean-shaped	60–140	Yellow-orange, 2 μm thick	Single or groups of 4–10		0.7 by 2.5–3.5	Grayish yellow		−
P. rugiseptum		Orange-red	Spherical or oval	85 average	2 layers: inner smooth, yellow; outer flaky, orange	Single					−
P. cellosum	0.8–1.2 by 3–10	See sorus color	Spherical to polygonal	20–30		4 to several hundred	Pale yellow to pink, orange, rusty red, brown or black	Length 1–3		+	−
P. sorediatum	0.8–1.2 by 3–6	See sorus color	Spherical to polygonal	5–15		Groups up to several hundred	Yellow-orange	0.8 by 3–4		−	−

Key to the genera of family **Polyangiaceae**

I. Sporangia sessile.
  A. Cells rod-shaped, never coccoid.
    Genus I. *Polyangium*, p. 92
  B. Cells become coccoid.
    Genus II. *Nannocystis*, p. 96
II. Sporangia stalked.
    Genus III. *Chondromyces*, p. 96

Genus I. **Polyangium** *Link 1809, 42*

Vegetative cells cylindrical, of uniform diameter **with blunt rounded ends.**

Sporangia sessile, solitary or in groups (sori) often bounded by a common envelope or layer of slime.

Myxospores resemble vegetative cells, not refractile or phase-dense, **lacking a slime capsule** in species examined.

Colonies of all species examined etch, erode and penetrate agar. Congo red is not adsorbed.

Cellulose may or may not be digested.

Type species is *Polyangium vitellinum* Link 1809, 42.

Note: There are 10 recognized species in this genus, of which only two (*P. fumosum* and *P. cellulosum*) have been grown in pure culture in recent studies. Table 2.1A summarizes the description of each species. The reader should also consult Plate 2.3.

Genus II. **Nannocystis** *Reichenbach 1970, 136*

Vegetative cells short, blunt ended rods or cocci. Sporangia solitary.

Type species is *Nannocystis exedens* Reichenbach 1970, 137.

TABLE 2.1B

Characteristics of the species of genus **Chondromyces**

Species	Vegetative rods (μm)	Sporangia				Stalks		
		Color	Shape	Size (μm)	Arrangement	Branching	Color	Length (μm)
C. crocatus	1.1–1.4 by 3–12	Straw-colored becoming golden or orange	Spindle-shaped, conical or spherical	10–25 by 15–30	In spherical clusters	+	Orange to brown	700 or more
C. apiculatus	1.1–1.4 3–14	Straw-colored to orange to brownish orange	Cylindrical to turnip-shaped with apical appendages	25–40 by 35–50	In spherical clusters of 60 or more	Seldom		700 or more
C. pediculatus	1.1–1.3 by 3–16	Pale yellow to orange becoming orange-red	Spherical, cylindrical, club-shaped or pyriform	25–40 by 35–60	In umbel-shaped groups of up to 60	—	Colorless to orange or brown when dry	To 750
C. catenulatus	1.0–1.3 by 4–6	Light yellow to orange	Fusiform, elliptical or irregular	18 by 20–50	In chains up to 300 μm long, sometimes branched	—	Orange to rust-colored	To 400
C. lanuginosis	0.9–1.0 by 3–8	White becoming yellow, pink then orange	Fused at base, with apical tuft of hair		In spherical or discoidal clusters of up to 80	Occasionally	Colorless becoming yellow	

Genus III. **Chondromyces** *Berkeley and Curtis in Berkeley 1874, 64*

Vegetative cells cylindrical, untapered rods, with blunt rounded ends.

Sporangia borne singly or in clusters **on** simple or branched **stalks.**

Myxospores lack capsules and resemble vegetative rods.

Vegetative swarms etch, erode and penetrate agar media. Vegetative slime does not adsorb congo red dye.

Aerobic.

Temperature range 18–37 C; optimum 28–30 C.

G + C content of species examined 69–70 moles % by T_m determinations.

Type species: *Chondromyces crocatus* Berkeley and Curtis in Berkeley 1874, 64.

Note: Table 2.1B summarizes the description of each species. The reader should also consult Plate 2.5.

ORDER II. CYTOPHAGALES *Leadbetter 1974, 99*

Cells rods, filaments. Hormogonia, gonidia or resting cells may be present. **Fruiting bodies are not produced. Motile by gliding in at least one morphological stage.** Gram-negative.

Chemolithotrophs, chemoorganotrophs or mixotrophs.

Key to the families of order **Cytophagales**

Key characteristics

Gliding motility of cells, filaments and gonidia	Presence of resting stages
Presence of carotenoid pigments	Presence of sulfur or calcium carbonate in cells
Cell shape	Filament shape and attachment

I. Cells motile with gliding motility; filaments not attached.
  A. Carotenoid pigments present; cells rods or filaments.
    Family I. *Cytophagaceae*, p. 99
  B. Carotenoid pigments absent; resting stages not known.
    1. Cells in cylindrical filaments; sulfur may or may not be present.
      Family II. *Beggiatoaceae*, p. 112
    2. Flat filaments, thus far found only in oral cavities of vertebrates.
      Family III. *Simonsiellaceae*, p. 116

II. Filaments generally attached at one end in natural habitats; only gonidia can glide.
    Family IV. *Leucotrichaceae*, p. 118

Familiae incertae sedis

I. Cells motile with gliding motility. Carotenoid pigments absent; resting stages not known.
  A. Cells spherical, ovoid or cylindrical, 5–33 by 15–125 μm, may or may not contain sulfur and calcium carbonate. (May have flagella.)
    Achromatiaceae, p. 120

II. Filaments, unbranched, occur singly or in bundles; some filaments are motile.
    Pelonemataceae, p. 122

FAMILY I. **CYTOPHAGACEAE** STANIER *1940, 630, emend. mut. char.*

Cells rods, single or filaments which may be helical, branched or sheathed. Resting stages known to be formed in one genus. Unsheathed cells or filaments **motile by gliding.** Gram-negative.

Chemoorganotrophs: Metabolism is respiratory or fermentative. **Cells or cell masses are colored** some shade of yellow, orange or red.

Type genus: *Cytophaga* Winogradsky 1929, 577.

Key to the genera of family **Cytophagaceae**

Key characteristics

Formation of microcysts
Cell shape
Decomposition of agar, cellulose or chitin
Presence of sheath

I. Resting stages (microcysts) not known.
  A. Cells, rods or filaments.
    1. Agar, cellulose or chitin attacked; cells not in sheaths.
      Genus I. *Cytophaga*, p. 101
    2. Agar, cellulose or chitin not attacked; cells not in sheaths.
      Genus II. *Flexibacter*, p. 105
    3. Agar and chitin not attacked; cellulose may be attacked; filaments may be in sheaths.
      Genus III. *Herpetosiphon*, p. 107
  B. Cells, filaments.
    1. Filaments may be sheathed, not helical.
      Genus IV. *Flexithrix*, p. 109
    2. Filaments helical, not sheathed.
      Genus V. *Saprospira*, p. 109
II. Microcysts formed.
      Genus VI. *Sporocytophaga*, p. 111

Genus I. Cytophaga *Winogradsky 1929, 577; Lewin 1969, 191 emend. mut. char.*

Single, short or elongate **flexible rods or filaments, not branched, sheathed** or **helical.** Cells 0.3–0.7 by 5–50 μm with round or tapered ends. Resting stages not known. Motile by gliding. Gram-negative.

Chemoorganotrophs: **Metabolism is respiratory,** using molecular oxygen as electron acceptor and, in one known instance, nitrate as an alternate electron acceptor, *or* **respiratory and fermentative,** with acetate, propionate and succinate among the end products. Characteristics are shown in Table 2.2. One or more **polysaccharides such as agar, cellulose or chitin may be decomposed,** or **carboxymethylcellulose depolymerized.** Organic growth factors are required by some species.

Contain yellow, orange or red carotenoid pigment(s).

Strict aerobes *or* facultative anaerobes. Temperature optimum 20–30 C, where examined.

The G + C content of the DNA of the species examined ranges from 33–42 moles %.

Type species: *Cytophaga hutchinsonii* Winogradsky 1929, 578.

Further Comments

Members of the genus are probably common in soil and in both fresh water and marine environments, but are often overlooked in isolation procedures: (a) employing media rich in nutrients and (b) unless particular effort is made to detect the often thin, spreading growth of these organisms. The selective isolation techniques as described by Warke and Dhala (1968) may prove useful.

In addition to properties noted in the generic description, members of this genus, in general, are less refractile than are most bacteria and often form at the colony periphery finger-like projections on the agar surface. When given sufficient carbon, some produce extracellular polymers which render liquid media rather viscous and colonies on agar mucilaginous.

Unless otherwise indicated the species descriptions have been drawn from the studies of Stanier (1940, 1941, 1942, 1947), Lewin (1969), Lewin and Lounsbery (1969), Mandel and Leadbetter (1965) and Mandel and Lewin (1969).

TABLE 2.2

Characteristics of the species of genus **Cytophaga**

For the carbon compounds listed, + or − refers to utilization, or not, of a given substrate. Where the data permit more explicit information is conveyed by: D = decomposed; L = liquefied; C = serve as (−C will not serve as) source of carbon and energy; H = hydrolyzed; −H, not hydrolyzed; d = some strains positive, some strains negative.

	Attack on									Catalase	NaCl required
	Agar	Cellulose	Chitin	Starch	Cellobiose	Glucose	Galactose	Xylose	Arabinose		
1. *C. hutchinsonii*	−H	C	−H		+	+	−	−	−	+	−
2. *C. rubra*	−H	C	−H		C	C	−	C	−	+	−
3. *C. johnsonae*	−	−	C	C	C	C	C	C	C	−	−
4. *C. krzemieniewskae*	L	D	−	D	+	+	+	+	−	+	+
5. *C. diffluens*	L	dD	−	d		d	d			−	+
6. *C. lytica*	H	−	−	H		+	+			+	+
7. *C. salmonicolor*	dH		−	+	+		+		+	+	+
8a. *C. fermentans* subsp. *fermentans*	H; −C	−C	−C	+	+	+	−	+		+	+
8b. *C. fermentans* subsp. *agarovorans*	H	−	−H	+	+	+	+	+			+

Genus II. **Flexibacter** *Soriano 1945, 92, Lewin 1969, 192 emend. mut. char.*

Flexible rods or **filaments** 0.5 by 5–100 μm. Resting stages not known. Motile by gliding. Gram-negative.

Cell masses are pink, red, orange or yellow due to carotenoid pigments.

Chemoorganotrophs: **Metabolism is usually respiratory,** utilizing molecular oxygen as terminal electron acceptor, but **one species is both respiratory and fermentative. Polymers such as agar, alginic acid, cellulose and chitin** are **not attacked.**

Characteristics of value in differentiating the species are given in Table 2.3.

Usually strict aerobes.

The G + C content of the DNA of the species examined ranges from 31.2–42.9 moles %.

Type species: *Flexibacter flexilis* Soriano 1945, 92.

TABLE 2.3

Differential characteristics of the species of genus **Flexibacter**

Reactions of *F. succinicans* and *F. columnaris* not known for tyrosine and dihydroxyphenylalanine degradation, indole or NH_3 production; negative for other species. Except for one strain of *F. succinicans*, all species fail to reduce nitrate to nitrite. + = 90% or more strains positive; − = 90% or more strains negative; d = some strains positive, some strains negative.

	Starch hydrolysis	Carboxymethylcellulose depolymerization	Carbon compounds utilized						Gelatin hydrolysis	H_2S production	Catalase presence
			Glucose	Galactose	Sucrose	Acetate	Lactate	Glycerol			
1. *F. flexilis*	−	−	d	d	+	d	−	−	+	+	−
2. *F. aggregans*	−	d	+	+	+	d	d	d	−	−	−
3. *F. giganteus*	+	−	+	d	−	−	−	−	+	d	−
4. *F. tractuosus*	d	−	+	d	d	d	−	d	+	d	−
5. *F. succinicans*	+		+	+	d		−	−			+
6. *F. columnaris*	−		+						+		

Genus III. **Herpetosiphon** *Holt and Lewin 1968, 2408; emend. mut. char.*

Unbranched, flexible, sheathed rods or **filaments** 0.7–1.5 by 5–150 μm or more (to several millimeters). Resting stages not known. **Unsheathed segments motile** by gliding. Gram-negative.

Chemoorganotrophs: **Metabolism is respiratory, using molecular oxygen as terminal electron acceptor.** Agar, alginic acid and chitin are not known to be attacked. Cellulose may be degraded, carboxymethylcellulose depolymerized or starch hydrolyzed (Table 2.4). Gelatin is liquefied. In all species, except *H. geysericola* which has not been examined, indole, ammonia or H_2S is not produced and nitrates are not reduced to nitrites. Cells possess yellow or orange carotenoid pigments.

Strict aerobes.

Temperature optima in the laboratory have not been determined. Where reported, pH optimum ca. 7.

Marine forms require sea water for growth.

Among the strains examined the G + C content of the DNA ranges from 44.9–53.1 moles % (Mandel and Lewin, 1969).

Type species: *Herpetosiphon aurantiacus* Holt and Lewin 1968, 2408.

TABLE 2.4

Differential characteristics of species of genus **Herpetosiphon**

	Color of cell mass	Cellulose hydrolyzed	Starch hydrolyzed	Tyrosine degraded	Catalase	Sea water required	Thermophilic	Nitrate as sole nitrogen source	G + C
									moles %
1. *H. aurantiacus*	Orange	−	+	+	+	−	−		48.1
2. *H. geysericola*	Orange	+	+		+	−	+		48.5
3. *H. cohaerens*	Orange	−	−	−	−	+	−	−	44.9
4. *H. persicus*	Orange	−	−	−	−	+	−	+	52.6
5. *H. nigricans*	Yellow	−	−	+	−	+	−	+	53.1

Genus IV. **Flexithrix** *Lewin 1970, 513*

Cells 0.3–0.5 by 5–15 μm, **usually as sheathed filaments** 0.5 μm in diameter and up to 500 μm long; may show false branching. Resting stages not known. Unsheathed cells 5–15 μm long, motile by gliding. Gram-negative.

Chemoorganotroph: Metabolism is respiratory, using molecular oxygen as terminal electron acceptor.

Strict aerobe.

The G + C content of the DNA of the single species examined is 37.2 moles % (Mandel and Lewin, 1969).

Type species: *Flexithrix dorotheae* Lewin 1970, 511.

Further Comments

Typical morphology is exhibited only in media in which the sugar concentration is low (e.g. 0.01%). In richer media, sheathed filaments are not produced, and the free, naked cells are then indistinguishable from those of species of *Cytophaga*. Under such conditions, *F. dorotheae* corresponds so closely to *Flexibacter aggregans*, in both cultural and nutritional features, that it might be regarded as a form variant of the latter species.

Genus V. **Saprospira** *Gross 1911, 190; Lewin 1962, 560 emend. mut. char.*

Helical filaments 0.8–1.5 by 10–500 μm; unbranched, unsheathed. Motile by gliding. Resting stages not known. Gram-negative.

Chemoorganotrophs: Metabolism is respiratory, using molecular oxygen as terminal electron acceptor. Agar, alginic acid, cellulose and chitin are not known to be attacked.

S. albida has not been studied extensively. The other three species can use tryptone but not glutamate or nitrate as sole nitrogen source, liquefy gelatin, do not produce indole or ammonia, do not reduce nitrate to nitrite and are catalase negative. Characteristics of differential value are given in Table 2.5.

Cell masses are colored, presumably due to carotenoid pigments.

Strict aerobes.

Both fresh water and marine forms are known.

The G + C content of the DNA of the species examined ranges from 35.4–48.3 moles % (Mandel and Lewin, 1969).

Type species: *Saprospira grandis* Gross 1911, 190.

TABLE 2.5

Differential characteristics of three species of genus **Saprospira**

	Color of cell mass	Carboxy-methyl-cellulose hydrolyzed	Starch hydrolyzed	Tyrosine degraded	Milk coagulated	Casamino acids as sole nitrogen source	Sources of carbon and energy					
							Acetate	Lactate	Galactose	Glucose	Sucrose	Glycerol
1. *S. grandis*	Orange	−	−	+	+	−	+	−	+	+	+	−
2. *S. thermalis*	Pink	−	+	−	−	+	−		−	+	−	+
3. *S. toviformis*	Orange-yellow	+	−	+	+	+	+	+	−	−	−	−
4. *S. albida*	Pale pink											

Genus VI. **Sporocytophaga** *Stanier 1940, 629*

Flexible rods with rounded ends, 0.3–0.5 by 5–8 μm, occurring singly. Spheroplasts and distorted cells occur in older cultures. A resting stage, the microcyst, is formed. Motile by gliding. Gram-negative.

Chemoorganotrophs: Metabolism is respiratory, using molecular oxygen as terminal electron acceptor. **Cellobiose, cellulose, glucose** and for some strains, **mannose, are the only known sources of carbon and energy.** Agar and chitin are not known to be attacked. Either ammonium or nitrate ions, or peptone, urea or yeast extract, can serve as sole nitrogen source. Amino acids, peptones, yeast extract or nutrient agar (Difco) cannot serve as sole carbon and energy sources. No organic growth factor requirements known.

Catalase positive.

Strict aerobe. Temperature optimum ca. 30 C.

The G + C content of the DNA of two strains examined was 36 moles % (buoyant density).

Type species: *Sporocytophaga myxococcoides* (Krzemieniewska) Stanier 1940, 629.

FAMILY II. **BEGGIATOACEAE** MIGULA 1894, 238

Colorless filaments containing cells in chains. Filaments are flexible and motile by gliding. Resting stages not known. Gram-negative.

Mixotrophs or chemoorganotrophs: Metabolism is respiratory, using molecular oxygen as terminal electron acceptor. Poly-β-hydroxybutyrate or volutin granules may occur intracellularly.

Aerobes or microaerophiles.

Carotenoid pigments not formed.

Key to the genera of family **Beggiatoaceae**

I. Filaments colorless, individual.
   A. Cells contain sulfur when grown in presence of H_2S.
      Genus I. *Beggiatoa*, p. 113
   B. Cells do not contain sulfur when grown in presence of H_2S.
      Genus II. *Vitreoscilla*, p. 114
II. One to many filaments in each sheath. Cells usually contain sulfur.
      Genus III. *Thioploca*, p. 115.

Genus I. **Beggiatoa** *Trevisan 1842, 56, Nom. cons.* Opin. 13, Jud. Comm. 1954, 152

Cells colorless, in unattached filaments, 1–30 by 4–20 μm. Motile by gliding. Resting stages not known. Gram-negative.

Mixotrophs: Metabolism is respiratory, using molecular oxygen as terminal electron acceptor. Organic growth factors are required by some strains.

Cells contain granules of sulfur when grown in the presence of hydrogen sulfide. **Intracellular granules** of **poly-β-hydroxybutyric acid** or **volutin may be present.**

Aerobic to microaerophilic.

Marine and fresh water forms known.

The G + C content of DNA of a strain regarded

as *B. leptomitiformis* is 37 moles % (Mandel and Lewin, 1969).

Type species: *Beggiatoa alba* (Vaucher) Trevisan 1845, 58.

Note: Species within this genus are separated solely on the basis of filament and cell size. Very few strains have been studied in pure culture. The reader is referred to the *Manual*, p. 113, for species descriptions.

Genus II. **Vitreoscilla** *Pringsheim 1949, 70; emend. mut. char.*

Colorless filaments, 1.2–2 by 3–70 μm, are composed of clearly delimited cylindrical or barrel-shaped cells. Reproduction is by fragmentation. Resting stages not known. Motile by gliding. Gram-negative.

Chemoorganotrophs: Metabolism is presumably solely respiratory. Proteins are not hydrolyzed. **Growth reported only on complex media;** minimal nutritional requirements unknown. **Intracellular sulfur granules not formed from hydrogen sulfide.**

The G + C content of two unnamed strains is 43.6 moles % (buoyant density) (Mandel and Lewin, 1969).

Type species: *Vitreoscilla beggiatoides* Pringsheim 1949, 70.

Note: Species within this genus are separated on the basis of cell size and shape and filament size. The reader is referred to the *Manual*, p. 114, for species descriptions.

Genus III. **Thioploca** *Lauterborn 1907, 242*

Flexible filaments made up of **numerous segments,** generally with **sulfur inclusions** (Plate 2.6, Fig. C), occur in parallel or braided **fascicles,** enclosed by a **common sheath** of variable width (Plate 2.6, Figs. A and B). The number of filaments embedded in one sheath is variable. Within one sheath, filaments may be of fairly uniform or of greatly differing diameters. The sheath is frequently encrusted with detritus (Plate 2.6, Fig. A). Individual filaments show **independent gliding** movement; their terminal segments are often tapered. Not isolated in pure culture.

Type species: *Thioploca schmidlei* Lauterborn 1907, 242.

TABLE 2.6

Differential characteristics of species of the genus **Thioploca**

	Diameter of		Length of	
	Individual filaments	Common sheath	Individual segments	Common sheath
	μm	*μm*	*μm*	*cm*
I. Filaments of fairly uniform diameter				
1. *T. schmidlei*	5.0–9.0	50–160	5.0–8.0	Several
2. *T. ingrica*	2.0–4.5	–80	1.5–8.0	–1.0
3. *T. minima*	0.8–1.5	–30	1.0–2.0	–0.6
II. Filaments of greatly differing diameter				
4. *T. mixta*	1.0 and 6.0–8.0			

FAMILY III. **SIMONSIELLACEAE** STEED 1962, 615

Cells arranged in apposition to form filaments which are flat, not cylindrical. Gliding motility when flat surface of filament is in contact with substrate, but not when filaments are coiled on edge as in colonies (Plate 2.7, Fig. 2)

Key to the genera of family **Simonsiellaceae**

I. Closely opposed cells form flat filaments which divide into hormogonia-like units with rounded terminal cells.

Genus I. *Simonsiella*

II. Cells arranged in pairs within flat filaments; terminal cells not rounded.

Genus II. *Alysiella*

Genus I. **Simonsiella** *Schmid 1922, 504*

Cells 0.4–0.7 by 2–4 μm, **closely apposed to form filaments with free faces of terminal cells rounded. Filaments are flat, not cylindrical,** and divide into hormogonia-like units (Plate 2.7, Figs. 1 and 2). **Gliding motility** when flat surface in contact with substrate. No spores. No capsules. Gram-negative.

Chemoorganotrophs: **Metabolism fermentative;** growth best in presence of blood or serum; minimal nutritional requirements not known.

Strict aerobes.

Temperature optimum 30–37 C.

Occur among **oral flora** in man and animals.

Type species: *Simonsiella muelleri* Schmid 1922, 504.

Note: Two species are recognized, *S. muelleri* and *S. crassa*, differentiated on the basis of cell and filament size.

Genus II. **Alysiella** *Langeron 1923, 116*

Cells 0.6 by 2–3 μm **arranged side by side in pairs to form filaments which are flat, not cylindrical; terminal cells are not rounded. Gliding motility** when flat surface is presented to substrate. No spores. No capsules. **Gram-negative.**

Chemoorganotrophs: **Metabolism fermentative;** growth best in presence of blood or serum; minimal nutritional requirements unknown.

Strict aerobes.

Temperature optimum 37 C.

Habitat: **Oral cavities** of animals.

Type species: *Alysiella filiformis* (Schmid) Langeron 1923, 118.

FAMILY IV. **LEUCOTRICHACEAE** BUCHANAN 1957, 850

Long filaments composed of short cylindrical or ovoid cells, cross-walls clearly in evidence, **colorless, unbranched;** filaments usually uniform in diameter throughout length although may taper from base to apex under some conditions. In nature filaments usually **attached to solid substrates by means of inconspicuous holdfasts, stalks absent. Filaments do not glide** but may wave sporadically from side to side. Dispersal by means of gonidia (single cells arising from cells of the filaments by rounding up, often released primarily from apices, but may also be intercalary); gonidia often but not always show jerky gliding motion on solid substrates. **Rosette formation is** a key diagnostic characteristic of the family, but is found rarely in nature, although frequently in culture. The rosettes may be formed of gonidia or, after conversion of gonidia to filaments, of several or more filaments attached at their bases. Filaments in culture often form true knots and knots also occur in nature although rarely. Resemble blue-green algae in many respects but differ from them in not forming photosynthetic pigments. **Strictly aerobic,** chemoorganotrophic or chemolithotrophic. **Aquatic.**

Genus I. **Leucothrix** *Oersted 1844, 44*

Morphological description as for family. **Sulfur granules are not formed,** even in habitats high in H_2S. **Marine,** requires NaCl for growth. Chemoorganotrophic.

Type species: *Leucothrix mucor* Oersted 1844, 44.

Genus II. **Thiothrix** *Winogradsky 1888, 39*

Morphologically very similar to *Leucothrix*, but **deposits sulfur granules inside the cells.** In H_2S-deficient habitats, *Thiothrix* may lose its sulfur granules and resemble very closely *Leucothrix*. Filaments usually attached to solid substrates. Dispersal by gonidia. Rosettes formed. Filaments non-motile. (May be confused with *Beggiatoa*, which also deposits sulfur, but the latter has motile gliding filaments.) *Thiothrix* is **found in both fresh water and marine habitats. It occurs commonly where H_2S concentrations are high.** In fresh water habitats it is found most frequently in sulfur springs and sewage plants and is especially adapted to flowing water. In marine environments, it is found where sulfur springs enter the sea or where H_2S is produced by sulfate reduction among rotting seaweeds, but usually where water flow occurs, and usually attached to marine algae. It apparently is an obligate autotroph but has not been consistently cultured.

Type species: *Thiothrix nivea* (Rabenhorst) Winogradsky.

FAMILIES AND GENERA OF UNCERTAIN AFFILIATION

Genus **Toxothrix** *Molisch 1925, 144*

Cells cylindrical, colorless, 0.5–0.75 by 3–6 μm, in filaments up to 400 μm long. A dense body often located at either end of cell (Plate 2.8, Fig. 1). Gram reaction not recorded.

Filaments often U-shaped (Plate 2.8, Figs. 1 and 2) and rotate while slowly moving forward with the rounded part in the lead; a mucoid substance, excreted from several sites on the trailing ends, is deposited as a double track ("railroad track") of twisted strings each 0.2 μm wide (Plate 2.8, Figs. 8–10). Fan-shaped structures may be deposited laterally along the track as the arms of the U move from side to side, and between the tracks as a result of the middle section being lifted and then touched down again (Plate 2.8, Fig. 7) (Krul *et al.*, 1970).

Oxidized iron may be deposited onto the mucoid threads, rendering them yellowish brown and brittle and giving them a diameter of 2.5 μm.

Have not been obtained in pure culture but heterotrophic and psychrophilic; have been maintained for long periods at 5 and 10 C. But filaments are extremely fragile during laboratory examination and explosive disintegration has been observed after short periods (Plate 2.8, Figs. 3–6). Grow attached to surfaces and develop best at reduced oxygen tensions (Hasselbarth and Lüdemann, 1967) and around neutrality (pH 6.5-7.2).

Originally found in water reservoir near the Biological Station on the river Dnieper in the U.S.S.R. Widely distributed in cold springs, bogs, ponds and lakes containing ferrous iron.

Type species: *Toxothrix trichogenes* (Cholodny) Beger in Beger and Bringmann 1953, 332.

FAMILY **ACHROMATIACEAE** MASSART 1901, 256

Cells spherical to ovoid or cylindrical. Division by constriction. Motile by slow, jerky (gliding?) movements. Resting stages not known. Cells may contain sulfur and calcium carbonate inclusions.

Genus **Achromatium** *Schewiakoff 1893, 1*

Cells **spherical** to **ovoid or cylindrical** with hemispherical ends (Plate 2.9, Fig. 1), 5–33 by 15–125 μm. Division by **constriction** in the middle. **Movements,** if any, are of a **slow rolling jerky type** dependent upon a substrate being present. **Photosynthetic pigments absent.** In their natural habitat cells may contain **sulfur droplets** and **large spherules** of **calcium carbonate.** Has not been cultivated in pure culture. **No resting stages known.**

Microaerophilic; apparently require sulfides.

Type species: *Achromatium oxaliferum* Schewiakoff 1893, 1.

Note: There are two species: *A. oxaliferum*, which contains calcium carbonate inclusions, and *A. volutans*, which does not.

FAMILY PELONEMATACEAE SKUJA 1956, 81

Cylindrical colorless cells, arranged in unbranched filaments, usually visible without staining. Filaments straight, flexuous or spirally coiled, with or without a thin slime-sheath, occurring singly or aggregated in bundles or bands. Single filaments may show a gliding movement combined with rotation around the longitudinal axis. Cells may contain gas vacuoles but not pigments or sulfur globules.

Reproduction by fragmentation of the filaments, rarely by arthrospore-like resting cells.

Characteristics differentiating the genera are given in Table 2.7.

Aquatic. Have not been cultivated.

TABLE 2.7
Characteristics differentiating the genera of family **Pelonemataceae**

	Pelonema	*Achroonema*	*Peloploca*	*Desmanthos*
Filaments				
Single	+	+	−	−
In bands or bundles	−	−	+	+
Gliding motility	∓	+	−	−
Holdfasts	−	−	−	+
Sheaths	−	D[a]	D	+
Gas vacuoles	+	D	D	−
Resting cells	−	D	−	−

[a] D, present in some species, not in others.

Genus I. **Pelonema** *Lauterborn 1916, 408*

Colorless cells with a delicate smooth wall; usually contain gas vacuoles. In multicellular, unbranched straight or loosely flexuous filaments of uniform thickness, which may or may not be constricted at the cross-walls. The filaments may occasionally show a slow gliding motility. Propagation by fragmentation of filaments. Resting cells not known.

Type species: *Pelonema tenue* Lauterborn 1916, 408.

Key to the species of genus **Pelonema**

I. Filaments usually slightly constricted at cross walls and 300–600 μm long.
  A. Cells considerably longer than wide.
    1. Cells 1.9–2.2 by 3–19 μm. Filaments usually not more than 300 μm long.
      1. *Pelonema tenue*
    2. Cells 1.0–1.7 by 5–16 μm. Filaments up to 600 μm or more in length.
      2. *Pelonema aphane*
  B. Cells isodiametrical or only slightly longer than wide.
    1. Cells 1.5–2.0 by 2–4 μm. Filaments up to 500 μm or more in length.
      3. *Pelonema pseudovacuolatum*

II. Filaments not constricted at cross-walls, relatively short.
  A. Cells 0.6–0.7 by 1.5–6 μm. Filaments up to 120 μm long.
      4. *Pelonema subtilissimum*
  B. Cells 2.0 by 2–6 μm. Filaments rarely more than 200 μm long.
      5. *Pelonema hyalinum*

Genus II. **Achroonema** *Skuja 1948, 30*

Cells have a delicate, smooth, sometimes distinctly porous cell wall; protoplasm homogeneous or granular, rarely with gas vacuoles. In multicellular, colorless, unbranched filaments, often constricted at cross-walls, straight, or nearly so or in very loose spirals. Multiply by fragmentation of filaments, rarely by formation of arthrospore-like resting cells.

Aquatic. Have not been cultivated.

Type species: *Achroonema spiroideum* Skuja 1948, 31.

Key to the species of genus **Achroonema**

Key characteristics

Cell and filament shape
Presence of constricted cross-walls
Filament size
Formation of resting cells
Color of wall of resting cell
Homogeneity of protoplast

I. Filaments more or less regularly coiled but in loose spirals.
1. *Achroonema spiroideum*

II. Filaments straight or only slightly and irregularly coiled.
A. Filaments not markedly constricted at cross-walls.
1. Filaments 2.5 μm or less in diameter.
a. Filaments 0.7–1.2 μm wide.
2. *Achroonema angustum*
b. Filaments 1.3–1.6 μm wide.
3. *Achroonema proteiforme*
c. Filaments ca. 2 μm wide.
4. *Achroonema profundum*
2. Filaments more than 2.5 μm in diameter.
5. *Achroonema splendens*
B. Filaments distinctly constricted at cross-walls.
1. Produce resting cells.
a. Wall of resting cells brown.
6. *Achroonema sporogenum*
b. Wall of resting cell colorless.
bb. Resting cell long and cylindrical.
7. *Achroonema lentum*
bbb. Resting cell short and more or less barrel-shaped.
8. *Achroonema inaequale*
2. Resting cells not known.
a. Cells usually isodiametrical.
9. *Achroonema simplex*
b. Cells usually longer than wide.
bb. Protoplasts have a darker centroplast.
10. *Achroonema articulatum*
bbb. Protoplasts more or less homogeneous.
i. Filaments 2.5–3 μm wide.
11. *Achroonema gotlandicum*
ii. Filaments 5–6.8 μm wide.
12. *Achroonema macromeres*

Genus III. **Peloploca** *Lauterborn 1913, 99*

Cells of variable length with homogeneous or granular protoplasm, with or without gas vacuoles. Cells arranged in uniform filaments with or without a gelatinous sheath; no constrictions at cross-walls or slightly constricted; cell walls delicate. Filaments bound together laterally to form rigid bundles or flat ribbons which may be straight, undulate in one plane or spirally wound; the ends may appear more or less fibrous. Not motile.

Propagation by fragmentation of bundles or bands.

Aquatic, free floating. Has not been cultivated.

Type species: *Peloploca undulata* Lauterborn 1913, 99.

Key to the species of genus **Peloploca**

Key characteristics

Presence of sheaths
Filaments in ribbon-like bands or bundles
Shape of bands or bundles
Fragility of bands
Presence of gas vacuoles

I. Separate filaments do not have a distinct gelatinous sheath; bind together in parallel to form a ribbon-like band.
  A. Bands have wavy shape in one plane.
    1. Filaments firmly bonded and bands not fragile; cells with gas vacuoles.
       1. *Peloploca undulata*
  B. Bands more or less flat.
    1. Filaments not strongly bound, bands fragile usually breaking into several pieces that remain together; cells without gas vacuoles.
       2. *Peloploca fibrata*
    2. Filaments firmly bonded and bands not fragile; cells with gas vacuoles.
       3. *Peloploca taeniata*

II. Separate filaments have a distinct gelatinous sheath; bind together in more or less spirally wound bundles.
  A. Bundles small, only slightly twisted or sigmoid; cells and sheath without gas vacuoles.
       4. *Peloploca ferruginea*
  B. Bundles longer, distinctly spirally wound; cells without gas vacuoles but with a vacuolated sheath.
       5. *Peloploca pulchra*

Genus IV. **Desmanthos** *Skuja 1958, 442*

Colorless cells in unbranched, more or less straight filaments thicker at the base than at the apex; filaments in bundles, the basic part of which is enclosed in a common hyaline, gelatinous sheath of variable thickness; the filaments at the top are free and divergent. Bundles attached by a holdfast or partially buried in bottom mud. Propagation by fragmentation of filaments and probably by longitudinal separation of the bundle-filaments. Motility doubtful.

Type species: *Desmanthos thiocrenophilum* Skuja 1958, 442.

PART 3

THE SHEATHED BACTERIA

Key to the genera of sheathed bacteria

Key characteristics

Motility of single cells
Encrustment of sheath with iron or manganese oxides
Sheaths attached
Filament shape at tip

I. Single cells motile by means of a polar flagellum or subpolar flagella.
  A. Sheaths rarely encrusted with iron and not encrusted with manganese oxides.
    Sphaerotilus, p. 128*
  B. Sheaths encrusted with iron or manganese oxides.
    Leptothrix, p. 129.
II. Single cells not motile by flagella.
  A. Sheaths not attached.
    1. Sheaths not encrusted with metal oxides.
      Streptothrix, p. 133.
    2. Sheaths may be encrusted with metal oxides.
      Lieskeella, p. 134.
  B. Sheaths attached.
    1. Sheaths not encrusted with metal oxides.
      Phragmidiothrix, p. 134.
    2. Sheaths encrusted with metal oxides.
      a. Filaments may be swollen at tip.
        Crenothrix, p. 135.
      b. Filaments taper at tip.
        Clonothrix, p. 136.

Note: Van Veen *et al.* (*Antonie van Leeuwenhoek Journal of Microbiology and Serology 39:* 207–216, 1973) have recently described a new genus and species, *Haliscomobacter hydrossis*, which are nonmotile, Gram-negative, aerobic, ensheathed rods which do not form holdfasts or become encrusted with metal oxides. The reader is referred to the above paper for a more complete description.

Genus **Sphaerotilus** *Kützing 1833, 386*

Straight rods, 0.7–2.4 by 3–10 μm, occurring in chains within a sheath of uniform width, which may be attached by means of a **holdfast** (Plate 3.1, Fig. 1) to submerged plants, stones, etc. **Sheaths usually thin without incrustation** by **ferric** and **manganic oxides** (Plate 3.1, Fig. 2). **Single cells motile** by means of a **bundle of subpolar flagella** (Plate 3.1, Fig. 3) often so inter-

* **All page numbers cited refer to** ***Bergey's Manual of Determinative Bacteriology*****, 8th edition, The Williams & Wilkins Company, Baltimore, 1974.**

twined as to give the appearance of a single large "unit flagellum." **Resting stages are not known. Gram-negative.**

Chemoorganotrophs: **metabolism respiratory,** never fermentative. Molecular oxygen is the universal electron acceptor. Alcohols, several organic acids and sugars are used as sources of carbon and energy. Inorganic nitrogen compounds (ammonium salts or nitrate) may serve as nitrogen sources in the presence of vitamin B_{12} or of methionine. Better growth is generally obtained with organic nitrogen sources: Casamino acids, peptone or a mixture of aspartic acid, glutamic acid and vitamin B_{12}. Gelatin is liquefied slowly.

Strict aerobes, but good growth can be obtained at low oxygen tensions. Temperature range: 15–37 C; optimum between 25 and 30 C.

The **G + C content** of the DNA of one strain of *S. natans* tested was **70.0 ± 0.5 moles %** (Mandel *et al.*, 1966).

The normal habitat is slowly running fresh water contaminated with sewage or waste water from paper or dairy industries. Long tassels of strongly cohering sheaths largely filled with chains of cells are formed, attached to submerged plants, stones, etc. Poorly precipitating, so-called bulking, activated sludge may also contain many filaments of this organism.

Type species: *Sphaerotilus natans* Kützing 1833, 386.

Genus Leptothrix *Kützing 1843, 198*

Straight rods, 0.6–1.5 by 3–12 μm, occurring in **chains** within a **sheath** or **free-swimming** as single cells, in pairs, and in some species as motile short chains containing up to eight cells. One species has well developed holdfasts. **Sheaths** have a pronounced tendency to become **impregnated** or **covered** with **hydrated ferric** or **manganic oxides. Free cells motile** by means of **one polar flagellum** (Plate 3.2, Fig. 1); one species has a subpolar tuft of several flagella (Table 3.1). Most species may contain globules of poly-β-hydroxybutyrate as intracellular reserve material (Plate 3.2, Fig. 3). Resting stages are not known. **Gram-negative.**

TABLE 3.1

Differential characteristics of species of **Leptothrix**[a]

	1. *L. ochracea*	2. *L. pseudo-ochracea*	3. *L. discophora*	4. *L. cholodnii*	5. *L. lopholea*
Cells					
Width, μm	0.8–1.0	0.8–1.3	0.6–0.8	0.9–1.3	1–1.4
Length, μm	2–3	5–12	2–4	2–5	3–6
Flagella					
Monotrichous, Polar		+	+	+	−
Polytrichous, Subpolar		−	−	−	+
Sheaths in:					
Fe^{++}-containing soil extract	s	s-u	u	u	s
Mn^{++} medium		g	e	e	g
Mn^{++} oxidation		+	+	+	+ (ret.)
Many holdfasts	−	−	−	−	+
Cohesion of sheaths in culture solution		+	−	−	−
Form and size (mm) of colonies on:					
Peptone-gluc. agar[b]		f–s(5)	s(0.1–0.3)	s(5)	f–s(1)
Mn-agar[c]		f(10)	s(0.5–2)	f–s(3–4)	s(2–3)
S-R dissociation by mutation		±	±	+	±
Response to nutrients		±	−	+	−

[a] Abbreviations: s, smooth; u, uneven, owing to precipitated ferric oxide; g, covered with small granules of MnO_2; e, encrusted with large masses of MnO_2; ret., retarded; f, filamentous.

[b] Basal medium containing 1 g of peptone and 1 g of glucose/liter.

[c] Nutrient medium containing 2 g of $MnCO_3$/liter.

Chemoorganotrophs: **metabolism respiratory,** never fermentative. Molecular oxygen is the universal electron acceptor. A variety of organic acids and sugars can be utilized as sources of carbon and energy but the number of assimilable compounds is less and the response usually poorer than with organisms of the genus *Sphaerotilus*. Most strains require organic compounds as the nitrogen source (peptone or casamino acids). A mixture of aspartic and glutamic acids can also be utilized when vitamin B_{12} is present. Few strains assimilate inorganic nitrogen compounds. For some strains a requirement for biotin and thiamine has been reported (Rouf and Stokes, 1964). All isolated species oxidize manganous compounds to manganic oxide in contrast to members of the genus *Sphaerotilus*.

Strict aerobes, but growth and manganese oxidation may proceed at low oxygen tensions.

Temperature range: 10–35 C; optimum between 20 and 25 C.

The **G + C content** of the DNA of one species tested was **69.5 ± 0.5 moles %** (Mandel *et al.*, 1966).

Type species: *Leptothrix ochracea* (Roth) Kützing 1843, 198.

Genus Streptothrix *Cohn 1875, 186, emend. mut. char. Migula 1895, 36*

Thin rods, 0.35–0.45 by 3.2–4.6 μm, **occurring in chains in hardly visible hyaline sheaths** of 0.5–0.8 μm width (Plate 3.2, Figs. 9 and 10). **Free cells** outside the sheath occur only sporadically. Motility and flagella of these cells have not been observed. **No ferric or manganic oxides** present in or on the sheaths. **Branching** of the filaments may incidentally occur in stationary cultures. **Lateral branches** generally very **short** as compared with main filament. **No resting stages known. Gram-negative.**

Colonies on poor agar media <0.5 mm, highly filamentous; on a peptone, sucrose, vitamin medium 1–3 mm, faintly pink with slightly filamentous edge. Fast-growing cultures in liquid media turn pink owing to the presence of carotenoid pigments.

Chemoorganotrophs: **metabolism respiratory,** never fermentative. Molecular oxygen is the universal electron acceptor. Glucose, lactose, sucrose and to a less extent mannitol are used as sources of carbon and energy. Nitrates and ammonium compounds are moderately good nitrogen sources. Better growth is obtained with organic nitrogen sources: glutamate or peptone. Thiamine and vitamin B_{12} are required for growth. Starch and gelatine are hydrolyzed.

Strict aerobes. Temperature range: 8–30 C; optimum: 25–27 C.

Widely distributed in fresh water and in activated sludge. Occurring in large masses in activated sludge loaded with residual water from meat industries or pig farms. Straight, sheathed chains of bacteria extending from the flocs interfere with settling of the sludge.

Type species: *Streptothrix hyalina* Migula 1895, 38.

Genus Lieskeella *Perfil'ev 1927, 335*

Cells rod-shaped with rounded ends, 0.6 by 2–3 μm, in chains. Show bipolar staining when treated with methylene blue. Usually two chains of cells (filaments) are wound around one another to give a double spiral, which is surrounded by a yellowish, slimy capsule, often with heavy deposits of small granules of ferric hydroxide. When the deposits are dissolved with dilute HCl, the cell chains fragment and the individual rods appear as a double zigzag band. Cell chains may aggregate in distinct layers or even in more or less solid skeins. The filaments separate rapidly upon removal from the normal environment.

There is a slow but incessant motion similar to that of cyanophytes.

Have not been obtained in pure culture. Gram-reaction not recorded.

Originally found in the upper layers of mud in bodies of water around Alt-Peterhof, Russia; also observed by Perfil'ev and Gabe (1961) in several other locations in western Russia throughout the year, usually as thin layers of up to 15 chains. Observed at depths of 0.5 to 1.2 meters in the littoral zone of Kristatellevyi Pond, Russia; also in the mud water interface of Lake Windermere, England (Skerman, personal communication).

Type species: *Lieskeella bifida* Perfil'ev 1927, 335.

Genus Phragmidiothrix *Engler 1883, 192*

Filaments are articulate, **unbranched,** colorless and over 100 μm long. They are **attached,** forming grayish white tufts. The free end may be of larger diameter than the base, but there is no tapering in either direction: the diameter varies between 3 and 6 μm. Surrounded by a very thin, delicate, gelatinous, colorless sheath, which is not encrusted with iron or manganese compounds. The

cells are of variable size, usually small and disc-shaped, the diameter being 4–6 times the thickness of the cell; cell walls are distinct and of even thickness throughout the filament. **Multiplication is by cross-septation and in certain regions of the filament by both cross- and longitudinal septation,** forming *Sarcina*-like aggregates of small, nearly cubical propagation cells. Septating cells may be of greater diameter causing localized swelling of the filament; in areas septating in both planes the filaments may be up to 6 μm in diameter (Plate 3.3*A*).

Has not been cultivated on artificial media in pure culture.

Originally found attached to the surface of living *Gammarus locusta* collected from the anaerobic, H_2S-containing, polluted area called "Weisser Grund" or "Todten Grund" in the Kieler Förde, Germany. Apparently not rare; have been reported on seaweed from polluted water of the Northern Adriatic Sea (Beger, 1957).

Type species: *Phragmidiothrix multiseptata* Engler 1883, 192.

Genus Crenothrix *Cohn 1870, 108*

Filaments up to 1 cm long, attached to a firm substrate and may be swollen at the free end. **Unbranched** but may show what appears to be false branching. The **very thin sheaths** surrounding the filaments may be colorless at the tip or encrusted with iron (or manganese) oxides at the base. **Cells cylindrical to disc-shaped, dividing by cross-septation in normal filaments and by cross- and longitudinal septation at the tips of enlarged filament ends.** Cells in enlarged filament ends are smaller and may round up; sometimes there is only one row of these present. Larger rod-shaped cells may also slip out of the sheath and form new filaments.

Gram-reaction has not been recorded. Not motile.

Has not been grown on artificial media in pure culture.

Found in stagnant and running waters containing organic matter and iron salts.

Type species: *Crenothrix polyspora* Cohn 1870, 131.

Genus Clonothrix *Roze 1896, 329*

Filaments up to 1.5 cm long, attached or free, surrounded by a **more or less distinct sheath** which may be encrusted with iron or manganese compounds giving a yellowish brown color. Filaments tapering; they may be single or **with false branches.** Cells cylindrical, colorless or bluish; cell masses may appear in shades of brown. **Reproduction by separation of individual cells,** followed by breakage of the sheath and release to the outside; such cells frequently attach themselves parallel to the older cell and produce new filaments giving rise to "pseudo-branching." Most of these morphological features were noted in Roze's original drawings (Plate 3.3*B*).

Have not been grown in pure culture on artificial media.

Type species: *Clonothrix fusca* Roze 1896, 330.

PART 4

BUDDING AND/OR APPENDAGED BACTERIA

Note: Only a partial key to the genera of this Part is presented in the *Manual*. The Part is very diverse, consisting of bacteria that reproduce by budding, produce appendages or do both. There are many active studies being carried out on this group, and future editions of the *Manual* should contain more useful keys. Separation of genera is on morphological grounds and the cultures are best viewed under phase contrast microscopy. The reader should consult Plates 4.1–4.4 for illustrations and the *Manual* for more extensive discussions of this Part. Attention is also drawn to the inclusion of budding or appendaged bacteria in other Parts.

Partial Key to Part 4

I. Bacteria in which products of binary fission are not equivalent.
  A. Prosthecate bacteria
    1. Prosthecae have a reproductive function; new cell formation by a budding process.
      Hyphomicrobium, p. 148*
      Hyphomonas, p. 150
      Pedomicrobium, p. 151
      (see also Part 1—*Rhodomicrobium*)
    2. Prosthecae have no reproductive function.
      Caulobacter, p. 153
      Asticcacaulis, p. 155
      Ancalomicrobium, p. 156
      Prosthecomicrobium, p. 157
    3. Prosthecae may have a reproductive function.
      Thiodendron, p. 158
  B. Non-prosthecate bacteria
    Reproduction by a budding process.
      Pasteuria, p. 158
      Blastobacter, p. 159
      Seliberia p. 160
      (see also Part 1—*Rhodopseudomonas;* Part 7—*Methylomonadaceae;* Part 12—*Nitrobacter*)
II. Bacteria with excreted appendages and holdfasts; products of binary fission may not be equivalent.
  A. Reproduce by binary fission only.
      Gallionella, p. 160
      Nevskia, p. 161
  B. Reproduce by a budding process.
      Planctomyces, p. 162
  Genera of uncertain affiliation
      Metallogenium, p. 163
      Caulococcus, p. 165
      Kusnezovia, p. 166

***All page numbers cited refer to *Bergey's Manual of Determinative Bacteriology*, 8th edition, The Williams & Wilkins Company, Baltimore, 1974.**

Note: Gossling and Moore (*International Journal of Systematic Bacteriology 25:* 202–207, 1975) have recently described a new genus of budding bacteria, isolated from fecal material, which they have named *Gemmiger*. They are strictly anaerobic, Gram-negative to Gram-variable carbohydrate fermenters which produce formic and butyric acids. Their taxonomic placement is very uncertain. The reader is referred to the above paper for a description and illustrations.

Also, Geitler (*Archives of Microbiology 103:* 231–236, 1975) has described a new genus and species, *Chamaesiphonsira cymbellicola*, which is epiphytic on certain diatoms. It produces long, conical, colorless filaments attached to the diatom by the broader base and which cut off up to 30 buds from the free apex. The organism is morphologically similar to the cyanobacterial genus *Chamaesiphon*.

Genus **Hyphomicrobium** *Stutzer and Hartleb 1898, 76*

Cells 0.5–1.0 by 1–3 μm; rod-shaped with pointed ends, oval, egg- or bean-shaped forms; produce mono- or bipolar filamentous outgrowths (hyphae) of varying length 0.3–0.4 μm in diameter when stained. The hyphae are not septate but may show true branching. Cells stain well with carbolfuchsin but only feebly with aqueous aniline dyes. Gram reaction not recorded.

Multiplication by budding at tip(s) of hyphae (Plate 4.1, Fig. 1); mature buds become motile, break off and often attach themselves to surfaces or other cells to form clumps. Motility is lost soon after attachment.

Liquid media are never turbid; growth occurs as a surface pellicle or ring, which in older cultures falls to the bottom of the vessel.

On solid media colonies are small, even after long incubation, reflect light from the shiny but granular surface which may show folds or concentric rings; dirty white changing to brownish with age.

Chemoorganotrophic. Carbon dioxide is required for growth. Oligocarbophilic, i.e. growth can occur in a mineral salts medium without added C sources. Growth is stimulated by humus or soil extracts, if the pH is maintained near neutral. Good growth with 0.2% (w/v) formate, acetate, propionate, isobutyrate, valerate, lactate, succinate or mannitol. Slow growth with 0.2% oxalate, glycerol or 1% Liebig's meat extract. No growth with 0.5–1.0% beef extract plus peptone in nutrient or plain gelatin. Fructose and sucrose not utilized; fructose is inhibitory in the presence of KNO_3.

Good growth with NH_4^+, NO_3^- or NO_2^- as N source. Grows in dilute urine or a solution of asparagine but not on asparagine agar.

Does not oxidize ammonia to nitrate. Anaerobic growth occurs in the presence of nitrate but accumulation of nitrite cannot be detected.

Aerobic.

Temperature range 15–30 C; optimum 25–30 C. Neutral or slightly alkaline media preferred.

Osmotolerant; grows in 5% NaCl, 4% $NaNO_2$ or 9% $NaNO_3$.

Widely distributed in soils of all continents, particularly in soil with nitrification potential.

Type species: *Hyphomicrobium vulgare* Stutzer and Hartleb 1898, 76.

Genus **Hyphomonas** *Pongratz 1957, 607*

Cells 0.6–1.0 μm in diameter and approximately 1.0 μm long, colorless, egg-shaped, oval- or pear-shaped, resembling human spermatozoa or cells of *Hyphomicrobium* spp. The narrow pole grows a tubelike and straight or wavy extension (prostheca) which may be branched like a fungal hypha and may be as long as 8 μm. Older cultures often contain abnormal cell types: giant cells, spindles or triangular forms, with the prosthecae (hyphae) highly branched.

Young cells and hyphae with dense, homogeneous cytoplasm; abnormal forms often vacuolated or granular. Multiplication by a budding process and *not* by binary fission. Slide cultures on nutrient agar and at 20 C show the development of a distal terminal knob on the hypha. In 1–2 hrs. a bud (daughter cell) is completed and separates from the stalked mother cell. While the bud becomes motile shortly after separation ("swarmer cell"), the mother cell stalk elongates slightly before growing the next bud. Intercalary buds have been found occasionally, forming a rosary-like chain of cells interconnected by short hyphae. Direct budding without the formation of a hypha has also been observed. Motility is due to a single, subpolarly inserted flagellum which is often retained after the development of a stalk and new daughter cell. The mother cell then tows the daughter cell.

Endospores, conidia or chlamydospores have not been found.

Staining: Gram-negative, not acid-fast. Stain well with anilin dyes, neutral red, gentian violet, etc. The use of Lugol's iodine as a mordant facilitates the staining of the narrow hyphae. Best results were obtained with silver impregnation.

Upon isolation on solid media, colonies appeared as smooth ("S") or rough ("R") types. The smooth colonies were round, convex, watery and translucent. They could be emulsified easily. The S organisms carried heavy capsules on the mother cells but usually not on hyphae or buds. Rough colonies were rare; they were smaller, foldy and dry, forming a central crater after several days. The organisms could not be readily emulsified and suspensions thus remained granular. They lacked capsules and were non-motile. While S forms were stable variants, R forms split into both S and R derivatives. The often voluminous capsules of the S form organisms were readily stainable with Ziehl's hot carbol fuchsin.

Growth on broth occurs in the form of a delicate veil.

Aerobic; catalase- and peroxidase-positive. Heterotrophic. Not proteolytic, hemolytic or saccharolytic. Gelatin is not digested, plasma is not coagulated. Pigments are not formed. Antibiotically not active against *Staphylococcus aureus*, *Streptococcus* sp., *Escherichia coli* or *Candida albicans*.

Optimum temperature 37 C. Growth slow at 18–20 C and better at 30 C. Optimum pH is 7.0–7.4, the range for growth is 6.5–8.5. Poor or no growth on acid media commonly used for the cultivation of fungi.

Growth is not influenced by light.

No growth on poor media such as tap water with KNO_3 and phosphate. Does not grow on mineral salts media (such as those of Korthof, van Niel or K. Boltjes) which contain $CaCO_3$, formate, acetate, propionate, lactate, methanol, ethanol, glycerol or glycine as the sole source of carbon.

Does not attack arabinose, rhamnose, xylose, glucose, fructose, galactose, lactose, saccharose, maltose, trehalose, melibiose, raffinose, amidon, inulin, dextrin, glycogen, esculin, glycerol, erythritol, mannitol, sorbitol or urea.

Good growth in coagulated serum, gelatin, blood agar, Loewenstein-Jensen medium, Dorset-agar and other rich media.

Reduce nitrate, nitrite, neutral red, methylene blue and janus green. Produce H_2S. Indol is formed from tryptophane. NH_3 is produced, the growth medium turns alkaline.

Resistant against therapeutic concentrations of penicillin, bacitracin or furadoine. Inhibited greatly by chloromycetin, streptomycin, neomycin and sulfonamides. The tetracyclines and erythromycin are slightly less inhibitory.

Not virulent against mice (5 ml), rats (1 ml), guinea pigs (1 ml) or rabbits (1 ml) if a suspension was injected subcutaneously or intraperitoneally which contained 5×10^9 cells/ml. The organisms could not be recovered from the treated animals.

Pure cultures survive at room temperature or in a refrigerator for several months if suspended in growth media.

Isolated only once from nasal mucus from a case of infectious sinusitis.

Type species: *Hyphomonas polymorpha* Pongratz 1957, 607.

Genus **Pedomicrobium** *Aristovskaya 1961, 957*

Cells spherical, oval, rod-shaped, pear- or bean-shaped, 0.4–2.0 μm wide, unevenly staining. **Multiplication is primarily by budding at the tips of cellular extensions (hyphae)** of constant diameter (0.15–0.3 μm). Daughter cells may remain attached to their mother hyphae or they may separate as uniflagellated "swarmers." After increasing in size, **mature swarmer cells grow from one to numerous hyphae from several sites of their cell surface.** Occasionally, division of single mother cells has been observed simultaneously with the budding process.

Gram-negative, microaerophilic to aerobic; **ferric and/or manganese depositions occur either on mother cells or on the hyphae.** Chemoheterotrophic, mesophilic; growth occurs on mineral salts media with simple organic compounds or with organomineral complexes of fulvenic acids and sesquioxides. Deposition of iron, manganese and slimes primarily on mother cells. Widely distributed in soils.

Type species: *Pedomicrobium ferrugineum* Aristovskaya 1961, 955 (translation p. 113).

Key to the species of genus **Pedomicrobium**

Key characteristics

Location of buds
Size of mother cell
Presence of iron or manganese compounds on mother cell

I. Buds form on hyphal tips only.
  A. Mother cell 0.6–2.0 μm wide.
    Ferric hydroxide deposited on mother cell.
      1. *Pedomicrobium ferrugineum*
  B. Mother cell 0.4 μm in diameter.
    Manganese compounds deposited on mother cell.
      2. *Pedomicrobium manganicum*
II. Buds form both on hyphal tips and directly on mother cells.
    Both iron and manganese deposited.
      3. *Pedomicrobium podsolicum*

Genus Caulobacter *Henrici and Johnson 1935, 83*

Cells rod-shaped, fusiform or vibrioid, 0.4–0.5 μm by 1–2 μm. Typically with a **stalk,** ca. 0.15 μm diameter, varying in length among isolates and with environmental conditions, **extending from one pole** as a continuation of the long axis of the cell. The **stalk consists of the cell wall layers, the cell membrane and a core of twisted membranes. A small mass of adhesive material is present at the distal end of the stalk.** Occur singly; in dense populations, **cells may adhere to each other in rosettes,** the distal ends of their stalks embedded in a common mass of adhesive material.

Cell division by transverse, binary, **asymmetrical fission of stalked cells.** At the time of separation one cell possesses a stalk, the other a single polar flagellum. Each appendage occurs at the cell pole opposite the one formed during fission. The flagellated cell secretes adhesive material at the base of the flagellum, develops a stalk at this site, and enters the immotile vegetative phase.

Resting stages not known.

Gram-negative. The cell wall consists of an outer component similar in thickness and organization to the cell membrane, and an inner component whose thickness varies among isolates; the inner component is soluble in lysozyme-EDTA.

Colonies circular, convex, glistening; colorless, pink, red, orange or yellow.

Chemoorganotrophic. Strictly respiratory. The principal pathway of intermediary carbon metabolism is the Entner-Doudoroff pathway. Most strains can store carbon as poly-β-hydroxybutyric acid; a few also store carbon as polysaccharide.

Pigmentation common, but not universal; pigmented isolates contain carotenoids. A few isolates produce both nondiffusible red or orange pigments and diffusible brown or red-brown pigments.

Requirements for organic growth factors vary among species. All strains can grow on peptone-yeast extract media.

Strict aerobes.

Temperature range of most isolates is 15–35 C; 20–25 C is optimal. A few isolates grow at 5–25 C; 15–20 C is optimal for these. Optimal pH for growth is around neutrality; pH range for all isolates is 6.0–9.0.

Fresh water and soil isolates typically grow well at low osmotic pressures, and multiplication can occur in distilled water. Growth is inhibited or cells become swollen and frequently lyse in media containing 1% (w/v) or more of organic material. Isolates from sea water require NaCl at 0.5 or 1% (w/v) for growth in organic media. Growth of most isolates is inhibited by streptomycin (0.1 mg/ml), penicillin G (1000 units/ml) and cycloserine (0.1 mg/ml).

Most species can be isolated from samples of fresh water containing low concentrations of organic materials and from soil. NaCl-requiring isolates have been obtained only from sea water.

The G + C content of the DNA is 64–67 mole % (buoyant density). Known bacteriophages, which include DNA- and RNA-containing viruses, are strain-specific among *Caulobacter* isolates. They do not lyse bacteria of other genera, nor have *Caulobacter* isolates been found susceptible to infection by phages lytic for other genera.

The type species is *Caulobacter vibrioides* Henrici and Johnson 1935, 84.

Key to the species of genus Caulobacter

Key characteristics
Taper and curvature of cells
Requirement for organic growth factors, e.g., Vitamins B_2, B_{12} and biotin
Requirement of NaCl in organic media
Inhibition by 4% (w/v) NaCl

I. Cells tapered.
  A. Long axis of the cell curved.
    1. Organic growth factors required.
      a. Vitamin B_2 necessary, but not sufficient.
        1. *C. vibrioides*
      b. Vitamin B_{12} necessary and sufficient.
        2. *C. henricii*
      c. Biotin necessary, but not sufficient.
        3. *C. intermedius*
      d. Growth not stimulated by B vitamins.
        4. *C. subvibrioides*
    2. Organic growth factors not required
        5. *C. crescentus*
  B. Long axis of the cell not curved.
    1. Organic growth factors required.
        6. *C. fusiformis*
    2. Organic growth factors not required.
        7. *C. leidyi*

II. Cells not tapered.
  A. NaCl not required for growth in organic media.
        8. *C. bacteroides*
  B. NaCl required for growth in organic media.
    1. Growth inhibited by 4% (w/v) NaCl.
        9. *C. halobacteroides*
    2. Growth not inhibited by 4% (w/v) NaCl.
        10. *C. maris*

Genus **Asticcacaulis** *Poindexter 1964, 282*

Cells rod-shaped (0.5–0.7 μm by 1–3 μm). **Typically with an appendage** ca. 0.15 μm in diameter, varying in length among isolates and with environmental conditions, **arising excentrically from one pole, or laterally along the cell.** In the latter case, there may be two appendages per cell. Although structurally similar to stalks of *Caulobacter*, these appendages are called pseudostalks since they lack adhesive material and **cannot function in attachment.** Each cell typically possesses a small mass of adhesive material at a subpolar (rarely polar) position on the cell. Occur singly; in dense populations, **cells may adhere to each other in rosettes,** the poles of the cells associated with a common mass of adhesive material.

Cell division by transverse, binary, **disparate, asymmetrical fission of pseudostalked cells.** At the time of separation, one cell is longer and possesses one or two pseudostalks; the shorter sibling has a single subpolar flagellum. The flagellated cell develops a pseudostalk and enters the immotile vegetative phase.

Resting stages not known.

Gram-negative. The cell wall consists of an outer component similar in thickness and organization to the cell membrane, and an inner component whose thickness varies among isolates; the inner component is soluble in lysozyme-EDTA.

Colonies circular, convex, glistening; colorless.

Chemoorganotrophic. Strictly respiratory. The principal pathway of intermediary carbon metabolism is the Entner-Doudoroff pathway. Carbon is stored as poly-β-hydroxybutyric acid.

Strict aerobes.

Temperature range is 15–35 C; 25–30 C is optimal. Optimal pH for growth is around neutrality; pH range is 6.0–9.0.

All isolates typically grow well at low osmotic pressures, and multiplication can occur in distilled water. Growth is inhibited or cells become swollen and frequently lyse in media containing 1% (w/v) or more of organic material. Isolates have not been obtained from marine sources.

Growth is inhibited by streptomycin (0.1 mg/ml) and by penicillin G (1000 units/ml).

Observations of these bacteria in natural materials have not been reported. Presumably, their distribution is similar to that of *Caulobacter*, since all known *Asticcacaulis* isolates have been obtained inadvertently during attempts at isolation of *Caulobacter*.

The G + C content of the DNA is 55–60 moles %

(buoyant density). Known bacteriophages, which are DNA viruses, are strain-specific among *Asticcacaulis* isolates. They do not lyse bacteria of other genera, nor have *Asticcacaulis* isolates been found susceptible to infection by phages lytic for other genera.

The type species is *Asticcacaulis excentricus* Poindexter 1964, 292.

Genus **Ancalomicrobium** *Staley 1968, 1940*

Unicellular, Gram-negative bacteria having from **two to eight prosthecae extending from the cell** (Plate 4.2, Fig. 1). **These prosthecae attain a length of about 3 μm at maturity,** about three times the diameter of the cell. Occasionally the prosthecae are bifurcated; the position of the bifurcation along the appendage is variable. **Cells reproduce by budding.** Buds are formed directly from one position on the mother cell, never from the prosthecae. Two to four prosthecae differentiate from the developing bud. Division occurs transversely when the bud has attained approximately the same size as the mother cell. Cells are **non-motile** and have **no holdfasts.** Gas vacuoles may be present in some cells. Found in natural waters.

Chemoorganotroph. Facultative anaerobe.

Type species: *Ancalomicrobium adetum* Staley 1968, 1940.

Genus **Prosthecomicrobium** *Staley 1968, 1940*

Unicellular, Gram-negative bacteria that have **many prosthecae extending in all directions from the cell** (Plate 4.2, Fig. 2). These prosthecae are normally **less than 2 μm long** and are **conical in shape,** tapering distally from the cell toward a blunt tip. Occasionally longer, non-tapering prosthecae are produced. **Cells divide by binary transverse fission.** Motile or non-motile; **motile strains exhibit a tumbling, circular motility.**

Chemoorganotroph, nonfermentative. Aerobic. Found in natural waters.

The G + C content of the DNA ranges from 65.8–69.9 moles % (buoyant density).

Type species: *Prosthecomicrobium pneumaticum* Staley 1968, 1940.

TABLE 4.1

Carbon sources utilized by species of **Ancalomicrobium** *and* **Prosthecomicrobium**

Utilized by all species: L-arabinose, cellobiose, D-fructose, L-fucose, D-galactose, D-glucose, lactose, D-lyxose, maltose, D-mannose, L-rhamnose, D-ribose, D-xylose.

Not utilized by any species: raffinose, inulin, pectin, starch, adipate, benzoate, phthalate, arabitol, erythritol, methanol.

Utilized by some species	1. *Ancalomicrobium adetum*	1. *Prosthecomicrobium pneumaticum*	2. *Prosthecomicrobium enhydrum*
D-Arabinose	+	+	?
Melibiose, glycogen	−	+	−
L-Sorbose, melezitose, arabin, dextrin	+	+	−
Xylan	?	−	−
Butyrate	−	−	+
Fumarate, succinate	+	−	+
Propionate	−	+	?
Adonitol	?	+	−
Dulcitol, ethanol	−	+	−
Glycerol, inositol, mannitol, sorbitol	+	+	−

? not tested or utilization uncertain.

Genus **Thiodendron** *Perfil'ev and Gabe 1961, 162*

Vibrio-shaped cells, spirally twisted and with both ends somewhat tapered, ca. 10–15 times longer than wide, bear, thin threads ("stalks," 0.15–0.25 μm) on either one or both cell poles. The threads are straight or more or less flexuous, often of considerable length and occasionally appear to be branched; they are arranged radially to a common center. The vibrio-shaped cells may be motile with flagella.

Colonies concentrically layered, ± globular and grayish to bluish white with alternating lighter and darker zonation. The appearance may be that of thalli of the alga *Padina pavonia*. The layering is assumed to reflect rhythmical deposition of colloidal sulfur as a result of H_2S oxidation. In natural sulfur spring water, the colonies grow to a size of 4 cm in 3–4 days.

Gram reaction not recorded. Presumably aerobic to microaerophilic. **Sulfur** in form of H_2S is required for growth.

Originally isolated in pure culture from a natural sulfur spring near Chokrakskoye in Russia. Also observed in peat mud, sand or various fresh water samples.

Type species: *Thiodendron latens* Perfil'ev and Gabe 1961, 162.

Genus **Pasteuria** *Metchnikoff 1888, 166, emend. mut. char. Hirsch 1972, 113*

Cells about 1–5 by 3–6 μm, **nearly spherical to pear-shaped** with one end broad and rounded, and the other pointed and attached. Older cells brighter near the rounded pole (DNA?). Occurs singly or in mass aggregations resembling grapes. **Multiplication is by budding from the top or side of the free rounded end** (Fig. 4A and Plate 4.1, Figs. 2–4) (Hirsch, 1972). Buds initially almost spherical. Not motile, not pigmented. Gram reaction not recorded.

Aerobic. Has not been cultured.

Originally isolated from *Daphnia pulex* and *Daphnia magna* and is probably pathogenic for them. Has recently been found in the body cavities and on antennae of *Daphnia* spp. from North America and England (Hirsch, 1972). Similar forms have been found attached to submerged surfaces in lake water (Henrici and Johnson, 1935; Hirsch, unpublished; Staley, personal communication).

Type species: *Pasteuria ramosa* Metchnikoff 1888, 166.

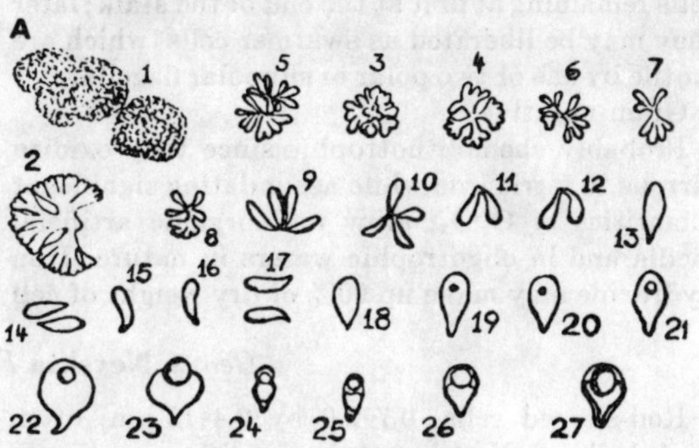

Fig. 4A. Stages in the life cycle of *Pasteuria ramosa* as redrawn from Plate I in the original description of Metchinkoff (1888). Stages 1–18 may be the crystals mentioned by Hirsch (1972). Stages 19–21 are also difficult to interpret. Stages 22–27 resemble the budding forms mentioned by Hirsch (1972). From Hirsch, P. 1972 Int. J. Syst. Bacteriol. *22:* 112–116.

Genus **Blastobacter** *Zavarzin 1961, 962*

Mature cells rod-shaped, wedge-shaped or club-shaped, often slightly curved, 0.7–1.0 (greatest width) by 2.0–4.5 μm (measurements from Fig. 13 of Zavarzin, 1961). Several cells attach by their narrow poles directly to a common base to form a rosette in the center of which there is usually a glistening corpuscle. Stalks absent. Multiplication by single buds appearing on the free cell pole. Buds spherical to oblong, 0.3 μm wide and non-motile.

Massive growth was observed in a cylinder containing iron water from a Northern Russian forest brook to which shreds of filter paper had been added. Growth was best in a zone with reduced iron and a pH of 6.2.

Gram reaction not recorded.

Not obtained in pure culture.

Originally found in isolated samples of fresh water ponds. Probably widely distributed in fresh water as part of the attached microflora.

Type species: *Blastobacter henricii* Zavarzin 1961, 962.

Genus **Seliberia** *Aristovskaya and Parinkina 1963, 56*

Rod-shaped, spirally twisted cells, 0.5–0.7 by 1–12 μm, forming **star-shaped figures** or rosettes (Plate 4.3, Figs. 1–4). Multiplies by transverse fission and by **budding**. When grown on soil media forms round to ovoid "generative" cells (Plate 4.3, Figs. 2 and 3). Resting stages not known. Gram reaction not determined.

Chemoorganotroph. Organic growth factors not required.

Facultatively anaerobic. Temperature optimum 24–25 C.

Type species: *Seliberia stellata* Aristovskaya and Parinkina 1963, 55.

Genus **Gallionella** *Ehrenberg 1838, 166 Nom. cons.* Opin. 9, Jud. Comm. 1954, 147

Cells kidney-shaped or rounded, on the ends of long "stalks" with the long axis of the cell transverse to that of the stalk. Stalks are formed of bundles of fibrils twisted around one another: they may be covered with iron hydroxide; manganese compounds are not deposited; fibrils may branch.

Multiplication is by fission of cells, the daughter cells remaining at first at the end of the stalk; later they may be liberated as swarmer cells, which are motile by one or two polar or subpolar flagella.

Gram-negative.

Probably chemolithotrophic since they oxidize ferrous to ferric iron while assimilating significant quantities of $C^{14}O_2$; grow in inorganic artificial media and in oligotrophic waters in nature. Iron hydroxide may make up 90% of dry weight of cell mass.

Microaerophilic, developing under oxygen concentrations of about 1 mg/liter. Optimum pH 6–7. Develop in cold water, often under snow, although a thermophilic strain has been reported.

Found in ferrous iron-containing waters and in soils; often associated with *Leptothrix ochracea* and with it is responsible for much iron oxide precipitation. Growth of these organisms may cause problems in water works.

Type species: *Gallionella ferruginea* Ehrenberg 1838, 166.

Note: There are two recognized species differentiated on the basis of the number of fibrils in the stalk. *G. ferruginea* has up to 90 fibrils, whereas *G. filamenta* has less than 12.

Genus **Nevskia** *Famintzin 1892, 484*

Rod-shaped cells, 0.7–2.0 by 2.4–12 μm, often slightly bent. Under certain conditions, cells are surrounded with slime but more is produced on one side forming acellular, hyaline stalks perpendicular to the long axis of the organism (Figs. 4.B and 4.C). The stalks branch dichotomously as a result of division of mature cells and leads to the formation of colonies up to 80 μm in size, which float on the surface of the water. They may be spherical, branched or cup-shaped and often form more or less dense pellicles. There is no metal encrustation on the stalks.

The slime is soluble in 1% KOH. Very dilute methyl violet stains the slime but not the cells.

Cells are usually packed with strongly refractile colorless globules. The globules are not sulfur or poly-β-hydroxybutyrate; they stain with Burdon's fat stain and may be dissolved with 70% ethanol.

Young cultures motile by means of one to three polar flagella.

Gram-negative. Aerobic.

Originally isolated from aquarium water. Occur on the surface of fresh water bodies (neuston communities). Have not been grown in pure culture.

Type species: *Nevskia ramosa* Famintzin 1892, 484.

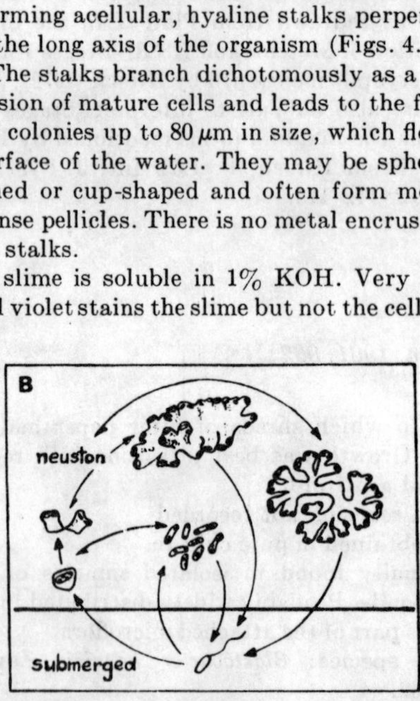

Fig. 4B. Life cycle of *Nevskia* sp. drawn schematically (Babenzien, 1967).

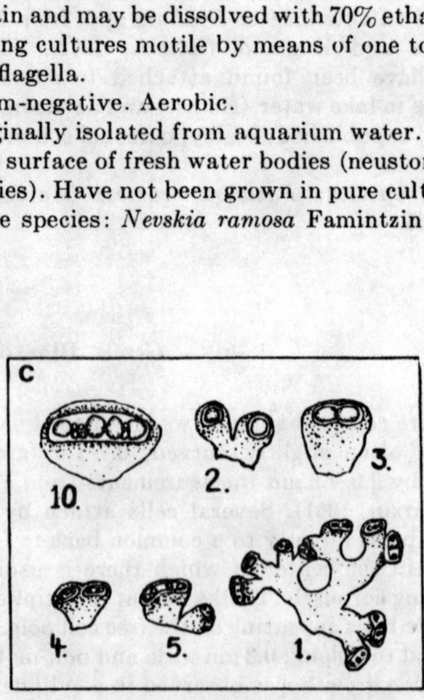

Fig. 4C. *Nevskia ramosa* as drawn by Famintzin (1892) (his Figures 1 thru 5 and 10).

Genus Planctomyces *Gimesi 1924, 4*

Cells spherical to oblong or pear-shaped, 0.3–1.7 μm in diameter with long and slender stalks when in the mature state. Stalks are 0.3–0.9 μm wide and up to 11 μm long and probably excreted; they often attach to a common holdfast and thus form rosettes of up to 50 cells. Older parts of the stalks (in the center of the rosette) may have depositions of iron which thus render the stalks thicker at the base. Multiplication is by terminal or lateral budding of mature or occasionally even smaller, immature cells. Planktonic bacteria that appear primarily in the autumn in aerobic surface layers of lakes.

Have not been grown in artificial media.

Type species: *Planctomyces bekefii* Gimesi 1924, 4.

Key to the species of genus Planctomyces

Key characteristics

Cell width and shape
Stalk width and length
Deposition of iron on stalks
Location of buds

I. Cells wider than 0.7 μm.

A. Cells spherical; stalks up to 10 μm long and 0.4–0.9 μm wide, often encrusted with ferric hydroxide.

1. *Planctomyces bekefii*

B. Cells globose, stalks very short or absent; cells attached to a central material containing iron; buds terminal.

2. *Planctomyces condensatus*

II. Cell diameter less than 0.7 μm.

A. Cells spherical to oblong, the long axis perpendicular to the main stalk axis. Stalks may be slightly curved, attached at the base to form rosettes. Iron may be deposited.

3. *Planctomyces kljasmensis*

B. Cells small and spherical; stalks thin and crsecent-shaped, up to 11 μm long; rosettes of few individuals. Iron deposition not reported:

4. *Planctomyces gracilis*

Genus Metallogenium *Perfil'ev and Gabe 1961, 50*

Cells coccoid, 0.5–1.5 μm in diameter (Plate 4.4, Fig. 1) and do not have a rigid cell wall; attach to surfaces. They either germinate directly or give rise, by repeated budding, to a group of round "elementary bodies." Germination of these bodies results in the formation of one to several flexible filaments with tapering ends (Plate 4.4, Figs. 2–9) and, according to Perfil'ev, occasional branching. In fixed and stained preparations the filaments or arais (Gr. *araios* thin strand) have an approximate diameter of 0.02–0.25 μm (Plate 4.4, Fig. 1) and may give a beaded appearance, similar to a rosary. The cells appear highly refractile when growing in the presence of manganese.

Microcolonies or coenobia growing in microcapillaries show one to several central cocci and the thin tapering arais radiating outward (Plate 4.4, Figs. 3–9). Coenobia may grow to a diameter of 10 μm (Dubinina, personal communication) or up to 1 mm (Perfil'ev and Gabe, 1961).

Multiplication is probably by a budding process. Cocci may also be formed at the end of arais and may remain attached to the filament, thus giving rise to daughter microcolonies around the periphery of the mother coenobium, much like strawberry plants with their runners and daughter plants around them (Plate 4.4, Figs. 5, 7, 9).

Microcolonies assume a brownish to black color in the presence of oxidizable manganese, beginning in the center of the colony. Manganese and/or iron oxides are then deposited onto the arais, increasing the width of the filaments from the base to the tip; they may thicken to a point where they join in the center to form dense lobate masses.

Gram reaction not recorded. For light microscopy and staining the metal deposits must be re-

moved by treatment with oxalic acid (0.2–1.0%) or EDTA, disodium salt (1–2%). Cells stain satisfactorily with carbol-gentian violet but poorly with carbol-erythrosine, methylene blue, aqueous or alcoholic gentian violet or crystal violet.

Heterotrophic. Growth in artificial medium is enhanced by the presence of living fungi. Growth of pure cultures in artificial culture media is slow; better growth is obtained in mixed populations. On ferrous ammonium citrate (1.5 g/liter) agar medium of pH 6.0–7.8 or on "Novogrudskii water agar" growth can be recognized after 4–6 weeks; manganese deposition (if present) is then meager. Colonies appear as irregularly shaped brownish spots. No growth on meat broth-peptone-KNO_3 agar at pH 7.0. Optimal pH: 6.8–7.0. Manganous compounds are not required but manganous carbonate and a low redox potential are stimulatory. Have been grown in pure culture in media containing serum such as is used for mycoplasmas.

Found in upper layers of bottom deposits in fresh water, planktonic in fresh water lakes and ponds and in microzones in soils of the northern hemisphere.

Type species: *Metallogenium personatum* Perfil'ev and Gabe 1961, 50.

Key to the species of genus **Metallogenium**

I. Tapering filaments contain minute cells which become visible after dissolution of the manganese oxides; grow in absence of living fungi although growth is stimulated by their presence.
  1. *Metallogenium personatum*

II. Filaments never contain minute cells. Multiplication by formation of single buds terminally or laterally on the arais. Grow only in presence of living fungi, yeasts or bacteria.
  2. *Metallogenium symbioticum*

Genus **Caulococcus** *Perfil'ev 1964, 43*

Coccoid cells ca. 0.5 µm diameter, which may be connected by fine threads of ca. 0.1 µm diameter. Multiply apparently by budding when in zoogloeal accumulations. Young cells motile, type of motility unknown. Microcolonies up to 0.3 mm with reticulate-ribbed surface structure; irregularly shaped and round, almost black (most common type), or radial-lobate (rare) or trichospherical (very rare). The margin of microcolonies often radially fringed by dichotomously branched filamentous cells.

The coccoid cells may be sparse (radial-lobate types of colonies) or frequent (round, heavily encrusted colonies).

Microaerophilic; deposits mainly Mn^{3+} (80–90%) and some Fe^{3+} (20–10%). Gram reaction not recorded.

Not cultured in the laboratory but can be transferred with mud samples (Perfil'ev and Gabe, 1961, 1964) and observed with the microcapillary technique.

Originally found abundantly in Lake Khepo-Yarvi (Karelian ASSR) especially in ore deposits; frequently observed in the upper layers of bottom mud deposits above the reducing horizon or in the bottom water over the mud surface. Also observed in several other lakes of that area.

Type species: *Caulococcus manganifer* Perfil'ev 1964, 43.

Genus **Kusnezovia** *Perfil'ev 1964, 40*

Cells coccoid, 0.5–1.5 µm in diameter, connected by filaments of ca. 0.1–0.2 µm diameter. Nonmotile, attached, multiplication by budding. Stainable with carbol gentian violet after removal of the often heavy encrustations of manganese with oxalic acid. Gram reaction not recorded.

Microcolonies and the rows of coccoid cells can be observed with submerged peloscopes (Perfil'ev and Gabe, 1961, 1964); they are attached by a stalk to the previously exposed capillary wall and hang freely downward into the capillary canal. The shapes of microcolonies are extremely variable; the following types of manganese-encrusted microcolonies have been observed: arrow-shaped shoots, open lily-of-the-valley leaves, goblets with a toothed margin, or compound, racemose structures resembling the lichen *Cetraria islandica*. In peloscopes, formations may also resemble clubs or candelabras, always pointing downward toward the reducing horizon with their wider parts. Layers of colonies in blackish brown horizons.

Budding can be observed in single-layered accumulations of the coccoid cells found in filamentous coenobia or leaflike formations.

Apparently microaerophilic. Development is stimulated by decomposing fungal hyphae. Manganese, and to much lesser extent iron, is oxidized and deposited.

Has not been cultured in the laboratory.

Originally found in some mud samples of Lake Ukshezero, Karelian ASSR; observed in secondary profiles in zones of manganese oxidation of mud samples stored in the laboratory. Rare.

Type species: *Kusnezovia polymorpha* Perfil'ev 1964, 40.

PART 5

THE SPIROCHETES

ORDER I. SPIROCHAETALES *BUCHANAN 1917, 163*

Slender, flexuous, helically coiled, unicellular bacteria 3–500 μm in length (span of helical organism) with one or more complete turns in the helix (Plate 5.1, Figs. 1–5). May occur in chains held together by the outer envelope, and may assume shape of a planar wave. Multiplication is by transverse fission.

Electron microscopic studies revealed that the cells consist of a protoplasmic cylinder intertwined with one or more axial fibrils which originate in approximately equal numbers from subterminal attachment discs located at either end of the protoplasmic cylinder. Both the protoplasmic cylinder and axial fibrils are enclosed by the outer envelope. The axial fibrils extend toward the end of the cell opposite the points of insertion and overlap one another. The unattached ends of the axial fibrils may extend beyond the ends of the protoplasmic cylinder giving the impression of external polar flagella; however, they are enclosed within the outer envelope.

Motile. Motility is of three types; rapid rotation about the long axis of the helix, flexion of the cells, and locomotion along a helical (corkscrew) or serpentine path.

Do not form endospores.

Larger cells are Gram-negative. Some spirochetes possess inclusions and show cross-striations in stained preparations. Those spirochetes with a transverse cell diameter near the limit of resolution achievable in bright-field microscopy are best observed by phase-contrast or dark-field microscopy.

Tendency to develop involution forms. Distention of the outer envelope may result in terminal or central swellings or spirochetal spheres in which the protoplasmic cylinder is coiled within the outer envelope. Such forms are probably degenerate. Rupture of the outer envelope may release the axial fibrils simulating flagella. Neither genetic transfer nor a life cycle has been established.

May be aerobic, facultatively anaerobic or anaerobic.

Chemoheterotrophic.

Free-living, commensal or parasitic. Some species are pathogenic.

FAMILY I. **SPIROCHAETACEAE** SWELLENGREBEL 1907, 581

Description as for the order.

Key to the genera of family **Spirochaetaceae**

Key characteristics

Size, shape of ends and degree of coiling of cells
Number of turns
Relationship to oxygen
Habitat
Presence of inclusions and bundles of axial filaments as seen in phase contrast microscopy
Catalase
Oxidase
Pathogenicity

I. 5–500 μm in length, 0.2–0.75 μm wide, anaerobic or facultatively anaerobic. Free-living in H_2S-containing fresh water and marine environments, in sewage and polluted water.

Genus I. *Spirochaeta*, p. 168*

II. 30–150 μm in length; 0.5–3.0 μm wide with 3 to 10 complete turns. In living specimens ovoid inclusion bodies and large bundles of axial fibrils may be seen by phase-contrast microscopy. Commensal, usually in mollusks.

Genus II. *Cristispira*, p. 171

III. 5–15 μm in length; 0.09–0.5 μm wide. Catalase and oxidase negative; anaerobic; commensal or parasitic. Some are pathogens.

Genus III. *Treponema*, p. 175

IV. 3–15 μm in length; 0.2–0.5 μm wide; anaerobic; parasitic. Some are pathogens, transmitted by ticks and lice.

Genus IV. *Borrelia*, p. 184

V. 6–20 μm in length; 0.1 μm wide; tightly coiled and may have bent or hooked ends; aerobic; free-living or parasitic. Some are pathogens.

Genus V. *Leptospira*, p. 190

Genus I. Spirochaeta *Ehrenberg 1835, 313*

Helical cells, 0.20–0.75 by 5–500 μm (Plate 5.1, Fig. 1). The cells have **axial fibrils.** The protoplasmic cylinder is wound around or together with the axial fibrils. The cells do not possess either terminal hooks or cross-striations. Motile, probably by means of the axial fibrils. **Free-living** in H_2S-containing mud, in sewage and polluted water Non-parasitic.

Of the five species described, only the type species has not been grown in pure culture. The four species which have been cultured are **chemoorganotrophs, obligate and facultative anaerobes, ferment carbohydrates** and have a **guanine + cytosine content** in their DNA ranging **from 50–66 moles %** (buoyant density). Cells of these four species possess two axial fibrils, one inserted near one end of the protoplasmic cylinder and the other near the opposite end. The two fibrils overlap in the central portion of the cell (1-2-1 arrangement). The protoplasmic cylinder and axial fibrils of these species are enclosed in a sheath or outer cell envelope.

One species is pigmented.

Type species: *Spirochaeta plicatilis* Ehrenberg 1835, 313.

Key to the species of genus Spirochaeta

Key characteristics

Size of cells
Relationship to oxygen
Products of fermentation of glucose
Growth requirement for Na^+, Cl^-, biotin, niacin and coenzyme A

I. Large, helical cells, presumably obligate or facultative anaerobes.
1. *Spirochaeta plicatilis*

II. Thin, helical cells.
A. Obligate anaerobes.
a. Ferment glucose to ethanol, acetic acid, lactic acid, CO_2 and H_2. Long cells (15–45 μm, up to 300 μm).

* All page numbers cited refer to *Bergey's Manual of Determinative Bacteriology*, 8th edition, The Williams & Wilkins Company, Baltimore, 1974.

2. *Spirochaeta stenostrepta*

b. Ferment glucose to acetic acid, lactic acid, succinic acid, CO_2 and H_2.

3. *Spirochaeta zuelzerae*

c. Ferment glucose to ethanol, acetic acid, CO_2, H_2 and small amounts of lactic, formic and pyruvic acids. Require exogenous supplements of Na^+, Cl^-, biotin, niacin and coenzyme A for growth.

4. *Spirochaeta litoralis*

B. Facultative anaerobes.

5. *Spirochaeta aurantia*

Genus II. Cristispira *Gross 1910, 44*

Cells single, flexuous, 0.5–3.0 μm in diameter and **helically coiled** or **planar wave-shaped** (Plate 5.1, Fig. 2). **Helix** spans **more than 30** and **up to 150 μm,** and has **2–10 complete turns. Ends** of cells either **rounded,** or **tapered,** or **pointed** with a rigid spicule. Divide by **transverse fission.**

The cells may contain 30–80 seriately aligned **ovoid inclusions** of unknown composition, which **displace** the **nuclear structure** so that it appears as **cross-striations** in stained preparations.

The normal movement is a **forward and backward locomotion** by means of an irrotational traveling helical wave. The helix may also rotate on its long axis or flex. There are **over 100 axial fibrils** intertwined with the protoplasmic cylinder; when grouped in a bundle, these may be visible by phase-contrast microscopy.

When removed from or disturbed in its habitat, **involutionary changes readily occur.** The axial fibrils may become detached from the protoplasmic cylinder within a distended spirochetal envelope and form the **so-called crista;** central and terminal bullae and spirochetal spheres ("cysts") may develop. Motility becomes irregular and cells may lyze.

Commensal: Widely distributed among **marine and fresh water** species of **bivalve and univalve mollusks,** usually found in the crystalline style (a rod-shaped structure composed of "mucus" secretions) or the fluid of the digestive tract. May also occur in other animals: large spirochetes have been observed in an echinoderm (Collier, 1921) and in tunicates (Hellmann, 1913). No species of *Cristispira* has thus far been grown in pure culture.

Proposed type species: *Cristispira pectinis* Gross 1910, 44 (Kuhn, 1970).

Genus III. Treponema *Schaudinn 1905, 1728*

Unicellular, helical rods 5–20 μm long, 0.09–0.5 μm wide, as determined by electron microscopy; with **tight regular or irregular spirals** (Plate 5.1 Fig. 3, 6–8). Cells have **one or more axial fibrils inserted at each end of the protoplasmic cylinder.**

Bullae and spirochetal spheres are seen in old cultures.

Motile. Gram-negative. Stain well with silver impregnation methods and Ryu's (1963) stain. Most species stain poorly with Giemsa's stain. Best **observed** under **dark-field or phase-contrast microscopy.** Most of the following part of the genus description refers only to the *in vitro* cultivatable species.

Chemoorganotrophs: **Metabolism fermentative,** using amino acids and/or carbohydrates.

Strict anaerobes. **Catalase and oxidase negative,** urease negative and Voges-Proskauer negative. Does not reduce nitrate. Not inhibited by iodoacetate.

Found in the oral cavity, intestinal tract and genital regions of man and animals. Some species are pathogenic.

The G + C content of the DNA (in the species examined) ranges from 32–50 moles % (T_m).

Type species: *Treponema pallidum* (Schaudinn and Hoffmann) Schaudinn 1905, 1728.

Key to the species of genus Treponema

Key characteristics

Pathogenicity
Products of metabolism
Mannitol fermentation
Width of cells
Utilization of lactate

I. Causes syphilis, yaws or pinta in man and animals.
   A. Cutaneous lesions in rabbits, no cutaneous lesions in hamsters or guinea pigs.
      1. *Treponema pallidum*
   B. Cutaneous lesions in rabbits and hamsters, but not in guinea pigs.
      2. *Treponema pertenue*
   C. Does not produce cutaneous lesions in rabbits, hamsters or guinea pigs.
      3. *Treponema carateum*
   D. Cutaneous lesions in rabbits and guinea pigs, not in hamsters.
      4. *Treponema paraluis-cuniculi*

II. Does not cause syphilis, yaws or pinta in man or animals.
   A. Ferments carbohydrates.
      1. Produces *n*-butyric, acetic and propionic acids; ethanol, *n*-propanol and *n*-butanol. Ferments mannitol.
         5. *Treponema phagedenis*
      2. Produces lactic, acetic and formic acids; traces of succinic acid. Does not ferment mannitol.
         6. *Treponema macrodentium*
   B. Does not ferment carbohydrates.
      1. Produces acetic acid and smaller amounts of succinic, lactic and formic acids.
         a. Cells 0.25–0.35 μm wide.
            7. *Treponema refringens*
         b. Cells 0.10–0.25 μm wide.
            8. *Treponema denticola*

TABLE 5.1[a]

Some identifying characteristics of cultivatable treponemes (Species 5–11)

	T. phagedenis biotype Reiter	*T. phagedenis* biotype Kazan	*T. macrodentium*	*T. refringens*	*T. refringens* biotype calligyrum	*T. denticola*	*T. denticola* biotype comandoni	*T. orale*	*T. scoliodontum*	*T. vincentii*
Cells 0.1–0.25 μm wide	−	−	+	−	−	+	+	+	+	+
Cells 0.25–0.35 μm wide	+	+	−	+	+	−	−	−	−	−
No. of axial fibrils	3	3	1	?	?	2	?	1	?	?
Carbohydrates required as energy source	−	−	+	−	−	−	−	−	−	−
Carbohydrates fermented										
Fructose	+	+	+	−	−	−	−	−	−	−
Glucose	+	+	+	−	−	−	−	−	−	−
Lactose	+	+	−	−	−	−	−	−	−	−
Mannitol	+	+	−	−	−	−	−	−	−	−
Sucrose	−	−	+	−	−	−	−	−	−	−
Esculin hydrolysis	−	+	?	+	+	+	+	?	−	−
Indole	+	+	−	+	+	+	−	+	−	+
H_2S	d	d	+	+	+	+	+	+	−	+
1% glycine	+	+	?	−	+	d	d	?	−	−
Convert lactate	−	−	−	−	−	−	−	+	−	−

[a] Symbols: + = Acid production, a pH of 6.3 or lower or a positive test in 90% or more strains.
− = Most (90% or more) strains negative.
d = Character inconstant and in one strain may sometimes be positive, sometimes negative.
? = Information not available.

2. Produces acetic and propionic acids.
   a. Uses lactate.

9. *Treponema orale*

3. Produces acetic and traces of propionic and *n*-butyric acids.
   a. Does not use lactate.

10. *Treponema scoliodontum*

4 Produces acetic and *n*-butyric acids.
   a. Does not use lactate.

11. *Treponema vincentii*

Genus IV. **Borrelia** *Swellengrebel 1907, 582*

Cells 0.2–0.5 by 3–20 μm, helical, with 3–10 or more coarse, uneven, often irregular coils, some of which may form obtuse angles (Plate 5.1, Fig. 4). Cells up to 1 μm wide and 25 μm long may occur; the measurements vary according to strain and the staining method used.

Electron microscopy shows a foamy, elastic envelope and a cytoplasmic membrane, between which there are 15–20 parallel fibrils coiled around the cell body. There is a central elongated fibrillar substance, no mitochondria and no undulating membrane. The fibrils constitute the locomotory apparatus, the coiled fibrils usually rotating in one direction, the cell body in the opposite direction but at the same rate. Motion is in forward and backward waves, laterally by bending and looping and corkscrew-like.

Gram-negative. Stain well with aniline dyes, particularly when phenol is used as a mordant.

Nutritional characteristics of most species unknown. Those studied possess fermentative metabolism. Growth of several species reported on complex media containing natural animal proteins or in developing chick embryos. Minimal nutritional requirements unknown.

Strict anaerobes. Growth observed at 20–37 C, optimum at 28–30 C. Preserved at −76 C.

Antigenic structure generally unstable resulting in considerable variations particularly in pathogenic strains during relapses.

No adequate system of species separation has yet been found.

Type species: *Borrelia anserina* (Sakharoff) Bergey *et al.* 1925, 435.

Key to the species of genus **Borrelia**

I. From birds.

1. *Borrelia anserina*

II. From animals other than birds.
  A. From man.
    1. Cause relapsing fever in man.
      a. Arthropod vector is the louse *Pediculus humanus* subsp. *humanus*.

2. *Borrelia recurrentis*

      b. Arthropod vector is a tick of the genus *Ornithodoros*.
      b1. Transmitted by *Ornithodoros erraticus erraticus*.

3. *Borrelia hispanica*

      b2. Transmitted by *Ornithodoros hermsi*.

4. *Borrelia hermsii*

      b3. Transmitted by *Ornithodoros moubata*.

5. *Borrelia duttonii*

      b4. Transmitted by *Ornithodoros parkeri*.

6. *Borrelia parkeri*

      b5. Transmitted by *Ornithodoros rudis*.

7. *Borrelia venezuelensis*

      b6. Transmitted by *Ornithodoros talaje*.

8. *Borrelia mazzottii*

      b7. Transmitted by *Ornithodoros tholozani*.

9. *Borrelia persica*

      b8. Transmitted by *Ornithodoros turicata*.

10. *Borrelia turicatae*

b9. Transmitted by *Ornithodoros verrucosus*.

11. *Borrelia caucasica*

B. From animals other than man.

1. From ticks (*Ornithodoros*).

a. From *O. brasiliensis*.

12. *Borrelia brasiliensis*

b. From *O. dugesi*.

13. *Borrelia dugesii*

c. From *O. graingeri*.

14. *Borrelia graingeri*

2. From animals other than arthropods.

a. From rodents.

a1. Transmitted by *Ornithodoros erraticus sonrai*.

15. *Borrelia crocidurae*

a2. Transmitted by *O. tartakovskyi*.

16. *Borrelia latyschewii*

a3. Transmitted by *O. zumpti*.

17. *Borrelia tillae*

b. From ruminants and horses.

18. *Borrelia theileri*

c. From primates.

19. *Borrelia harveyi*

Genus V. **Leptospira** *Noguchi 1917, 755*

Cells **single, flexuous, helical,** 6–20 μm or more by 0.1 μm in diameter; coils 0.2–0.3 μm in overall diameter, pitch 0.3–0.5 μm (Plate 5.1, Fig. 5). **One or both ends may be bent**—more or less at right angles to the long axis—**or hooked.** Some strains may lose their hooked shape during subculture in certain media and retain the straight form in that medium.

Motile by means of axial filaments, comprising the axistyle, which are now regarded as flagellar analogues; no external flagella. **Movements are of three basic sorts: shunting** in either direction of the long axis (translation without polar differentiation); **rapid rotation** or oscillation about the long axis; and **flexion.** Leptospires may be motionless for a few moments and at such times the tightly coiled appearance, suggestive of a tiny rope, is usually visible as are the shape and size of the hooks if these are present. Translation is on a straight or gently curved course in liquid milieux; but in more viscous milieux (e.g. semi-solid agar) the course is usually sinuous or serpentine. When a hook-ended organism rotates rapidly around its long axis a variety of shapes may result, commonly like an hourglass, the figure 8, a tennis racquet or the capital letter T, depending on the number and form of the hooks. Flexion of the body may occur spontaneously when the leptospire is stationary and also when it meets an obstruction; it is usual during translation in viscous media; and it is also associated with transverse division of the organism, when it may be very vigorous.

Not readily visible in preparations treated with Giemsa stain or the Gram-reaction, but revealed, often distorted, by silver impregnation methods. **Readily visible by dark-field microscopy** at magnifications of 100 × or greater: less clearly observed by phase-contrast; not visible by ordinary illumination.

The structure of *Leptospira* as revealed by electron microscopy consists of a cellular body wound around an axistyle with a common external sheath enveloping both these elements. The cellular body is roughly circular in transverse section. Nuclear material, cytoplasm and a limiting cytoplasmic membrane can be differentiated. The axistyle seems to be a single structure, about 0.01–0.02 μm in diameter, comprised of two filaments each of which is inserted subterminally, by means of a disc or knob, at one end of the body with their free ends near the middle of the body.

Chemoorganotrophs: Metabolism respiratory. Terminal electron acceptors are cytochromes a, c and c_1 (for strains investigated; Baseman and Cox, 1969). Energy-yielding substrates should include fatty acids as the source of carbon. Parasitic strains require unsaturated fatty acids (containing 15–16 carbon atoms or more) or the corresponding polysorbate (Tween). This requirement is increased at temperatures above or below the optimum. Saturated fatty acids are not used unless suitable unsaturated fatty acids are also present (Johnson, Harris and Walby, 1969). Inorganic NH_4 salts are suitable as the source of nitrogen. Other minimal nutritional requirements include thiamine, vitamin B_{12}, and various metallic salts (see Ellinghausen and McCullough, 1965; Shenberg, 1967). Some strains seem to need other, undetermined, nutritional factors.

Can be cultured on various media which are based on a solution of inorganic salts, buffered with phosphates. Some media are supplemented with pooled rabbit serum, 7–20% v/v (Fletcher, 1928; Korthof, 1932; Stuart, 1948); others with bovine albumin, fraction V, and Tween 80 (Ellinghausen and McCullough, 1965; Johnson and Harris, 1967) usually without rabbit serum. Pooled normal rabbit serum added (2–5%, v/v) to bovine albumin Tween 80 medium enhances the rate of multiplication of certain strains—notably those which belong to serotypes which did not multiply in Shenberg's synthetic medium (Shenberg, 1967). Solid media are not recommended for routine isolation of strains because satisfactory multiplication is much less certain than in liquid or semi-solid (containing 0.2% agar, w/v) media.

Strict **aerobes** utilizing atmospheric oxygen.

Temperature optimum 28–30 C. More rapid multiplication may sometimes result from incubation at 37 C for 1–2 days. Exposure to 40 C is deleterious (Turner, 1966); and at 56 C it is lethal (for pathogenic strains investigated). Strains of free-living, presumed saprophytic, leptospires will multiply at lower temperatures (e.g. 13 C) than parasitic and pathogenic strains (Johnson and Harris, 1967).

Optimal pH requirement 7.2–7.4.

The G + C content of the DNA, for the few strains reported, ranges from 36–39 ± 1% (T_m and buoyant density, Haapala *et al.*, 1969).

Some strains parasitic or pathogenic in vertebrates; others free-living.

Type species: *Leptospira interrogans* (Stimson) Wenyon 1926, 1281.

PART 6

SPIRAL AND CURVED BACTERIA

Note: Spiral and curved organisms other than those included in the family *Spirillaceae* may be found on pages 75 and 76 among the Genera of Uncertain Affiliation. These may be differentiated on the basis of parasitism of bacteria (*Bdellovibrio*); cells in ring formation (*Microcyclus*); cells in sigmoid aggregates (*Pelosigma*); and bow-shaped, gas-vacuolated cells (*Brachyarcus*). Other Parts of the *Manual* (1, 2, 5, 8, 9, 12, 17, 18 and 19) also contain curved organisms; however, these are usually differentiated on the basis of some other major criterion, such as phototrophism.

FAMILY I. SPIRILLACEAE MIGULA 1894, 237

Rigid, helically curved rods with less than one complete turn to many turns. Cell diameter, 0.2–1.7 μm; length of helix 0.5–60 μm. Motile, swimming in straight lines with a characteristic corkscrew-like motion. Cells may possess only a single polar flagellum or a fascicle of several polar flagella. Flagella may be at one or both poles. Some species are strictly aerobic or obligately microaerophilic, with oxygen required as the terminal electron acceptor. Others are anaerobic but can also grow under microaerophilic conditions. Chemoorganotrophs. Incapable of fermenting carbohydrates, although a few species can oxidize a limited variety. Usually no alcohols or fatty acids are detected by gas chromatography of culture supernatants; when detected they are present in only trace amounts. Oxidase positive. Indole negative. Some produce a yellowish green water-soluble fluorescent pigment. Some grow in a simple mineral salts medium containing a single carbon source and ammonium sulfate; other require a complex peptone-based medium. Some are free living in fresh water or marine environments. Others are saprophytic or parasitic. Some are pathogenic.

Key to the genera of family **Spirillaceae**

I. Cells 0.25–1.7 μm wide and 2–60 μm long. Polytrichous polar flagella, usually at both poles. Polyhydroxybutyrate granules usually present in cytoplasm. Strictly aerobic or obligately microaerophilic. The G + C content of the DNA ranges from 38–65 moles % (T_m).

Genus I. *Spirillum*, p. 196*

II. Cells 0.2–0.8 μm wide and 0.5–5 μm long. Single polar flagellum at one or both poles. Polyhydroxybutyrate granules not present. Microaerophilic to anaerobic. The G + C content of the DNA ranges from 30–35 moles % (T_m).

Genus II. *Campylobacter*, p. 207

Genus I. Spirillum *Ehrenberg 1832, 38*

Rigid, helical cells, 0.25–1.7 μm in diameter, with less than one turn to many turns. Intracellular granules of polyhydroxybutyrate present in most species. Motile by means of polar polytrichous flagella. All but one species (*S. delicatum*) exhibit bipolar flagellation. Gram-negative.

Chemoorganotrophs, having a strictly respiratory metabolism with oxygen as the terminal

* All page numbers cited refer to *Bergey's Manual of Determinative Bacteriology*, 8th edition, The Williams & Wilkins Company, Baltimore, 1974.

electron acceptor. A few species can grow anaerobically with nitrate.

Acid reactions are usually not produced from sugars; in the few positive cases the reactions are evident only with a low peptone concentration. Sugars usually do not serve as sole carbon sources.

Casein and hippurate not hydrolyzed. Indole, sulfatase, amylase and phenylalanine deaminase negative. Gelatin usually not attacked. Hydrogen sulfide usually produced from cysteine. A few species are urease positive.

Oxidase and catalase positive; the latter is sometimes weak. Phosphatase usually positive.

Yellowish green fluorescent water-soluble pigment produced by nearly half of the species.

The nutrition is simple; most species do not require amino acids, vitamins, purines or pyrimidines. Various organic acids, alcohols or amino acids can be utilized as sole carbon sources. Ammonium ions can usually be utilized as a sole nitrogen source. Incorporation of sea water into the media is required for marine species.

The type species is obligately microaerophilic. All other species can grow aerobically, although some may possibly prefer microaerophilic conditions. Optimum temperature 30 C, with no growth at 10 or 45 C. Fresh water species do not grow in NaCl concentrations greater than 3%, and some fail to grow in concentrations as low as 0.3%.

Found in fresh and salt waters containing organic matter.

The G + C content of the DNA ranges from 38–65 moles %.

Type species: *Spirillum volutans* Ehrenberg 1832, 38 (Stiles, 1905; Hylemon *et al.*, 1973).

Further Comments

In order to provide a suitable basis for comparison with the descriptions, the media and methods described in the dissertations by Wells (1970) and Hylemon (1971) should be used for the determination of the characteristics of spirilla; the methods are summarized in Hylemon, Wells, Krieg and Jannasch in the October 1973 issue of the *International Journal of Systematic Bacteriology.*

Editorial Note. Hylemon, Wells, Krieg and Jannasch (The genus *Spirillum:* A taxonomic study. Int. J. Syst. Bacteriol. *23:* 340–380, 1973) propose a division of the genus into three genera:

1. **Spirillum** with the obligately microaerophilic *S. volutans* as the type and only species.

2. **Aquaspirillum,** a new genus for the aerobic

TABLE 6.1

Differential characteristics of the fresh water species of genus **Spirillum**

	1. *S. volutans*	2. *S. giesbergeri*	3. *S. graniferum*	4. *S. anulus*	5. *S. metamorphum*	6. *S. putridiconchylium*	7. *S. itersonii*	8. *S. peregrinum*	9. *S. serpens*	10. *S. sinuosum*	11. *S. gracile*	12. *S. delicatum*	13. *S. polymorphum*
Microaerophilic	+	−	−	−	−	−	−	−	−	−	−	−	−
Cell diameter													
1.5 μm	+												
0.5–1.5 μm		+	+	+	+	+	+	+	+	+			
0.25–0.4 μm											+	+	+
Succinate as sole carbon source		−	−	−	+	+			−	−			
H_2S from cysteine		+	−	+	+	+			+	+			
Urease		+	+	−	−	−			d	+			
Gelatin liquefied		−	−	−	+	−			d	−			
Anaerobic growth from nitrate		−	−	−	−	−	+	−	−	−			
Fructose acid		−	−	−	−	−	+	+	−	−			
Glutamate as sole carbon source		−	−	+	+	+	+	+	+	−			
Glucose acid											+	−	−
Phosphatase											+	+	−
Krebs cycle acids as sole carbon source											−	+	−
G + C moles %	38	58	57	59	63	52	62	62	50	57	65	63	62

TABLE 6.2

Other characteristics of species of genus **Spirillum**

Reactions are determined in peptone-succinate-salts (PSS) medium (see "Description of the species of genus *Spirillum*"). The peptone must be decreased to 0.2% to demonstrate acid reactions from sugars. Urease is determined using heavy washed suspensions in a BES-buffered phenol red medium. W = weak reaction; d = varies among strains.

Test	Species																	
	1. *S. volutans*	2. *S. giesbergeri*	3. *S. graniferum*	4. *S. anulus*	5. *S. metamorphum*	6. *S. putridiconchylium*	7. *S. itersonii*	8. *S. peregrinum*	9. *S. serpens*	10. *S. sinuosum*	11. *S. gracile*	12. *S. delicatum*	13. *S. polymorphum*	14. *S. lunatum*	15. *S. minutulum*	16. *S. beijerinckii*	17. *S. linum*	18. *S. japonicum*
Catalase	W	+	+	+	+	W	+	+	+	+	+	+	+	+	+	W	W	W
Phosphatase	+	+	+	+	+	+	+	+	+	+	+	+	−	+	−	+	+	+
H_2S from cysteine	+	+	−	+	+	+	+	+	+	+	+	+	+	+	−	+	d	−
Urease	−	+	+	−	−	−	−	+	d	+	+	−	−	−	−	−	−	−
Esculin hydrolysis	−	−	−	−	−	−	+	+	−	−	−	−	+	−	−	−	−	−
Gelatin liquefaction	−	−	−	−	+	−	−	−	d	−	−	−	−	−	−	−	−	−
Nitrate reduction to nitrite or more reduced products	−	−	−	−	−	−	+	−	d	−	+	+	+	+	+	−	−	+
Selenite reduction	−	−	−	−	−	−	d	+	−	−	−	−	W	+	+	−	W	W
Polyhydroxybutyrate granules	+	+	+	+	+	+	+	+	+	+	−	+	+	−	+	+	+	+
DNase	−	+	+	−	−	−	+	−	d	+	+	−	−	+	−	+	−	−
RNase	−	+	+	d	−	+	+	−	d	−	+	+	−	+	−	+	d	−
Acid reaction from sugars	−	−	−	−	−	−	+	+	−	−	+	−	−	+	−	−	−	−
Growth in presence of:																		
a. 1% bile	−	+	+	−	+	+	+	+	+	+	+	−	+	+	+	+	+	+
b. 1% glycine	−	−	−	−	−	−	−	−	−	−	−	−	−	+	+	−	+	−
Growth on:																		
a. Triple sugar iron slants	−	−	−	−	+	+	+	+	d	−	−	+	−	+	+	−	+	−
b. EMB agar	−	−	−	−	+	+	+	+	+	−	+	−	+	+	−	−	−	−
c. MacConkey agar	−	−	−	−	−	−	+	−	d	−	−	−	+	+	−	−	−	−
d. MRVP broth	−	+	+	−	−	−	+	−	d	−	−	+	−	+	−	−	+	−
e. Seller's slants	−	−	−	−	+	−	+	+	d	−	−	−	−	+	−	−	−	−
Pigment produced from:																		
a. Phenylalanine	−	−	−	−	−	−	−	+	−	−	−	−	−	−	−	−	−	−
b. Tyrosine	−	−	−	−	−	−	+	−	−	−	−	−	−	+	−	+	+	−
c. Tryptophan	−	−	−	−	−	−	+	−	−	−	−	−	−	+	d	−	d	−
Alkaline reaction in litmus milk	−	−	−	−	−	−	−	+	−	−	−	−	−					
Water-soluble fluorescent pigment	−	−	−	−	+	−	+	+	d	−	d	−	+	+	−	−	+	−
Anaerobic growth with nitrate	−	−	−	−	−	−	+	−	−	−	−	−	−	+	−	−	−	−
Coccoid bodies predominant in older cultures	−	−	−	−	−	−	+	+	−	−	−	−	+	−	+	+	+	−

TABLE 6.3

Sole carbon sources for species of genus **Spirillum**

The following are not used as sole carbon sources by any species of *Spirillum:* D-ribose, L-xylose, 2-ketogluconate, D-galactose, lactose, sucrose, maltose, sedoheptulose, D-raffinose, tryptophan, methionine, threonine. Propionate, ethanol, *n*-butanol, L-ornithine, L-citrulline, L-arginine, L-lysine, L-cysteine, L-leucine, L-isoleucine and L-valine are used by some strains of *S. itersonii*.

A washed standardized inoculum is used. + = the production of at least 10 Klett units of turbidity (blue filter) in the second 72-hr subculture in defined mineral salts basal medium containing carbon source, using 16-mm cuvettes.

Sole carbon source	2. *S. giesbergeri*	3. *S. graniferum*	4. *S. anulus*	5. *S. metamorphum*	6. *S. putridiconchylium*	7. *S. itersonii*	8. *S. peregrinum*	9. *S. serpens*	10. *S. sinuosum*	11. *S. gracile*	12. *S. delicatum*	13. *S. polymorphum*	14. *S. lunatum*	15. *S. minutulum*	16. *S. beijerinckii*	17. *S. linum*	18. *S. japonicum*
Citrate	−	−	−	−	−	−	−	−	−	−	−	−	+	−	−	−	−
Aconitate	−	−	−	−	+	+	+	−	−	−	+	−	+	−	−	−	−
Isocitrate	−	−	−	−	+	−	−	−	−	−	+	−	+	−	−	−	−
α-Ketoglutarate	−	−	−	+	+	+	+	d	−	−	+	−	+	d	−	−	−
Succinate	−	−	−	+	+	+	+	−	−	−	+	−	+	+	−	−	−
Fumarate	−	+	−	+	+	+	+	−	−	−	+	−	+	+	−	−	−
L-Malate	−	−	−	+	+	+	+	−	+	−	+	−	+	+	+	−	+
Oxaloacetate	−	−	−	+	+	d	+	d	+	−	+	−	+	+	−	−	+
Pyruvate	+	+	+	+	+	d	+	d	+	−	+	−	+	+	+	−	+
L-Lactate	−	−	−	+	+	d	+	−	−	−	+	−	+	+	−	−	+
Malonate	−	−	−	+	+	d	+	−	−	−	+	+	+	−	−	−	−
Acetate	−	−	−	−	+	−	+	−	−	−	+	+	+	+	−	d	−
β-Hydroxybutyrate	−	−	−	−	+	+	+	−	−	−	−	−	+	−	−	−	−
n-Propanol	−	−	−	−	−	+	−	−	−	−	−	−	+	−	−	−	−
L-Histidine	−	−	−	−	+	+	−	−	−	−	−	−	−	−	−	−	−
L-Tyrosine	−	−	−	−	−	−	−	−	−	−	−	−	+	−	−	−	−
L-Phenylalanine	−	−	−	−	−	+	−	−	−	−	−	−	−	−	−	−	−
L-Alanine	−	−	−	+	+	d	+	d	−	−	−	−	+	−	−	−	+
L-Glutamate	−	−	d	+	+	+	+	+	−	−	−	+	+	+	−	−	+
L-Aspartate	−	−	−	+	+	+	+	+	−	−	−	+	−	−	−	−	−
L-Asparagine	−	−	−	+	+	+	+	d	−	−	−	−	+	−	−	−	−
L-Proline	−	−	−	−	+	+	+	d	−	−	−	+	+	+	−	−	−
L-Hydroxyproline	−	−	−	−	−	d	+	−	−	−	−	−	−	−	−	−	−
L-Serine	−	−	−	−	−	−	−	d	−	−	−	−	−	−	−	−	−
Glycine	−	−	−	−	−	−	−	−	−	−	−	−	+	−	−	−	−
Putrescine	−	−	−	−	−	d	−	−	−	−	+	−	−	−	−	−	−
D-Glucose	−	−	−	−	−	−	−	−	−	−	−	−	+	−	−	−	−
D-Gluconate	−	−	−	−	−	−	−	−	−	−	−	−	+	−	−	−	−
D-Mannitol	−	−	−	−	−	−	−	−	−	−	−	−	+	−	−	−	−
D-Fructose	−	−	−	−	−	d	+	−	−	−	−	−	−	−	−	−	−
Glycerol	−	−	−	−	−	d	−	−	−	−	−	−	+	−	−	−	−

fresh water forms containing species 2, 4–13 recognized herein and two new species *A. dispar* (type strain: ATCC 27510) and *A. aquaticum* (type strain: ATCC 11330); *A. serpens* is the type species. *S. graniferum* (No. 3) is considered a synonym of *A. giesbergeri* (No. 2).

3. **Oceanospirillum,** a new genus for the marine forms that do not attack carbohydrates, with *O. linum* (No. 17) as the type species plus species 15, 16, 18 and a new species *O. maris* (type strain: ATCC 27509).

S. lunatum (No. 14) was not included in genus *Oceanospirillum* because of its higher G + C content and its ability to attack a variety of carbohydrates. The authors note that the authenticity of the type strain was in question; several characteristics are not consistent with the original description.

TABLE 6.4

Differential characteristics of sea water-requiring species of genus **Spirillum**

	14. *S. lunatum*	15. *S. minutulum*	16. *S. beijerinckii*	17. *S. linum*	18. *S. japonicum*
Acid reaction from sugars	+	−	−	−	−
Phosphatase		−	+	+	+
Nitrate reduction		+	−	−	−
Selenite reduction		+	−	+	+
Krebs cycle acids and amino acids as sole carbon sources		+		−	+
G + C moles %	63	42	47	48	45

Genus II. Campylobacter *Sebald and Véron* 1963, 907

Slender **non-spore-forming spirally curved rods,** 0.2–0.8 μm wide and 0.5–5 μm long. The rods may have one or more spirals and can be as long as 8 μm. They also appear S-shaped and gull-winged when two cells form short chains. Cells in old cultures form spherical or coccoid bodies.

Motile with a characteristic corkscrew-like motion. They have a **single polar flagellum at one or both ends of the cell.** Flagella may be 2–3 times the length of the cells. **Gram-negative.**

Chemoorganotrophs: Carbohydrates are neither fermented nor oxidized. Metabolism of those species studied is respiratory. No acid or neutral end products produced.

Do not require serum for growth. Energy from amino acids or tricarboxylic acid cycle intermediates not carbohydrates. Gelatin and urea not hydrolyzed. Methyl red and Voges Proskauer negative. No lipase activity. Oxidase positive. Pigments are not produced.

Microaerophilic to anaerobic. Some species are microaerophilic requiring an oxygen concentration of between 3 and 15%. Other species are anaerobic or can grow under either microaerophilic or anaerobic conditions. Occasionally a few strains may grow slightly under aerobic conditions.

Some species are pathogenic for man and animals. Found in the reproductive organs, intestinal tract and oral cavity of animals and man.

The G + C content of the DNA (of those species studied) ranges from 30–35 moles % (T_m).

Type species: *Campylobacter fetus* (Smith and Taylor) Sebald and Véron 1963, 907.

Further Comments

Most of the organisms in this genus have been studied using Albimi Brucella broth (Pfizer Diag-

TABLE 6.5

Some differential characteristics of species of the genus **Campylobacter**

	Catalase	Nitrite reduction	H_2S on TSI	H_2S, lead acetate strips	1% glycine	3.5% NaCl	25 C
1.a. *C. fetus* subsp. *fetus*	+[a]	−	−	−	−	−	+
b. *C. fetus* subsp. *intestinalis*	+	−	−	+	+	−	+
c. *C. fetus* subsp. *jejuni*	+	−	−	+	+	−	−
2.a. *C. sputorum* subsp. *sputorum*	−	+	+	+	+	−	+
b. *C. sputorum* subsp. *bubulus*	−	+	+	+	+	+	d
3. *C. fecalis*	+	?	+	+	+	d	−

[a] + = most (90% or more) strains positive for this characteristic; − = most (90% or more) strains negative; d = some (less than 90%) strains positive, some negative; ? = reaction not known.

nostics, Flushing, New York) or agar as the basal medium. Most tests were made in semisolid Brucella broth (0.16% agar) which provides microaerophilic conditions. For tests in liquid media or on agar plates, the oxygen content must be reduced to 3–15%. Casein, gelatin, DNA and RNA hydrolyses were made on Brucella agar plates incubated in an atmosphere of 5% oxygen and 95% nitrogen. Many of the methods are described in detail by Smibert (1965 and 1969).

GENERA OF UNCERTAIN AFFILIATION

Genus **Bdellovibrio** *Stolp and Starr 1963, 243*

Cells are single, small, curved, motile rods in the parasitic state; motility by means of a polar, sheathed flagellum, usually monotrichous. Cells elongate to helical (spiral) forms in the non-parasitic stage of their life cycle. Gram-negative.

May or may not be obligately parasitic on bacteria. Parasitic strains attach to and penetrate into the host; may exhibit a host-range specificity.

Chemoorganotrophs, metabolism respiratory, not fermentative. Parasitic strains require the presence of suitable host bacteria for intracellular multiplication; only facultative strains or host-independent (H-I mutants) derived from parasitic strains are able to grow in complex media.

Aerobic.

Temperature optimum generally 30 C; growth extremely poor above 37 C.

The G + C content of the DNA of most cultures of the type species is 50.4 ± 0.9 moles % (T_m, buoyant density). A second group of isolates of facultatively parasitic strains has a G + C content of 43 moles % (Seidler *et al.*, 1969, 1972).

Types species: *Bdellovibrio bacteriovorus* Stolp and Starr 1963, 243.

Key to the species of genus **Bdellovibrio**

Key characteristics

DNA base ratio
Catalase
Sensitivity to vibriostat 0/129
Degree of parasitism
Protease activity
Host range

I. High GC ratio (50.4 ± 0.9 moles %). Catalase positive; sensitive to vibriostat 0/129. Obligately parasitic or obligately host-independent. Relative low protease activity.

1. *Bdellovibrio bacteriovorus*

II. Low GC ratio (42–43 moles %). Facultatively parasitic or host-independent.

1. Catalase positive; sensitive to vibriostat 0/129. High protease activity.

2. *Bdellovibrio stolpii*

2. Catalase negative; resistant to vibriostat 0/129. Moderate protease activity; host-range restricted to pseudomonads.

3. *Bdellovibrio starrii*

Genus **Microcyclus** *Ørskov 1928, 183*

Curved rods, 0.5–2.0 by 1.0 to more than 10 μm long. **Rings,** 1.5–10.0 μm outer diameter, **formed** when cells elongate prior to division. **Non-motile.** Resting stages not known. **Gram-negative.**

Chemoorganotrophs; **metabolism respiratory,** never fermentative. **Strict aerobes.**

Temperatures optimum 22–37 C; maximum usually below 37 C, for one strain 43 C; minimum 5 C.

Isolated from fresh water sources, including ponds, lakes and wells.

The G + C content of the DNA ranges from 39.5–68.4 moles % (buoyant density).

Type species: *Microcyclus aquaticus* Ørskov 1928, 183.

Note: Three species are recognized in the *Manual*; their differentiation is based on DNA base ratio, pigment formation and morphology. However, as the *Manual* points out, there may be more than one genus represented in *Microcyclus*, and species definition is uncertain at this time.

Genus **Pelosigma** *Lauterborn 1913, 100*

Cells S-shaped, slender filaments, 0.23–0.35 × 9–30 μm, colorless or pale gray, usually with a slight spiral twist. Generally arranged side by side in flat sigmoid aggregates of four (Plate 6.1 Fig. 6) or multiples of four; multiple aggregates of considerable thickness have been observed. The cells may be held together throughout their length by a mucoid substance, or only at one end forming a point, the other end being wider and spread out to reveal the individual filaments.

The aggregate may be motile by an organelle, visible with the light microscope and located at the pointed end. This organelle is interpreted by the author as a tuft of flagella.

Multiplication is presumably by synchronous cross-division of the aggregate and sudden rapid separation of the daughter aggregates.

Gram stain not recorded; have not been obtained in pure culture.

Found in and on mud in fresh and brackish waters.

Types species: *Pelosigma cohnii* (Warming) Lauterborn 1913, 100.

Note: Two species, *P. cohnii* and *P. palustre*, are recognized in the *Manual*, based on very old observations, and differentiation is on size. The reader is referred to Plate 6.1, Fig. 6 and the *Manual*, pp. 215–216.

Genus **Brachyarcus** *Skuja 1964, 19*

Rod-shaped cells, bent like a bow, 1 by 1.5–2.5 μm, colorless, with several gas vesicles of reddish tinge arranged in the center, and occasionally with some minute sulfur granules. Cells arranged in groups (coenobia) as a result of polar growth and median cross-division combined with tight attachment to surfaces by means of a mucoid substance. Coenobia of two (rings) or four (clover leaf appearance) or more cells quite common. Delayed cell division resulting in the formation of pretzel-shaped cells. Division in the coenobia may or may not be synchronous. Coenobia of 2–10 cells may measure 3–6 μm in diameter, secondary families of irregularly humped agglomerates up to 100 μm or more in size. The mucoid "capsule" thin and of uncertain (unsharp) limitation.

Non-motile. Microaerophilic or anaerobic.

Not obtained in pure culture.

Originally found in Lake Vuolep Njakajaure, Swedish Lappland, in April in 12–13.5 m depth. Hirsch (unpublished) has observed this organism in several Michigan marl lakes and ponds throughout the year, as well as in Lake Plussee (Germany).

Type species: *Brachyarcus thiophilus* Skuja 1964, 20.

PART 7

GRAM-NEGATIVE AEROBIC RODS AND COCCI

Note: There is no key to the families given in the *Manual* so a synopsis of the five recognized families is provided. For the identification of Gram-negative aerobic rods or cocci which do not seem to fit in any of the five families the reader is referred to either Part 10 or the six Genera of Uncertain Affiliation of this Part. Three of these Genera of Uncertain Affiliation are well-known pathogenic types (*Brucella, Bordetella* and *Francisella*); one is thermophilic (*Thermus*); one is capable of oxidizing ethanol to acetic acid (*Acetobacter*); and one (*Alcaligenes*) contains peritrichously flagellated respiratory rods.

Synopsis of recognized families of Part 7

I. Rods, motile by polar flagella; catalase-positive; do not fix dinitrogen or use one-carbon compounds as carbon source.

Family I. *Pseudomonadaceae*, p. 217*

II. Cells large, rod-shaped to oval; fix dinitrogen in a nitrogen-free medium with an organic carbon source.

Family II. *Azotobacteraceae*, p. 253

III. Rods; incite nodules on roots of legumes or gall hypertrophies on roots and stems of diverse plants (one exception). Are difficult to identify if isolated from soil.

Family III. *Rhizobiaceae*, p. 261

IV. Rods or cocci; utilize only one-carbon compounds as a carbon source.

Family IV. *Methylomonadaceae*, p. 267

V. Rods or cocci; require high concentrations (ca. 12%) NaCl for growth.

Family V. *Halobacteriaceae*, p. 269

FAMILY I. PSEUDOMONADACEAE WINSLOW, BROADHURST, BUCHANAN, KRUMWIEDE, ROGERS AND SMITH 1917, 555

Straight or curved rods, motile by polar flagella. Gram-negative.

Chemoorganotrophs: Metabolism respiratory, never fermentative. Do not fix nitrogen. Able to use other than single-carbon compounds as sole source of carbon.

Strict aerobes.

Catalase positive; oxidase usually positive.

Growth occurs from 4 or lower to 43 C.

The G + C content of the DNA ranges between 58–70 moles %.

Type genus: *Pseudomonas* Migula 1894, 237.

Note: The makeup of this family is such that the type genus *Pseudomonas* is used as a base against which the other genera are differentiated on the basis of unique structural, physiological or ecological properties. Yellow colony producers which are normally pathogenic for plants are included in *Xanthomonas*. Members of the genus *Zoogloea* aggregate into macroscopic flocs in their natural habitat. Organisms which oxidize ethanol to acetic acid are included in a fourth genus, *Gluconobacter*.

Genus I. Pseudomonas *Migula 1894, 237 Nom. cons.* Opin. 5, Jud. Comm. 1952, 121

Cells single, **straight or curved rods,** but not helical. Dimensions, generally 0.5–1 μm by 1.5–4 μm. Motile by **polar flagella;** monotrichous or multitrichous. Do not produce sheaths or pros-

* All page numbers cited refer to *Bergey's Manual of Determinative Bacteriology*, 8th edition, The Williams & Wilkins Company, Baltimore, 1974.

thecae. No resting stages known. **Gram-negative.**

Chemoorganotrophs: Metabolism **respiratory, never fermentative. Some are facultative chemolithotrophs,** able to use H_2 or CO as energy source. Molecular oxygen is the universal electron acceptor; **some can denitrify,** using nitrate as an alternate acceptor.

Strict aerobes, except for those species which can use denitrification as a means of anaerobic respiration. Catalase positive.

The G + C content of the DNA of those species examined ranges from 58–70 moles %.

Type species: *Pseudomonas aeruginosa* (Schroeter) Migula 1900, 884.

Key to the sections of genus **Pseudomonas**

Note: See Stanier *et al.* (1966) for methodology.

I. Growth factors not required. Capable of growth through several transfers in a purely mineral medium with the addition of an appropriate single carbon source.

(Acetate is used by all species, pyruvate (pure, filter-sterilized) and succinate by most. It is recommended that tests be also performed with glucose, lactate and L-alanine, all at 0.5% concentration.)

A. Poly-β-hydroxybutyrate is not accumulated as an intracellular carbon reserve.

(For this test DL-β-hydroxybutyrate is the preferred carbon source in a nitrogen-deficient medium (Stanier *et al.*, 1966). Most other species can use this substrate so, without evidence

TABLE 7.1

Characteristics of species 1–10 of genus **Pseudomonas**

	1. *P. aeruginosa*	2. *P. putida*	3–5. See Table 7.2	6. *P. syringae*	7. *P. cichorii*	8. *P. stutzeri*	9. *P. mendocina*	10. *P. alcaligenes*
No. of flagella	1	>1	>1	>1	>1	1[a]	1[a]	1
Fluorescent pigments	d[b]	+[b]	+	+	+	−[b]	−	−
Pyocyanine	d	−	−	−	−	−	−	−
Carotenoids	−	−	−	−	−	−	+	−
Growth at 41 C	+	−	−	−	−	d	+	+
Levan formation from sucrose	−	−	d	d	−	−	−	−
Arginine dihydrolase	+	+	+	−	−	−	+	+
Oxidase reaction	+	+	+	−	+	+	+	+
Denitrification	+	−	d	−	−	+	+	−
Hydrolysis of:								
Gelatin	+	−	+	d	−	−	−	+
Starch	−	−	−	−	−	+	−	−
Poly-β-hydroxybutyrate	−	−	−	−	−	−	−	−
Carbon sources for growth:								
Glucose	+	+	+	+	+	+	+	−
Trehalose	−	−	+	−	−	−	−	−
2-Ketogluconate	+	+	+	−	−	−	−	−
meso-Inositol	−	−	+	d	d	−	−	−
Geraniol	+	−	−	−	−	−	+	−
L-Valine	+	+	+	−	−	+	+	−
β-Alanine	+	+	+	−	−	−	+	+
DL-Arginine	+	+	+	d	+	−	+	+

[a] Lateral flagella of short wave length may also be produced under certain conditions.

[b] + = positive for 90% or more of strains; − = negative for 90% or more of strains; d = positive for more than 10% but less than 90% of all strains studied.

TABLE 7.2

Characteristics of species 3 (biotypes), 4 and 5 of genus **Pseudomonas**

	3. *P. fluorescens*					4. *P. chloro-raphis*	5. *P. aureo-faciens*
	Biotype I	Biotype II	Biotype III	Biotype IV	Mis-cella-neous strains		
P. fluorescens biotypes according to Stanier *et al.* (1966)[a]	A	B	C	F	G	D	E
Non-fluorescent pigments:							
Green (chlororaphin)	−[b]	−	−	−	−	+[b]	−
Orange (phenazine-1-carboxylate)	−	−	−	−	−	−	+
Blue, non-diffusible	−	−	−	+	−	−	−
Levan formation from sucrose	+	+	−	+	−	+	+
Denitrification	−	+	+	+	−	+	d[b]
Carbon sources for growth:							
L-Arabinose	+	+	−	+	d	−	+
Sucrose	+	+	−	+	d	+	d
Saccharate	+	+	−	+	d	+	+
Propionate	+	+	d	−	d	+	+
Butyrate	−	d	−	+	d	+	d
Sorbitol	+	+	d	+	d	−	−
Adonitol	+	d	+	−	d	−	−
Propylene glycol	−	+	d	−	d	−	−
Ethanol	−	+	d	−	d	+	−

[a] For more complete comparative tables, the original paper should be consulted.

[b] + = positive for 90% or more of strains; − = negative for all strains studied; d = positive for more than 10% but less than 90% of all strains studied.

to the contrary, a strain incapable of growing with β-hydroxybutyrate but capable of growing with other carbon sources may be presumed not to accumulate the polymer.)

Section I, p. 221

B. Poly-β-hydroxybutyrate is accumulated as an intracellular carbon reserve.

1. DL-Arginine and betaine are used as sole carbon sources. Grow in complex media at 40 C; most strains grow at 41 C.

Section II, p. 227

2. DL-Arginine is not used as sole carbon source and betaine is used by very few strains. Do not grow at 41 C, with the exception of *P. lemoignei*.

Section III, p. 230

II. Growth factors required; not capable of growth through several transfers in mineral media with a single carbon source.

Section IV, p. 235

Key to the species of Section I

(See also Tables 7.1 and 7.2)

A. Do not accumulate poly-β-hydroxybutyrate intracellularly. Do not hydrolyze poly-β-hydroxybutyrate extracellularly.

1. Fluorescent pigments produced by most strains.

a. Arginine dihydrolase present. Saprophytic.

1. *P. aeruginosa*

2. *P. putida*

3. *P. fluorescens*
4. *P. chlororaphis*
5. *P. aureofaciens*

b. Arginine dihydrolase absent. Phytopathogenic.

6. *P. syringae*
7. *P. cichorii*

2. Fluorescent pigments not produced.

a. Nitrate denitrified.

8. *P. stutzeri*
9. *P. mendocina*

b. Do not denitrify.

10. *P. alcaligenes*

TABLE 7.3

Characteristics of species 11–16 of genus **Pseudomonas**

	11. *P. pseudoalcaligenes*	12. *P. pseudomallei*	13. *P. mallei*	14. *P. caryophylli*	15. *P. cepacia*	16. *P. marginata*
No. of flagella	1	>1	0	>1	>1	>1
Diffusible pigments[a]	–[b]	–	–	–	+[b]	d[b]
Arginine dihydrolase	+	+	+	+	–	–
Denitrification	–	+	+	+	–	–
Hydrolysis of:						
Starch	–	+	d	–	–	–
Extracellular poly-β-hydroxybutyrate	–	+	d	–	–	–
Gelatin	–	+	+	–	d	+
Carbon sources for growth:						
Glucose	–	+	+	+	+	+
D-Xylose	–	–	+	+	d	+
D-Ribose	–	+	–	+	+	+
L-Rhamnose	–	–	–	+	d	–
Saccharate	–	–	–	+	+	+
Levulinate	–	+	–	–	+	–
Citraconate	–	–	–	–	+	+
Mesaconate	+	–	–	–	–	+
D(–)-Tartrate	–	–	–	–	–	+
Mesotartrate	–	–	–	+	+	+
Erythritol	–	+	–	–	–	–
Adonitol	–	d	–	–	+	+
2,3-Butylene glycol	–	–	–	+	+	–
meta-Hydroxybenzoate	–	–	–	–	+	–
Tryptamine	–	–	–	–	+	–
α-Amylamine	–	+	+	–	+	–

[a] Strains of *Pseudomonas cepacia* may produce non-fluorescent yellowish, greenish, brownish, violet or purple pigments that may be associated with the colony and/or diffuse into the medium. Diffusible yellow-green non-fluorescent pigments may also be produced by *Pseudomonas marginata* and *Pseudomonas caryophylli*.

[b] + = positive for 90% or more of strains; – = negative for 90% or more of strains; d = positive for more than 10% but less than 90% of all strains studied.

Key to the species of Section II
(See also Table 7.3.)

A. Accumulate poly-β-hydroxybutyrate as a carbon reserve. Utilize arginine and betaine as sole carbon sources. Grow at 40 C in complex medium; most strains grow at 41 C.
   1. Arginine dihydrolase present.
      11. *P. pseudoalcaligenes*
      12. *P. pseudomallei*
      13. *P. mallei*
      14. *P. caryophylli*
   2. Arginine dihydrolase not present.
      15. *P. cepacia*
      16. *P. marginata*

Key to the species of Section III
(See also Table 7.4)

A. Accumulate poly-β-hydroxybutyrate as a carbon reserve. Arginine and betaine not utilized; arginine dihydrolase absent. Do not grow at 41 C, with the exception of *P. lemoignei*.
   1. Cannot grow autotrophically with hydrogen.
      17. *P. lemoignei*
      18. *P. testosteroni*
      19. *P. acidovorans*
      20. *P. delafieldii*
      21. *P. solanacearum*

TABLE 7.4
Characteristics of species 17–26 of genus **Pseudomonas**

	17. *P. lemoignei*	18. *P. testosteroni*	19. *P. acidovorans*	20. *P. delafieldii*	21. *P. solanacearum*	22. *P. facilis*	23. *P. saccharophila*	24. *P. ruhlandii*	25. *P. flava*	26. *P. palleronii*
No. of flagella	1	>1	>1	1	>1	1	1	1	1[a]	1[a]
Carotenoid pigments	−[b]	−	−	−	−	−	−	−	+[b]	+
Autotrophic growth with H_2	−	−	−	−	−	+	+	+	+	+
Hydrolysis of:										
Starch	−	−	−	−	−	−	+	−	−	−
Poly-β-hydroxybutyrate	+	d[b]	−	+	−	+	−	−	−	−
Gelatin	−	−	−	−	−	+	−	−	−	−
Carbon sources for growth:										
DL-Lactate	−	+	+	+	d	+	+	+	+	+
Glucose	−	−	−	+	+	+	M[b]	M	+	M
Fructose	−	−	+	+	+	+	M	−	+	−
L-Arabinose	−	−	−	+	−	+	+	−	+	−
Sucrose	−	−	−	−	+	−	+	−	+	−
Malonate	−	−	+	+	−	+	−	−	−	−
Ethanol	−	−	+	−	−	−	−	+	+	+
Glycollate	−	+	+	−	d	−	−	−	−	+
para-Hydroxybenzoate	−	+	+	−	d	−	−	−	−	+
Testosterone	−	+	−	−	−	−	−	−	−	−

[a] With a tendency toward multitrichous flagellation and subpolar flagellar insertion.
[b] + = positive for all strains studied; − = negative for all strains studied; d = positive for more than 10% but less than 90% of all strains studied; M = positive for all strains, but may require mutation in strains isolated from nature; > = one or more.

2. Can grow autotrophically with hydrogen.
   22. *P. facilis*
   23. *P. saccharophila*
   24. *P. ruhlandii*
   25. *P. flava*
   26. *P. palleronii*

Key to the species of Section IV

(See also Table 7.5)

A. Require one or more growth factors
   1. Do not accumulate poly-β-hydroxybutyrate as intracellular growth reserve
      27. *P. maltophila*
   2. Accumulate poly-β-hydroxybutyrate
      28. *P. vesicularis*
      29. *P. diminuta*

TABLE 7.5

Characteristics of species 27–29 of genus **Pseudomonas**

	27. *P. maltophilia*	28. *P. vesicularis*	29. *P. diminuta*
Flagellation	Multitrichous	Monotrichous	Monotrichous
Color of colonies	Yellow	Yellow	None
Growth factors	Methionine	Pantothenate Biotin Cyanocobalamin	Pantothenate Biotin Cyanocobalamin Cystine
Oxidase reaction	Negative	Weak	Strong
Hydrolysis of:			
Starch	−[a]	−	−
Gelatin	+[a]	−	−
Tween 80	+	−	−
Poly-β-hydroxybutyrate	−	−	−
Carbon sources used for growth:			
Glucose	+	+	−
Cellobiose	+	+	−
DL-β-Hydroxybutyrate	−	+	+
Histidine	+	−	+
Pantothenate	−	−	+

[a] + = positive for 90% or more of strains; − = negative for 90% or more of strains.

Genus II. **Xanthomonas** *Dowson 1939, 187*

Cells single, **straight rods.** Dimensions 0.2–0.8 by 0.6–2.0 μm, usually *ca.* 0.4–1.0 μm. Motile by means a **polar flagellum.** Do not produce sheaths. No resting stages known. **Gram-negative.**

Growth on agar media usually yellow.

Chemoorganotrophs. **Metabolism respiratory,** never fermentative. Molecular oxygen is the electron acceptor. **Oxidase reaction negative or weak.** Catalase positive.

In a weakly buffered medium (Dye, 1962), acid is produced in small amounts from many carbohydrates but not from rhamnose, inulin, adonitol, dulcitol, inositol or salicin and rarely from sorbitol. **No acid produced in purple milk.** Acetate, citrate, malate, propionate and succinate are utilized but generally not benzoate, oxalate or tartrate; gluconate may be utilized after a delay. Most species hydrolyze starch and Tween 80 rapidly.

Nitrates not reduced. Hydrogen sulfide produced from cysteine and by most species from thiosulfate and peptone. Acetoin and indole not produced, sodium hippurate not hydrolyzed. **Minimal growth requirements are complex** and usually include methionine, glutamic acid and nicotinic acid in various combinations. **Asparagine not utilized as a sole source of carbon and nitrogen. Growth** on nutrient agar **inhibited** by 0.1% and **usually by 0.02% triphenyltetrazolium chloride.**

Strict aerobes.

Temperature optimum 25–27 C; none grow at 5 C, most grow at 7 C but some unable to grow below 9 C; all grow at 30 C, none grow at 40 C.

The G + C content of the DNA of 29 nomenspecies of the *X. campestris* group **ranges from 63.5–69.2 moles % (mostly T_m).**

Type species: *Xanthomonas campestris* (Pammel) Dowson 1939, 190.

TABLE 7.6

Characters differentiating the species of genus **Xanthomonas**

	1. *X. campestris*	2. *X. fragariae*	3. *X. albilineans*	4. *X. axonopodis*	5. *X. ampelina*
Growth at 35°C	+[a]	+	+	+	−[a]
Hydrolysis of esculin[b]	+	+	+	+	−
Mucoid growth on nutrient glucose agar	+	+	−	−	−
Gelatin liquefaction	+	+	d[a]	−	−
Proteolysis of milk	+	−	−	−	−
H_2S from peptone	+	−	−	+	d
Urease	−	−	−	−	+
Tolerance of NaCl, %	2.0–5.0	0.5–1.0	<0.5	1.0	1.0
Acid production[c] from:					
Arabinose	+	−	−	−	+
Glucose	+	+	+	+	−
Mannose	+	+	+	−	−
Galactose	+	−	d	−	+
Trehalose	+	−	−	+	−
Cellobiose	+	−	−	−	−

[a] + = 90% or more of strains positive in this character; − = 90% or more of strains negative in this character; d = some (less than 90%) strains positive, some negative.

[b] Some reactions delayed.

[c] Within 21 days by the methods of Dye (1962).

Genus III. **Zoogloea** Itzigsohn 1868, 30

Cells rod shaped, range 0.5–1.0 by 1.0–3.0 μm, **no spores or cysts. Young cells actively motile with polar monoflagellation.** In natural waters and certain media cells later **aggregate in macroscopic flocs,** free floating (Plate 7.1, Fig. 1*B*) or attached to surfaces (Plate 7.1, Fig. 2*A–C*). **On the flocs** there are **finger-like or dendritic outgrowths** (Plate 7.1, Fig. 1*A*). Capsules not demonstrable on the neotype strain but reported on one other strain. The flocs are firm, not soft or slimy; recent studies have shown **extracellular fibrils interlacing the cells in the flocs** (Plate 7.1, Fig. 3). **Pleomorphic with age;** old cells usually **distended with granules of poly-β-hydroxybutyric acid.** The amount of polymer and extent of flocculation are influenced by the age of culture and nature of medium (Crabtree *et al.*, 1965).

Gram-negative.

Cells occur naturally (in natural waters and sewage) in flocs as described above. In laboratory media, growth is **flocculent when the carbon: nitrogen ratio is >10:1** and is **dispersed when ratio is <5:1.** In broth cultures containing 0.05% arginine and 0.5% glucose, there develops a fragile pellicle with lacy tapelike flocs adhering to the sidewalls and eventually extending to the bottom of the tube (Plate 7.1, Fig. 2*A–C*). In shake cultures, starlike flocs are formed (Plate 7.1, Fig. 1*B*).

Chemoorganotrophs. **Metabolism is respiratory, never fermentative. Molecular oxygen** is the **terminal electron acceptor.** Xylose, fructose, glucose, mannose, ethanol and glycerol and certain amino acids are oxidized. Dulcitol, melibiose, melezitose, raffinose, glycogen, soluble starch and inulin are not oxidized. Cellulose and native corn starch are not hydrolyzed.

Not proteolytic. Gelatin, coagulated egg albumin and casein not hydrolyzed. Litmus milk not peptonized. Indole and hydrogen sulfide not produced. Nitrates not reduced to nitrites or to gaseous nitrogen. Ammonia not produced from peptone and amino acids. Cytochrome oxidase, catalase positive. Citrate not utilized. Non-pigmented.

Vitamin B_{12} required, biotin stimulatory. With vitamins provided, arginine, aspartic acid, glutamic acid, histidine, ornithine, citrulline, glutamine, pantothenate, glucosamine, adenine, uracil and thiamine may serve as sole sources of carbon and nitrogen. Ammonia may also serve, when both vitamins are provided. Acetate, pyruvate, malate, fumarate and oxalate are utilized as carbon sources.

Strict aerobe. Optimum temperature 28–30 C; slow growth at 10 C and none at 45 C. Optimum pH 7.0–7.5; no growth at pH 4.5 or 9.6. No distinctive ionic or osmotic requirement but will grow with 3% NaCl, not with 6.5% NaCl.

Base composition of DNA. Not determined.

Type species: *Zoogloea ramigera* Itzigsohn 1868, 30. Characteristics differentiating species of this genus are given in Table 7.6 A.

TABLE 7.6A
Characteristics of the species of genus **Zoogloea**

Characteristic	*Z. ramigera*	*Z. filipendula*
Carbohydrate oxidation		
Arabinose	+	−
Galactose	+	−
Sorbose	+	−
Rhamnose	+	−
Fucose	+	−
Maltose	+	−
Sucrose	+	−
Lactose	+	−
Trehalose	+	−
Cellobiose	+	−
Salicin	+	−
Inositol	+	−
Mannitol	+	−
Sorbitol	+	−
Arginine dihydrolase produced	+	−
Tyrosine decarboxylase produced	+	−
Urea hydrolysis	−	+

Genus IV. **Gluconobacter** *Asai 1935, 689, emend. mut. char. Asai, Iizuka and Komagata 1964, 100*

Cells ellipsoidal to rod-shaped, 0.6–0.8 by 1.5–2.0 μm, occurring singly, in pairs or in chains. Involution forms may occur in old cultures. **Motile with three to eight polar flagella,** rarely one flagellum, **or non-motile.** Endospores not formed. Gram-negative; weakly Gram-positive in older cultures.

Chemoorganotrophs: **Metabolism respiratory,** never fermentative; oxygen is terminal electron acceptor.

Characteristics differentiating the genus from *Acetobacter* and *Pseudomonas* are given in Table 7.7.

Oxidize ethanol to acetic acid, sometimes weakly, **at** neutral and **acid reactions (pH 4.5). Do not oxidize acetate or lactate to CO_2.** Usually pronounced **ketogenesis and acid formation from sugars.** Many strains produce 2- and 5-ketogluconic acids, water-soluble brown pigments and γ-pyrones. Usually strongly **catalase positive.**

Strict aerobes.

Temperature optimum 25–30 C; range tolerated 7–41 C, although many strains do not grow above 37 C. pH optimum 5.5–6.0; growth and acetic acid production occur at pH 4–4.5; slight growth in weakly alkaline media.

Occur in flowers, souring fruits, vegetables, beer, South African "kaffir beer," cider, wine, wine vinegar, baker's yeast, garden soil.

The G + C content of the DNA ranges from 60–64 moles %, with one exception of 57 (T_m, chemical analysis).

Type species: *Gluconobacter oxydans* (Henneberg) De Ley 1961, 47 (proposed by Asai, Iizuka and Komagata 1964, 118).

TABLE 7.7

Characteristics differentiating the genera **Gluconobacter, Acetobacter** *and* **Pseudomonas**

	Gluconobacter	*Acetobacter*	*Pseudomonas*
Flagellation	Polar or none	Peritrichous or none	Polar
Growth at pH 4.5	+	+	−
Oxidation of:			
Ethanol to acetic acid at pH 4.5	+(M)[a]	+(S)	−
Acetic acid to CO_2	−	+	d
Lactate to CO_2	−	+	+
Glucose to gluconate	+	d	d
Amino acids by resting cells	−	+	+
Krebs cycle	−	+	+
Production of 5-ketogluconate	+	d	−
Ketogenesis	+	d	−
Quinones Q_{10}	+	−	
Q_9	−	+	
Hydrolysis of:			
Lactose and starch	−	−	d
Gelatin	−/W	−	d
Greenish and/or fluorescent pigments	−	−	d

[a] M, moderate; S, strong; W, weak.

FAMILY II. AZOTOBACTERACEAE PRIBRAM 1933, 5

Cells **large,** predominantly **bluntly rod-shaped** to oval, but **changes dramatically in morphology** with time or changes in growth conditions. Cells often in pairs. **Motile by** peritrichous or polar **flagella** or non-motile.

Gram-negative but may be Gram-variable.

Endospores not produced but **some species form cysts.**

Heterotrophic. **Capable of fixing molecular nitrogen** in a nitrogen-free medium with an organic carbon source.

Normally fix (with 1–2% carbohydrate) **at least 10 mg of atmospheric nitrogen/g of carbohydrate consumed.** Under these conditions some *Beijerinckia* strains fix less (6–9 mg of N/g of carbohydrate consumed) because of copious polysaccharide (slime) production. Efficiency of nitrogen fixation appreciably increased at lower carbohydrate levels (Becking, 1971) or lower oxygen tension (Meyerhoff and Burk, 1928). Organic growth factors not required for growth but for nitrogen fixation trace elements, in particular molybdenum (a specific catalyst of nitrogen fixation) required.

Grows on media with or without combined nitrogen. However, some representatives (*Azomonas agilis*, *Beijerinckia* sp.) are unable to utilize nitrate as sole source of nitrogen or utilize it very poorly. Strains of *Beijerinckia* do not grow on peptone agar. Most strains capsulated and produce copious slime. Fluorescent pigments produced by some strains.

Strict aerobes, but also able to grow and fix nitrogen under reduced oxygen pressure. **Catalase positive or negative** (*Derxia*).

Mesophilic inhabitants of soil, water and leaf surfaces (phyllosphere).

Type genus: *Azotobacter* Beijerinck 1901, 567.

Key to the genera of family **Azotobacteraceae**

Key characteristics

Cell shape
Extracellular slime production
Growth rate
Catalase
Cyst formation
Presence of lipoidal granules
DNA base ratio (not essential)

I. Large ovoid cells; most species produce extracellular slime. Rapid growth; catalase positive.
  A. Cysts formed; G + C 63–66%.
  I. *Azotobacter*
  B. Cysts not formed; G + C 53–59%.
  II. *Azomonas*

II. Small rods producing very tenacious extracellular slime or gum, and conspicuous internal globular lipoid bodies. Slow growth; catalase may or may not be produced.
  A. Lipoid bodies bipolar; G + C 55–59%. Catalase positive.
  III. *Beijerinckia*
  B. Lipoid bodies numerous; G + C 70%. Catalase negative.
  IV. *Derxia*

Genus I. **Azotobacter** *Beijerinck 1901, 567*

Large ovoid cells, 2 μm or more in diameter, of varying length down to coccoid morphology. Occur singly, in pairs or irregular clumps, and rarely in chains of more than four cells. Marked pleomorphism. Do not produce endospores, but **form thick-walled cysts.** May produce copious amounts of capsular slime. Motile with peritrichous flagella or non-motile. Gram-reaction negative, with marked variability.

Some strains elaborate a water-soluble fluorescent pigment which appears **green under ultraviolet light** (Table 7.8).

Generally fix at least 10 mg of atmospheric nitrogen non-symbiotically/g of carbohydrate consumed. Growth media contain carbohydrates (usually glucose), alcohols or organic acids. Molybdenum required for nitrogen fixation; may be replaced by vanadium.

Not proteolytic but can utilize nitrate, ammonia and amino acids as sources of nitrogen.

Catalase positive.

Grow well aerobically but can also grow under reduced oxygen tension.

Optimum temperature between 20 and 30 C. Growth range pH 5.5–8.5; optimum 7.0–7.5.

Found in soil and water.

The G + C content of the DNA ranges from 63–66 moles %.

Type species: *Azotobacter chroococcum* Beijerinck 1901, 567.

TABLE 7.8

Characters of the species of genera **Azotobacter** *and* **Azomonas**

	1. *Azotobacter chroococcum*	2. *Azotobacter beijerinckii*	3. *Azotobacter vinelandii*	4. *Azotobacter paspali*	1. *Azomonas agilis*	2. *Azomonas insignis*	3. *Azomonas macrocytogenes*
Pigments							
Water-soluble fluorescence	None	None	Green	Green	White	None	White
Water-insoluble in cells	Black	Cinnamon					
Carbohydrate utilization							
Starch	+	−	−	−	−	−	−
Mannitol	+	−	+	−	−	−	+
Rhamnose	−	−	+	−	−	−	−
Motility	+	−	+	+	+	+	+
Flagella							
Peritrichous	+		+	+	+		
Lophotrichous						+	
Monotrichous							+
Cysts formed	+	+	+	+	−	−	−
Capsular slime produced	+	+	+	+	+	−	+
Habitat usually soil	+	+	+	+	−	−	+
Habitat limited to fresh water					+	+	
DNA base ratio GC%	65–66	66	66	63–65	53–54	57–58	58–59

Genus II. **Azomonas** *Winogradsky 1938, 391*

Large ovoid cells 2 μm or more in diameter, of varying length down to coccoid morphology. Occur singly, in pairs or irregular clumps. Possess marked pleomorphism. **Do not produce** endospores or **cysts.** May produce copious amounts of capsular slime. Motile with polar or peritrichous flagella. Gram-negative or variable.

Strains that elaborate a water soluble fluorescent pigment **appear white under ultraviolet light** (see Table 7.8 for distinguishing characteristics).

Generally fix at least 10 mg of atmospheric nitrogen non-symbiotically/g of carbohydrate consumed. Growth media usually contain glucose or sucrose.

Not proteolytic. Catalase positive.

Grow well aerobically but can also grow under reduced oxygen tension.

Optimum temperature between 20 and 30 C. pH range 4.5–9.0; optimum 7.0–7.5.

Found in soil and water.

The G + C content of the DNA ranges from 53–59 moles %.

Type species: *Azomonas agilis* (Beijerinck) Winogradsky 1938, 400.

Note: See Table 7.8 for the characteristics of the species of *Azomonas.*

Genus III. **Beijerinckia** *Derx 1950, 145*

Cells single, straight, slightly curved or pear-shaped rods, 0.5–1.5 by 1.7–4.5 μm with distinctly rounded ends. Sometimes large misshapen cells, 3 by 5–6 μm, that are occasionally branched or forked. Characteristic large, highly refractile, lipoid (poly-β-hydroxybutyrate) bodies occur at each end of the cell (Plate 7.3, Fig. 4). Under certain conditions some strains show rounded, coccoid cells without the terminal lipoid globules. Motile or non-motile (Table 7.9). Flagella peritrichous, but flagellation has been studied in only a few strains (Hofer, 1944).

Cysts (enclosing one cell) and capsules (enclosing several cells) found in some species (*B. mobilis* and *B. fluminensis*) (Plate 7.3, Fig. 5). Gram-negative.

In liquid media no surface pellicle formed, but the whole medium becomes a homogenous, highly viscous semitransparent mass. In some species (*B. fluminensis*) the whole medium becomes opalescent and turbid, and adhering slime is not produced. On agar medium copious slime is produced and giant colonies develop with a smooth, folded or plicated surface (Plate 7.3, Figs. 2 and 3). The slime is extremely tough, tenacious or elastic and makes it difficult to remove part of a colony with a loop. In some strains the slime is of a more granular consistency (*B. mobilis, B. fluminensis*) resembling that of *Azotobacter* and therefore easier to remove. The polysaccharide slime consists of glucose, galactose, mannose, glucuronic acid and galacturonic acid, but no heptose. The main component, 51.8% of the dry weight of the polysaccharide, is glucose (López and Becking, 1968).

Atmospheric nitrogen is fixed in a nitrogen-deficient medium. **Molybdenum is required for nitrogen fixation, but, unlike Azotobacter, this cannot be replaced by vanadium.** Also in contrast to *Azotobacter*, calcium is not required for growth and nitrogen fixation.

Glucose, fructose and sucrose utilized by all strains; a wide variability exists in the ability to attack other carbon substrates (see Table 7.10). Oxidize organic substrates to CO_2 and a small amount of acetic acid so that the pH of medium is lowered during growth.

No growth on peptone agar or broth. Glutamate utilized poorly or not at all.

Cells highly acid tolerant, growth between pH 3 and 9.5 or 10.0. In neutral and alkaline media acid is produced, in very acid media an alkaline substance which increases the pH of the medium is produced (Becking, 1961).

Cells contain cytochromes with absorption peaks at 415–424 nm (Sorêt); 480, 518 nm (β); 525–527 nm, 551–556 nm (αc); and 604, 630 nm ($a_1aa_3a_2$) (Moss and Tchan, 1958). Therefore, the cells contain cytochrome *c* (λ_{max} 524 and 552 nm) and cytochrome *a* (λ_{max} 590 and 630 nm).

Strict aerobe, but also grows and fixes nitrogen under reduced oxygen pressure.

Catalase positive.

Temperature range, 10–35 C; optimum for growth 20–30 C, no growth at 37 C. Cells frost resistant; no reduction of viability after storage for 3–4 months at −4 C (Becking, 1961).

The **G + C content** of DNA ranges from 54.7–60.7 moles % (T_m) (De Ley and Park, 1966). In DNA-hybridization experiments with ^{14}C-DNA fragments from *Pseudomonas fluorescens* and *P. putida* the percentage DNA bound relative to these two species is 28 and 19–29, respectively (De Ley and Park, 1966).

Type species: *Beijerinckia indica* (Starkey and De) Derx 1950, 146.

TABLE 7.9

Differential characteristics of species of genus **Beijerinckia**

	B. inaica	*B. mobilis*	*B. fluminensis*	*B. derxii*
Water-soluble green fluorescent pigment	−	−	−	+
Colony color after aging	Fulvus or pink	Amber brown	Fulvus or pink	Buff
Motility of cells	+[a] or ±	+++	+[a] or ±	−
Growth on casein agar	−/w	100[b]	15	100
Growth with nitrate as N source	±[c]	++	−/w	−/w
Growth when C and N source is asparagine	6	60	−/w	−/w
Growth on carbon sources:				
Lactose	70	0	20	100
Erythritol	0	45	0	0
Propanol	50	100	0	0
Acetate, butyrate, fumarate, lactate, malate	50–95	100	20–80	−/w
Benzoate	6	100	0	0

[a] A minority of cells, mostly in young stages.

[b] Numbers represent percentage of strains growing on the medium.

[c] ± About 40–50% of strains tested positive; ++ good growth; −/w negative or weak reaction.

TABLE 7.10

Utilization of carbon compounds by **Beijerinckia** *species*

	1. *B. indica*	2. *B. mobilis*	3. *B. fluminensis*	4. *B. derxii*
	(48)[a]	(7)	(5)	(3)
Arabinose	85[b]	100	100	−/w
Galactose	85	45	100	100
Inulin	65	15	0	−/w
Maltose	90	70	100	100
Rhamnose	35	30	0	−/w
Sorbose	70	100	100	100
Starch	55	30	20	−/w
Xylose	70	45	100	100
Butanol	80	100	40	0
Ethanol	90	100	80	100
Glycerol	70	100	100	0
Inositol	55	100	40	0
Mannitol	90	85	40	100

	1. *B. indica*	2. *B. mobilis*	3. *B. fluminensis*	4. *B. derxii*
Acetate[c]	50	100	20	−/w
Butyrate	85	100	40	−/w
Citrate	40	0	0	−/w
Formate	0	70(w)	0	−/w
Fumarate	80	100	60	−/w
Lactate	95	100	80	−/w
Malate	85	100	20	−/w
Malonate	50	0	40	−/w
Oxalate	2	0	0	−/w
Propionate	0	0	0	−/w
Succinate	80	100	40	100
Tartrate	15	0	0	−/w
Valerate	8	70(w)	0	−/w
Salicylate	6	0	0	0

[a] Figures in parentheses at top of table indicate number of strains tested.

[b] Numbers in the table indicate percentage of strains tested that utilize the compound. See also Table 7.9. All strains utilize glucose, fructose and sucrose; none can grow on amyl alcohol. With the exception of *B. indica* the percentages are indicative only, because of the small number of strains available for test. Tests made on nitrogen-free agar containing K_2HPO_4 0.8 g, KH_2PO_4 0.2 g, $MgSO_4 \cdot 7H_2O$ 0.5 g, $FeCl_3 \cdot 6H_2O$ 0.1 g, $Na_2MoO_4 \cdot 2H_2O$ 0.005 g, distilled water 1000 ml (pH 6.7). −/w, negative or weak reaction; (w), weak growth.

[c] Tested as Na- or Ca-salt; formate and oxalate tested only as Na-salt.

Genus IV. **Derxia** *Jensen, Petersen, De and Bhattacharya 1960, 193*

Cells rod-shaped with rounded ends, 1.0–1.2 by 3.0–6.0 μm, occurring singly or in short chains. Cells rather pleomorphic, some cells occasionally becoming very large. Young cells have a homogenous cytoplasm; older cells show typical large refractive bodies throughout the whole cell. Motile by a short polar flagellum; motile cells numerous in liquid (glucose) medium containing combined nitrogen, but rare on nitrogen-deficient solid media. Resting stages not known. Gram-negative.

Agar colonies at first slimy and semitransparent; later massive and opaque, highly raised with a wrinkled surface (Plate 7.3, Fig. 1); older colonies develop a dark mahogany brown color. Liquid media turn into a gelatinous mass, but growth near the surface is more luxuriant forming a thick, tough pellicle.

A wide range of sugars, alcohols and organic acids are oxidized mostly to CO_2 and a small amount of acid, probably acetic, when growing in an alkaline medium.

Atmospheric nitrogen fixed in a nitrogen-deficient medium. Molybdenum required for nitrogen fixation; vanadium cannot replace molybdenum in this process.

Strict aerobe, also growing and fixing nitrogen under reduced oxygen pressure. Catalase negative.

Growth slow at 15 C, optimal at 25–35 C, feeble at 40 C, no growth at 50 C. Acid tolerant.

Found in tropical soils (Asia, Africa, South America).

The G + C content of the DNA is 70.4 moles %.

Type species: *Derxia gummosa* Jensen, Petersen, De and Bhattacharya 1960, 193.

FAMILY III. **RHIZOBIACEAE** CONN 1938, 321

Cells without endospores, normally **rod-shaped. Motile;** one polar or subpolar flagellum, or two to six peritrichous flagella. **Aerobic. Gram-negative. Many carbohydrates utilized.** Considerable extracellular slime usually produced during growth on carbohydrate-containing media. Some strains of rhizobia and agrobacteria show a close relationship in DNA base composition.

All species, with the exception of *Agrobacterium radiobacter*, **incite cortical hypertrophies on plants. Nodules** are **incited on roots of leguminous species** (family: *Leguminosae*) by strains of rhizobia (symbionts). **Gall hypertrophies** are produced **on roots and stems of** diverse **plant species** by strains of agrobacteria (tumorigenic phytopathogens). Can be isolated from nodules or galls; not easily identified when isolated from soil. Confirmation of isolates should be made by proper plant inoculation tests.

Key to the genera of the family **Rhizobiaceae**

I. Cells stimulate nodule production on roots of leguminous plants. Fix free nitrogen when in the symbiotic stage within root nodules.
Do not utilize citrate.
Do not produce 3-ketolactose.
Genus I. *Rhizobium*, p. 262

II. Cells do not stimulate root nodule production on leguminous plants but most species do produce other types of hypertrophies on many plants.
Do not fix free nitrogen.
Utilize citrate.
Genus II. *Agrobacterium*, p. 264

Genus I. **Rhizobium** *Frank 1889, 338. Nom. gen. cons.* Opin 34, Jud. Comm. 1970, 11

Rods 0.5–0.9 by 1.2–3.0 μm. **Commonly pleomorphic under adverse growth conditions.** Often contain granules of poly-β-hydroxybutyrate which are refractile under phase contrast, stainable with Sudan black B and soluble in chloroform. **Motile** by two to six peritrichous flagella or by a polar or subpolar flagellum. **Non-sporing. Gram-negative. Growth on carbohydrate media usually accompanied by** copious extracellular, polysaccharide **slime.**

Chemoorganotrophs: **Metabolism respiratory.** Molecular oxygen is the terminal electron acceptor. **3-Ketoglycosides not produced** (Bernaerts and DeLey, 1963). Gelatin is not liquefied or liquefied slowly within 2 months. Casein and agar not hydrolyzed. Produce little or no hydrogen sulfide on bismuth sulfite agar. Colonies colorless to white; exceptions are the pink to red colonies produced by strains from *Lotononis bainesii* Baker. Utilize a wide range of carbohydrates, without gas formation, but not cellulose or starch. Ammonium salts, nitrates and most amino acids can serve as nitrogen sources. Some strains will grow in a simple mineral-salts medium containing only vitamin-free casein hydrolysate as the sole source of both carbon and nitrogen. Some strains require water-soluble vitamins.

Aerobic; often able to produce excellent growth

under oxygen tensions less than 0.01 atmosphere (Wilson, 1940). Temperature optima 25–30 C. pH range 5.0–8.5.

Members of this genus characteristically **able to invade root hairs of leguminous plants** (family: *Leguminosae*) **and incite production of root nodules,** wherein the bacteria occur as intracellular symbionts. All strains exhibit host range affinities (host "specificity"). Bacteria usually present in nodule as pleomorphic forms (bacteroids) enclosed singly or in small groups within plant-produced membranous sacs. **Nodule bacteroids** characteristically **involved in fixing molecular nitrogen** into combined forms utilizable by the host plant.

Plant tests are essential for identification.

The **G + C content** of the DNA ranges from **59.1–65.5** moles %, T_m (DeLey and Rassel, 1965).

Type species: *Rhizobium leguminosarum* (Frank) Frank 1889, 338.

Further Comments

Members of the genus *Rhizobium* are common soil inhabitants. Identification is relatively easy if isolated from host plant nodules but difficult if isolated from the soil or if a non-infective variant. The ability to cause nodule formation (infectiveness) is more stable than the ability to fix nitrogen in symbiosis (effectiveness). Strains tend to lose effectiveness after serial cultivation on media containing certain amino acids, especially DL- or D-forms, and after many years of storage on laboratory media. Infectiveness and effectiveness represent discrete phenomena and vary within wide limits, depending upon genetic factors present in both bacterial strain and host plant; occasional strains nodulate certain hosts but are ineffective.

Cells are commonly pleomorphic (swollen and either globular, ellipsoidal or branched) in old cultures, within root nodules, or when subjected to other conditions in which protein synthesis is partially inhibited. Such conditions vary with the strains but may include extremes of temperature and pH, low oxygen tension, and growth in laboratory media containing excessive amounts of various amino acids, or certain alkaloids, glycosides, dyes or antibiotics.

Most satisfactory growth is provided by media **containing yeast or other plant extracts. Yeast extract mineral salts media containing mannitol or glucose are among the most conventional. Pentoses are preferred by some species.**

Species of *Rhizobium* can generally be distinguished from those of *Agrobacterium* by a combination of tests (see key to *Rhizobiaceae* and species description of *Agrobacterium radiobacter*). In addition, rhizobia have negative hypertrophy-initiating ability as shown by the Pinto bean (*Phaseolus vulgaris* L.) leaf test (Lippincott and Heberlein, 1965) and the carrot disc assay (Klein and Tenenbaum, 1955), as used by Lippincott and Lippincott (1969).

Key to the groups of species of the genus **Rhizobium**

I. 2–6 peritrichous flagella; rapid growth on yeast extract media.

Group I.
1. *R. leguminosarum*
2. *R. phaseoli*
3. *R. trifolii*
4. *R. meliloti*

II. Polar or subpolar flagellum; slow growth on yeast extract media.

Group II.
5. *R. japonicum*
6. *R. lupini*

Note: Species in each group are separated on the basis of host plant.

Genus II. **Agrobacterium** *Conn 1942, 359. Nom. gen. cons.* Opin. 33, Jud. Comm. 1970, 10

Rods 0.8 by 1.5–3.0 μm. Motile by one to four peritrichous flagella; if only one flagellum, lateral attachment is more common than polar. **Nonsporing. Gram-negative.** Fimbriae are common. **Growth on carbohydrate-containing media usually accompanied by** copious extracellular, polysaccharide **slime.** Colonies are non-pigmented and usually smooth tending to become striated with age, but rough colonies are produced by many strains.

Chemoorganotrophs: **Metabolism respiratory.** Molecular oxygen is the terminal electron acceptor. **3-Ketolactose commonly produced by** *A. tumefaciens* and *A. radiobacter* (DeLey *et al.*, 1966). Gelatin slowly liquefied in several weeks or not at all. Casein not hydrolyzed. Utilize a wide range of simple carbohydrates, organic acids and amino acids as energy sources but not cellulose, starch, agar or chitin. The release of either organic acids and/or CO_2 into the medium indicating utilization of carbohydrate usually produces an acidic reaction at the surface of synthetic media low in organic nitrogen compounds but the reaction may be hidden by the release of basic substances in media containing peptone or other materials rich in amino acids. Generally produce hydrogen sulfide on bismuth sulfite agar. Rapid rate of growth on meat extract or yeast extract peptone media; turbidity with pellicle usually within 24–36 hr; exception, strains of *A. rubi*. Ammonium salts, nitrates and some amino acids serve as sole nitrogen sources for some species; others require amino acids with additional growth factors (Starr, 1946). Normally oxidase positive (Table 7.11).

Aerobic, but able to grow under reduced oxygen tensions in plant tissue. Temperature optima 25–30 C. pH range 4.3–12.0. Optimum pH range 6.0–9.0.

With the exception of *A. radiobacter*, members of this genus **initiate stem hypertrophies on diverse plants** wherein the bacteria occur as intercellular parasites. **Bacteria enter host tissue through pre-existing lesions or abrasions; ability of strains to invade host plant tissue directly is lacking.** Host range affinities are not sharply defined. **Tissue puncture inoculation tests are essential to confirm strain identification** based on routine biochemical tests.

The **G + C content** of the DNA ranges **from 59.6–62.8** moles %, T_m (De Ley *et al.*, 1966).

Type species: *Agrobacterium tumefaciens* (Smith and Townsend) Conn 1942, 359.

Key to the species of the genus **Agrobacterium**

A. Amino acids, nitrates and ammonium salts used as sole source of nitrogen. 3-Ketolactose produced.
   1. Produce galls.
      1. *A. tumefaciens*
   2. Do not produce galls.
      2. *A. radiobacter*

B. Do not utilize amino acids, nitrates and ammonium salts as sole source of nitrogen. 3-Ketolactose not produced.
   1. Produce hairy root of nursery stock.
      3. *A. rhizogenes*
   2. Produce galls on raspberries.
      4. *A. rubi*

TABLE 7.11

Differential characteristics of the species of genus **Agrobacterium**

	1. *A. tumefaciens*,[a] 2. *A. radiobacter*	3. *A. rhizogenes* 4. *A. rubi*
Asparagine used as sole source of carbon and nitrogen	+	–
Nitrites produced from nitrates	+	–
Ammonium salts, nitrates or amino acids utilized as sole nitrogen source	+	–
Growth factors and amino acids required	–	+
Calcium, or sodium, glycerophosphate mannitol nitrate agar:		
Growth	+	–/S[b]
Halo or browning	+	–

	1. *A. tumefaciens*,[a] 2. *A. radiobacter*	3. *A. rhizogenes* 4. *A. rubi*
Starr's and Lippincotts' basal salts media, growth	+	–
Congo red and aniline blue mannitol agars:		
Growth	+	M
Dye absorbed	P	F
Sodium selenite-yeast water-glucose agar:		
Growth	+	S[c]
Selenite reduced	+	–
Litmus milk:		
Reaction	N/Alk	A
Serum zone formation	+	F
3-Ketolactose production	+	–

[a] Since this table was prepared it has been shown that there are two biotypes of these species, but the description given here applies only to biotype I (Keane, Kerr and New. 1970 Australian J. Biol. Sci. *23:* 585–595.

[b] S = scant growth; M = moderate growth; F = faint; A = acid; Alk = alkaline; N = neutral; P = pronounced.

[c] Some strains of *A. rubi* will not grow at all on this medium.

FAMILY IV. METHYLOMONADACEAE LEADBETTER 1974, 267

Bacteria utilizing only one-carbon organic compounds, such as methane or methanol, as a carbon source. Gram-negative.

Genus I. Methylomonas *Leadbetter 1974, 268*

Cells **single, straight, curved** or **branched rods,** but **not helical.** Dimensions 0.5–1 by 1–4 μm. **Motile** by monotrichous polar flagellum. Sheaths or prosthecae not known. Resting stages not known. **Gram-negative.**

Chemoorganotrophs: **Metabolism is respiratory, using molecular oxygen as terminal electron acceptor. Methane** and **methanol** are the **only known sources of carbon** and **energy.** Organic growth factors are not required.

Strict aerobes. Temperature range 20–35 C. The **G + C content** of the DNA of the type species examined is **52.1 moles %** (buoyant density).

Type species: *Methylomonas methanica* (Söhngen) Leadbetter 1975.

Note: The reader is referred to the *Manual* for descriptions of the three species included in this genus.

Genus II. Methylococcus *Leadbetter 1974, 269*

Cells spherical, usually occurring in pairs. Non-motile. Resting stages not known. **Gram-negative.**

Chemoorganotrophs: **Metabolism is respiratory, using molecular oxygen as terminal electron acceptor. Methane and methanol** are **only known sources of carbon and energy.** No organic growth factors are required.

Strictly aerobic.

The **G + C content** of the DNA of the type strain is **62.5 moles** % (buoyant density) (Foster and Davis, 1966).

Type species: *Methylococcus capsulatus* Leadbetter 1975.

FAMILY V. HALOBACTERIACEAE GIBBONS 1974, 269

Rods and cocci which require high concentrations (>2 M or *ca.* 12%) of sodium chloride for growth. Reproduce by binary fission. Resting stages not known.

Cell walls do not contain diaminopimelic acid or muramic acid. Cell walls largely lipoprotein, the proteins being acidic and probably more acidic than those of other Gram-negative organisms. Lipids mostly non-saponifiable phospholipid and glycolipid derivatives of a glycerol di-ether. The di-ether linkage is unusual and may be unique, as is the presence of di-hydrophytol groups rather than fatty acids (Kates *et al.*, 1966; Kates, 1972) (ether-linked lipids have been reported in *Thermoplasma acidophilum* (Langworthy *et al.*, 1972) and seem to be an adaptation to extreme environments). Lipid synthesis is by the mevalonic pathway, rather than the more usual malonate pathway (Kates *et al.*, 1968). A fatty acid synthetase has been found in one species but is inhibited by the concentration of salt in the cell (Pugh *et al.*, 1971).

Chemoorganotrophs: Metabolism respiratory, never fermentative. Energy is obtained from amino acids. Little is known about the utilization of carbohydrates.

Contain carotenoids; the main pigment is bacterioruberin (Kelly and Liaaen-Jensen, 1967). Colonies are of various shades of red: pink, orange-red, vermilion, mauve-red.

Found wherever NaCl and other required ions occur in adequate concentrations; salterns (where salt is prepared by solar evaporation of sea water), some salt lakes, the Dead Sea and proteinaceous material preserved with solar salts, such as fish, intestines (sausage casings) and hides.

The DNA shows a major and minor component. The minor component makes up 10–30% of the DNA; such a high content of a minor component has so far been found only in the family *Halobacteriaceae.*

The G + C content of the major component ranges from 66–68 moles %, of the minor component 57–60 moles % (buoyant density: Joshi *et al.*, 1963; Moore and McCarthy, 1969).

Type genus: *Halobacterium* Elazari-Volcani 1957, 207.

Genus I. Halobacterium *Elazari-Volcani 1957, 207*

Cells rod-shaped, 0.6–1.0 by 1–6 μm, occurring singly; may be pleomorphic, especially in deficient media. Reproduce by binary fission. Motile by a tuft of polar flagella or non-motile. Resting stages not known. Gram-negative, although Gram-reaction and usual staining methods not satisfactory because of the high salt concentrations required for growth. Best observed by phase-contrast microscopy.

Mature cells retain their rod shape in 3.5 M

NaCl; at lower concentrations pleomorphic forms appear and at 1.5 M the cells are spherical, because of loss of cell wall. At this concentration suspensions are viscous as a result of partial lysis and at 0.5 M few, if any, cells can be detected. Cell structure is maintained by $MgCl_2$ and $CaCl_2$.

Sodium, chloride and magnesium ions are required to maintain cell structure and rigidity. As their concentration is reduced the cell wall dissolves and the cell membrane breaks up into tiny fragments.

The internal ion is mainly potassium which is required to maintain the structural integrity of the ribosomes and for protein synthesis, as well as being one of the ions required to maintain activity of intracellular enzymes.

The cell wall does not contain diaminopimelic acid or muramic acid. The wall is largely lipoprotein, being acid and probably more acidic than that of other Gram-negative bacteria. The cell envelope contains about 20% lipid, mostly non-saponifiable derivatives of a glycerol di-ether. The di-ether linkage is unusual as is the presence of dihydrophytol groups rather than fatty acids. Lipid synthesis seems to be by the mevalonic pathway, rather than the more usual malonate pathway.

Colonies on agar are small (<2 mm), round, convex, entire, translucent and pinkish mauve to bright red or vermilion in color.

Chemoorganotrophs: Metabolism respiratory, never fermentative. Energy obtained from amino acids. Carbohydrates are used only slightly, if at all; acid is not produced in sugar-containing media which become alkaline as the result of deamination and/or decarboxylation of amino acids. Some carbohydrates, such as glucose and glycerol, may be stimulatory under certain conditions and with 2% glycerol acid production has been noted (Gochnauer and Kushner, 1969). Some sugars may be utilized as indicated by manometric techniques. Starch not hydrolyzed.

Contain carotenoids; the main pigment is bacterioruberin, a 50-carbon carotenoid having four hydroxyl groups. This pigment is not restricted to the halophiles, since it occurs in *Corynebacterium poinsettiae*. A few colorless (white) strains have been isolated, probably mutants.

Squalene, dihydrosqualene, tetrahydroxysqualene and vitamin MK8 have been found (Tornabene *et al.*, 1969).

Proteins and proteoses preferred for growth in complex media (peptones not satisfactory), supplemented by at least 3 M NaCl plus Mg and K ions. A useful medium is 0.5% each of proteose-peptone and tryptone (Difco), 25% NaCl, 2% $MgSO_4 \cdot 7H_2O$ and 0.2% KCl. Will grow in media saturated with NaCl. The ions are required both for growth and maintenance of cell structure. Will not grow much below 3 M NaCl; will grow in 2.5 M NaCl + 1.5 M KCl. For details of cultivation see Eimhjellen (1965) or Gibbons (1969).

Oxidase and catalase positive. Gelatin usually liquefied. H_2S produced from peptone and thiosulfate, usually from cysteine. Indole produced in 25 and 30% NaCl. Arginine dihydrolase present. Urease negative. Some biochemical reactions affected by salt concentration in the growth medium.

Strict aerobes, growing on solid surfaces exposed to air, in shallow layers of liquid medium or in well aerated liquid cultures.

Grows well at 30–50 C; optimum for rate of growth 40 C, for cell production 35 C. pH range 5.5–8.0; optimum 7.2–7.4.

Most strains sensitive to polymyxin (300 units); not sensitive to aureomycin, chloromycetin, penicillin, terramycin.

Found where NaCl and other required ions occur in adequate concentrations: salterns, the Dead Sea and some other salt lakes, and heavily salted proteinaceous materials. Survives in solar salt for several years.

The DNA shows a major and a minor component, the latter making up 10–30% of the DNA. The G + C content of the major component ranges from 66–68 moles %, of the minor component 57–60 moles % (buoyant density; Joshi *et al.*, 1963; Moore and McCarthy, 1969).

Type species: *Halobacterium salinarium* (Harrison and Kennedy) Elazari-Volcani 1957, 208.

Note: There are two species recognized, *H. salinarium* and *H. halobium*, which are separated on the basis of DNA homology.

Genus II. **Halococcus** *Schoop 1935, 817*

Cocci, 0.6–1.5 μm in diameter, occurring in pairs, tetrads and irregular clusters of tetrads. Nonmotile. Gram-negative.

Grow only in the presence of 2.5 M or higher concentrations of NaCl. Cells retain their shape in lower concentrations of salt and seem more resistant to osmotic damage although the cell walls resemble those of *Halobacterium* in not containing diaminopimelic acid or muramic acid.

Colonies small (<2 mm) opaque, convex, pink to red in color. Cells contain carotenoids.

Chemoorganotrophs: Metabolism respiratory, never fermentative.

Aerobic.

The G + C content of the DNA is 67 moles % for the major component and 59 moles % for the minor component, which makes up 31% of the total DNA (Moore and McCarthy, 1969).

Type species: *Halococcus morrhuae* (Farlow) Kocur and Hodgkiss 1973, 154.

GENERA OF UNCERTAIN AFFILIATION

Genus **Alcaligenes** *Castellani and Chalmers 1919, 936*

Cells rods, coccal rods or cocci 0.5–1.2 μm by 0.5–2.6 μm usually occurring singly. Motile with one to four (occasionally up to eight) peritrichous or degenerate peritrichous flagella. No resting stages known. Gram-negative.

Chemoorganotrophs: Metabolism respiratory, never fermentative. Molecular oxygen is the final electron acceptor.

Strict aerobes. Some strains capable of anaerobic respiration in the presence of nitrate or nitrite, which acts as an alternate electron acceptor.

Most strains (except *A. aquamarinus*) have simple nitrogenous nutritional requirements and produce very turbid growth in liquid media with ammonium or nitrate salts as the sole nitrogen source; some strains require organic nitrogen compounds (amino acids and/or vitamins). Do not fix gaseous nitrogen.

Not actively proteolytic in casein or gelatin media. Cellulose, chitin and agar not hydrolyzed. Oxidase positive (Kovacs' test).

Optimum temperatures between 20 and 37°C. All species grow quickly at pH 7.0.

The G + C content of the DNA ranges from 57.9–70 moles %.

Members of the genus *Alcaligenes* are common, apparently saprophytic, inhabitants of the intestinal tract of vertebrates. They occur in dairy products, rotting eggs and other foods and in fresh water, marine and terrestrial environments in which they are involved in decomposition and mineralization processes. Not known to enter into either pathogenic or symbiotic associations with plants or animals.

Type species: *Alcaligenes faecalis* Castellani and Chalmers, 1919, 936.

Note: Insufficient information is given in the descriptions of the four species to construct a useful key. The reader is referred to the *Manual*, pp. 274–275, for species descriptions.

Genus **Acetobacter** *Beijerinck 1898, 215*

Cells ellipsoidal to rod-shaped, straight or slightly curved, 0.6–0.8 by 1.0–3.0 μm, occurring singly, in pairs or in chains. Involution forms frequent in some species and may be spherical, elongated, swollen, club-shaped, curved, branched or filamentous. Motile by peritrichous flagella or non-motile. Endospores not formed. Young cells Gram-negative, in older cultures some strains

TABLE 7.12

Characteristics differentiating the species and subspecies of genus **Acetobacter**

	1. *A. aceti*				2. *A. pasteurianus*					3. *A. peroxydans*
	1a. subsp. *aceti*	1b. subsp. *orleanensis*	1c. subsp. *xylinum*	1d. subsp. *liquefaciens*	2a. subsp. *pasteurianus*	2b. subsp. *lovaniensis*	2c. subsp. *estunensis*	2d. subsp. *ascendens*	2e. subsp. *paradoxus*	
Catalase	(+)	(+)	(+)	(+)	(+)	(+)	(+)	(+)	−	−
Ketogenesis in glycerol or erythritol	+[a]	+[a]	+[a]	+[a]	−[b]	−[b]	−[b]	−[b]	−	−
Formation of										
5-Ketogluconate	(+)	(+)	(+)	(+)	−	−	−	−	−	−
2-Ketogluconate	(+)	(+)	(+)	(+)	d	−	−	−	−	−
Gluconate	+	+	+	+	(+)	+	+	−	−	−
Growth on ethanol[c]	+	−	(−)	+	(−)	+	+	−	−	+
Produces										
Cellulose	−	−	+		−	−	+	−	−	−
γ-Pyrones	−	−	−	+	−	−	−	−	−	−
Brown pigment	−	−	−	+	−	−	−	−	−	−
G + C, moles %	59–65	60–61	62–63	64	55–62		62		55–56	61–64

[a] Moderate to strong.

[b] Negative or very weak; ( ) usually positive or negative.

[c] In Hoyer-Frateur medium: ethanol, 3% (v/v); $(NH_4)_2SO_4$, 0.1%; K_2HPO_4, 0.01%; KH_2PO_4, 0.09%; $MgSO_4 \cdot 7H_2O$, 0.025%; $FeCl_3$, 0.0005%.

become Gram variable.

Chemoorganotrophs: Metabolism respiratory, never fermentative; oxygen is the terminal acceptor.

Characteristics differentiating the genus from *Gluconobacter* and *Pseudomonas* are given in Table 7.7.

Oxidize ethanol to acetic acid at neutral and acid reactions (pH 4.5). Acetate and lactate are oxidized to CO_2 and H_2O ("overoxidizers"). Ethanol, lactate are good carbon sources. Hexoses and glycerol are used as carbon sources by most strains; mannitol and glutamate are used poorly or not at all; lactose, dextrin and starch are not hydrolyzed. γ-Pyrones usually not produced.

Many amino acids readily oxidized by resting cells.

Generally no pigments produced, but cell masses may be pink due to porphyrins; some strains produce a brown, water-soluble pigment.

Growth reported on simple and complex media; most strains do not require vitamins.

Strict aerobes.

Optimum temperature for growth *ca.* 30 C; temperature range 5–42 C. pH optimum 5.4–6.3; growth occurs at pH 4.0–4.5; little growth at pH 7–8.

Found on fruits and vegetables, souring fruit juices, vinegar, alcoholic beverages.

The G + C content of the DNA ranges from 55–64 moles % (T_m, chemical analysis).

Type species: *Acetobacter aceti* (Pasteur) Beijerinck 1898, 215.

Genus **Brucella** *Meyer and Shaw 1920, 173*

Coccobacilli or short rods, 0.5–0.7 by 0.6–1.5 μm, arranged singly, more rarely in short chains. No capsules. **Non-motile.** Do not form endospores. **Gram-negative.** Do not show bipolar staining.

Chemoorganotrophs: Metabolism respiratory. Growth of some strains can occur in a chemically defined medium with ammonium ion as the sole source of nitrogen (McCullough and Dick, 1943); most strains show better growth, and from a smaller inoculum, using multiple amino acids (Rode, Oglesby and Schuhardt, 1958; Gerhardt, 1958).

The following vitamins are required for growth: thiamin, niacin and biotin; Ca pantothenate often stimulates growth; **hemin (X factor) and coenzyme I (V factor) not required.**

Catalase positive; oxidase usually positive but *B. neotomae* and *B. ovis* are oxidase negative. Urea hydrolyzed to a variable extent. Nitrates reduced to nitrites (except for *B. ovis*). Citrate not utilized; indole not produced; methyl red and Voges-Proskauer tests negative. Litmus milk no change.

Strict aerobes; some require 5–10% added CO_2 for growth especially on initial isolation.

Temperature range 20–40 C; optimum 37 C. Optimum pH 6.6–7.4.

Mammalian parasites and pathogens; facultatively intracellular with a relatively wide host range.

The G + C content of the DNA ranges from 56–58 moles % (buoyant density) (Hoyer and McCullough, 1968); members of the genus comprise a closely knit and sharply demarcated genetic group as defined by DNA hybridization studies.

Type species: *Brucella melitensis* (Hughes) **Meyer and Shaw 1920, 179 (see Table 7.13).**

Genus **Bordetella** *Moreno-López 1952, 178*

Minute coccobacilli, 0.2–0.3 μm by 0.5–1.0 μm, arranged singly or in pairs, more rarely in short chains. **Non-motile** or **motile** by lateral polytrichous flagella. **Gram-negative,** bipolar.

Colonies on potato-glycerol-blood agar (Bordet and Gengou, 1906) medium, smooth, convex, pearly, glistening, nearly transparent, surrounded by zone of hemolysis without definite periphery.

Chemoorganotrophs: **Metabolism respiratory,** never fermentative. Contain catalase. No indole reaction. No gelatin liquefaction. Litmus milk made alkaline. **Require nicotinic acid, cysteine** and **methionine** (most strains); **utilize oxidatively** alanine, asparagine, glutamic acid, proline and serine (Rowatt, 1957; Meyer and Cameron, 1957; **hemin** (X factor) and **coenzyme** I (V factor) **not required.**

Strict aerobes.

Temperature optimum 35–37 C.

Mammalian parasites and **pathogens** of the **respiratory tract:** cause whooping cough or whooping cough-like disease in man; one species causes bronchopneumonia in dogs, guinea pigs and other animal species.

Have a **genus specific heat stable O antigen, antigenic heat-labile dermonecrotic toxin,** and the **common heat-labile agglutinogen no. 7. Each species** has a **specific agglutinogen;** 10 other agglutinogens occur among the species.

Type species: *Bordetella pertussis* (Bergey *et al.*) **Moreno-López 1952, 178 (see Table 7.14).**

TABLE 7.13

Differential characters of the species and biotypes of the genus **Brucella**

Biotype	CO_2 required	H_2S produced	Growth on dye media[a]			Agglutination in			Lysis by phage Tb, RTD[b]	Substrates metabolized oxidatively										
			Basic fuchsin b.	Thionin a.	Thionin b.	Monospecific sera A	Monospecific sera M	Anti-rough serum		L-Alanine	L-Asparagine	L-Glutamic acid	L-Arabinose	D-Galactose	D-Ribose	D-Glucose	L-Erythritol	D-Xylose	L-Arginine[c]	L-Lysine
1. *B. melitensis*																				
1	−	−	+	−	+	−	+	−	−	+	+	+	−	−	−	+	+	−	−	−
2	−	−	+	−	+	+	−	−	−	+	+	+	−	−	−	+	+	−	−	−
3	−	−	+	−	+	+	+	−	−	+	+	+	−	−	−	+	+	−	−	−
2. *B. abortus*																				
1	d	+	+	−	−	+	−	−	+	+	+	+	+	+	+	+	+	−	−	−
2	d	+	−	−	−	+	−	−	+	+	+	+	+	+	+	+	+	−	−	−
3	d	+	+	+	+	+	−	−	+	+	+	+	+	+	+	+	+	−	−	−
4	d	+	+	−	−	−	+	−	+	+	+	+	+	+	+	+	+	−	−	−
5	−	−	+	−	+	−	+	−	+	+	+	+	+	+	+	+	+	−	−	−
6	−	−	+	−	+	+	−	−	+	+	+	+	+	+	+	+	+	−	−	−
7	−	d	+	−	+	+	+	−	+	+	+	+	+	+	+	+	+	−	−	−
8	+	−	+	−	+	−	+	−	+	+	+	+	+	+	+	+	+	−	−	−
9	−	+	+	−	+	−	+	−	+	+	+	+	+	+	+	+	+	−	−	−
3. *B. suis*																				
1	−	+	−	+	+	+	−	−	−	−	−	−	+	+	+	+	+	+	+	+
2	−	−	−	−	+	+	−	−	−	−	+	+	+	+	+	+	+	+	+	−
3	−	−	+	+	+	+	−	−	−	−	−	+	−	−	+	+	+	+	+	+
4	−	−	+	+	+	+	+	−	−	−	−	+	−	−	+	+	+	+	+	+
4. *B. neotomae*	−	+	−	−	+	+	−	−	−	−	+	+	+	+	d	+	+	+	−	−
5. *B. ovis*	+	−	+	+	+	−	−	+	−	+	+	+	−	−	−	−	−	−	−	−
6. *B. canis*	−	−	−	+	+	−	−	+	−	−	−	−	−	−	+	+	d	−	+	+

[a] Certified dyes (National Aniline Division, Allied Chemical and Dye Co., New York) at concentrations a. = 1:25,000; b. = 1:50,000.

[b] RTD, routine test dilution

[c] Same reactions with DL-citrulline and DL-ornithine.

TABLE 7.14
Differential characteristics of species of genus **Bordetella**

	1. *B. pertussis*	2. *B. parapertussis*	3. *B. bronchiseptica*
Flagella	−[a]	−	+[a]
Nitrate → nitrite	−	−	+
Citrate utilized	−	+	+
Urease produced	−	+	+
Peptone agar			
Growth	−	+	+
Browning	−	+	−
Bordet-Gengou agar			
Growth in 1–2 days	−	+	+
Growth in 3–4 days	+		
Litmus milk alkaline			
In 1–4 days	−	+	+
In 12–14 days, M[b]	+		
Sensitize mice to histamine	+	−	−
Specific heat-labile antigen			
Factor 1	+	−	−
Factor 12	−	−	+
Factor 14	−	+	−
G + C content, moles %, T_m	61	61	66

[a] − = All strains negative; + = all strains positive.
[b] M = modulated from Phase I to Phase II–IV.

Genus **Francisella** *Dorofe'ev 1947, 176*

Very small, coccoid to ellipsoidal pleomorphic rods, which often show bipolar staining by special methods; many pass through Berkefeld filters. **Non-motile. Gram-negative.**

Acid without gas produced in media containing some carbohydrates. Catalase negative, H_2S produced. Cells soluble in sodium lauryl sulfate and sodium ricinoleate.

No growth on ordinary media without enrichment.

Strictly aerobic, no growth anaerobically. Optimum temperature 37 C.

Frequently found in natural waters; can be parasitic on man, other mammals, birds and arthropods.

Type species: *Francisella tularensis* (McCoy and Chapin) Dorofe'ev 1947, 178.

Note: There are two species recognized; *F. tularensis* produces acid from sucrose and *F. novicida* does not. They are also differentiated on pathogenicity and by serological methods.

Genus **Thermus** *Brock and Freeze 1969, 295*

Rods and filaments, 0.5–0.8 μm in diameter. Rods 5–10 μm in length, filaments variable length from 20 to greater than 200 μm. **Large spheres** 10–20 μm in diameter derived from the association of individual cells usually formed in old cultures. **Gram-negative. Non-motile. Flagella and endospores absent.**

Compact slowly spreading colonies. Many strains produce **yellow to bright orange colonies** although non-pigmented strains have also been isolated (Ramaley and Hixson, 1970). Surface pellicle often formed in unshaken liquid cultures.

Chemoorganotrophs. **Obligately aerobic.** Growth on amino acids, sugars, organic acids.

Temperature optimum 70–72 C; maximum 79 C, minimum 40 C. pH optimum 7.5–7.8, maximum 9.5, minimum 6.

The **G + C content** of the DNA ranges from **64–67 moles %** (buoyant density) among five strains tested.

Type species: *Thermus aquaticus* Brock and Freeze 1969, 295.

Note: Jackson *et al.* (*International Journal of Systematic Bacteriology 23:* 28–36, 1973) have recently described a new genus of thermophilic, Gram-negative bacteria named *Thermomicrobium*. This genus is differentiated from *Thermus* by the production of short, irregularly shaped rods.

PART 8

GRAM-NEGATIVE FACULTATIVELY ANAEROBIC RODS

Key to families in Part 8

I. Motile by peritrichous flagella. Oxidase-negative.
Family I. *Enterobacteriaceae*

II. Motile by polar flagella. Oxidase-positive.
Family II. *Vibrionaceae*

Synopsis of genera of uncertain affiliation in Part 8

I. Ferments glucose to ethanol and carbon dioxide.
Genus *Zymomonas*

II. Produces violet colonies.
Genus *Chromobacterium*

III. Produces yellow, orange, red or brown colonies. No acid or gas from carbohydrates.
Genus *Flavobacterium*

IV. Parasites. Require X and/or V factors from blood.
Genus *Haemophilus*

V. Parasites. Cells rod-shaped or ovoid showing bipolar staining. Catalase- and oxidase-positive.
Genus *Pasteurella*

VI. Cells rod-shaped with interspersed coccal forms giving a Morse Code form.
Genus *Actinobacillus*

VII. Pleomorphic rods. Catalase-negative. Indole- and oxidase-positive. Nitrates not reduced.
Genus *Cardiobacterium*

VIII. Parasites. Rods in long chains or unbranched filaments. Require body fluids for growth.
Genus *Streptobacillus*

IX. Pathogenic for humans. Pleomorphic rods showing a bipolar condensation of chromatin.
Genus *Calymmatobacterium*

X. Pathogenic for humans and animals causing a follicular type of conjunctivitis. Optimum temperature for growth 28–30 C.
Genus *Noguchia* (this genus was inadvertently omitted from the eighth edition of the *Manual* and the description is taken from the seventh edition)

FAMILY I. **ENTEROBACTERIACEAE** RAHN 1937, 281 *Nom. gen. cons.* Opin. 15, Jud. Comm. 1958, 73

Small **Gram-negative rods; motile by peritrichate flagella** or non-motile. Capsulated or non-capsulated. Not spore-forming; not acid-fast. Aerobic and **facultatively anaerobic.** Grow readily on meat extract media but some members have special growth requirements. **Chemoorganotrophic; metabolism respiratory and fermentative. Acid is produced** from the fermentation of **glucose,** other carbohydrates and alcohols; **usually aerogenic** but anaerogenic groups and mutants occur. **Catalase positive** with the exception of one serotype of *Shigella;* **oxidase negative. Nitrates are reduced to nitrites** except by some strains of *Erwinia*. G + C content of DNA: 39–59 moles %.

Type genus: *Escherichia* Castellani and Chalmers 1919, 941. Designated type genus Opinion 15 Jud. Comm. 1958, 73.

TABLE 8.1

Distinguishing characteristics of the five primary groups (tribes)

	Tribe I. *Escherichieae*	Tribe II. *Klebsielleae*	Tribe III. *Proteeae*	Tribe IV. *Yersinieae*	Tribe V. *Erwinieae*
Fermentation pattern	Mixed acid	2,3-Butanediol		Mixed acid	Mixed acid & 2,3-butanediol
M.R.	+	D	+	+	
V.P.	−	D	D	−	D
Phenylalanine deamination	−	−	+	−	D
Nitrate reduction	+	+	+	+	D
Urease	−	D	D	D	−
KCN, growth in	D	+	+	−	D
Optimal temp. for growth	37 C	37 C	37 C	30–37 C	27–30 C
G + C, %	50–53	52–59	39–42	45–47	50–58

Tribe I. Escherichieae

Genus I. **Escherichia** *Castellani and Chalmers 1919, 491 Nom. cons.* Opin. 15, Jud. Comm. 1958, 73

Straight rods, 1.1–1.5 by 2.0–6.0 μm (living) or 0.4–0.7 by 1.0–3.0 μm (dried and stained) (Luria, 1960). Singly or in pairs. Motile by peritrichous flagella or non-motile.

Grow readily on simple nutrient media. Colonies on nutrient agar may be smooth (S), low convex, moist, shiny surface, entire edge, gray and easily emulsified in saline, or rough (R), dry and do not emulsify well in saline. There are intergrading forms between these extremes. Mucoid forms may occur.

Acetate can be used but citrate cannot be used as sole carbon source. Glucose and other carbohydrates are fermented with the production of pyruvate, which is further converted into lactic, acetic and formic acids. Part of the formic acid is split by a complex hydrogenlyase enzyme system into equal amounts of CO_2 and H_2 (de Ley, 1962). Some strains are anaerogenic. Lactose is fermented by most strains, but this may be delayed or absent.

Other biochemical characteristics are shown in Tables 8.2 and 8.3.

G + C content of DNA: 50–51 moles % (buoyant density and T_m).

Type species: *Escherichia coli* (Migula) Castellani and Chalmers 1919, 941.

Genus II. **Edwardsiella** *Ewing and McWhorter 1965, 37*

Motile peritrichously flagellated **rods:** not encapsulated.

Citrate and malonate cannot be used as a sole C source. Late fermentation sometimes occurs in citrate.

H_2S is produced abundantly **from TSI agar,** but poorly producing biotypes may occur. Indole is produced. Lysine and ornithine decarboxylases are present.

Mannitol is not fermented by most strains. No action on *d*-tartrate and mucate.

Type species: *Edwardsiella tarda* Ewing and McWhorter 1965, 37.

Genus III. **Citrobacter** *Werkman and Gillen 1932, 173*

Motile peritrichously flagellated **rods,** not encapsulated. Grow readily on ordinary media and **can use citrate as sole C source.** Glucose and other carbohydrates fermented with the production of acid and gas (CO_2:H_2 in the ratio 1:1). Lactose fermentation may be delayed or absent. **Trimethyleneglycol formed from glycerol.** Growth not inhibited by KCN.

Able to grow on Muller's tetrathionate broth, Leifson's selenite broth, sodium deoxycholate-citrate agar, Wilson and Blair's bismuth sulfite medium and Kristensen's brilliant green-phenol red agar, all of which inhibit or retard the growth of *E. coli*.

Other characteristics are given in Tables 8.2, 8.3 and 8.4.

Found in water, food, feces and urine, but their pathogenicity is problematical. They seem to be normal intestinal inhabitants and are found constantly in healthy people. Certain serotypes seem to occur in sporadic and mass alimentary infections, and in infections of the urinary tract, gall bladder, middle ear and meninges.

Type species: *Citrobacter freundii* (Braak) Werkman and Gillen 1932, 173.

TABLE 8.2

Main biochemical characters of primary groups I to IV

	Group I					Group II				Group III. *Proteus*	Group IV. *Yersinia*
	Escherichia	*Edwardsiella*	*Citrobacter*	*Salmonella*	*Shigella*	*Klebsiella*	*Enterobacter*	*Hafnia*	*Serratia*		
Catalase	+	+	+	+	D[a]	+	+	+	+	+	+
Oxidase	−	−	−	−	−	−	−	−	−	−	−
β-Galactosidase	+	−	+	D	d	+	+	+	+	−	+
Gas from glucose at 37 C	+	+	+	+	−	d	+	+	d	D	−
KCN (growth on)	−	−	+	D	−	+	+	+	+	+	−
Mucate (acid)	+	−	+	D	−	d	d	−	−	+	−
Nitrate reduced	+	+	+	+	+	+	+	+	+	+	+
G + C, moles %	50–51			50–53		52–56	52–59	52–57	53–59	39–42 (one species = 50)	45–47
Carbohydrates (acid from)											
Adonitol	−	−	−	−	−	d	+	−	d	D	D
Arabinose	+	−	+	+		+	+	+	−	−	+
Dulcitol	d	−	d	D	d	d	−	−	−	−	−
Esculin	d	−	d	−			D	−	−	d	D
Inositol	−	−	−	d	−	+	D	−	d	D	−
Lactose	+ or ×	−	+ or ×	D	D	D	+	−	−	−	−
Maltose	+	+	+	+		+	+	+	+	D	+
Mannitol	+	−	+	+	D	+	+	+	+	D	+
Salicin	d	−	d	−	−	+	+	−	+	d	D
Sorbitol	+	−	+	+		+	+	−		−	D
Sucrose	d	−	d	−	D	+	+	− or ×	+	D	D
Trehalose	+	−	+	+		+	+	+	+	d	+
Xylose	d	−	+	+	D	+	+	+	d	D	D
Related C sources											
Citrate	−	−	+	+	−	d	+	+	+		−
Gluconate		−		−	−	+	+	+	+		
Malonate	−	−	d	D	−	D	+		−		−
d-Tartrate	d	−	+	D		d	−	−			D
M.R.	+	+	+	+	+	D	−	−	−	+	+
V.P.	−	−	−	−	−	D	+	+	D	d	−
Protein reactions											
Arginine	d	−	d	+	−	−	D	−	−	−	−
Gelatin hydrolysis	−	−	−	D	−	(d)	(+)	−	+	D	−
H_2S from TSI	−	+	D	+	−	−	−	−	−	D	D
Indole	+	+	D	−	D	d	−	−	−	D	D
Lysine decarboxylated	+	+	−	+	−	d	D	+	+	d	−
Ornithine	d	+	d	+	d	−	+	+	+	D	D
Urea hydrolyzed	−	−	(+)	−	−	d	(d)	−	−	D	D
Glutamic acid	−	−	−	−	−	−				+	
Phenylalanine	−	−	−	−	−	−	−	−	−	+	−

[a] D = different reactions given by different species of a genus; d = different reactions given by different strains of a species or serotype; × = late and irregularly positive (mutative).

TABLE 8.3

Distinguishing characteristics of the genera of the tribe **Escherichieae** *(group I)*

	Escherichia	*Edwardsiella*	*Citrobacter*	*Salmonella*	*Shigella*
Motility	+	+	+	+	−
Indole	+	+	D	−	d
H_2S on TSI	−	+	D	+	−
β-Galactosidase	+	−	+	D	d
Lactose	+/×	−	+/×	D	−/×
Mucate	+	−	+	D	−
KCN	−	−	d	D	−
Gelatin	−	−	−	D	−
Malonate	−	−	d	D	−
d-Tartrate	d	−	−	D	−
Dulcitol	d	−	d	D	d

TABLE 8.4

Distinguishing characters between **Citrobacter freundii** *and biotypes of* **C. intermedius**

	H_2S	Indole	Malonate	KCN	Other designations
1. *C. freundii*	+	−	−	+	*Salmonella coli*
2. *C. intermedius* biotype a	−	+	−	+	*Padlewskia, Levinea amalonatica*
C. intermedius biotype b	−	+	+	−	*C. koseri, Levinea malonatica*

Genus IV. **Salmonella** *Lignières 1900, 389*

Rods, usually motile by peritrichous flagella; non-motile mutants may occur and one type (*S. gallinarum*, *S. pullorum*) is always non-motile. Colonies are generally 2–4 mm in diameter but certain types (*S. abortus-equi*, *S. typhi-suis* and *Salmonella abortus-ovis*) produce colonies of about 1 mm. Most strains will grow on defined media without special growth factors, and they **can use citrate as C source. Most strains are aerogenic** but *S. typhi*, an important exception, never produces gas. Anaerogenic variants of normally gas-producing serotypes are found in nature, and this is particularly common with *S. dublin*.

The common biochemical characters are shown in Table 8.2. Subdivision by biochemical characters into the so-called subgenera of Kauffmann (1960, 1963, 1964) is shown in Table 8.5. These subdivisions correspond more closely to species or subspecies in other groups of bacteria. Whatever rank is assigned to them, the worthiness of the subdivisions was confirmed by Rohde (1965, 1966, 1967).

DNase and lipase are not produced.

G + C content of DNA: 50–53 moles % (by chemical analysis, T_m and buoyant density (Hill, 1966)).

Type species: *Salmonella cholerae-suis* (Smith) Weldin.

Note: The *Manual* contains descriptions of some well-known species of *Salmonella* arranged by sub-genus.

TABLE 8.5

Distinguishing characters of the subgenera of **Salmonella**

	I	II	III	IV
Dulcitol	+	+	−	−
Lactose	−	−	+ or ×[a]	−
β-Galactosidase	−	×	+	−
d-Tartrate	+	−	−	−
Mucate	+	+	d	−
Malonate	−	+	+	−
Gelatin	−	(+)	(+)	(+)
KCN	−	−	−	+

[a] × = late and irregularly positive; (+) = late, but always positive.

TABLE 8.6

Kauffmann-White scheme including subgenus III (S. arizonae) and serotypes described up to December 1970

Species or serotype	Antigen		
	O	H	
		Phase 1	Phase 2
Group A			
S. paratyphi A	*1*,2,12	a	—
S. nitra	2,12	g,m	—
S. kiel	*1*,2,12	g,p	—
Group B			
**S. makoma*	4,12	a	—
S. kisangani	*1*,4,[5],12	a	1,2
S. hessarek	4,12,*27*	a	1,5
S. fulica	4,[5],12	a	1,5
S. arechavaleta	4,[5],12	a	[1,7]
S. bispebjerg	*1*,4,[5],12	a	e,n,x
S. abortus-equi	4,12	—	e,n,x
S. tinda	*1*,4,12,*27*	a	e,n,z_{15}
S. nakuru	*1*,4,12,*27*	a	z_6
S. schottmuelleri (*S. paratyphi B*)	*1*,4,[5],12	b	1,2
S. java[1]	*1*,4,[5],12	b	[1,2]
S. limete	*1*,4,12,*27*	b	1,5
S. canada	4,12	b	1,6
S. uppsala	4,12,*27*	b	1,7
**S. sofia*	*1*,4,12,*27*	b	[e,n,x]
S. abony	*1*,4,[5],12	b	e,n,x
S. abortus-bovis	*1*,4,12,*27*	b	e,n,x
S. wagenia	*1*,4,12,*27*	b	e,n,z_{15}
S. schleissheim	4,12,*27*	b	—
S. wien	*1*,4,12,*27*	b	l,w
S. legon	4,12	c	1,5
S. abortus-ovis	4,12	c	1,6
S. altendorf	4,12,*27*	c	1,7
S. jericho	*1*,4,12,*27*	c	e,n,z_{15}
S. bury	4,12,*27*	c	z_6
S. stanley	*1*,4,[5],12,*27*	d	1,2
S. eppendorf	*1*,4,12,*27*	d	1,5
S. schwarzengrund	*1*,4,12,*27*	d	1,7
**S. kluetjenfelde*	4,12	d	e,n,x
S. sarajane	4,12,*27*	d	e,n,x
S. duisburg	*1*,4,12,*27*	d	e,n,z_{15}
S. salinatis	4,12	d,e,h	d,e,n,z_{15}
S. mons	*1*,4,12,*27*	d	l,w
S. ayinde	4,12,*27*	d	z_6
S. saint-paul	*1*,4,[5],12	e,h	1,2
S. reading	4,[5],12	e,h	1,5
S. kaapstad	4,12	e,h	1,7
S. chester	4,[5],12	e,h	e,n,x
S. san-diego	4,[5],12	e,h	e,n,z_{15}
**S. makumira*	*1*,4,12,*27*	e,n,x	1,7
**Salmonella*	4,12	(f),g	—
S. derby	*1*,4,[5],12	f,g	[1,2]
S. agona	*1*,4,12	f,g,s	—
S. essen	4,12	g,m	—
**S. caledon*	4,12	g,m,t	e,n,x
S. hato	4,[5],12	g,m,s	—
S. california	4,12	g,m,t	—
S. kingston	*1*,4,[5],12,*27*	g,s,t	—
S. budapest	*1*,4,12	g,t	—
**S. bechuana*	4,12,*27*	g,t	—
S. travis	4,[5],12	g,z_{51}	1,7
S. banana	4,[5],12	m,t	—
S. typhimurium	*1*,4,[5],12	i	1,2
S. lagos	*1*,4,12	i	1,5
S. agama	4,12	i	1,6
S. gloucester	*1*,4,12,*27*	i	l,w
S. massenya	*1*,4,12,*27*	k	1,5
S. neumuenster	*1*,4,12,*27*	k	1,6
**Salmonella*	*1*,4,12,*27*	k	1,6
S. ljubljana	4,12,*27*	k	e,n,x
S. texas	4,[5],12	k	e,n,z_{15}
S. azteca	4,[5],12	l,v	1,5
S. clackamas	4,12	l,v	1,6
S. bredeney	*1*,4,12,*27*	l,v	1,7
S. fyris	4,5,12	l,v	1,2
S. kimuenza	*1*,4,12,*27*	l,v	e,n,x
S. brandenburg	*1*,4,12	l,v	e,n,z_{15}
S. togo	4,12	l,w	1,6
**S. kilwa*	4,12	l,w	e,n,x
S. ayton	*1*,4,12,*27*	l,w	z_6
S. vom	4,12,*27*	l,z_{13},z_{28}	e,n,z_{15}
S. kunduchi	*1*,4,[5],12,*27*	l,z_{28}	1,2
S. tyresoe	4,12	l,z_{28}	1,5
S. heidelberg	*1*,4,[5],12	r	1,2
S. bradford	4,12,*27*	r	1,5
S. remo	*1*,4,12,*27*	r	1,7
S. bochum	4,[5],12	r	l,w
S. africana	4,12	r,i	l,w
S. coeln	4,[5],12	y	1,2
S. trachau	4,12,*27*	y	1,5
S. teddington	4,12,*27*	y	1,7
S. ball	*1*,4,[5],12,*27*	y	e,n,x
S. jos	*1*,4,12,*27*	y	e,n,z_{15}
S. kamoru	4,12,*27*	y	z_6
S. shubra	4,[5],12	z	1,2
**Salmonella*	*1*,4,12,*27*	z	1,5
S. kiambu	4,12	z	1,5
S. indiana	*1*,4,12	z	1,7
**S. nordenham*	*1*,4,12,*27*	z	e,n,x
S. koenigstuhl	*1*,4,12	z	e,n,z_{15}
S. preston	*1*,4,12	z	l,w
S. entebbe	*1*,4,12,*27*	z	z_6
S. stanleyville	*1*,4,[5],12,*27*	z_4, z_{23}	[1,2]
S. kalamu	4,12	z_4,z_{24}	[1,5]
S. haifa	*1*,4,[5],12	z_{10}	1,2

This table was prepared by the World Health Organization International Reference Centre for Salmonella, Pasteur Institute, Paris.

Definition of symbols. * Subgenus II, † Subgenus III, original Arizona formula (Edwards *et al.*, 1965) given in brackets, ‡ Subgenus IV; ( ), atypical, weakly agglutinable by ordinary sera; [ ], may be lacking. O antigens numbered 1–65; when italicized present only when organism is lysogenized by converting bacteriophage. H antigens: Phase 1 lettered a–z, z_1–z_{59}; Phase 2 originally numbered but because of crossed reactions now includes many with Phase 1 designations. Groups: A group contains one or more O antigens in common. Listing within groups is alphabetically by Phase 1 and/or numerically by Phase 2 antigens.

[1] d-tartrate positive contrary to *S. paratyphi* B.

TABLE 8.6—*Continued*

Species or serotype	Antigen		
	O	H	
		Phase 1	Phase 2
S. ituri	*1*,4,12	z_{10}	1,5
S. tudu	4,12	z_{10}	1,6
S. albert	4,12	z_{10}	e,n,x
S. tokoin	4,12	z_{10}	e,n,z_{15}
S. mura	*1*,4,12	z_{10}	l,w
S. fortune	*1*,4,12,*27*	z_{10}	z_6
S. vellore	*1*,4,12,*27*	z_{10}	z_{35}
S. brancaster	*1*,4,12,*27*	z_{29}	—
**S. helsinki*	*1*,4,12	z_{29}	[e,n,x]
S. tafo	*1*,4,12,*27*	z_{35}	1,7
S. sloterdijk	*1*,4,12,*27*	z_{35}	z_6
S. yaounde	*1*,4,12,*27*	z_{35}	e,n,z_{15}
S. tejas	4,12	z_{36}	—
S. wilhelmsburg	*1*,4,[5],12,*27*	z_{38}	—
**S. durbanville*	*1*,4,12,*27*	z_{39}	1,[5],7
Group C_1			
S. san-juan	6,7	a	1,5
S. umhlali	6,7	a	1,6
S. austin	6,7	a	1,7
S. oslo	6,7	a	e,n,x
S. denver	6,7	a	e,n,z_{15}
S. coleypark	6,7	a	l,w
**Salmonella*	6,7	a	z_6
**S. calvinia*	6,7	a	z_{42}
S. brazzaville	6,7	b	1,2
S. edinburg	6,7	b	1,5
S. koumra	6,7	b	1,7
S. georgia	6,7	b	e,n,z_{15}
S. ohio	6,7	b	l,w
S. leopoldville	6,7	b	z_6
S. kotte	6,7	b	z_{35}
**S. bloemfontein*	6,7	b	[e,n,x]:z_{42}
S. hirschfeldii (*S. paratyphi C*)	6,7,[Vi]	c	1,5
S. cholerae-suis	6,7	[c]	1,5
S. typhi-suis	6,7	c	1,5
S. birkenhead	6,7	c	1,6
S. kisii	6,7	d	1,2
S. isangi	6,7	d	1,5
S. kivu	6,7	d	1,6
S. amersfoort	6,7	d	e,n,x
S. gombe	6,7	d	e,n,z_{15}
S. livingstone	6,7	d	l,w
S. wil	6,7	d	l,z_{13},z_{28}
S. larochelle	6,7	e,h	1,2
S. lomita	6,7	e,h	1,5
S. norwich	6,7	e,h	1,6
S. braenderup	6,7	e,h	e,n,z_{15}
S. rissen	6,7	f,g	—
S. afula	6,7	f,g,t	e,n,x
S. montevideo	6,7	g,m,s,[p]	—
S. othmarschen	6,7	g,m,[t]	—
S. menston	6,7	g,s,t	—
S. riggil	6,7	g,t	—
**Salmonella*	6,7	g,t	e,n,x:z_{42}
S. alamo	6,7	g,z_{51}	1,5
S. haelsingborg	6,7	m,p,t,[u]	—
S. oranienburg	6,7	m,t	—
S. augustenborg	6,7	i	1,2
S. oritamerin	6,7	i	1,5
S. garoli	6,7	i	1,6
S. norton	6,7	i	l,w
S. galiema	6,7	k	1,2
S. thompson	6,7	k	1,5
S. daytona	6,7	k	1,6
S. baiboukoum	6,7	k	1,7
S. singapore	6,7	k	e,n,x
S. escanaba	6,7	k	e,n,z_{15}
†*Salmonella arizonae* (Ar.27:22:31)	6,7	(k)	z
**Salmonella*	6,7	k	[z_6]
S. concord	6,7	l,v	1,2
S. irumu	6,7	l,v	1,5
S. bonn	6,7	l,v	e,n,x
S. potsdam	6,7	l,v	e,n,z_{15}
†*Salmonella arizonae* (Ar.27:23:31)	6,7	l,v	z
S. gdansk	6,7	l,v	z_6
†*Salmonella arizonae* (Ar.27:23:25)	6,7	l,v	z_{53}
S. gabon	6,7	l,w	1,2
S. colorado	6,7	l,w	1,5
S. ness-ziona	6,7	l,z_{13}	1,5
S. kenya	6,7	l,z_{13}	e,n,x
S. neukoelln	6,7	l,z_{13},z_{28}	e,n,z_{15}
S. makiso	6,7	l,z_{13},z_{28}	z_6
**S. heilbron*	6,7	l,z_{28}	1,5:[z_{42}]
S. virchow	6,7	r	1,2
S. infantis	6,7	r	1,5
S. nigeria	6,7	r	1,6
S. colindale	6,7	r	1,7
S. papuana	6,7	r	e,n,z_{15}
S. richmond	6,7	y	1,2
S. bareilly	6,7	y	1,5
S. gatow	6,7	y	1,7
S. hartford	6,7	y	e,n,x
S. mikawasima	6,7	y	e,n,z_{15}
**S. tosamanga*	6,7	z	1,5
S. oakland	6,7	z	1,6,7
S. businga	6,7	z	e,n,z_{15}
**Salmonella*	6,7	z	z_6
**S. oysterbeds*	6,7	z	z_{42}
S. goma	6,7	z_4,z_{23}	z_6
S. obogu	6,7	z_4,z_{23}	1,5
S. aequatoria	6,7	z_4,z_{23}	e,n,z_{15}
‡*S. roterberg*	6,7	z_4,z_{23}	—
S. somone	6,7	z_4,z_{24}	—
‡*S. kralendyk*	6,7	z_4,z_{24}	—
†*Salmonella arizonae* (Ar.27:1,6,7:-)	6,7	z_4,z_{32}	—
**S. cape*	6,7	z_6	1,7
S. menden	6,7	z_{10}	1,2
S. inganda	6,7	z_{10}	1,5
S. eschweiler	6,7	z_{10}	1,6
S. ngili	6,7	z_{10}	1,7
S. djugu	6,7	z_{10}	e,n,x
S. mbandaka	6,7	z_{10}	e,n,z_{15}
**Salmonella*	6,7	z_{10}	z_{35}
S. tennessee	6,7	z_{29}	—
**Salmonella*	6,7	z_{29}	—
‡*S. argentina*	6,7	z_{36}	—
**S. bacongo*	6,7	z_{36}	z_{42}
S. lille	6,7	z_{38}	—

TABLE 8.6—*Continued*

Species or serotype	Antigen O	Antigen H Phase 1	Antigen H Phase 2
**S. gilbert*	6,7	z_{39}	1,5,7
S. hillsborough	6,7	z_{41}	l,w
S. tamilnadu	6,7	z_{41}	z_{35}
**S. sullivan*	6,7	z_{42}	1,7
**Salmonella*	6,7	z_{42}	e,n,x:1,6
Group C_2			
S. doncaster	6,8	a	1,5
S. curacao	6,8	a	1,6
S. nordufer	6,8	a	1,7
S. narashino	6,8	a	e,n,x
S. leith	6,8	a	e,n,z_{15}
**S. tulear*	6,8	a	z_{52}
S. nagoya	6,8	b	1,5
S. stourbridge	6,8	b	1,6
S. gatuni	6,8	b	e,n,x
S. presov	6,8	b	e,n,z_{15}
S. bukuru	6,8	b	l,w
S. banalia	6,8	b	z_6
S. wingrove	6,8	c	1,2
S. utah	6,8	c	1,5
S. bronx	6,8	c	1,6
S. belfast	6,8	c	1,7
S. belem	6,8	c	e,n,x
S. quiniela	6,8	c	e,n,z_{15}
S. muenchen	6,8	d	1,2
S. manhattan	6,8	d	1,5
S. sterrenbos	6,8	d	e,n,x
S. herston	6,8	d	e,n,z_{15}
S. newport	6,8	e,h	1,2
S. kottbus	6,8	e,h	1,5
S. tshiongwe	6,8	e,h	e,n,z_{15}
S. sandow	6,8	f,g	e,n,z_{15}
S. chincol	6,8	g,m,s	e,n,x
**Salmonella*	6,8	g,m,t	[e,n,x]
**S. baragwanath*	6,8	m,t	1,5
**S. germiston*	6,8	m,t	e,n,x
S. lindenburg	6,8	i	1,2
S. takoradi	6,8	i	1,5
S. warnow	6,8	i	1,6
S. bonariensis	6,8	i	e,n,x
S. aba	6,8	i	e,n,z_{15}
S. blockley	6,8	k	1,5
S. schwerin	6,8	k	e,n,x
S. litchfield	6,8	l,v	1,2
S. loanda	6,8	l,v	1,5
S. manchester	6,8	l,v	1,7
S. holcomb	6,8	l,v	e,n,x
S. edmonton	6,8	l,v	e,n,z_{15}
S. fayed	6,8	l,w	1,2
S. breukelen	6,8	l,z_{13}	e,n,z_{15}
S. bovis-morbificans	6,8	r	1,5
S. akanji	6,8	r	1,7
S. hidalgo	6,8	r	e,n,z_{15}
S. gold-coast	6,8	r	l,w
S. tananarive	6,8	y	1,5
S. praha	6,8	y	e,n,z_{15}
S. mowanjum	6,8	z	1,5
S. kuru	6,8	z	l,w
S. lezennes	6,8	z_4,z_{23}	1,7
S. chailey	6,8	z_4,z_{23}	e,n,z_{15}
S. duesseldorf	6,8	z_4,z_{24}	—
S. tallahassee	6,8	z_4,z_{32}	—
S. zerifin	6,8	z_{10}	1,2
S. mapo	6,8	z_{10}	1,5
S. cleveland	6,8	z_{10}	1,7
S. hadar	6,8	z_{10}	e,n,x
S. glostrup	6,8	z_{10}	e,n,z_{15}
S. wippra	6,8	z[illegible]	z_6
S. uno	6,8	z_{29}	—
S. yarm	6,8	z_{35}	1,2
Group C_3			
S. djelfa	8	b	1,2
S. korbol	8,*20*	b	1,5
S. sanga	8	b	1,7
S. shipley	8,*20*	b	e,n,z_{15}
S. alexanderpolder	8	c	l,w
S. virginia	8	d	1,2
S. labadi	8,*20*	d	z_6
S. atakpame	8,*20*	e,h	1,7
S. bardo	8	e,h	1,2
S. ferruch	8	e,h	1,5
S. rechovot	8,*20*	e,h	z_6
S. emek	8,*20*	g,m,s	—
S. yokoe	8	m,t	—
S. kentucky	8,*20*	i	z_6
S. kentucky var. *jerusalem*	8	i	z_6
S. haardt	8	k	1,5
S. pakistan	8	l,v	1,2
S. amherstiana	8	l,v	1,6
S. hindmarsh	8	r	1,5
S. pikine	8,*20*	r	z_6
S. altona	8,*20*	r,i	z_6
S. cocody	8,*20*	r,i	e,n,z_{15}
S. giza	8	y	1,2
S. brunei	8,*20*	y	1,5
S. alagbon	8	y	1,7
S. sunnycove	8	y	e,n,x
S. kralingen	8,*20*	y	z_6
S. corvallis	8,*20*	z_4,z_{23}	—
S. albany	8,*20*	z_4,z_{24}	—
S. paris	8,*20*	z_{10}	1,5
S. molade	8,*20*	z_{10}	z_6
S. istanbul	8	z_{10}	e,n,x
S. tamale	8,*20*	z_{29}	[e,n,z_{15}]
S. angers	8,*20*	z_{35}	z_6
S. diogoye	8,*20*	z_{41}	z_6
S. apeyeme	8,*20*	z_{38}	—
Group C_4 (*Salmonella* of group C_1 lysogenized by "phage 14")			
S. nienstedten	6,7,*14*	b	[l,w]
S. kaduna	6,7,*14*	c	e,n,z_{15}
S. hissar	6,7,*14*	c	1,2
S. omderman	6,7,*14*	d	e,n,x
S. eimsbuettel	6,7,*14*	d	l,w
S. nieukerk	6,7,*14*	d	z_6
S. ardwick	6,7,*14*	f,g	—
S. thielallee	6,7,*14*	m,t	—
S. gelsenkirchen	6,7,*14*	l,v	z_6

TABLE 8.6—*Continued*

Species or serotype	Antigen O	Antigen H Phase 1	Antigen H Phase 2
S. jerusalem	6,7,14	z_{10}	l,w
S. bornum	6,7,14	z_{38}	—
Group D_1			
S. sendai	1,9,12	a	1,5
S. miami	1,9,12	a	1,5
*Salmonella	9,12	a	1,5
S. os	9,12	a	1,6
S. saarbruecken	1,9,12	a	1,7
S. loma-linda	9,12	a	e,n,x
S. durban	9,12	a	e,n,z_{15}
S. onarimon	1,9,12	b	1,2
S. frintrop	1,9,12	b	1,5
*S. mjimwema	1,9,12	b	e,n,x
*S. blankenese	1,9,12	b	z_6
*S. suederelbe	1,9,12	b	z_{39}
S. goeteborg	9,12	c	1,5
S. ipeko	9,12	c	1,6
S. alabama	9,12	c	e,n,z_{15}
S. ridge	9,12	c	z_6
S. typhi	9,12[Vi]	d	—
S. ndolo	1,9,12	d	1,5
S. tarshyne	9,12	d	1,6
*S. rhodesiense	9,12	d	e,n,x
S. zega	9,12	d	z_6
S. jaffna	1,9,12	d	z_{35}
S. bournemouth	9,12	e,h	1,2
S. eastbourne	1,9,12	e,h	1,5
S. israel	9,12	e,h	e,n,z_{15}
*Salmonella	9,12	e,n,x	1,6
*S. lindrick	9,12	e,n,x	1,[5],7
S. berta	1,9,12	f,g,t	—
S. enteritidis	1,9,12	g,m	—
S. blegdam	9,12	g,m,q	—
*S. muizenberg	9,12	g,m,s,t	1,5
*S. kuilsrivier	1,9,12	g,m,s,t	e,n,x
S. manica	1,9,12	g,m,s,t	z_{42}
*S. hamburg	1,9,12	g,m,t	—
S. dublin	1,9,12[Vi]	g,p	—
S. naestved	1,9,12	g,p,s	—
S. rostock	1,9,12	g,p,u	—
S. moscow	9,12	g,q	—
*S. neasden	9,12	g,s,t	e,n,x
S. new-mexico	9,12	g,z_{51}	1,5
S. pensacola	1,9,12	m,t	—
S. seremban	9,12	i	1,5
S. claibornei	1,9,12	k	1,5
S. goverdhan	9,12	k	1,6
S. mendoza	9,12	l,v	1,2
S. panama	1,9,12	l,v	1,5
S. kapemba	9,12	l,v	1,7
*Salmonella	9,12	l,v	e,n,x
S. goettingen	9,12	l,v	e,n,z_{15}
S. victoria	1,9,12	l,w	1,5
†S. dar-es-salaam (S. salamae)	1,9,12	l,w	e,n,x
S. miyazaki	9,12	l,z_{13}	1,7
S. napoli	1,9,12	l,z_{13}	e,n,x
S. javiana	1,9,12	l,z_{28}	1,5
S. jamaica	9,12	r	1,5
S. lomé	9,12	r	z_6
S. lawndale	1,9,12	z	1,5
*S. stellenbosch	1,9,12	z	1,7
*S. angola	1,9,12	z	z_6
*S. hueningen	9,12	z	z_{39}
S. wangata	1,9,12	z_4,z_{23}	[1,7]
S. portland	9,12	z_{10}	1,5
*S. canastel	9,12	z_{29}	1,5
S. penarth	9,12	z_{35}	z_6
S. elomrane	1,9,12	z_{38}	—
*S. wynberg	1,9,12	z_{39}	1,7
S. gallinarum	1,9,12	—	—
Group D_2[1]			
S. baildon	9,46	a	e,n,x
S. zadar	9,46	b	1,6
*S. lundby	9,46	b	e,n,x
S. bamboye	9,46	b	l,w
S. itutaba	9,46	c	z_6
S. strasbourg	9,46	d	1,7
S. plymouth	9,46	d	z_6
S. bergedorf	9,46	e,h	1,2
S. wernigerode	9,46	f,g	—
*S. duivenhoks	9,46	g,m,s,t	e,n,x
S. gateshead	9,46	g,s,t	—
S. mathura	9,46	i	e,n,z_{15}
S. potto	9,46	i	z_6
S. marylebone	9,46	k	1,2
S. ceyco	9,46	k	z_{35}
S. india	9,46	l,v	1,5
S. geraldton	9,46	l,v	1,6
S. shoreditch	9,46	r	e,n,z_{15}
S. mayday	9,46	y	z_6
S. benin	9,46	y	1,7
*S. haarlem	9,46	z	e,n,x
S. ekotedo	9,46	z_4,z_{23}	—
*S. maarssen	9,46	z_4,z_{24}	z_{39}:z_{42}
S. lishabi	9,46	z_{10}	1,7
S. inglis	9,46	z_{10}	e,n,x
*Salmonella	9,46	z_{10}	z_6
S. ouakam	9,46	z_{29}	—
S. hillegersberg	9,46	z_{35}	1,5
S. fresno	9,46	z_{38}	—
Group D_3			
*S. zuerich	1,9,12,(46),27	c	z_{39}
*Salmonella	1,9,12,(46),27	y	z_{39}
Group E_1			
S. aminatu	3,10	a	1,2
S. goelzau	3,10	a	1,5
S. oxford	3,10	a	1,7
*S. matroosfontein	3,10	a	e,n,x
S. galil	3,10	a	e,n,z_{15}
S. kalina	3,10	b	1,2
S. butantan	3,10	b	1,5
S. allerton	3,10	b	1,6
S. huvudsta	3,10	b	1,7

[1] The serotypes of this group also contain the factors O:3 and (10) the latter not very well developed. They can be lysogenized by phages ϵ_{15} and ϵ_{34} and in the case of double lysogenization become strongly agglutinable, like strains of group E_2, by antisera against O:34 and O:12_2.

TABLE 8.6—*Continued*

Species or serotype	Antigen		
	O	H	
		Phase 1	Phase 2
S. benfica	3,10	b	e,n,x
S. yaba	3,10	b	e,n,z_{15}
S. epicrates	3,10	b	l,w
S. pramiso	3,10	c	1,7
S. agege	3,10	c	e,n,z_{15}
S. anderlecht	3,10	c	l,w
S. okefoko	3,10	c	z_6
S. stormont	3,10	d	1,2
S. shangani	3,10	d	1,5
S. onireke	3,10	d	1,7
S. souza	3,10	d	e,n,x
S. madjorio	3,10	d	e,n,z_{15}
S. birmingham	3,10	d	l,w
S. weybridge	3,10	d	z_6
S. maron	3,10	d	z_{35}
S. vejle	3,10	e,h	1,2
S. muenster	3,10	e,h	1,5
S. anatum	3,10	e,h	1,6
S. nyborg	3,10	e,h	1,7
S. newlands	3,10	e,h	e,n,x
S. meleagridis	3,10	e,h	l,w
S. sekondi	3,10	e,h	z_6
**S. chudleigh*	3,10	e,n,x	1,7
S. regent	3,10	f,g	—
S. suberu	3,10	g,m	—
S. amsterdam	3,10	g,m,s	—
S. westhampton	3,10	g,s,t	—
**S. islington*	3,10	g,t	—
S. southbank	3,10	m,t	
**S. stikland*	3,10	m,t	e,n,x
S. amounderness	3,10	i	1,5
S. falkensee	3,10	i	e,n,z_{15}
S. yeerongpilly	3,10	i	z_6
S. wimborne	3,10	k	1,2
S. zanzibar	3,10	k	1,5
S. marienthal	3,10	k	e,n,z_{15}
S. new-rochelle	3,10	k	l,w
S. nchanga	3,10	l,v	1,2
S. sinstorf	3,10	l,v	1,5
S. london	3,10	l,v	1,6
S. give	3,10	l,v	1,7
S. ruzizi	3,10	l,v	e,n,z_{15}
**S. fuhlsbuettel*	3,10	l,v	z_6
S. uganda	3,10	l,z_{13}	1,5
S. fallowfield	3,10	l,z_{13},z_{28}	e,n,z_{15}
S. joal	3,10	l,z_{28}	1,7
**S. westpark*	3,10	l,z_{28}	e,n,x
**Salmonella*	3,10	l,z_{28}	z_{39}
S. seegefeld	3,10	r,i	1,2
S. ughelli	3,10	r	1,5
S. elisabethville	3,10	r	1,7
S. simi	3,10	r	e,n,z_{15}
S. weltevreden	3,10	r	z_6
S. amager	3,10	y	1,2
S. orion	3,10	y	1,5
S. mokola	3,10	y	1,7
S. ohlstedt	3,10	y	e,n,x
S. bolton	3,10	y	e,n,z_{15}
S. langensalza	3,10	y	l,w
S. stockholm	3,10	y	z_6
**S. alexander*	3,10	z	1,5
**S. finchley*	3,10	z	e,n,x
S. clerkenwell	3,10	z	l,w
**S. tafelbaai*	3,10	z	z_{39}
S. adabraka	3,10	z_4,z_{23}	[1,7]
S. florian	3,10	z_4,z_{24}	—
S. okerara	3,10	z_{10}	1,2
S. lexington	3,10	z_{10}	1,5
S. coquilhatville	3,10	z_{10}	1,7
S. kristianstad	3,10	z_{10}	e,n,z_{15}
S. biafra	3,10	z_{10}	z_6
S. jedburgh	3,10	z_{29}	—
S. cairina	3,10	z_{35}	z_6
S. macallen	3,10	z_{36}	—
S. bolombo	3,10	z_{38}	—
**S. mpila*	3,10	z_{38}	z_{42}
**S. winchester*	3,10	z_{39}	1,7
Group E_2 (*Salmonella* of group E_1 lysogenized by phage ϵ_{15})			
S. clichy	3,*15*	a	1,5
S. rosenthal	3,*15*	b	1,5
S. pankow	3,*15*	d	1,5
S. eschersheim	3,*15*	d	e,n,x
S. goerlitz	3,*15*	e,h	1,2
S. new-haw	3,*15*	e,h	1,5
S. newington	3,*15*	e,h	1,6
S. selandia	3,*15*	e,h	1,7
S. cambridge	3,*15*	e,h	l,w
S. drypool	3,*15*	g,m,s	—
**S. parow*	3,*15*	g,m,s,t	—
S. halmstad	3,*15*	g,s,t	—
S. nancy	3,*15*	l,v	1,2
S. portsmouth	3,*15*	l,v	1,6
S. new-brunswick	3,*15*	l,v	1,7
S. kinshasa	3,*15*	l,z_{13}	1,5
S. lanka	3,*15*	r	z_6
S. tuebingen	3,*15*	y	1,2
S. binza	3,*15*	y	1,5
S. tournai	3,*15*	y	z_6
S. manila	3,*15*	z_{10}	1,5
Group E_3 (*Salmonella* of group E_1 lysogenized by phages ϵ_{15} and ϵ_{34})			
S. khartoum	3,*15*,*34*	a	1,7
S. arkansas	3,*15*,*34*	e,h	1,5
S. minneapolis	3,*15*,*34*	e,h	1,6
S. wildwood	3,*15*,*34*	e,h	l,w
S. canoga	3,*15*,*34*	g,s,t	—
S. menhaden	3,*15*,*34*	l,v	1,7
S. thomasville	3,*15*,*34*	y	1,5
S. illinois	3,*15*,*34*	z_{10}	1,5
S. harrisonburg	3,*15*,*34*	z_{10}	1,6
Group E_4			
S. gwoza	1,3,19	a	e,n,z_{15}
S. gnesta	1,3,19	b	1,5
S. visby	1,3,19	b	1,6
S. broughton	1,3,19	b	l,w
S. accra	1,3,19	b	z_6

TABLE 8.6—*Continued*

Species or serotype	Antigen		
	O	H	
		Phase 1	Phase 2
S. madiago	1,3,19	c	1,7
S. ahmadi	1,3,19	d	1,5
S. liverpool	1,3,19	d	e,n,z_{15}
S. tilburg	1,3,19	d	l,w
S. niloese	1,3,19	d	z_6
S. vilvoorde	1,3,19	e,h	1,5
S. sankt-marx	1,3,19	e,h	1,7
S. sao	1,3,19	e,h	e,n,z_{15}
S. calabar	1,3,19	e,h	l,w
S. rideau	1,3,19	f,g	—
S. maiduguri	1,3,19	f,g,t	e,n,z_{15}
S. senftenberg	1,3,19	g,[s],t	—
S. cannstatt	1,3,19	m,t	—
S. stratford	1,3,19	i	1,2
S. machaga	1,3,19	i	e,n,x
S. avonmouth	1,3,19	i	e,n,z_{15}
S. zuilen	1,3,19	i	l,w
S. taksony	1,3,19	i	z_6
S. ngor	1,3,19	l,v	1,5
S. westerstede	1,3,19	l,z_{13}	[1,2]
S. lokstedt	1,3,19	l,z_{13},z_{28}	1,2
S. bedford	1,3,19	l,z_{13},z_{28}	e,n,z_{15}
S. yalding	1,3,19	r	e,n,z_{15}
S. fareham	1,3,19	r,i	l,w
S. krefeld	1,3,19	y	l,w
S. korlebu	1,3,19	z	1,5
S. schoeneberg	1,3,19	z	e,n,z_{15}
S. carno	1,3,19	z	l,w
S. dallgow	1,3,19	z_{10}	e,n,z_{15}
S. llandoff	1,3,19	z_{29}	—
S. chittagong	(1),3,10,(19)	b	z_{35}
S. bilu	(1),3,10,(19)	f,g,t	1,(2),7
S. ilugun	(1),3,10,(19)	z_4,z_{23}	z_6
S. dessau	(1),3,*15*,(19)	g,s,t	—
S. cannonhill	(1),3,*15*,(19)	y	e,n,x
Group F			
S. toowong	11	a	1,7
S. marseille	11	a	1,5
S. luciana	11	a	e,n,z_{15}
**S. glencairn*	11	a	z_6:z_{42}
S. leeuwarden	11	b	1,5
**Salmonella*	11	b	1,7
**S. srinagar*	11	b	e,n,x
S. pharr	11	b	e,n,z_{15}
S. chandans	11	d	e,n,x
**S. montgomery*	11	d(a)	d,e,n,z_{15}
S. findorff	11	d	z_6
S. chingola	11	e,h	1,2
S. adamstua	11	e,h	1,6
S. redhill	11	e,h	l,z_{13},z_{28}
**S. grabouw*	11	g,m,s,t	z_{39}
‡*S. mundsburg*	11	g,z_{51}	—
**S. lincoln*	11	m,t	e,n,x
S. aberdeen	11	i	1,2
S. brijbhumi	11	i	1,5
S. heerlen	11	i	1,6
S. veneziana	11	i	e,n,x
S. pretoria	11	k	1,2
S. abaetetuba	11	k	1,5
S. sharon	11	k	1,6
S. colobane	11	k	1,7
S. kisarawe	11	k	e,n,x
S. amba	11	k	l,z_{13},z_{28}
†*Salmonella arizonae* (Ar.17:29:25)	11	k	z_{53}
S. stendal	11	l,v	1,2
S. maracaibo	11	l,v	1,5
S. fann	11	l,v	e,n,x
S. bullbay	11	l,v	e,n,z_{15}
S. glidji	11	l,w	1,5
S. osnabrueck	11	l,z_{13},z_{28}	e,n,x
**S. huila*	11	l,z_{28}	e,n,x
S. senegal	11	r	1,5
S. rubislaw	11	r	e,n,x
S. volta	11	r	l,z_{13},z_{28}
S. solt	11	y	1,5
S. herzliya	11	y	e,n,x
S. nyanza	11	z	z_6
**S. soutpan*	11	z	z_{39}
†*Salmonella arizonae* (Ar.17:1,2,5:-)	11	z_4,z_{23}	—
‡*S. parera*	11	z_4,z_{23}	—
S. remete	11	z_4,z_{23}	1,6
S. etterbeek	11	z_4,z_{23}	e,n,z_{15}
S. yehuda	11	z_4,z_{24}	—
‡*Salmonella*	11	z_4,z_{32}	—
S. wentworth	11	z_{10}	1,2
S. straengnaes	11	z_{10}	1,5
S. tel-hashomer	11	z_{10}	e,n,x
S. lene	11	z_{38}	—
S. maastricht	11	z_{41}	1,2
**Salmonella*	11	—	1,5
Group G_1			
†*Salmonella arizonae* (Ar.18:-:-)	13,22	—	—
S. mim	13,22	a	1,6
S. ibadan	13,22	b	1,5
S. vaertan	13,22	b	e,n,x
S. bahati	13,22	b	e,n,z_{15}
S. haouaria	13,22	c	e,n,x,z_{15}
S. friedenau	13,22	d	1,6
S. diguel	*1*,13,22	d	e,n,z_{15}
S. willemstad	*1*,13,22	e,h	1,6
S. raus	13,22	f,g	e,n,x
**Salmonella*	13,22	(f),g,t	—
S. bron	13,22	g,m	[e,n,z_{15}]
**S. limbe*	*1*,13,22	g,m,t	[1,5]
**S. rotterdam*	*1*,13,22	g,t	1,5
S. lovelace	13,22	l,v	1,5
S. borbeck	13,22	l,v	1,6
S. tanger	*1*,13,22	y	1,6
S. poona	*1*,13,22	z	1,6
S. bristol	13,22	z	1,7
S. roodepoort	*1*,13,22	z_{10}	1,5
S. agoueve	13,22	z_{29}	—
**S. clifton*	13,22	z_{29}	1,5
**S. goodwood*	13,22	z_{29}	e,n,x
S. mampong	13,22	z_{35}	1,6
S. leiden	13,22	z_{38}	—
**Salmonella*	13,22	z_{39}	1,5,(7)

TABLE 8.6—*Continued*

Species or serotype	Antigen		
	O	H	
		Phase 1	Phase 2
Group G_2			
S. chagoua	1,13,23	a	1,5
**S. tygerberg*	1,13,23	a	z_{42}
S. atlanta	13,23	b	—
S. mississippi	1,13,23	b	1,5
**S. acres*	1,13,23	b	[1,5]:z_{42}
S. bracknell	13,23	b	1,6
S. ullevi	1,13,23	b	e,n,x
S. durham	13,23	b	e,n,z_{15}
S. mishmar-haemek	1,13,23	d	1,5
S. grumpensis	13,23	d	1,7
S. tel-el-kebir	13,23	d	e,n,z_{15}
**Salmonella*	13,23	d	e,n,x
S. putten	13,23	d	l,w
S. isuge	13,23	d	z_6
S. wichita	1,13,23	d	[z_{37}]
S. tschangu	1,13,23	e,h	1,5
**S. epping*	1,13,23	e,n,x	1,7
S. havana	1,13,23	f,g,[s]	—
S. agbeni	13,23	g,m	—
**S. kraaifontein*	1,13,23	g,m,t	[e,n,x]
S. okatie	13,23	g,s,t	—
**S. luanshya*	13,23	g,m,s,t	—
S. congo	13,23	g,m,s,t	—
**S. gojenberg*	1,13,23	g,t	1,5
†*Salmonella arizonae* (Ar.18:13,14:-)	1,13,23	g,z_{51}	—
S. kintambo	13,23	m,t	—
**S. katesgrove*	1,13,23	m,t	1,5
**S. worchester*	1,13,23	m,t	e,n,x
**S. boulders*	13,23	m,t	z_{42}
S. kedougou	1,13,23	i	l,w
S. idikan	13,23	i	1,5
S. jukestown	13,23	i	e,n,z_{15}
**Salmonella*	13,23	l,z_{28}	z_6
**S. vredelust*	1,13,23	l,z_{28}	z_{42}
S. adjame	13,23	r	1,6
S. linton	13,23	r	e,n,z_{15}
S. yarrabah	13,23	y	1,7
S. ordonez	1,13,23	y	l,w
S. tunis	1,13,23	y	z_6
**S. nachshonim*	1,13,23	z	1,5
S. farmsen	13,23	z	1,6
S. worthington	1,13,23	z	l,w
S. ajiobo	13,23	z_4,z_{23}	—
S. romanby	13,23	z_4,z_{24}	—
†*Salmonella arizonae* (Ar.18:1,3,11:-)	1,13,23	z_4,z_{24}	—
S. demerara	13,23	z_{10}	l,w
S. cubana	1,13,23	z_{29}	[z_{37}]
S. fanti	13,23	z_{38}	—
**S. stevenage*	1,13,23	[z_{42}]	1,7
Group H			
S. garba	1,6,14,25	a	1,5
S. ferlac	1,6,14,25	a	e,n,x
S. tucson	[1],6,14,[25]	b	[1,7]
**Salmonella*	6,14	b	e,n,x,z_{15}
S. blijdorp	1,6,14,25	c	1,5
S. kassberg	1,6,14,25	c	1,6
S. minna	1,6,14,25	c	l,w
S. heves	6,14,24	d	1,5
S. finkenwerder	[1],6,14,[25]	d	1,5
S. florida	1,6,14,25	d	1,7
S. lindern	6,14,24	d	e,n,x
S. charity	1,6,14,25	d	e,n,x
S. teko	1,6,14,25	d	e,n,z_{15}
S. encino	1,6,14,25	d	l,z_{13},z_{28}
S. albuquerque	1,6,14,24	d	z_6
S. bahrenfeld	6,14,24	e,h	1,5
S. onderstepoort	1,6,14,[25]	e,h	1,5
S. magumeri	1,6,14,25	e,h	1,6
S. beaudesert	[1],6,14,[25]	e,n	1,7
S. warragul	1,6,14,25	g,m	—
S. caracas	1,6,14,25	g,m,s	—
S. kaitaan	1,6,14,25	m,t	—
**S. rooikrantz*	1,6,14	m,t	1,5
**S. emmerich*	6,14	[m,t]	e,n,x
S. mampeza	1,6,14,25	i	1,5
S. buzu	1,6,14,25	i	1,7
S. schalkwijk	6,14,24	i	e,n,..
S. harburg	1,6,14,25	k	1,5
**Salmonella*	6,14	k	[e,n,x]
†*Salmonella arizonae* (Ar.7a,7c:29:31)	6,14	k	z
**Salmonella*	1,6,14	k	z_6:z_{42}
†*Salmonella arizonae* (Ar.7a,7c:29:25)	6,14	k	z_{53}
S. boecker	[1],6,14,[25]	l,v	1,7
S. horsham	1,6,14,25	l,v	e,n,x
S. aflao	1,6,14,25	l,z_{28}	e,n,x
S. surat	1,6,14,25	[r],i	e,n,z_{15}
S. carrau	6,14,24	y	1,7
S. madelia	1,6,14,25	y	1,7
S. fischerkietz	1,6,14,25	y	e,n,x
S. homosassa	1,6,14,25	z	1,5
S. soahanina	6,14,24	z	e,n,x
S. sundsvall	1,6,14,25	z	e,n,x
S. poano	1,6,14,25	z	l,z_{13},z_{28}
S. bousso	1,6,14,25	z_4,z_{23}	[e,n,z_{15}]
S. uzaramo	1,6,14,25	z_4,z_{24}	—
S. nessa	1,6,14,25	z_{10}	1,2
**S. simonstown*	1,6,14	z_{10}	1,5
**S. bornheim*	1,6,14,25	z_{10}	1,(2),7
†*Salmonella arizonae* (Ar.7a,7c:27:31)	6,14	z_{10}	z
†*Salmonella arizonae* (Ar.7a,7c:27:31:38)	6,14	z_{10}	z:z_{56}
**S. slangkop*	1,6,14	z_{10}	z_6:z_{42}
†*Salmonella arizonae* (Ar.7a,7c:27:38)	6,14	z_{10}	z_{56}
S. sara	1,6,14,25	z_{38}	[e,n,x]
**Salmonella*	1,6,14	z_{42}	1,6
Group I (O 16)			
S. hannover	16	a	1,2
S. brazil	16	a	1,5
S. amunigun	16	a	1,6
S. nyeko	16	a	1,7
S. fischerhuette	16	a	e,n,z_{15}
S. heron	16	a	z_6
S. hull	16	b	1,2
S. wa	16	b	1,5
S. glasgow	16	b	1,6

TABLE 8.6—*Continued*

Species or serotype	Antigen O	Antigen H Phase 1	Antigen H Phase 2
S. hvittingfoss	16	b	e,n,x
S. sangera	16	b	e,n,z_{15}
S. malstatt	16	b	z_6
**Salmonella*	16	b	z_{42}
S. vancouver	16	c	1,5
S. shamba	16	c	e,n,x
S. oldenburg	16	d	1,2
S. gaminara	16	d	1,7
S. barranquilla	16	d	e,n,x
S. nottingham	16	d	e,n,z_{15}
S. barmbek	16	d	z_6
S. malakal	16	e,h	1,2
S. saboya	16	e,h	1,5
S. weston	16	e,h	z_6
**S. bellville*	16	e,n,x	1,7
S. tees	16	f,g	—
S. adeoyo	16	g,m	—
S. nikolaifleet	16	g,m,s	—
**S. mobeni*	16	g,m,[s],t	—
**S. merseyside*	16	g,t	1,5
**S. rowbarton*	16	m,t	[z_{42}]
S. amina	16	i	1,5
S. wisbech	16	i	1,7
S. frankfurt	16	i	e,n,z_{15}
S. abobo	16	i	z_6
†*Salmonella arizonae* (Ar.25:33:21)	16	i	z_{35}
S. szentes	16	k	1,2
S. nuatja	16	k	e,n,x
S. orientalis	16	k	e,n,z_{15}
†*Salmonella arizonae* (Ar.25:29:31)	16	k	z
†*Salmonella arizonae* (Ar.25:29:25)	16	k	z_{53}
†*Salmonella arizonae* (Ar.25:22:21)	16	(k)	z_{35}
†*Salmonella arizonae* (Ar.25:23:30)	16	l,v	1,5,7
S. shanghai	16	l,v	1,6
S. welikade	16	l,v	1,7
S. salford	16	l,v	e,n,x
S. burgas	16	l,v	e,n,z_{15}
†*Salmonella arizonae* (Ar.25:23:31)	16	l,v	z
S. los-angeles	16	l,v	z_6
†*Salmonella arizonae* (Ar.25:23:21)	16	l,v	z_{35}
†*Salmonella arizonae* (Ar.25:23:25)	16	l,v	z_{53}
S. lomnava	16	l,w	e,n,z_{15}
**S. noordhoek*	16	l,w	z_6
S. mandera	16	l,z_{13}	e,n,z_{15}
S. enugu	16	l,z_{13},z_{28}	—
**S. sarepta*	16	l,z_{28}	z_{42}
S. annedal	16	r,i	e,n,x
S. zwickau	16	r,i	e,n,z_{15}
S. avignon	16	y	e,n,z_{15}
S. saphra	16	y	1,5
S. akuafo	16	y	1,6
S. kikoma	16	y	e,n,x
S. lingwala	16	z	1,7
**S. louwbester*	16	z	e,n,x

Species or serotype	Antigen O	Antigen H Phase 1	Antigen H Phase 2
S. kibi	16	z_4,z_{23}	—
**S. haddon*	16	z_4,z_{23}	—
‡*S. ochsenzoll*	16	z_4,z_{23}	—
‡*S. chameleon*	16	z_4,z_{32}	—
†*Salmonella arizonae* (Ar.25:27:30)	16	z_{10}	1,5
S. lisboa	16	z_{10}	1,6
†*Salmonella arizonae* (Ar.25:27:28)	16	z_{10}	e,n,x,z_{15}
S. redlands	16	z_{10}	e,n,z_{15}
**S. jacksonville*	16	z_{29}	—
**S. woodstock*	16	z_{42}	1,(5),7
**S. elsiesrivier*	16	z_{42}	1,6
Group J (O 17)			
S. bonames	17	a	1,2
S. jangwani	17	a	1,5
S. kinondoni	17	a	e,n,x
S. kirkee	17	b	1,2
**S. hillbrow*	17	b	e,n,x,z_{15}
S. victoriaborg	17	c	1,6
**S. woerden*	17	c	z_{39}
S. berlin	17	d	1,5
S. niamey	17	d	l,w
**S. verity*	17	e,n,x,z_{15}	1,6
**S. bleadon*	17	(f),g,t	[e,n,x,z_{15}]
**Salmonella*	17	k	—
S. irenea	17	k	1,5
S. matadi	17	k	e,n,x
S. morotai	17	l,v	1,2
S. michigan	17	l,v	1,5
S. carmel	17	l,v	e,n,x
†*Salmonella arizonae* (Ar.12:23:28)	17	l,v	e,n,x,z_{15}
†*Salmonella arizonae* (Ar.12:23:21)	17	l,v	z_{35}
S. gori	17	z	1,2
**S. constantia*	17	z	l,w:z_{42}
†*Salmonella arizonae* (Ar.12:1,2,5:-) (Ar.12:1,2,6:-)	17	z_4,z_{23}	—
†*Salmonella arizonae* (Ar.12:1,3,11:-)	17	z_4,z_{24}	—
†*Salmonella arizonae* (Ar.12:1,6,7:-) (Ar.12:1,6,7,9:-) (Ar.12:1,7,8:-)	17	z_4,z_{32}	—
S. djibouti	17	z_{10}	e,n,x
†*Salmonella arizonae* (Ar.12:27:28)	17	z_{10}	e,n,x,z_{15}
†*Salmonella arizonae* (Ar.12:27,(35):28)	17	z_{10},(a)	e,n,x,z_{15}
†*Salmonella arizonae* (Ar.12:27:31)	17	z_{10}	z
S. kandla	17	z_{29}	—
†*Salmonella arizonae* (Ar.12:17,20:-)	17	z_{36}	—
Group K (O 18)			
S. fluntern	*6,14*,18	b	1,5
S. usumbura	18	d	1,7
†*Salmonella arizonae* (Ar.7a,7b:13,14:-)	18	g,z_{51}	—

TABLE 8.6—*Continued*

Species or serotype	Antigen O	Antigen H Phase 1	Antigen H Phase 2
S. langenhorn	18	m,t	—
**Salmonella*	18	m,t	1,5
S. memphis	18	k	1,5
†*Salmonella arizonae* (Ar.7a,7b:22:34)	18	(k)	z_{54}
†*Salmonella arizonae* (Ar.7a,7b:24:31)	18	r	z
**Salmonella*	18	y	e,n,x,z_{15}
S. cerro	*6,14*,18	z_4,z_{23}	[1,5]
†*Salmonella arizonae* (Ar.7a,7b:1,2,5:-) (Ar.7a,7b:1,2,6:-)	18	z_4,z_{23}	—
S. blukwa	18	z_4,z_{24}	—
†*Salmonella arizonae* (Ar.7a,7b:1,7,8:-)	18	z_4,z_{32}	—
S. carnac	18	z_{10}	z_6
**S. zeist*	18	z_{10}	z_6
**S. beloha*	18	z_{36}	—
S. sinthia	18	z_{38}	—
Group L (O 21)			
S. assen	21	a	—
S. ghana	21	b	1,6
S. minnesota	21	b	e,n,x
S. rhône	21	c	e,n,x
**Salmonella*	21	c	e,n,x
†*Salmonella arizonae* (Ar.22:32:28)	21	c	e,n,x,z_{15}
S. spartel	21	d	1,5
S. magwa	21	d	e,n,x
S. good	21	f,g	e,n,x
†*Salmonella arizonae* (Ar.22:13,14:-)	21	g,z_{51}	—
S. diourbel	21	i	1,2.
†*Salmonella arizonae* (Ar.22:33:28)	21	i	e,n,x,z_{15}
†*Salmonella arizonae* (Ar.22:29:28)	21	k	e,n,x,z_{15}
†*Salmonella arizonae* (Ar.22:23:31)	21	l,v	z
S. ruiru	21	y	e,n,x
S. baguida	21	z_4,z_{23}	—
‡*S. soesterberg*	21	z_4,z_{23}	—
**S. gwaai*	21	z_4,z_{24}	—
†*Salmonella arizonae* (Ar.22:27:28)	21	z_{10}	e,n,x,z_{15}
†*Salmonella arizonae* (Ar.22:27:31)	21	z_{10}	z
**S. wandsbek*	21	z_{10}	z_6
†*Salmonella arizonae* (Ar.22:16,17,18:-)	21	z_{29}	—
S. gambaga	21	z_{35}	e,n,z_{15}
Group M (O 28)			
S. solna	28	a	1,5
S. dakar	28	a	1,6
S. seattle	28	a	e,n,x
S. honelis	28	a	e,n,z_{15}
S. moero	28	b	1,5
S. ashanti	28	b	1,6
S. bokanjac	28	b	1,7
S. langford	28	b	e,n,z_{15}
**S. kaltenhausen*	28	b	z_6
S. hermannswerder	28	c	1,5
S. eberswalde	28	c	1,6
S. halle	28	c	1,7
S. dresden	28	c	e,n,x
S. wedding	28	c	e,n,z_{15}
S. techimani	28	c	z_6
S. amoutive	28	d	1,5
S. mundonobo	28	d	1,7
S. mocamedes	28	d	e,n,x
S. patience	28	d	e,n,z_{15}
S. kpeme	28	e,h	1,7
**Salmonella*	28	e,n,x	1,7
S. friedrichsfelde	28	f,g	—
S. abadina	28	g,m	[e,n,z_{15}]
S. croft	28	g,m,s	—
S. ona	28	g,s,t	—
**S. llandudno*	28	g,s,t	1,5
S. vinohrady	28	m,t	—
**Salmonella*	28	m,t	—
S. cotham	28	i	1,5
S. volkmarsdorf	28	i	1,6
S. kuessel	28	i	e,n,z_{15}
S. guildford	28	k	1,2
S. ilala	28	k	1,5
S. adamstown	28	k	1,6
S. taunton	28	k	e,n,x
S. ank	28	k	e,n,z_{15}
S. leoben	28	l,v	1,5
S. vitkin	28	l,v	e,n,x
S. nashua	28	l,v	e,n,z_{15}
S. chicago	28	r	1,5
S. bassadji	28	r	1,6
S. kibusi	28	r	e,n,x
**S. oevelgoenne*	28	r	e,n,z_{15}
S. sankt-georg	28	r,i	e,n,z_{15}
S. oskarshamn	28	y	1,2
S. nima	28	y	1,5
S. pomona	28	y	1,7
S. kitenge	28	y	e,n,x
S. tel-aviv	28	y	e,n,z_{15}
S. shomolu	28	y	l,w
S. ezra	28	z	1,7
S. brisbane	28	z	e,n,z_{15}
**S. ceres*	28	z	z_{39}
S. teltow	28	z_4,z_{23}	1,6
S. babelsberg	28	z_4,z_{23}	[e,n,z_{15}]
S. rogy	28	z_{10}	1,2
S. umbilo	28	z_{10}	e,n,x
S. luckenwalde	28	z_{10}	e,n,z_{15}
S. moroto	28	z_{10}	l,w
S. djermaia	28	z_{29}	—
S. aderike	28	z_{38}	—
Group N (O 30)			
S. doulassame	30	a	e,n,z_{15}
S. overvecht	30	a	1,2
S. zehlendorf	30	a	1,5
**S. odijk*	30	a	z_{39}
S. louga	30	b	1,2
S. aschersleben	30	b	1,5

TABLE 8.6—*Continued*

Species or serotype	Antigen		
	O	H	
		Phase 1	Phase 2
S. urbana	30	b	e,n,x
S. messina	30	d	1,5
**S. slatograd*	30	f,g,t	—
S. godesberg	30	g,m	—
S. giessen	30	g,m,s	—
S. sternschanze	30	g,s,t	[z_{59}]
S. wayne	30	g,z_{51}	—
S. landau	30	i	1,2
S. morehead	30	i	1,5
S. soerenga	30	i	l,w
S. hilversum	30	k	1,2
S. ramat-gan	30	k	1,5
S. aqua	30	k	1,6
S. angoda	30	k	e,n,x
S. odozi	30	k	e,n,x,z_{15}
**Salmonella*	30	k	e,n,[x],z_{15}
S. ligeo	30	l,v	1,2
S. donna	30	l,v	1,5
S. morocco	30	l,z_{13},z_{28}	e,n,z_{15}
S. gege	30	r	1,5
S. matopeni	30	y	1,2
S. steinplatz	30	y	1,6
S. baguirmi	30	y	e,n,x
S. bietri	30	y	1,5
S. bodjonegoro	30	z_4,z_{24}	—
S. kumasi	30	z_{10}	e,n,z_{15}
S. aragua	30	z_{29}	—
S. ago	30	z_{38}	—
**Salmonella*	30	z_{39}	1,7
Group O (O 35)			
S. umhlatazana	35	a	e,n,z_{15}
S. tchad	35	b	—
S. yolo	35	c	—
S. dembe	35	d	l,w
S. gassi	35	e,h	z_6
S. adelaide	35	f,g	—
S. ealing	35	g,m,s	—
**Salmonella*	35	g,m,s,t	—
S. ebrie	35	g,m,t	—
S. anecho	35	g,s,t	—
S. agodi	35	g,t	—
†*Salmonella arizonae* (Ar.20:13,14:-)	35	g,z_{51}	—
S. monschaui	35	m,t	—
†*Salmonella arizonae* (Ar.20:33:28)	35	i	e,n,x,z_{15}
S. gambia	35	i	e,n,z_{15}
S. bandia	35	i	l,w
†*Salmonella arizonae* (Ar.20:33:31)	35	i	z
†*Salmonella arizonae* (Ar.20:29:31)	35	k	z
†*Salmonella arizonae* (Ar.20:29:25)	35	k	z_{53}
†*Salmonella arizonae* (Ar.20:22:31)	35	(k)	z
†*Salmonella arizonae* (Ar.20:23:30)	35	l,v	1,5,7
†*Salmonella arizonae* (Ar.20:23:21)	35	l,v	z_{35}
**Salmonella*	35	l,z_{28}	—
†*Salmonella arizonae* (Ar.20:24:28)	35	r	e,n,x,z_{15}
S. massakory	35	r	l,w
†*Salmonella arizonae* (Ar.20:24:21)	35	r	z_{35}
†*Salmonella arizonae* (Ar.20:24:41)	35	r	z_{58}
S. alachua	35	z_4,z_{23}	—
†*Salmonella arizonae* (Ar.20:1,2,6:-)	35	z_4,z_{23}	—
S. westphalia	35	z_4,z_{24}	—
S. camberene	35	z_{10}	1,5
S. enschede	35	z_{10}	l,w
S. ligna	35	z_{10}	z_6
†*Salmonella arizonae* (Ar.20:27:21)	35	z_{10}	z_{35}
†*Salmonella arizonae* (Ar.20:16,17,18:-)	35	z_{29}	—
**S. utbremen*	35	z_{29}	e,n,x
†*Salmonella arizonae* (Ar.20:26:30)	35	z_{52}	1,5,7
†*Salmonella arizonae* (Ar.20:26:28)	35	z_{52}	e,n,x,z_{15}
†*Salmonella arizonae* (Ar.20:26:21)	35	z_{52}	z_{35}
Group P (O 38)			
S. sheffield	38	c	1,5
S. kidderminster	38	c	1,6
**S. carletonville*	38	d	1,5
S. thiaroye	38	e,h	1,2
S. kasenyi	38	e,h	1,5
S. korovi	38	g,m,s	—
**S. foulpointe*	38	g,t	—
†*Salmonella arizonae* (Ar.16:13,14:-)	38	g,z_{51}	—
S. mgulani	38	i	1,2
S. lansing	38	i	1,5
†*Salmonella arizonae* (Ar.16:33:25)	38	i	z_{53}
S. njala	38	k	e,n,x
S. echa	38	k	1,2
S. inverness	38	k	1,6
†*Salmonella arizonae* (Ar.16:29:31)	38	k	z
†*Salmonella arizonae* (Ar.16:22:30)	38	(k)	1,5,7
†*Salmonella arizonae* (Ar.16:22:31)	38	(k)	z
†*Salmonella arizonae* (Ar.16:22:21)	38	(k)	z_{35}
†*Salmonella arizonae* (Ar.16:22:34)	38	(k)	z_{54}
†*Salmonella arizonae* (Ar.16:22:35)	38	(k)	z_{55}
S. alger	38	l,v	1,2
S. kimberley	38	l,v	1,5
S. roan	38	l,v	e,n,x
†*Salmonella arizonae* (Ar.16:23:31)	38	l,v	z
†*Salmonella arizonae* (Ar.16:23:21)	38	l,v	z_{35}

TABLE 8.6—*Continued*

Species or serotype	Antigen O	Antigen H Phase 1	Antigen H Phase 2
†*Salmonella arizonae* (Ar.16:23:25)	38	l,v	z_{53}
†*Salmonella arizonae* (Ar.16:23:34)	38	l,v	z_{54}
S. lindi	38	r	1,5
S. emmastad	38	r	1,6
†*Salmonella arizonae* (Ar.16:24:31)	38	r	z
†*Salmonella arizonae* (Ar.16:24:31:40a, 40b)	38	r	z:z_{57}
†*Salmonella arizonae* (Ar.16:24:21)	38	r	z_{35}
S. freetown	38	y	1,5
S. colombo	38	y	1,6
S. perth	38	y	e,n,x
‡*Salmonella*	38	z_4,z_{23}	—
S. yoff	38	z_4,z_{23}	1,2
†*Salmonella arizonae* †(Ar.16:27:31)	38	z_{10}	z
†*Salmonella arizonae* (Ar.16:27:25)	38	z_{10}	z_{53}
Salmonella arizonae (Ar.16:39:25)	38	z_{47}	z_{53}
Group Q (O 39)			
S. wandsworth	39	b	1,2
S. logone	39	d	1,5
S. mara	39	e,h	1,5
S. hofit	39	i	1,5
S. champaign	39	k	1,5
S. kokomlemle	39	l,v	e,n,x
**S. mondeor*	39	l,z_{28}	e,n,x
S. anfo	39	y	1,2
S. windermere	39	y	1,5
Group R (O 40)			
S. shikmonah	40	a	1,5
S. greiz	40	a	z_6
**S. springs*	40	a	z_{39}
**Salmonella*	40	b	—
S. riogrande	40	b	1,5
S. johannesburg	*1*,40	b	e,n,x
S. duval	*1*,40	b	e,n,z_{15}
S. benguella	40	b	z_6
**S. suarez*	*1*,40	c	e,n,x,z_{15}
**S. ottershaw*	40	d	—
S. driffield	*1*,40	d	1,5
S. tilene	*1*,40	e,h	1,2
**Salmonella*	*1*,40	(f),g	e,n,x,z_{15}
**S. boksburg*	40	g,m,s,t	e,n,x
**S. alsterdorf*	*1*,40	g,m,t	—
†*Salmonella arizonae* (Ar.10a,10b:13, 14:-)	40	g,z_{51}	—
‡*S. seminole*	*1*,40	g,z_{51}	—
**Salmonella*	*1*,40	m,t	z_{42}
†*Salmonella arizonae* (Ar.10a,10b:33:30)	40	i	1,5,7
S. goulfey	*1*,40	k	1,5
S. allandale	*1*,40	k	1,6
S. hann	40	k	e,n,x
**S. sunnydale*	*1*,40	k	e,n,x,z_{15}
†*Salmonella arizonae* (Ar.10a,10b:29:31: 40a,40c)	40	k	z:z_{57}
S. millesi	*1*,40	l,v	1,2
†*Salmonella arizonae* (Ar.10a,10b:23:25)	40	l,v	z_{53}
S. bukavu	*1*,40	l,z_{28}	1,5
S. santhiaba	40	l,z_{28}	1,6
**S. bulawayo*	*1*,40	z	1,5
S. nowawes	40	z	z_6
†*Salmonella arizonae* (Ar.10a,10b:1,2, 6:-)	40	z_4,z_{23}	
‡*S. sachsenwald*	*1*,40	z_4,z_{23}	—
†*Salmonella arizonae* (Ar.10a,10b:1,3, 11:-)	40	z_4,z_{24}	—
‡*Salmonella*	40	z_4,z_{24}	—
**S. degania*	40	z_4,z_{24}	z_{39}
‡*S. bern*	*1*,40	z_4,z_{32}	—
†*Salmonella arizonae* (Ar.10a,10b:1,7, 8:-)	40	z_4,z_{32}	—
‡*Salmonella*	40	z_4,z_{32}	—
**Salmonella*	*1*,40	z_6	1,5
S. trotha	40	z_{10}	z_6
†*Salmonella arizonae* (Ar.10a,10c:27:21)	40	z_{10}	z_{35}
S. omifisan	40	z_{29}	—
†*Salmonella arizonae* (Ar.10a,10b:16,17, 18:-)	40	z_{29}	—
**S. fandran*	*1*,40	z_{35}	e,n,x,z_{15}
†*Salmonella arizonae* (Ar.10a,10b:17, 20:-)	40	z_{36}	—
**S. grunty*	*1*,40	z_{39}	1,6
S. karamoja	*1*,40	z_{41}	1,2
**Salmonella*	40	—	1,7
Group S (O 41)			
S. vietnam	41	b	
**Salmonella*	41	b	
S. egusi	41	d	[1,5]
**S. hennepin*	41	d	z_6
**S. lethe*	41	g,t	—
†*Salmonella arizonae* (Ar.13:13,14:-)	41	g,z_{51}	—
**Salmonella*	41	k	—
†*Salmonella arizonae* (Ar.13:22:-)	41	(k)	—
**S. dubrovnik*	41	z	1,5
‡*Salmonella*	41	z_4,z_{23}	—
S. waycross	41	z_4,z_{23}	—
†*Salmonella arizonae* (Ar.13:1,2,5:-) (Ar.13:1,2,6:-)	41	z_4,z_{23}	—
S. ipswich	41	z_4,z_{24}	—
†*Salmonella arizonae* (Ar.13:1,3,11:-)	41	z_4,z_{24}	—

TABLE 8.6—*Continued*

Species or serotype	Antigen O	Antigen H Phase 1	Antigen H Phase 2
†*Salmonella arizonae*	41	z_4,z_{32}	—
(Ar.13:1,6,7:-)			
(Ar.13:1,7,8:-)			
**S. negev*	41	z_{10}	1,2
S. leipzig	41	z_{10}	1,5
S. landala	41	z_{10}	1,6
S. inpraw	41	z_{10}	e,n,x
**S. lurup*	41	z_{10}	[e,n,x,z_{15}]
**S. lichtenberg*	41	z_{10}	z_6
†*Salmonella arizonae*	41	z_{29}	—
(Ar.13:16,17,18:-)			
†*Salmonella arizonae*	41	z_{36}	—
(Ar.13:17,20:-)			
S. offa	41	z_{38}	—
Group T (O 42)			
S. faji	1,42	a	e,n,z_{15}
**S. chinovum*	42	b	1,5
**S. uphill*	42	b	e,n,x,z_{15}
S. egusitoo	1,42	b	z_6
S. kampala	1,42	c	z_6
**S. fremantle*	42	(f),g,t	—
†*Salmonella arizonae*	42	g,z_{51}	—
(Ar.15:13,14:-)			
S. maricopa	1,42	g,z_{51}	1,5
S. waral	1,42	m,t	—
**Salmonella*	42	m,t	e,n,x,z_{15}
S. kaneshie	1,42	i	l,w
S. middlesbrough	1,42	i	z_6
†*Salmonella arizonae*	42	k	—
(Ar.15:29:-)			
S. haferbreite	42	k	1,6
†*Salmonella arizonae*	42	(k)	z_{35}
(Ar.15:22:21)			
†*Salmonella arizonae*	42	l,v	1,5,7
(Ar.15:23:30)			
**S. portbech*	42	l,v	e,n,x,z_{15}
S. sipane	1,42	r	e,n,z_{15}
**S. nairobi*	42	r	—
†*Salmonella arizonae*	42	r	z
(Ar.15:24:31)			
†*Salmonella arizonae*	42	r	z_{50}
(Ar.15:24:42)			
†*Salmonella arizonae*	42	r	z_{53}
(Ar.15:24:25)			
S. harvestehude	1,42	y	z_6
**S. detroit*	42	z	1,5
S. ursenbach	42	z	1,6
**S. rand*	42	z	e,n,x,z_{15}
**S. nuernberg*	42	z	z_6
†*Salmonella arizonae*	42	z_4,z_{23}	—
(Ar.15:1,2,6:-)			
S. gera	1,42	z_4,z_{23}	1,6
S. toricada	1,42	z_4,z_{24}	—
†*Salmonella arizonae*	42	z_4,z_{24}	—
(Ar.15:1,3,11:-)			
†*Salmonella arizonae*	42	z_{10}	e,n,x,z_{15}
(Ar.15:27:28)			
†*Salmonella arizonae*	42	z_{10}	z
(Ar.15:27:31)			
S. loenga	1,42	z_{10}	z_6
S. kahla	1,42	z_{35}	1,6
S. weslaco	42	z_{36}	—
**Salmonella*	42	—	1,6
S. taset	1,42	z_{41}	—
Group U (O 43)			
S. graz	43	a	1,2
S. berkeley	43	a	1,5
**S. kommetje*	43	b	z_{42}
**Salmonella*	43	e,n,x,z_{15}	1,(5),7
**Salmonella*	43	e,n,x,z_{15}	1,6
S. milwaukee	43	f,g	—
**Salmonella*	43	f,g,t	1,5
**S. mosselbay*	43	g,m,s,t	z_{42}
**S. veddel*	43	g,t	—
S. mbao	43	i	1,2
S. ahuza	43	k	1,5
†*Salmonella arizonae*	43	k	z
(Ar.21:29:31)			
†*Salmonella arizonae*	43	l,v	z_{53}
(Ar.21:23:25)			
†*Salmonella arizonae*	43	r	e,n,x,z_{15}
(Ar.21:24:28)			
S. farcha	43	y	1,2
S. kingabwa	43	y	1,5
**Salmonella*	43	z	1,5
‡*S. houten*	43	z_4,z_{23}	—
(*S. houtenae*)			
†*Salmonella arizonae*	43	z_4,z_{24}	—
(Ar.21:1,3,11:-)			
‡*S. tuindorp*	43	z_4,z_{32}	—
‡*Salmonella*	43	z_{29}	—
**Salmonella*	43	z_{29}	z_{42}
S. ahepe	43	z_{35}	1,6
†*Salmonella arizonae*	43	z_{36}	—
(Ar.21:17,20:-)			
‡*S. volksdorf*	43	z_{36},z_{38}	—
S. irigny	43	z_{38}	—
**S. bunnik*	43	z_{42}	1,5,7
Group V (O 44)			
S. niakhar	44	a	1,5
S. niarembe	44	a	l,w
S. sedgwick	44	b	e,n,z_{15}
S. madigan	44	c	1,5
S. bobo	44	d	1,5
S. fischerstrasse	44	d	e,n,z_{15}
S. vleuten	44	f,g	—
S. gamaba	44	g,m,s	—
‡*Salmonella*	44	g,z_{51}	—
S. muguga	44	m,t	—
S. lawra	44	k	e,n,z_{15}
S. uhlenhorst	44	z	l,w
S. kua	44	z_4,z_{23}	—
**Salmonella*	44	z_4,z_{23}	—
†*Salmonella arizonae*	44	z_4,z_{23}	—
(Ar.13:1,2,5:-)			
(Ar.13:1,2,6:-)			
S. christiansborg	44	z_4,z_{24}	—
‡*Salmonella*	44	z_4,z_{24}	—
†*Salmonella arizonae*	44	z_4,z_{32}	—
(Ar.13:1,6,7,9:-)			
(Ar.13:1,7,8:-)			
(Ar.13:1,2,10:-)			

TABLE 8.6—*Continued*

Species or serotype	Antigen		
	O	H	
		Phase 1	Phase 2
‡*S. lohbruegge*	44	z_4, z_{32}	—
S. guinea	44	z_{10}	[1,7]
‡*Salmonella*	44	z_{36}, [z_{38}]	—
S. koketime	44	z_{38}	—
**S. clovelly*	1,44	z_{39}	e,n,x,z_{15}
Group W (O 45)			
**S. vrindaban*	45	a	e,n,x,z_{15}
**S. ejeda*	45	a	z_{10}
S. riverside	45	b	1,5
S. deversoir	45	c	e,n,x
S. dugbe	45	d	1,6
S. karachi	45	d	e,n,x
S. suelldorf	45	f,g	—
S. tornow	45	g,m	—
**S. bremen*	45	g,m,s,t	e,n,x
**S. windhoek*	45	g,t	1,5
†*Salmonella arizonae* (Ar.11:13,14:-)	45	g,z_{51}	—
‡*Salmonella*	45	g,z_{51}	—
S. apapa	45	m,t	—
**S. perinet*	45	m,t	e,n,x,z_{15}
S. casablanca	45	k	1,7
S. cairns	45	k	e,n,z_{15}
**S. klapmuts*	45	z	z_{39}
‡*Salmonella*	45	z_4, z_{23}	—
S. jodhpur	45	z_{29}	—
†*Salmonella arizonae* (Ar.11:16,17,18:-)	45	z_{29}	—
S. lattenkamp	45	z_{35}	1,5
Group X (O 47)			
**S. bilthoven*	47	a	1,5
S. saka	47	b	—
**S. phoenix*	47	b	1,5
**S. khami*	47	b	[e,n,x,z_{15}]
†*Salmonella arizonae* (Ar.23:32:28) (Ar.28:32:28:[40_a, 40_c])	47	c	e,n,x,z_{15}: [z_{57}]
†*Salmonella arizonae* (Ar.28:32:31)	47	c	z
S. stellingen	47	d	e,n,x
**S. quimbamba*	47	d	z_{39}
S. sljeme	1,47	f,g	—
S. luke	1,47	g,m	—
S. mesbit	47	m,t	e,n,z_{15}
†*Salmonella arizonae* (Ar.23:33:28:[42])	47	i	e,n,x,z_{15}: [z_{50}]
S. bergen	47	i	e,n,z_{15}
†*Salmonella arizonae* (Ar.28:33:31)	47	i	z
†*Salmonella arizonae* (Ar.23:33:21) (Ar.28:33:21)	47	i	z_{35}
†*Salmonella arizonae* (Ar.23:33:25) (Ar.28:33:25:[40_a, 40_c]	47	i	z_{53}: [z_{57}]
S. bootle	47	k	1,5
S. dahomey	47	k	1,6
†*Salmonella arizonae* (Ar.28:29:28)	47	k	e,n,x,z_{15}
S. lyon	47	k	e,n,z_{15}
†*Salmonella arizonae* (Ar.28:29:31)	47	k	z
†*Salmonella arizonae* (Ar.23:23:30:42)	47	l,v	1,5,7: [z_{50}]
†*Salmonella arizonae* (Ar.28:23:28)	47	l,v	e,n,x,z_{15}
†*Salmonella arizonae* (Ar.28:23:21)	47	l,v	z_{35}
†*Salmonella arizonae* (Ar.28:23:25)	47	l,v	z_{53}
S. teshie	1,47	l,z_{13},z_{28}	e,n,z_{15}
†*Salmonella arizonae* (Ar.23:24:31)	47	r	z
†*Salmonella arizonae* (Ar.23:24:25)	47	r	z_{53}
**Salmonella*	47	r	z_{53}
S. moualine	47	y	1,6
S. mount-pleasant	47	z	1,5
S. kaolack	47	z	1,6
**S. chersina*	47	z	z_6
S. bere	47	z_4, z_{23}	z_6
**Salmonella*	47	z_6	1,6
†*Salmonella arizonae* (Ar.28:27:30)	47	z_{10}	1,5,7
†*Salmonella arizonae* (Ar.28:27:31)	47	z_{10}	z
†*Salmonella arizonae* (Ar.28:16,17,18:-)	47	z_{29}	—
S. alexanderplatz	47	z_{38}	—
S. quinhon	47	z_{44}	—
†*Salmonella arizonae* (Ar.28:26:30)	47	z_{52}	1,5,7
†*Salmonella arizonae* (Ar.28:26:28)	47	z_{52}	e,n,x,z_{15}
†*Salmonella arizonae* (Ar.28:26:31)	47	z_{52}	z
†*Salmonella arizonae* (Ar.28:26:21)	47	z_{52}	z_{35}
Group Y (O 48)			
S. hisingen	48	a	1,5,7
**Salmonella*	48	a	z_6
†*Salmonella arizonae* (Ar.5:35:21)	48	a	z_{35}
**S. hagenbeck*	48	d	z_6
S. fitzroy	48	e,h	1,5
**S. hammonia*	48	e,n,x,z_{15}	z_6
**S. erlangen*	48	g,m,t	—
†*Salmonella arizonae* (Ar.5:13,14:-)	48	g,z_{51}	—
‡*S. marina*	48	g,z_{51}	—
**S. sydney*	48	i	z
†*Salmonella arizonae* (Ar.5:33:25)	48	i	z_{53}
†*Salmonella arizonae* (Ar.5:29:30)	48	k	1,5,7
†*Salmonella arizonae* (Ar.5:29:28)	48	k	e,n,x,z_{15}
S. dahlem	48	k	e,n,z_{15}
†*Salmonella arizonae* (Ar.5:29:21)	48	k	z_{35}

TABLE 8.6—*Continued*

Species or serotype	Antigen		
	O	H	
		Phase 1	Phase 2
**S. sakaraha*	48	k	z_{39}
†*Salmonella arizonae* (Ar.5:29:25)	48	k	z_{53}
†*Salmonella arizonae* (Ar.5:22:25)	48	(k)	z_{53}
†*Salmonella arizonae* (Ar.5:23:30:[39])	48	l,v	1,5,7:[z_{47}]
†*Salmonella arizonae* (Ar.5:24:28)	48	r	e,n,x,z_{15}
S. toucra	48	z	1,5:[z_{58}]
†*Salmonella arizonae* (Ar.5:1,2,5:-) (Ar.5:1,2,5,6:-) (Ar.5:1,6:-)	48	z_4,z_{23}	—
S. djakarta	48	z_4,z_{24}	—
†*Salmonella arizonae* (Ar.5:1,3,11:-)	48	z_4,z_{24}	—
†*Salmonella arizonae* (Ar.5:1,6,7:-) (Ar.5:1,7,8:-)	48	z_4,z_{32}	—
‡*Salmonella*	48	z_4,z_{32}	—
**S. ngozi*	48	z_{10}	[1,5]
†*Salmonella arizonae* (Ar.5:27:28)	48	z_{10}	e,n,x,z_{15}
†*Salmonella arizonae* (Ar.5:16,17,18:-)	48	z_{29}	—
S. bongor	48	z_{35}	—
†*Salmonella arizonae* (Ar.5:17,20:-)	48	z_{36}	—
†*Salmonella arizonae* (Ar.5:26:31)	48	z_{52}	z
Group Z (O 50)			
**Salmonella*	50	b	z_6
S. rochdale	50	b	e,n,x
**S. krugersdorp*	50	e,n,x	1,7
**S. namib*	50	g,m,s,t	1,5
†*Salmonella arizonae* (Ar.9a,9b:13,14:-)	50	g,z_{51}	—
‡*S. wassenaar*	50	g,z_{51}	—
**S. atra*	50	m,t	z_6:z_{42}
†*Salmonella arizonae* (Ar.9a,9c:33:30)	50	i	1,5,7
†*Salmonella arizonae* (Ar.9a,9c:33:28)	50	i	e,n,x,z_{15}
†*Salmonella arizonae* (Ar.9a,9c:33:31)	50	i	z
†*Salmonella arizonae* (Ar.9a,9c:29:30)	50	k	1,5,7
†*Salmonella arizonae* (Ar.9a,9c:29:28)	50	k	e,n,x,z_{15}
†*Salmonella arizonae* (Ar.9a,9b:29:31) (Ar.9a,9c:29:31)	50	k	z
†*Salmonella arizonae* (Ar.9a,9b:22:31)	50	(k)	z
**S. seaforth*	50	k	z_6
†*Salmonella arizonae* (Ar.9a,9b:29:21)	50	k	z_{35}
†*Salmonella arizonae* (Ar.9a,9b:22:21)	50	(k)	z_{35}
†*Salmonella arizonae* (Ar.9a,9c:29:25)	50	k	z_{53}
†*Salmonella arizonae* (Ar.9a,9b:23:28)	50	l,v	e,n,x,z_{15}
†*Salmonella arizonae* (Ar.9a,9c:23:31)	50	l,v	z
†*Salmonella arizonae* (Ar.9a,9c:23:21)	50	l,v	z_{35}
**Salmonella*	50	l,w	e,n,x,z_{15}: z_{42}
**Salmonella*	50	l,z_{28}	z_{42}
†*Salmonella arizonae* (Ar.9a,9b:24:30)	50	r	1,5,7
†*Salmonella arizonae* (Ar.9a,9b:24:31)	50	r	z
†*Salmonella arizonae* (Ar.9a,9b:24:21)	50	r	z_{35}
†*Salmonella arizonae* (Ar.9a,9b:24:25)	50	r	z_{53}
S. dougi	50	y	1,6
**S. greenside*	50	z	e,n,x
†*Salmonella arizonae* (Ar.9a,9b:1,2,5:-) (Ar.9a,9b:1,2,6:-)	50	z_4,z_{23}	—
‡*S. flint*	50	z_4,z_{23}	—
‡*Salmonella*	50	z_4,z_{24}	—
†*Salmonella arizonae* (Ar.9a,9b:1,3,11:-)	50	z_4,z_{24}	—
†*Salmonella arizonae* (Ar.9a,9b:1,6,7:-) (Ar.9a,9b:1,7,8:-)	50	z_4,z_{32}	—
‡*S. bonaire*	50	z_4,z_{32}	—
†*Salmonella arizonae* (Ar.9a,9c:27:31)	50	z_{10}	z
**S. hooggraven*	50	z_{10}	z_6:z_{42}
†*Salmonella arizonae* (Ar.9a,9b:16,17, 18:-)	50	z_{29}	—
†*Salmonella arizonae* (Ar.9a,9b:17,20:-)	50	z_{36}	—
**S. faure*	50	z_{42}	1,7
†*Salmonella arizonae* (Ar.9a,9b:26:30)	50	z_{52}	1,5,7
†*Salmonella arizonae* (Ar.9a,9b:26:31) (Ar.9a,9c:26:31)	50	z_{52}	z
†*Salmonella arizonae* (Ar.9a,9b:26:21) (Ar.9a,9c:26:21)	50	z_{52}	z_{35}
†*Salmonella arizonae* (Ar.9a,9b:26:25) (Ar.9a,9c:26:25)	50	z_{52}	z_{53}
Group 51			
S. tione	51	a	e,n,x
S. gokul	1,51	d	[1,5]
S. meskin	51	e,h	1,2
†*Salmonella arizonae* (Ar.1,2:13,14:-)	51	g,z_{51}	—
S. kabete	51	i	1,5
†*Salmonella arizonae* (Ar.1,2:33:21)	51	i	z_{35}
S. dan	51	k	e,n,z_{15}
S. overschie	51	l,v	1,5

TABLE 8.6—*Continued*

Species or serotype	Antigen		
	O	H	
		Phase 1	Phase 2
†*Salmonella arizonae*	51	l, v	z
(Ar.1,2:23:31)			
**S. askraal*	51	l, z_{28}	—
S. antsalova	51	z	1,5
S. treforest	*1*,51	z	1,6
†*Salmonella arizonae*	51	z_4, z_{23}	—
(Ar.1,2:1,2,5:-)			
‡*S. harmelen*	51	z_4, z_{23}	—
†*Salmonella arizonae*	51	z_4, z_{24}	—
(Ar.1,2:1,3,11:-)			
**Salmonella*	51	z_{29}	e, n, x, z_{15}
†*Salmonella arizonae*	51	z_{52}	z
(Ar.1,2:26:31)			
**S. roggeveld*	51	—	1,7
Group 52			
S. flottbek	52	b	—
S. utrecht	52	d	1,5
**Salmonella*	52	d	e, n, x, z_{15}
S. butare	52	e, h	1,6
S. sainte-marie	52	g, t	—
†*Salmonella arizonae*	52	k	z_{35}
(Ar.31:29:21)			
†*Salmonella arizonae*	52	l, v	z_{53}
(Ar.31:23:25)			
**Solmonella*	52	z_{44}	1,5,7
Group 53			
†*Salmonella arizonae*	53	g, z_{51}	—
(Ar.1,4:13,14:-)			
†*Salmonella arizonae*	53	i	z
(Ar.1,4:33:31)			
†*Salmonella arizonae*	53	k	e, n, x, z_{15}
(Ar.1,4:29:28)			
†*Salmonella arizonae*	53	k	z
(Ar.1,4:29:31)			
†*Salmonella arizonae*	53	(k)	z_{35}
(Ar.1,4:22:21)			
†*Salmonella arizonae*	53	l, v	e, n, x, z_{15}
(Ar.1,4:23:28)			
**S. midhurst*	53	l, z_{28}	z_{39}
†*Salmonella arizonae*	53	r	z
(Ar.1,4:24:31)			
†*Salmonella arizonae*	53	r	z_{35}
(Ar.1,4:24:21)			
†*Salmonella arizonae*	53	r	z_{56}
(Ar.1,4:24:38)			
†*Salmonella arizonae*	53	z	1,5,7
(Ar.1,4:31:30)			
**Salmonella*	53	z	z_6
†*Salmonella arizonae*	53	z_4, z_{23}	—
(Ar.1,4:1,2,5:-)			
(Ar.1,4:1,2,6:-)			
**S. humber*	53	z_4, z_{24}	—
†*Salmonella arizonae*	53	z_4, z_{24}	—
(Ar.1,4:1,3,11:-)			
†*Salmonella arizonae*	53	z_4, z_{32}	—
(Ar.1,4:1,6,7:-)			
(Ar.1,4:1,6,7,9:-)			
†*Salmonella arizonae*	53	z_{10}	z_{35}
(Ar.1,4:27:21)			
†*Salmonella arizonae*	53	z_{29}	—
(Ar.1,4:16,17,18:-)			
‡*S. bockenheim*	1,53	z_{36}, z_{38}	—
†*Salmonella arizonae*	53	z_{52}	z_{35}
(Ar.1,4:26:21)			
†*Salmonella arizonae*	53	z_{52}	z_{53}
(Ar.1,4:26:25)			
Group 54			
S. tonev	54	b	e, n, x
S. rossleben	54	e, h	1,6
S. uccle	54	g, s, t	—
S. poeseldorf	54	i	z_6
S. ochsenwerder	54	k	1,5
S. steinwerder	(3), (15), 54	y	1,5
Group 55			
**S. tranoroa*	55	k	z_{39}
Group 56			
**S. artis*	56	b	—
**Salmonella*	56	d	—
**Salmonella*	56	e, n, x	1,7
†*Salmonella arizonae*	56	z_4, z_{23}	—
(Ar.14:1,2,5:-)			
(Ar.14:1,2,6:-)			
†*Salmonella arizonae*	56	z_4, z_{32}	—
(Ar.14:1,6,7,9:-)			
Group 57			
**Salmonella*	57	g, m, s, t	z_{42}
†*Salmonella arizonae*	57	i	e, n, x, z_{15}
(Ar.34:33:28)			
†*Salmonella arizonae*	57	i	z
(Ar.34:33:31)			
**S. locarno*	57	z_{29}	z_{42}
**S. manombo*	57	z_{39}	e, n, x, z_{15}
**S. tokai*	57	z_{42}	1,6:z_{53}
Group 58			
**Salmonella*	58	a	—
†*Salmonella arizonae*	58	l, v	e, n, x, z_{15}
(Ar.1,33:23:28)			
†*Salmonella arizonae*	58	l, v	z_{35}
(Ar.1,33:23:21)			
**S. basel*	58	l, z_{13}, z_{28}	1,5
†*Salmonella arizonae*	58	r	e, n, x, z_{15}
(Ar.1,33:24:28)			
†*Salmonella arizonae*	58	r	z_{53}:[z_{47}]:[z_{57}]
(Ar.1,33:24:25:			
[39]:[40_a, 40_c])			
†*Salmonella arizonae*	58	z_{52}	z
(Ar.1,33:26:31)			
†*Salmonella arizonae*	58	z_{52}	z_{35}
(Ar.1,33:26:21)			
Group 59			
†*Salmonella arizonae*	59	c	e, n, x, z_{15}
(Ar.19:32:28)			
†*Salmonella arizonae*	59	i	z
(Ar.19:33:31)			
†*Salmonella arizonae*	59	i	z_{35}
(Ar.19:33:21)			
**S. betioky*	59	k	(z)
†*Salmonella arizonae*	59	k	z_{53}
(Ar.19:29:25)			

TABLE 8.6—*Continued*

Species or serotype	Antigen O	Antigen H Phase 1	Antigen H Phase 2
†*Salmonella arizonae* (Ar.19:22:28)	59	(k)	e,n,x,z_{15}
†*Salmonella arizonae* (Ar.19:22:31)	59	(k)	z
†*Salmonella arizonae* (Ar.19:22:21)	59	(k)	z_{35}
†*Salmonella arizonae* (Ar.19:1,2,5,:-) (Ar.19:1,2,6:-)	59	z_4,z_{23}	—
†*Salmonella arizonae* (Ar.19:27:25)	59	z_{10}	z_{53}
†*Salmonella arizonae* (Ar.19:27:40a,40c)	59	z_{10}	z_{57}
†*Salmonella arizonae* (Ar.19:16,17,18:-)	59	z_{29}	—
†*Salmonella arizonae* (Ar.19:17,20:-)	59	z_{36}	—
†*Salmonella arizonae* (Ar.19:26:-)	59	z_{52}	—
Group 60			
**S. setubal*	60	g,m,t	z_6
†*Salmonella arizonae* (Ar.24:33:21)	60	i	z_{35}
†*Salmonella arizonae* (Ar.24:29:28)	60	k	e,n,x,z_{15}
†*Salmonella arizonae* (Ar.24:29:31)	60	k	z
†*Salmonella arizonae* (Ar.24:29:21)	60	k	z_{35}
†*Salmonella arizonae* (Ar.24:22:25)	60	(k)	z_{53}
†*Salmonella arizonae* (Ar.24:23:31)	60	l,v	z
†*Salmonella arizonae* (Ar.24:24:28)	60	r	e,n,x,z_{15}
†*Salmonella arizonae* (Ar.24:24:31)	60	r	z
†*Salmonella arizonae* (Ar.24:24:25)	60	r	z_{53}
**S. luton*	60	z	e,n,x
†*Salmonella arizonae* (Ar.24:27:31)	60	z_{10}	z
†*Salmonella arizonae* (Ar.24:26:21)	60	z_{52}	z_{35}
†*Salmonella arizonae* (Ar.24:26:25)	60	z_{52}	z_{53}
Group 61			
†*Salmonella arizonae* (Ar.26:32:30)	61	c	1,5,7
†*Salmonella arizonae* (Ar.26:32:21)	61	c	z_{35}
†*Salmonella arizonae* (Ar.26:33:28)	61	i	e,n,x,z_{15}
**S. eilbek*	61	i	z
†*Salmonella arizonae* (Ar.26:33:31)	61	i	z
†*Salmonella arizonae* (Ar.26:33:21)	61	i	z_{35}
†*Salmonella arizonae* (Ar.26:33:25)	61	i	z_{53}
†*Salmonella arizonae* (Ar.26:29:30)	61	k	1,5,7
†*Salmonella arizonae* (Ar.26:22:25)	61	(k)	z_{53}
†*Salmonella arizonae* (Ar.26:23:30:[40_a, 40_b])	61	l,v	1,5,7:[z_{57}]
†*Salmonella arizonae* (Ar.26:23:31)	61	l,v	z
†*Salmonella arizonae* (Ar.26:23:21)	61	l,v	z_{35}
†*Salmonella arizonae* (Ar.26:23:25)	61	l,v	z_{53}
†*Salmonella arizonae* (Ar.26:24:30)	61	r	1,5,7
†*Salmonella arizonae* (Ar.26:24:25)	61	r	z_{53}
†*Salmonella arizonae* (Ar.26:26:25)	61	z_{52}	z_{53}
Group 62			
†*Salmonella arizonae* (Ar.6:13,14:-)	62	g,z_{51}	—
†*Salmonella arizonae* (Ar.6:1,2,5:-)	62	z_4,z_{23}	—
†*Salmonella arizonae* (Ar.6:1,7,8:-)	62	z_4,z_{32}	—
Group 63			
†*Salmonella arizonae* (Ar.8:13,14:-)	63	g,z_{51}	—
†*Salmonella arizonae* (Ar.8:1,7,8:-)	63	z_4,z_{32}	—
†*Salmonella arizonae* (Ar.8:17,20:-)	63	z_{36}	—
Group 64			
†*Salmonella arizonae* (Ar.29:32:31)	64	c	z
†*Salmonella arizonae* (Ar.29:33:31)	64	i	z
†*Salmonella arizonae* (Ar.(5),29:33:[21]:[40a,40b])	64	i	[z_{35}]:[z_{57}]
†*Salmonella arizonae* (Ar.(5),29:29:30)	64	k	1,5,7
**Salmonella*	64	k	e,n,x,z_{15}
†*Salmonella arizonae* (Ar.29:29:25)	64	k	z_{53}
†*Salmonella arizonae* (Ar.29:23:31)	64	l,v	z
†*Salmonella arizonae* (Ar.29:24:31) (Ar.(5),29:24:31)	64	r	z
†*Salmonella arizonae* (Ar.29:27:31)	64	z_{10}	z
**Salmonella*	64	z_{29}	—
†*Salmonella arizonae* (Ar.(5),29:17,20:-)	64	z_{36}	—
†*Salmonella arizonae* (Ar.29:26:28)	64	z_{52}	e,n,x,z_{15}

TABLE 8.6—*Continued*

Species or serotype	Antigen O	Antigen H Phase 1	Antigen H Phase 2
Group 65			
†*Salmonella arizonae* (Ar.30:32:25)	65	c	z_{53}
†*Salmonella arizonae* (Ar.30:22:31)	65	(k)	z
†*Salmonella arizonae* (Ar.30:22:21)	65	(k)	z_{35}
†*Salmonella arizonae* (Ar.30:22:25)	65	(k)	z_{53}
†*Salmonella arizonae* (Ar.30:23:30)	65	l, v	1,5,7
†*Salmonella arizonae* (Ar.30:23:31)	65	l, v	z
†*Salmonella arizonae* (Ar.30:23:25)	65	l, v	z_{53}
†*Salmonella arizonae* (Ar.30:27:28)	65	z_{10}	e, n, x, z_{15}
†*Salmonella arizonae* (Ar.30:26:31)	65	z_{52}	z
†*Salmonella arizonae* (Ar.30:26:21)	65	z_{52}	z_{35}
†*Salmonella arizonae* (Ar.30:26:25)	65	z_{52}	z_{53}

Genus V. **Shigella** *Castellani and Chalmers 1919, 936*

Non-motile rods, not encapsulated. Grow well on nutrient media and do not need special growth factors; growth inhibited by bismuth sulfite in Wilson and Blair's medium. **Cannot use citrate** or malonate **as sole C source. Growth inhibited by KCN.** H_2S is not produced.

Glucose and other carbohydrates are **fermented with the production of acid but not gas** (except for aerogenic biotypes of one serotype). Adonitol, inositol and salicin are not fermented.

Catalase is usually produced, but exceptions occur in one species.

Other biochemical characters are shown in Tables 8.2 and 8.7.

Type species: *Shigella dysenteriae* (Shiga) Castellani and Chalmers.

TABLE 8.7

Distinguishing characters of the **Shigella** *species*

	1. *S. dysenteriae*	2. *S. flexneri*	3. *S. boydii*	4. *S. sonnei*
Lactose (acid)	–	–	–	(+)
Mannitol	–	+	+	+
Sucrose	–	–	–	(+)
Dulcitol	d	–	d	–
Xylose	–	–	d	d
Indole	d	d	d	–
Ornithine	–	–	–	+
Arginine	–	–	–	d

Tribe II. **Klebsielleae**

TABLE 8.8

Distinguishing characteristics of the genera of the tribe **Klebsielleae**

	Klebsiella	*Enterobacter*	*Hafnia*	*Serratia*
Motility	–	+	+	+
Ornithine decarboxylase	–	+	+	+
Sorbitol	+	+	–	–
Red pigment	–	–	–	D
DNase	–	–	–	+
Arginine dihydrolase	–	D	–	–

Genus VI. **Klebsiella** *Trevisan 1885, 105 Nom. cons.* Opin. 13, Jud. Comm. 1954, 152

Non-motile, capsulated **rods,** 0.3–1.5 μm by 0.6–6.0 μm, arranged singly, in pairs or short chains. Grow on meat extract media producing more-or-less dome-shaped, glistening colonies of varying degrees of stickiness, depending on the strain and the composition of the medium.

No special growth requirements and most strains **can use citrate and glucose as sole C source,** and ammonia as N source. Glucose is fermented with the production of acid and gas (more CO_2 than H_2), but anaerogenic strains occur. Most strains produce 2,3-butanediol as a major end product of the fermentation of glucose and the V.P. reaction is usually positive; lactic, acetic and formic acids are formed in smaller amounts and ethanol in larger amounts than in a mixed acid fermentation. H_2S is not produced from TSI; gelatinase and indole are usually not produced.

Other characters are shown in Tables 8.2 and 8.8.

Optimal temperature for growth 35–37 C; optimal pH about 7.2.

G + C content of DNA: 52–56 moles % (T_m and buoyant density).

Klebsiella strains are resistant to penicillin in standard doses but may be sensitive to high concentrations; sensitivity to the following drugs may vary: ampicillin, cephalosporins, streptomycin, chloramphenicol, tetracycline, neomycin, kanamycin, polymyxin B, sulfonamides and nitrofurantoin. The proportion of resistant strains is increasing steadily, which may be due to mutations but in many cases is probably caused by transfer of plasmids carrying resistance determinants to a varying number of drugs.

Type species: *Klebsiella pneumoniae* (Schroeter) Trevisan.

TABLE 8.9

Differentiation within the genus **Klebsiella**

Character or test	1. *K. pneumoniae*	2. *K. ozaenae*	3. *K. rhinoscleromatis*
Gas from glucose	+	d	–
Lactose (acid)	+	(+)	–
Dulcitol (acid)	d		
Methyl red	–	+	+
Voges-Proskauer	+	–	–
Citrate utilization	+	d	–
Urease	+	d	–
Malonate	+	–	+
Lysine decarboxylase	+	d	–
Mucate	+	d	
Organic acids			
Citrate	d	d	–
d-Tartrate	d	d	–

Genus VII. **Enterobacter** *Hormaeche and Edwards 1960, 72 Nom. cons.* Opin. 28, Jud. Comm. 1963, 38

Motile rods, peritrichously flagellated; some strains encapsulated. **Citrate and acetate can be used as sole C source.**

At 37 C glucose is fermented with the production of acid and gas (CO_2:H::2:1). At 44.5 C gas is not produced from glucose. Dulcitol is usually not attacked. The Voges-Proskauer test is usually positive; the methyl red test is negative.

Alginate is not utilized; pectate is not decomposed.

Gelatin is liquefied slowly by most strains; H_2S is not produced on TSI agar.

Other characteristics are given in Tables 8.2, 8.8 and 8.10.

G + C content of DNA: 52–59 moles %.

Type species: *Enterobacter cloacae* (Jordan) Hormaeche and Edwards 1960, 72.

TABLE 8.10

Differentiation between **Enterobacter cloacae** *and* **E. aerogenes**

	1. *E. cloacae*	2. *E. aerogenes*
Gas produced from		
Glycerol	–	+
Inositol	–	d
Adonitol	d	+
Malonate as C source	d	+
Gelatin liquefaction	(+)[a]	((+))
Esculin hydrolysis	–	+
Lysine decarboxylase	–	+
Arginine dihydrolase	+	–

[a] (+) slow liquefaction; ((+)) very slow liquefaction.

Genus VIII. **Hafnia** *Møller 1954, 272*

Motile, peritrichously flagellated **rods;** not encapsulated. **Citrate and acetate can be used as sole C source.**

Glucose is fermented with the production of acid and gas. Acid is not produced from citrate, *d*-tartrate, and mucate. The M.R. test is usually negative at 37 C; the V.P. test is usually positive at 22 C, but may be negative at 37 C.

Alginate is not utilized; pectate is not decomposed. H_2S is not produced from TSI agar, but is produced weakly in peptone iron agar by many strains. Lipase is not produced; DNase is not produced.

Other characteristics are given in Tables 8.2 and 8.8.

G + C content of DNA is 52–57 moles %.

(See Sakazaki and Namioka (1957) for a review of the literature.)

Type species: *Hafnia alvei* Møller 1954, 272.

Genus IX. **Serratia** *Bizio 1823, 288*

Motile peritrichously flagellated **rods;** some strains are capsulated.

Citrate and acetate can be used as sole C source. **Many strains produce pink, red or magenta pigment.**

Glucose is fermented with or without the production of a small volume of gas; cellobiose, inositol and glycerol are fermented without gas production. The fermentation of xylose and adonitol are variable characters. The M.R. test is negative, the V.P. is usually positive.

Malonate and alginate are not utilized; pectate is not decomposed. Urea is usually not decomposed, but some strains attack it weakly. **DNase is produced.** For other characteristics see Tables 8.2 and 8.8.

When it is present, the characteristic feature of the genus is a non-diffusible pigment, prodigiosin. But pigmentation is a variable characteristic of strains and it is influenced by cultural conditions and medium; many strains fail to produce it.

G + C content of DNA: 53–59 moles %.

Type species: *Serratia marcescens* Bizio 1823, 288.

Tribe III. **Proteeae**

Genus X. **Proteus** *Hauser 1885, 12*

Usually straight rods, 0.4–0.6 by 1.0–3.0 μm; coccoid and irregular involution forms, filaments and spheroplasts are frequent under certain conditions. May occur in pairs or chains. Not encapsulated. Motile by peritrichous flagella; swimming motility most pronounced at 20 C and often absent at 37 C. Non-pigmented.

Acid is formed regularly and rapidly from glucose, and more slowly and less regularly from fructose, galactose and glycerol; the methyl red test is usually positive. Gas in small amounts may or may not be formed from glucose. With rare exceptions, acid is not formed from L-arabinose, dextrin, dulcitol, glycogen, inulin, lactose, melibiose, raffinose, sorbitol, sorbose or starch. Acid formation from other sugars varies (Table 8.12). Pectolytic and alginolytic activity not present.

Nitrates reduced to nitrites; Pichinoty's type A reductase present in all species but some strains of *P. inconstans* have type B reductase (Pichinoty *et al.*, 1966). Sulfite reductase not present.

Phenylalanine deaminated to phenylpyruvic acid. Indole formed except by one species. Glutamic acid decarboxylated; arginine not decarboxylated. Growth not inhibited by KCN in Møller's test.

Some species grow in a mineral medium with ammonium ion as sole N source and glucose as sole energy and C source; other species require growth factors (Table 8.11).

Temperature range for growth about 10–43 C.

The G + C content of the DNA ranges from 38–42 moles % (T_m) for four species and is about 50 for *P. morganii*.

Type species: *Proteus vulgaris* Hauser 1885, 12.

TABLE 8.11

Differential characteristics of species within the genus **Proteus**

	1. *P. vulgaris*	2. *P. mirabilis*	3. *P. morganii*	4. *P. rettgeri*	5. *P. inconstans*
Gas from glucose	+[a]	+	+	d[b]	d
Urease	+	+	+	+	−
Indole production	+	−[c]	+	+	+
Gelatin liquefaction	+	+	−	−	−
H_2S production	+	+	−	−	−
Ornithine decarboxylase	−	+	+	−	−
Growth requirement					
Nicotinic acid	+	+	+	−	−
Pantothenic acid	−	−	+	−	−
Moles % G + C	39.3	39.3	50.0	39.0	41.5
±	1.2	1.4	0.7	1.5	0.6

[a] + = more than 90% strains have this character.

[b] d = 89–11% positive.

[c] − = less than 10% of strains positive for this character.

TABLE 8.12

Fermentation reactions within the genus **Proteus** *(all known sugars giving differential reactions are included)*

	1. *P. vulgaris*	2. *P. mirabilis*	3. *P. morganii*	4. *P. rettgeri*	5. *P. inconstans* A	5. *P. inconstans* B
Adonitol	−	−	−	+	+	−
Amygdalin[a]	+	−	−	−	(d)[b]	(d)
Arabitol[a]	−	−	−	(+)	(d)?	(d)?
Arbutin[a]	d	−	−	+	−	−
Cellobiose	(d)	(d)	−	(d)	−	(d)
Erythritol	−	−	−	+	−	−
Esculin	d	−	−	+	−	−
Inositol	−	−	−	+	−	+
Maltose	+	−	−	−	−	−
Mannitol	−	−	−	+	−	−
Mannose[a]	−	−	+	+	+	+
Melezitose[a]	(+)	−	−	−	−	−
Rhamnose	−	−	−	d	−	−
Salicin	d	(d)	−	d	−	−
Sucrose	+	(+)	−	d	(d)	(+)
Trehalose	(+)	+	d	−	−	d
Xylose	d	+	−	d	−	−

[a] Results for these sugars are based on a small number of strains.

[b] Results in parentheses mean that the reaction is usually delayed, i.e. occurring later than 48 hr.

Tribe IV. **Yersinieae**

Genus XI. **Yersinia** *van Loghem 1944, 15*

Cells ovoid or rods, 0.5–1.0 by 1–2 μm; non-motile at 37 C; at temperatures below 37 C two species motile with peritrichous flagella. Not encapsulated.

Lactose usually not fermented; β-galactosidase produced. Fructose, glucose, glycerol, maltose, mannitol, mannose and trehalose fermented without the production of gas. Dulcitol, erythritol, fucose, glycogen, inositol, melezitose and raffinose not fermented. Methyl red test positive; Voges-Proskauer negative at 37 C for *Y. pestis* and *Y. pseudotuberculosis*.

Grows well on meat extract agar (1–2 days); cannot utilize citrate as sole source of C. Malonate not utilized. Tetrathionate not reduced.

Gelatin not hydrolyzed; nitrates reduced, except by one variety. Lysine not decarboxylated, phenylalanine not deaminated. Arginine dihydrolase reaction not present. Indole usually not produced by *Y. pestis* and *Y. pseudotuberculosis*.

Temperature range said to be between −2 C and 45 C; optimal 30–37 C.

The G + C content of the DNA ranges from 45.8–46.8 moles % (T_m).

Type species: *Yersinia pestis* (Lehmann and Neumann) van Loghem 1944, 15.

TABLE 8.13

Distinguishing characters between the species of **Yersinia,** *and a comparison of these characters with those of* **Salmonella** *and* **Pasteurella multocida**

	1. *Y. pestis*	2. *Y. pseudotuberculosis*	3. *Y. enterocolitica*	*Salmonella*	*Pasteurella*
Motility					
22 C	−	+	+	+	−
37 C	−	−	−	+	−
Urea hydrolyzed	−[a]	+	+	−	−
Esculin	+	+	−	−	−
Rhamnose	(d)	+	−	+	−
Salicin	d	+	−	−	−
Sucrose	−	−	+	−	+
Gas from glucose	−	−	−	+	−
β-Galactosidase	+	+	+	d	−
Ornithine decarboxylase	−	−	+	+	
Citrate utilized	−	−	−	+	−
Indole	−	−	d	−	d

[a] = sometimes positive in freshly isolated strains.

TABLE 8.14

Characters, other than those of special value in differentiating species **of Yersinia**

	1. Y. pestis[a]	*2. Y. pseudo-tuberculosis*	*3. Y. entero-colitica*
Adonitol	−	−	−
Amygdalin	(+)[b]	−	−
Arabinose	+	+	d
Arabitol		−	−
Arbutin	−	+	+
Cellobiose	−	−	+
Dextrin	+	+	d
Galactose	(+)	+	+
Inulin	w[c]	−	−
Lactose	−	−	−
Melibiose	d	+	−
Sorbitol	d	−	+
Sorbose	−	−	+
Starch	w	+	d
Xylose	+	+	d
KCN (growth in)	−	−	d
d-Tartrate	+	−	−
H_2S production	+	−	−

[a] The reactions given for *Y. pestis* are based on the examination of only a small number of strains.

[b] ( ) = delayed reaction.

[c] w = weak reaction

Tribe V. **Erwinieae**

Genus XII. **Erwinia** *Winslow, Broadhurst, Buchanan, Krumwiede, Rogers and Smith 1920, 209*

Cells predominantly single, straight rods, 0.5–1.0 by 1.0–3.0 μm. **Motile** (one exception) by peritrichous flagella. Gram-negative.

Produce **acid from** fructose, glucose, galactose, **β-methylglucoside and sucrose,** usually from mannose and ribose, but **rarely from adonitol, dulcitol or melezitose.** Utilize acetate, fumarate, gluconate, malate and succinate but **not benzoate, oxalate or propionate, as carbon- and energy-yielding sources. Gas production comparatively weak or absent.** Fermentation end products from glucose are CO_2 and different combinations of succinate, lactate, formate and acetate; some form 2,3-butanediol and some ethanol (White and Starr, 1971). **Do not hydrolyze starch** beyond dextrins.

Do not decarboxylate glutamic acid. Decarboxylases for arginine, lysine or ornithine can be detected by Møller's method (Møller, 1955) in only a few (less than 5%) of strains of the *E. carotovora* group and in none of other species.

Rarely produce urease or lipases.

Optimum temperature for growth 27–30 C; maximum varies between 32 C and at least 40 C.

Facultatively anaerobic but anaerobic growth by some species is weak. Oxidase negative, catalase positive.

Associated with plants as pathogens, saprophytes or as constituents of the epiphytic flora. One species has also been isolated from animal and human hosts.

The G + C content of DNA ranges from 50–58 moles % (buoyant density and T_m).

Type species: *Erwinia amylovora* (Burrill) Winslow *et al.* 1920, 209.

TABLE 8.15

Acid production from organic compounds[a] *by* **Erwinia** *species and their varieties*[b]

	Amylovora group						Herbicola group				Carotovora group				
	1. *E. amylovora*	2. *E. salicis*	3. *E. tracheiphila*[c]	4. *E. nigrifluens*	5. *E. quercina*	6. *E. rubrifaciens*	7a. *E. herbicola* var. *herbicola*	7b. *E. herbicola* var. *ananas*	8. *E. stewartii*	9. *E. uredovora*	10a. *E. carotovora* var. *carotovora*	10b. *E. carotovora* var. *atroseptica*	11. *E. chrysanthemi*[d]	12. *E. cypripedii*	13. *E. rhaponici*
Arabinose	d[e]	−	−	+	−	+	+	+	+	+	+	+	+	+	+
Mannitol	−	+	−	+	+	+	+	+	+	+	+	+	+	+	+
Salicin	−	+	−	+	+	−	d	+	−	d	+	+	+	+	+
α-Methyl glucoside	−	−	−	−	+	+	−	−	−	−	−	+	−	−	d
Xylose	−	−	−	+	−	−	+	+	+	+	+	+	+	+	−
Raffinose	−	+	−	+	−	−	d	+	+	+	+	+	d	−	+
Dulcitol	−	−	−	−	−	−	−	−	−	−	−	−	−	−	+
Inositol	−	+	−	+	−	−	−	+	−	+	d	−	−	+	+
Lactose	−	−	−	−	−	−	d	+	+	+	+	+	−	−	+
Melezitose	−	−	−	−	−	−	−	−	−	+	−	−	−	−	+
Melibiose	−	+	−	+	−	−	−	+	+	+	+	+	d	d	+
Maltose	−	−	−	−	−	−	+	+	−	+	d	+	−	d	+
Adonitol	−	−	−	−	−	−	−	−	−	+	−	−	−	−	−
Cellobiose	−	−	−	−	−	−	−	+	−	+	+	+	+	+	+
Dextrin	−	−	−	−	−	−	+	−	−	+	−	−	−	−	−
Esculin	−	+	−	+	+	−	d	d	−	d	+	+	+	+	+
Glycerol	−	d	−	+	+	d	−	+	−	+	−	−	−	−	+
Mannose	−	+	−	+	+	+	+	+	+	+	+	+	+	+	+
Rhamnose	−	−	−	+	−	−	+	d	−	+	+	+	+	+	+
Ribose	+	+	−	+	+	+	+	+	+	+	+	+	+	+	+
Sorbitol	d	+	−	+	+	+	+	+	+	+	+	+	+	+	+

[a] After 7 days growth at 27 C in unshaken aqueous solution of 1% organic compound, 1% peptone with bromcresol purple as an indicator. Data mostly from Dye (1968, 1969).

[b] Characters in boxes are those most useful for differentiation within groups. For invariant characters see generic description.

[c] *E. tracheiphila* grows very slowly in the above medium.

[d] Many strains of *E. chrysanthemi* are delayed lactose fermenters.

[e] + = 80% or more of strains positive; − = 20% or less of strains positive; d = 21–79% of strains positive.

TABLE 8.16

***Biochemical, physiological and cultural characters of* Erwinia *species and their varieties*[a]**

	Amylovora group						Herbicola group				Carotovora group				
	1. *E. amylovora*	2. *E. salicis*	3. *E. tracheiphila*	4. *E. nigrifluens*	5. *E. quercina*	6. *E. rubrifaciens*	7a. *E. herbicola* var. *herbicola*	7b. *E. herbicola* var. *ananas*	8. *E. stewartii*	9. *E. uredovora*	10a. *E. carotovora* var. *carotovora*	10b. *E. carotovora* var. *atroseptica*	11. *E. chrysanthemi*	12. *E. cypripedii*	13. *E. rhapontici*
Anaerobic growth	±[b]	±	±	+	+	+	+	+	+	+	+	+	+	+	+
Growth factors required[c]	+	+	+	−	+	−	−	−	−	−	−	−	−	−	−
H_2S from cysteine[d]	−	+	+	+	+	+	+	d	−	−	+	+	+	+	+
Urease[d]	−	−	−	+	−	−	−	−	−	−	−	−	−	−	−
Growth at 36 C	−	−	−	+	+	+	+	+	d	+	+	−	+	+	−
Gluconate oxidation[e]	−	−	+	−	−	−	d	+	d	+	−	−	−	+	+
Sucrose, reducing compounds[f]	+	+	d	−	+	−	d	+	d	+	−	+	d	−	+
Pectate degradation[g]	−	+	−	−	−	+	−	−	−	−	+	+	+	−	−
Pink diffusible pigment[h]	−	−	−	−	−	+	−	−	−	−	−	−	−	−	+
Acetoin[d]	+	+	d	+	+	−	+	+	−	+	+	+	+	−	+
Mucoid growth[i]	+	+	−	−	+	+	d	+	+	−	d	−	d	d	d
Symplasmata[j]							d	−	−	d	.	.	.	.	.
Motility	+	+	+	+	+	+	+	+	−	+	+	+	+	+	+
Nitrate reduction[d]	−	−	−	−	−	−	+	−	−	+	+	+	+	+	+
DNase[k]	−	−	−	−	−	−	−	−	−	+	−	−	−	−	−
Gelatin liquefaction[d]	+	−	−	−	−	−	+	+	−	+	+	+	+	−	−
Phenylalanine deaminase[l]	−	−	−	−	−	−	+	−	−	−	−	−	−	+	d
Indole[m]	−	−	−	−	−	−	−	+	−	+	−	−	+	−	−
Gas from glucose[n]	−	−	−	−	−	−	−	−	−	−	d	d	+	+	−
Casein hydrolysis[d]	−	−	−	−	−	−	−	−	−	−	+	d	d	−	−
Blue pigment[o]	−	−	−	−	−	−	−	−	−	−	−	−	+	−	−
Yellow pigment[p]	−	−	−	−	−	−	+	+	+	+	−	−	−	−	−
Cottonseed oil hydrolysis[d]	−	−	−	−	−	−	−	−	−	−	d	−	d	+	d
KCN inhibition	+	+	+	+	+	+	+	+	+	+	d	d	d	+	−
Maximum growth temperature[q]	33–34	35	32–34	38	38	38	37–40	40	35–39	37–39	37–40	35	37–40	38	34
Growth in 5% NaCl											+	+	−	+	+
Lecithinase[r]											−	−	+	−	−
Phosphatase[s]											−	−	+	+	+
Sensitivity to erythromycin (50 μg)											−	−	+	+	+

[a] Characters in boxes are those most useful for differentiation within groups. For invariant characters see generic description.

[b] ± = weak growth; + = 80% or more of strains positive; − = 20% or less of strains positive; d = 21–79% of strains positive.

[c] *E. amylovora* requires nicotinic acid. Other growth factor requiring species will grow in an inorganic salts medium with utilizable C source and yeast extract; their exact requirements are not known.

[d] By the methods of Dye (1968).

[e] After 4 days shake culture at 27 C in the medium of Shaw and Clarke (1955). The production of an orange or brown color (with or without precipitate) with an equal volume of Benedict's quantitative reagent after 10 min in a boiling water bath constitutes a positive reaction.

[f] After 2 days shake culture at 27 C in 4% sucrose, 1% peptone, 0.5% beef extract broth and tested and read as footnote *e* above.

[g] In 3 days at 27 C on Paton's medium (Paton, 1959).

[h] On 1% yeast extract, 1% D-glucose, 2% ppt. chalk agar (YDC). Pigment production by *E. rhapontici* is more consistent on media containing (%): sucrose, 2; peptone, 0.5; K_2HPO_4, 0.05; $MgSO_4$, 0.025; agar, 2; pH 7.2–7.4.

[i] On 5% sucrose nutrient agar.

[j] See Graham and Hodgkiss (1967).

[k] On DNase test agar after 2 days at 27 C (Graham and Hodgkiss, 1967).

[l] On phenylalanine agar after 2-3 days at 27 C (Report of Enterobacteriaceae Subcommittee, 1958). The reaction is weaker than that given by *Proteus* spp.

[m] After 2 and 5 days at 27 C in 1% tryptone, 0.1% tryptophan broth and tested with Kovacs' reagent. *E. chrysanthemi* probably converts tryptophan to α-methyl indole and not indole (Lelliott, 1956).

[n] In the sealed tube of Hugh and Leifson's (1953) O/F medium. *E. quercina* and *E. rubrifaciens* produce small amounts of gas (possibly from peptone) on some other media.

[o] On YDC (see footnote *h* above) after 5–10 days at 27 C.

[p] On nutrient agar. Non-pigmented strains of *E. herbicola* occur (Billing and Baker, 1963) but their frequency in relation to pigmented forms is not known.

[q] Data from Dye (1968, 1969); not tested for growth above 40 C. Range indicates strain variation.

[r] On egg yolk agar after 7 days at 27 C.

[s] As Cowan and Steel (1965) but on 0.05% sodium phenolphthalein diphosphate agar after 2 days at 27 C.

TABLE 8.17

Utilization of some organic compounds[a] *as a source of carbon and energy by* **Erwinia** *species and their varieties*[b]

	Formate	Malonate	Tartrate	Lactate	Citrate	Galacturonate
1. *E. amylovora*	+[c]	−	−	+	+	−
2. *E. salicis*	−	−	−	−	−	−
3. *E. tracheiphila*	d	−	−	−	d	−
4. *E. nigrifluens*	+	−	+	+	−	−
5. *E. quercina*	+	−	−	+	+	−
6. *E. rubrifaciens*	+	−	+	+	+	−
7a. *E. herbicola* var. *herbicola*	+	d	d	+	+	−
7b. *E. herbicola* var. *ananas*	+	−	+	+	+	d
8. *E. stewartii*	+	−	+	+	+	−
9. *E. uredovora*	+	−	+	+	+	−
10. *E. carotovora*	+	−	−	+	+	d
11. *E. chrysanthemi*	+	+	+	+	+	d
12. *E. cypripedii*	+	d	+	+	+	+
13. *E. rhapontici*	+	+	−	+	+	−

[a] In 21 days at 27 C in OY medium (Dye, 1968).

[b] For invariant characters see generic description.

[c] + = 80% or more of strains positive; − = 20% or less of strains negative; d = 21–79% of strains positive. Data mostly from Dye (1968, 1969).

FAMILY II. **VIBRIONACEAE** VÉRON 1965, 5245

Rigid Gram-negative rods, straight or curved; usually motile by polar flagella but some cells may have, in addition, lateral flagella produced under certain growth conditions.

Chemoorganotrophs, metabolism both fermentative and respiratory. Oxidase positive. Several species produce butylene glycol from glucose, some are proteolytic, and some produce indole.

Facultative anaerobes without exacting nutritional requirements.

Usually found in fresh or sea water, occasionally in fish or man.

The G + C content of the DNA ranges from 39 to about 63 moles %.

Type genus: *Vibrio* Pacini 1854, 411.

TABLE 8.18

Differential characteristics of the genera of family **Vibrionaceae** *and genus* **Pseudomonas**

Genus	Flagella	Oxidase	0/129 sensitivity	Luminescence	Carbohydrate metabolism (O/F medium)	Gas production	Diffusible pigment
Vibrio	Polar	+	+	d	Fermentative	−	−
Aeromonas	Polar	+	−	−	Fermentative	d	− or brown
Plesiomonas	Polar (lophotrichous)	+	+	−	Fermentative and respiratory	−	−
Photobacterium	Polar	− or +	+ (occasionally −)	+	Fermentative	+	−
Lucibacterium	Usually peritrichous	+	−	+	Fermentative	−	−
Pseudomonas	Polar	+	−	−	Respiratory or not metabolized	−	Green fluorescent or—(occasionally red, brown or orange pigments produced)

Genus I. **Vibrio** *Pacini 1854, 411*

Short asporogenous **rods, axis curved or straight,** 0.5 by 1.5–3.0 μm, single or occasionally united into S shapes or spirals. **Motile by a single polar flagellum,** or, in some species, two or more flagella in one polar tuft; very occasionally non-motile. In some species the flagellum has a central core with an outer sheath (visible in electron microscope preparations). Spheroplasts frequently present, usually formed in adverse environmental conditions. **Gram-negative.** Not acid-fast. No capsules. Grow well and rapidly on standard nutrient media.

Chemoorganotrophs, **metabolism is both respiratory** (oxygen is utilized) **and fermentative.** Metabolism of carbohydrates is fermentative with mixed products but no CO_2 or H_2. **Oxidase-positive.** Non-pigmented or yellow. Generally able to grow on simple mineral ammonium media with a simple carbon source; glutamate and succinate are oxidizable substrates, probably universal within the genus, but the range of substrates utilized is relatively limited. Frequently V.P. positive. Nitrites usually formed from nitrates. Acid but no gas formed from glucose. Urease negative.

Facultatively anaerobic. Temperature optima range from 18–37 C. pH range 6.0–9.0. Optimum NaCl requirement usually 3.0%, some strains fail to grow in the absence of sodium chloride. Usually **sensitive to 2,4-diamino-6,7-diisopropyl pteridine** (0/129) and novobiocin.

The G + C content of the DNA (of those species examined) ranges from 40–50 moles %.

Found in fresh and salt water, and in the alimentary canal of man and animals; some species are pathogenic for man and other vertebrates (fish).

Type species: *Vibrio cholerae* Pacini 1854, 411.

TABLE 8.19

Characteristics of the species of genus **Vibrio**

	Vibrio cholerae	*Vibrio parahaemolyticus*	*Vibrio anguillarum*	*Vibrio fischeri*	*Vibrio costicola*
Flagella					
1 polar	+[a]	+	+	d	+
lophotrichous	−	−	−	d	−
Indole	+	+	+	−	−
Methyl red	d	d	d	+	−
Voges-Proskauer	d	d	d	−	d
Citrate utilization	d	+	−	−	−
O/129 sensitivity	+	(+)	+	+	d
Novobiocin sensitivity	+	(+)	+	+	+
Gelatin liquefaction	+	+	+	+	d
Casein hydrolysis	+	+	+	+	−
Tween 80 hydrolysis	+	+	d	+	d
Lecithinase (egg yolk)	+	+			d
Hydrogen sulfide	−	−	−	−	−
Acid from arabinose	−	d	−	−	−
Acid from mannose	d	+	+	+	+
Acid from sucrose	+	d	+	d	+
Acid from mannitol	+	+	+	d	d
Acid from inositol	−	−	−	−	−
Acid from salicin	−	−	−	d	−
Hydrolysis of starch	+	+	+	d	−
Growth without added NaCl	+	−	+	−	−
Growth in 7.0% NaCl	d	+	d	+	+
Growth in 10.0% NaCl	−	d	−	−	+
Growth at 5 C	−	−	+	+	+
Growth at 37 C	+	+	d	−	d
Growth at 42 C	d		−	−	−
Møller's media:					
Arginine–alkaline reaction	−	−	+	−	+
Lysine decarboxylase	+	+	−	+	−
Ornithine decarboxylase	+	+	−	−	−
Hemolysis: horse blood agar	d	d	d		−
Luminescence	−[b]	−	−	d	−
Pathogenicity	+	+	+	−	−
DNA base ratio–moles % G + C	45–49	42–45	44–45	40–46	50

[a] + positive; − negative; (+) weak positive; d more than 10% of strains positive or negative.

[b] *Vibrio cholerae* biotype *albensis* is luminous, but only one strain is known and has been examined.

TABLE 8.20

Characters that distinguish the biotypes of **Vibrio cholerae**

	cholerae	eltor	proteus	albensis
Number of flagella	1	1	1	1 or more
Hemolysis (in tube)	−	d	+	−
Fermentation of arabinose	−	−	−	−
Fermentation of mannose	+	+	+	−
Fermentation of sucrose	+	+	+	+
V.P. at 22 C	−	+	d	+
V.P. at 37 C	d	+	d	+
M.R. at 22 C	+	d	d	−
M.R. at 37 C	+	+	d	d
Nitrates reduced to nitrites	+	+	−	+
Indole produced at 22 C	+	+	+	(+)[a]
Indole produced at 37 C	d	d	(+)	−
Cholera red reaction	+	+	−	−
Lysine decarboxylated	+	+	−	+
Growth at 5 C	−	−	+	−
Growth at 37 C	+	+	+	+
Growth at 42 C	d	d	−	−
Growth in NaCl 0%	+	+	−	+
Growth in NaCl 4%	+	+	+	+
Growth in NaCl 7%	d	d	+	−
Growth in NaCl 8%	−	−	(+)	−
Growth at pH 10	+	+	+	+
Growth at pH 11	d	d	+	−
Agglutinated by O:1 antiserum	+	+	−	−
Luminescence	−	−	−	+

[a] (+) = delayed.

Genus II. **Aeromonas** *Kluyver and van Niel 1936, 398*

Cells straight, rod-shaped with rounded ends to coccoid 1.0–4.4 μm, occasionally forming filaments up to 8 μm long; occurs singly, in pairs or chains; **motile by polar flagella, generally monotrichous;** some species non-motile. No resting stages known. **Gram-negative.**

Chemoorganotrophs: **metabolism both respiratory and fermentative;** carbohydrates broken down to acid or acid and gas (CO_2 and H_2). Some species show a 2,3-butanediol fermentation (produce acetoin from glucose).

Some species grow on a mineral medium with ammonia as the sole source of nitrogen and one of the following as the sole source of carbon: glucose, arginine, asparagine or histidine; all species grow in a mixture of L-arginine, asparagine, leucine and methionine.

Glucose, fructose, maltose and trehalose are broken down; adonitol, dulcitol, inositol, inulin, melezitose, sorbose and xylose are not fermented. Starch, dextrin and glycerol hydrolyzed.

Casein hydrolyzed; gelatin liquefied; deoxyribonuclease, arginine dehydrogenase and phosphatase produced; glutamic acid not decarboxylated; urea not broken down. Nitrate reduced to nitrite. Cytochrome oxidase, **oxidase** and catalase **positive.**

Facultative anaerobes.

Some species show no growth at 37 C, for others the maximum growth temperature is 38–41 C. Minimum growth temperature 0–5 C. Most biochemical tests are performed at the optimum temperature of 30 C, although some are best performed at 20 C. Growth is restricted to the pH range 5.5–9.0.

Not sensitive to the vibriostatic agent 2,4-diamino-6,7-diisopropyl pteridine (O/129).

The G + C content of the DNA ranges from 57–63 moles %.

Proposed type species: *Aeromonas hydrophila* (Chester) Stanier 1943, 213.

TABLE 8.21

Characteristics differentiating the species, subspecies and biotypes of genus **Aeromonas**

	1. *A. hydrophila* subsp.			2. *A. punctata* subsp.		3. *A. salmonicida* subsp.		
	hydrophila	*anaerogenes*	*proteolytica*	*punctata*	*caviae*	*salmonicida*	*achromogenes*	*masoucida*
Growth in nutrient broth at 37 C (water bath)	+[a]	+	+	+	+	−	−	−
Brown pigment (water-soluble) on trypticase soy agar	−	−	−	−	−	+	−[b]	−[b]
Ammonium ion and glucose as sole source of nitrogen and carbon	+[c]	+[c]	+	+	+[c]	−	−	−
Amino acids as sole source of carbon								
L-Arginine	+	+	+	+	+	−	−	−
L-Asparagine	+	+	+	+	+	−	−	−
L-Histidine	+	+	+	+	+	−	−	−
L-Glutamic acid	+	+[c]	+	+	+	−	−	−
L-Serine	+[c]	d	+	+	+	−	−	−
L-Alanine	d	d	+	d	+	−	−	−
Growth in nutrient broth containing 7.5% NaCl	−	−	+	−	−	−	−	−
Butanediol dehydrogenase	+	+	+	−	−	+	+	+
Production of:								
Gas from glycerol	+	−	−	−	−	+	−	+
Gas from glucose	+	−	−	+	−	+	−	+
Indole in tryptone broth containing 0.1% tryptophan	+[c]	+[c]	+	+	+	−	+	+
Hydrogen sulfide from 2.5% peptone water	+	+	−	+	+	−	−[b]	+
KCN broth (Møller technique)	+	+	+	+	+	−	−	−
Lysine decarboxylase reaction (Møller technique)	−	−	+	−	−	−	−	+
biotypes	1	2						
Voges-Proskauer reaction	+ −	+ −	+	−	−	−	−	+
Gluconate oxidase test	+ −	+ −	+	−	−	−	−	+
Breakdown of								
Galactose	+	+	−	+	+	+	+	+
Sucrose	d	+	−	d	+	−	+	+
Mannitol	+	+	+	d	+	+	−	+
Arabinose	d	+	−	d	+	+	−	+
Esculin	d	d	−	d	+	+	−	+

[a] + = positive; − = negative; d = some strains positive, some negative.
[b] May be a delayed positive reaction.
[c] Aberrant strains occur.

Genus III. **Plesiomonas** *Habs and Schubert 1962, 324*

Cells round-ended, straight, rod-shaped 0.8–1.0 by 3.0 μm; growing singly in pairs or short chains, motile by polar flagella, generally lophotrichous. Resting stages not known. Gram-negative.

Chemoorganotrophs: **metabolism both respiratory and fermentative;** breakdown of carbo-

TABLE 8.22

Features which distinguish **Photobacterium** *from other heterotrophic Gram-negative bacteria*

	Photobacterium	*Lucibacterium*	*Vibrio*	*Aeromonas*	*Pseudomonas*	*Alcaligenes*	*Enterobacteriaceae*
Flagella	Polar (occasionally nonmotile, no flagella)	Usually peritrichous	Polar (one or more per cell)	Polar (usually one per cell)	Polar (one or more per cell)	Peritrichous	Peritrichous
Oxidase	d	+	+	+	+	+	−
Sensitivity to 2,4-diamino-6,7-diisopropyl pteridine	+	−	+	−	−	−	−
Luminescence	+	+	d	−	−	−	−
Carbohydrate metabolism	Fermentative	Fermentative	Fermentative	Fermentative	Respiratory, or not metabolized	Respiratory, or not metabolized	Fermentative
Gas production	+	−	−	d	−	−	d
DNA base ratio, moles % G + C	39–42	45–46	40–50	57–63	58–70	58–70	39–59

hydrates by **production of acid but without gas.** Most strains grow on mineral media containing ammonia as sole source of nitrogen and glucose as sole source of carbon. Cytochrome oxidase, **oxidase** and **catalase** reactions **positive. Absence of exoferments** like diastase, lipase, deoxyribonuclease and proteinases; breakdown of inositol; positive lysine, ornithine decarboxylase and **arginine dehydrogenase** reactions (Møller technique).

Facultative anaerobe. Optimum growth at 30 C, good growth at 37 C; maximum growth temperature ranges in between 39–41 C. No growth in nutrient broth containing 7.5% sodium chloride.

Most strains are **sensitive to** the vibriostatic agent **2,4-diamino-6,7-diisopropyl pteridine** (0/129).

The G + C content of the DNA is 51 moles %.

Type species: *Plesiomonas shigelloides* (Bader) Habs and Schubert 1962, 324.

Genus IV. Photobacterium *Beijerinck 1889, 401*

Cells **coccobacilli** or occasionally rods, 1.0–2.5 by 0.4–1.0 μm, **asporogenous,** axis straight, **occurring** singly and occasionally in pairs. Pleomorphic forms occur in adverse conditions of growth. **Motile** by one or more **polar flagella,** or occasionally non-motile. Strains may appear non-motile as only a small proportion of the cells in a culture are seen to move actively and detection of flagella is difficult. **Gram-negative.** No capsules. Some strains do not grow well on nutrient media unless the NaCl concentration is 2.0–3.0%.

Chemoorganotrophs, metabolism is both respiratory and fermentative. **Carbohydrate metabolism is fermentative** and **gas is usually produced.** In peptone media **acid** and usually gas produced from **hexoses** but pentoses and more complex carbohydrates are not generally attacked. Nitrites produced from nitrates. Ammonia produced from peptone. **Indole not produced.** Methyl red positive. **Acetylmethylcarbinol usually produced.** Trimethylamine oxide generally reduced to trimethylamine. Possess a lysine decarboxylase but not an ornithine decarboxylase, alkaline products usually produced from arginine. **Starch not hydrolyzed.** Chitin may be digested. Gelatin may occasionally be slowly liquefied.

Sensitive to chloramphenicol, streptomycin, polymyxin B and **2,4-diamino-6,7-diisopropyl pteridine.** Insensitive to penicillin.

Facultatively anaerobic. Optimum temperature usually 20–30 C. Optimum NaCl concentration usually 3.0%, growth occurs in nutrient media with 0.5–5.0% NaCl, no growth without NaCl. Growth at pH 6–9. Usually luminescent in suitable conditions, but luminescence may be lost after long periods in artificial culture.

The G + C content of the DNA (in the strains examined) ranges from 39–42 moles %.

Habitat: Sea water, surface and alimentary tract of some marine fishes and the luminous organs of some fish and cephalopods.

Type species: *Photobacterium phosphoreum* (Cohn) Ford 1927, 615.

Features which distinguish the genus from other Gram-negative, motile heterotrophic bacteria are given in Table 8.22.

TABLE 8.23

Characters that distinguish species of genus **Photobacterium**

	P. phosphoreum	*P. mandapamensis*
Oxidase (Kovacs)	−	+
Maltose fermentation	+	−
Voges-Proskauer	+	+
2,3-Butanediol (Bullock's method)	+	−
Growth at 5 C	+	−
Growth at 37 C	−	+

Genus V. Lucibacterium *Hendrie, Hodgkiss and Shewan 1970, 166*

Cells **asporogenous rods,** axis straight or slightly curved, 0.5–1.0 by 1.2–2.5 μm. **Motile** by **peritrichous flagella.** Some strains also have a thick, sheathed polar flagellum in addition to the finer lateral flagella. Fimbriae may also be present. Occasionally only the polar flagellum can be detected. Pleomorphic forms occur in adverse cultural conditions. No capsules. **Gram-negative.**

Chemoorganotrophs, metabolism is both respiratory and fermentative. **Carbohydrate metabolism is fermentative without gas** production. A wide range of carbohydrates are fermented but pentoses, lactose, raffinose, cellulose, inositol, dulcitol and inulin are not usually attacked. Nitrites produced from nitrates. **Indole produced.** Methyl red positive. **Acetylmethylcarbinol not produced.** Trimethylamine oxide reduced to trimethylamine. Ammonia produced from peptone. **Oxidase** (Kovacs') and catalase **positive. Starch hydrolyzed. Gelatin liquefied.** Chitin may be digested. Possesses lipases and lecithinase. Usually possesses decarboxylases for lysine and ornithine but does not generally give rise to alkaline products from arginine.

Sensitive to chloramphenicol, streptomycin and polymyxin B. **Insensitive** to penicillin and **2,4-diamino-6,7-diisopropyl pteridine.**

Facultatively anaerobic. Optimum temperature 25–30 C. Growth in nutrient media containing 0.5–5.0% NaCl, optimum NaCl concentration 2.0–3.0%, **no growth without NaCl.** Growth at pH 6–9. Usually luminescent in suitable growth conditions. In some strains luminescence is transient (i.e. it fades rapidly).

The G + C content of the DNA (of the strains examined) is 45–46 moles %.

Habitat: Sea water. Also found on the surfaces of dead marine animals.

Type species: *Lucibacterium harveyi* (Johnson and Shunk) Hendrie, Hodgkiss and Shewan 1970, 166.

Features which distinguish the genus from other heterotrophic, motile, Gram-negative bacteria are given in Table 8.22.

GENERA OF UNCERTAIN AFFILIATION

Genus **Zymomonas** *Kluyver and van Niel 1936, 399*

Plump rod-shaped cells with rounded ends, occurring singly or in pairs, motile by means of lophotrichous flagella. No resting stages known. Gram-negative.

Chemoorganotrophs showing vigorous fermentative metabolism of glucose or fructose which yields nearly equimolar quantities of ethanol and CO_2 by the Entner-Doudoroff pathway. Although not an end product, acetaldehyde accumulates in the culture medium. Arabinose, rhamnose, xylose, galactose, lactose, maltose, raffinose, mannitol and dulcitol are not fermented.

Produces catalase and H_2S. Gelatin is not liquefied, indole not produced, nitrate is not reduced to nitrite, and the methyl red test is negative (Millis, 1956).

Anaerobes, but able to tolerate some oxygen.

Optimum temperature 30 C.

The G + C content of the DNA is 47–48 moles %.

Type species: *Zymomonas mobilis* (Lindner) Kluyver and van Niel 1936, 399.

Note: There are two species recognized; *Z. mobilis* ferments sucrose and *Z. anaerobia* does not.

Genus **Chromobacterium** *Bergonzini 1881, 153. Nom. gen. cons.* Opin 16, Jud. Comm., 1958, 152

Rods with rounded ends, sometimes slightly curved, occurring singly, occasionally with some pairs or short chains. Dimensions 0.6–1.2 by 1.5–6 μm, without definite capsules although sometimes with intercellular slime. Motile by both one **polar flagellum** and usually one to four **subpolar or lateral flagella. No resting stages known. Gram-negative,** often with barred or bipolar staining and lipid inclusions.

Produce low convex round **violet colonies,** and in nutrient broth a **violet ring** at the junction of the liquid surface and the container wall.

Chemoorganotrophs: **metabolism either respiratory (oxidative) or fermentative,** yielding organic acids from glucose and some other carbohydrates, but no visible gas. Oxidize lactate to CO_2.

Usually oxidase positive by the method of Kovacs (1956), although pigment may interfere with reading. Catalase positive, but **highly sensitive to hydrogen peroxide.** Indole negative, V.P. negative. Nitrate and nitrite reduced, sometimes with visible gas production. Produce ammonia from peptone. Phosphatase positive, arylsulfatase negative.

Violet pigment (violacein) produced on suitable media, not diffusing readily, **soluble in ethanol but not in water or chloroform.** Violacein has antibiotic properties (DeMoss 1967).

Grow on ordinary peptone media. Utilize citrate and NH_3 as sole carbon and nitrogen sources for growth. Without distinctive organic growth factor requirements.

TABLE 8.24

Main differential characters of species

	C. violaceum	*C. lividum*
Growth at 4 C (7 days)	−	+
Growth at 37 C (7 days)	+[a]	−
HCN production	+	−
Turbidity from egg yolk	+	−
Acid from trehalose	+	−
Acid from arabinose	−	+
Acid from xylose	−	+
Casein hydrolysis	+	− or w
Esculin hydrolysis	−	+

[a] + = positive; w = weak; − = negative.

Strict aerobes or facultative anaerobes.

Grow at 25 C, but species differ in optimum and in maximum and minimum temperatures. Do not grow below pH 5, optimum pH 7–8. Do not grow in media containing 6% or more of NaCl.

Resistant to benzylpenicillin, 10 μg/ml; sensitive to tetracycline, 30 μg/ml. **Resistant to 0/129** (2,4-diamino-6,7-diisopropyl pteridine) disc method, 30 μg/disc.

Soil and water organisms, occasionally causing infections of mammals or food spoilage.

The **G + C content** of the DNA ranges from **63–72 moles %** (T_m method).

Type species: *Chromobacterium violaceum* Bergonzini 1881, 153.

Genus Flavobacterium *Bergey et al. 1923, 97*

Cells vary from coccobacilli to slender rods. **Motile with peritrichous flagella or non-motile.** The latter **do not show gliding movement** (flexibacterial) **or swarming** (cytophagal) **growth on a nutrient agar.** Endospores not formed. Gram-negative.

TABLE 8.25

Characteristics of the species of genus **Flavobacterium**

Character	Section I						Section II					
	1. *F. aquatile*	2. *F. breve*	3. *F. meningosepticum*	4. *F. ferrugineum*	5. *F. halmephilum*	6. *F. uliginosum*	7. *F. capsulatum*	8. *F. lutescens*	9. *F. rigense*	10. *F. indoltheticum*	11. *F. tirrenicum*	12. *F. devorans*
Motile	−	−	−	−	−	−	−	−	+	+	+	+
Pigmentation	yo[a]	y	y	yo	y	yo	y	y	yo	yo	r	y
Specific attribute basis for isolation	−	−	−	+	+	−	+	−	−	+	+	+
Grows at 37 C	−	0	+	+	−	−	−	+	+	−	−	−
Halotolerant	−	+	+	+	+	−	+	+	+	+	+	+
Require added NaCl	−	−	−	−	+	+	−	−	−	+	+	−
Hydrolysis of												
Gelatin	+	−	+	+	+	+	−	+	+	+	−	+
Casein	+	−	+	0	0	+	0	+	−	+	0	+
Starch	−	0	0	+	−	−	−	+	0	+	−	−
Agar	−	0	0	0	0	+	0	0	0	0	0	−
Cellulose	−	0	0	0	0	0	+	0	0	0	0	−
Chitin	0	0	0	0	0	0	0	0	0	+	0	−
Acid from												
Glucose	+	+	+	+	−	+		+	+	+	−	+
Lactose[b]	−	−	+	+	−	+	+	−	0	−	−	−
Sucrose	+	0	−	+	−	+	+	−	0	+	−	+
Maltose	+	+	+	+	−	+	+	−	0	+	−	+
Methyl red	−	−	0	0	0	0	0	−	0	0	0	−
Citrate	−	−	V	0	0	0	0	+	0	−	0	+
Indole	−	−	+	0	−	−	−	0	−	+	−	−
H_2S	−	0	V	0	0	0	0	0	0	+	0	0
Urease	−	0	−	0	0	−	0	−	0	−	0	−
NO_3 to NO_2	−	0	−	0	0	+	+	+	+	−	−	−
Catalase	+	0	+	0	0	0	+	0	0	0	0	+
Isolated from marine environment	−	−	−	−	+	+	−	+	−	+	+	−
Isolated from soil or fresh water	+	+	−	+	−	−	+	−	+	−	−	+
Human pathogen	−	−	+	−	−	−	−	−	−	−	−	−
Moles % G + C	33	26	36	43	0	32	63	65	69	70	0	69

[a] Pigmentation: y = yellow; yo = yellow to orange; r = red.

[b] Slowly; V = variable; 0 = no data recorded.

Growth on solid media is pigmented yellow, orange, red or brown and **hue may vary with media and temperature.** Color most pronounced on potato, gelatin or milk-containing media and at lower temperatures (15–20 C). Often light is required for maximum pigmentation. **Pigments not soluble in media** and have not been characterized but generally presumed to be carotenoid; however, non-carotenoid pigmentation is probable for at least two of the species. Colonies typically translucent, smooth and entire, or occasionally opaque.

Chemoorganotrophs. Cultures often difficult to maintain following primary isolation and fastidiousness has been associated with nitrogen requirements, need for exogenous metabolites such as B complex vitamins or physical conditions such as concentration of agar in the medium.

Metabolism respiratory. Fermentative behavior, not conspicuous and **neither acid nor gas reactions common** in carbohydrate-containing broth. Acid reactions are common when such cultures are shaken during incubation and concentrations of peptone in media are low. A few cultures are facultatively anaerobic.

Characteristics for the species are given in Table 8.25.

Incubation temperatures below 30 C preferable and growth may be inhibited by higher temperatures. A few species grow at 37 C.

Widely distributed in soil and fresh and marine waters. Commonly found on vegetables during commercial processing and in dairy products. Unidentified pigmented isolates from human infections have been assigned to the genus and one pathogenic species has been included.

Two distinct ranges of DNA nucleotide base ratios (G + C moles %) have been reported: 30–42 and 63–70.

Type species: *Flavobacterium aquatile* (Frankland and Frankland) Bergey *et al.* 1923, 100.

Genus Haemophilus *Winslow, Broadhurst, Buchanan, Krumwiede, Rogers and Smith 1917, 561*

Minute to medium-sized, **coccobacillary to rod-shaped** cells which sometimes form **threads and filaments,** and may show marked **pleomorphism.** Non-motile. Gram-negative. Strict parasites, **requiring provision of growth factors present in blood, especially X and/or V factors.** Aerobic, facultatively anaerobic. May or may not be pathogenic. Optimum temperature usually 37 C; range usually 25–43 C. Occur in various lesions and secretions as well as on normal mucous membranes of vertebrates.

The G + C content of the DNA (in the species examined) is from 38–42 moles %.

Type species: *Haemophilus influenzae* (Lehmann and Neumann) Winslow *et al.* 1917, 561.

TABLE 8.26

Differential characteristics of the species of genus **Haemophilus**[a]

	1. *H. influenzae*	2. *H. suis*	3. *H. haemolyticus*	4. *H. parainfluenzae*	5. *H. parasuis*	6. *H. parahaemolyticus*	7. *H. haemoglobinophilus*	8. *H. influenzaemurium*	9. *H. gallinarum*	10. *H. paraphrophilus*	11. *H. paragallinarum*	12. *H. paraphrohaemolyticus*	13. *H. aphrophilus*	14. *H. ducreyi*	Species incertae sedis: 15. *H. ovis*	16. *H. putoriorum*	17. *H. citreus*	18. *H. piscium*	19. *H. aegyptius*
V requirement	+	+	+	+	+	+	−	−	+	+	+	+	−	−	−	?[b]	?	−†	+
X requirement	+	+	+	−	−	−	+	+	+	−	−	−	+	+	w	?	?	−†	+
Hemolysis	−	−	+	−	−	+	−	−	−	−	−	+	−	(+)	−	−	d	+	−
Increased CO_2 requirement	−	−	−	−	−	−	−	−	+	+	+	+	+	+	−	−	−	−	−
Serum or other additional enrichment required	−	+	−	−	+	−	−	−	+	−	+	−	−	+	−	−	−	−	−
Nitrites from nitrates	+	+	+	+	+	+	+	?	+	+	+	+	+	?	+	?	+	−	+
Indole	d	−	d	d	−	−	+	−	−	−	−	d	−	?	−	?	+	−	−
Catalase	+	+	+	+	+	+	+	−	?	+	−	+	(+)	?	?	?	?	?	?
δ-ALA utilization	−	−	?	+	+	+	?	?	?	?	+	?	?	?	?	?	?	?	?

[a] Species may be identified by the first five requirements listed.

[b] ?, not determined, not known; †, *H. piscium* requires diphosphothiamine or ATP; w, weak reaction or poor growth, in *H. ovis* X requirement is demonstrable in freshly isolated cultures; (+), delayed; d, some (less than 90%) strains positive, some negative.

Species incertae sedis.

Haemophilus vaginalis* Gardner and Dukes 1955, 963

Description from authors mentioned in text and from: Dukes and Gardner, 1961; Lutz, Wurch and Grootten, 1956; Lutz, Grootten and Wurch, 1956; Lapage, 1961; Redmond and Kotcher, 1963; Park *et al.*, 1968; Dunkelberg and McVeigh, 1969.

Bacilli and coccobacilli, 0.3–0.6 by 1–2 μm. Pleomorphism not marked, filaments not seen. Arranged singly or in pairs lying end to end; palisades, cells lying at sharp angles and clubs may be present. Arrangement suggestive of *Corynebacterium* rather than *Haemophilus*. No capsules, spores or flagella.

Gram-variable; uneven staining of cells, some cells retain Gram stain which may be more marked in young (8–12 hrs) cultures. Frankly Gram-positive on inspissated serum (Zinnemann and Turner, 1963). Metachromatic granules with Albert's or Neisser's stain. Sudanophilic inclusions found. Not acid-fast.

Electron microscopy shows cell walls and septa to be of Gram-positive type (Reyn *et al.*, 1966). Vickerstaff and Cole (1969) found 6-deoxytalose and no arabinose in the wall of the type strain.

Type strain contained lysine, glutamic and aspartic acids but diaminopimelic acid not demonstrated (Lapage, unpublished).

No growth on usual diagnostic media. No growth on nutrient agar, and poor if any growth on blood agar; on Casman's agar (1947) colonies hardly visible in 24 hrs, 0.1–0.2 mm diameter in 24–48 hrs; convex, circular, entire, homogeneous, smooth, translucent and colorless. On further incubation, become dull and more opaque and may reach 1 mm diameter, no further differentiation occurs.

Reports on hemolysis vary from α,β to nonhemolytic. Differs with species of origin of blood. On Casman's medium with human or rabbit blood, present author finds α-lysis at first then β, followed by conversion of the medium to a chocolate color. Little action on horse or sheep blood.

In stabs in semi-solid media, no growth on surface or in the top few millimeters of the medium, then growth to the base of the medium. Light inocula may form discrete colonies.

In liquid media turbidity and deposit. May be discrete colonies.

Chemoorganotroph, fermentative. Principal product of fermentation acetic acid (Moss and Dunkelberg, 1969). Methyl red test may be positive in glucose-containing media. There are differences in the published results on the fermentation patterns, perhaps due to the different media used. Acid is produced but no gas, and fermentation may be weak especially of the sugars less frequently attacked. Most strains produce acid but no gas from starch, glycogen and dextrin; glucose and maltose are commonly fermented; arabinose, fructose, galactose, rhamnose and xylose may be attacked. Some authors report strains which ferment sucrose, glycerol and inulin. Do not ferment dulcitol, inositol, lactose, mannitol, raffinose, salicin, sorbitol or trehalose.

Catalase and cytochrome oxidase not produced, nitrates not reduced. Peroxidase production variable (Edmunds, 1962). Does not produce acetylmethylcarbinol, H_2S, indole or urease. Fails to liquefy gelatin or coagulated serum; litmus milk no change. Sensitive to hydrogen peroxide by inhibition test on an agar plate (Dunkelberg *et al.*, 1970). Does not decarboxylate lysine, ornithine or arginine, oxidize ethanol, digest cellulose or form starch from maltose (Dunkelberg *et al.*, 1970).

Exacting growth needs, but does not need hemin (X factor), niacin adenine nucleotide (V factor) or similar coenzyme-like substances. A need for blood has been found by most workers and Edmunds (1960) found that a factor in red cell stroma was necessary. Dunkelberg *et al.* (1970) have grown the organism on a semi-synthetic medium containing enzymic digest of casein, carbohydrates, vitamins, bases, salts and trace metals. They considered that the strains tested required particular commercial proteose peptones as nitrogen, vitamin and purine pyrimidine base sources, five B vitamins: thiamine, riboflavin, nicotinic acid, folic acid and biotin, and one or more nucleic acid bases; the type strain required adenine; the amino acid requirements are not known. Growth improved by fermentable carbohydrates (Zinnemann and Turner, 1963; Dunkelberg *et al.*, 1970).

Facultative anaerobe. Optimum temperature 36–37 C (Dukes and Gardner, 1961). Optimum pH 6–6.5, little growth at pH 4.5, none at pH 8 (Edmunds, 1960). Added CO_2 may improve growth on some media, e.g. chocolate blood agar, but have little effect on others, e.g. Casman's medium. Viability poor, cultures on solid media may die after 24 hrs but semi-solid cultures may live 7 days or longer. Viability good when freeze-dried.

Reports vary on the presence of growth on tellurite media, grows in the presence of 0.01% potas-

* **Editorial Note**. This species does not belong in the genus *Haemophilus* but its taxonomic position has not yet been settled. It is put here as the place in which readers will most probably look for it.

sium cyanide; but does not grow in the presence of 0.5% NaCl or 0.5% bile in peptone starch glucose medium (Dunkelberg *et al.*, 1970).

Reports vary but most strains seem to be sensitive to penicillin, chloramphenicol, tetracycline, bacitracin, erythromycin, neomycin. Sensitivity to streptomycin varies.

Seven serological groups by precipitin tests (Edmunds, 1962), common antigen by tube agglutination and fluorescent antiserum (Redmond and Kotcher, 1963). Some strains show hemagglutination (cold) probably not fimbrial, of human, horse, ox and fowl red cells, but not sheep or rabbit (Edmunds, 1962).

Found in the human genital tract, and has been isolated in America, Europe and Australia. Reported as cause of 'non-specific' vaginitis and urethritis. Masses of bacterial cells may be seen to cover the epithelial cells in the discharge, the so-called 'clue-cells.' Transmission has been claimed in human volunteer experiments (Gardner and Dukes, 1955). Not pathogenic for laboratory animals.

The G + C content of the DNA is 42 ± 1 moles % (strain Dukes 594).

Type strain (holotype): 594, Gardner and Dukes; NCTC 10287; ATCC 14018.

Genus **Pasteurella** *Trevisan 1887, 94 Nom. cons.* Opin. 13, Jud. Comm. 1954, 153

Cells ovoid or rod-shaped, 1.4 ± 0.4 by 0.4 ± 0.1 μm, singly or less frequently in pairs or short chains. **Non-motile;** do not form endospores. **Gram-negative; bipolar staining** common, especially in preparations made from infected animal tissues stained with Giemsa or methylene blue.

Chemoorganotrophic, growing best on media containing blood; minimal nutritional requirements variable. **Metabolism fermentative;** glucose and other fermentable substances including fructose attacked with the production of small amounts of acid but not gas.

Catalase positive; almost always **oxidase positive;** reduce nitrates; gelatinase negative; M.R. and V.P. negative. Aerobic, **facultatively anaerobic;** temperature range 22–42 C, optimum 37 C.

Parasitic on mammals (including man) and birds.

G + C content of DNA of those members tested is in the range 36.5–43.0 moles %.

Type species: *Pasteurella multocida* (Lehmann and Neumann) Rosenbusch and Merchant 1939, 85.

TABLE 8.27

Characters of the species of genus **Pasteurella**

Characteristic	1. *P. multocida*	2. *P. pneumotropica*	3. *P. haemolytica*	4. *P. ureae*
Hemolysis on blood agar	–[a]	–	+	+
Growth on MacConkey's agar	–	–	+	–
Indole	+	+	–	–
Urease	–	+	–	+
H_2S	+	+	d	–
Ornithine decarboxylase	+	d	d	–
Mannitol	+	–	+	+
Sorbitol	+	–	d	+
Trehalose	d	+	d	–
Xylose	d	d	+	–
Lactose	–	d	d	–

[a] – = most (90% or more) strains negative; + = most (90% or more) strains positive; d = some (less than 90%) strains positive, some negative.

TABLE 8.28

Characteristics of the biotypes of **Pasteurella haemolytica**

	A	T
Acid (within 10 days) from		
Arabinose	+	–
Xylose	+	–
Lactose	d	–
Mannose	–	d
Salicin	–	d
Trehalose	–	+
Cultures	Die out rapidly	Survive longer
Penicillin	Sensitive	Resistant
Capsule serotypes	1, 2, 5, 6, 7, 8, 9, 11, 12	3, 4, 10
Somatic serotypes	A, B	C, D
Found as		
Pathogen	Pneumonia in cattle and sheep Septicemia in newborn lambs	Septicemia in lambs over 3 months
Commensal	Nasopharynx of cattle and sheep	–

Genus **Actinobacillus** *Brumpt 1910, 849*

Cells spherical, oval or rod-shaped, 0.4 ± 0.1 by 1.0 ± 0.4 μm with mostly **bacillary** but **interspersed with coccal elements** which often lie at the pole of a bacillus **giving a characteristic Morse code form;** occasionally longer forms up to 6 μm, more common on media containing glucose or maltose; arranged singly, in pairs or more rarely in chains. **Non-motile.** Do not form endospores. **Gram-negative.** Irregular staining. Small amounts of extracellular slime may be demonstrated in wet india ink preparations.

Cultures very sticky especially on primary isolation; colonies may be difficult to remove completely from the agar surface. Surface cultures are of low viability, dying out in 5–7 days.

Chemoorganotrophs: metabolism fermentative. Acid but no gas within 24 hrs from glucose, fructose and xylose. Dulcitol, inositol and inulin are not fermented. Other carbohydrates may be fermented with acid but no gas. **β-Galactosidase positive.** Methyl red test negative.

Hydrogen sulfide produced by most strains (tested by lead acetate paper). **Nitrates reduced to nitrites. Indole not produced. Urease positive.**

Growth reported only on complex media; minimal nutritional requirements unknown. Grows on MacConkey's medium.

Facultative anaerobes.

Temperatures range 20–42 C; optimum 37 C.

Not inhibited by novobiocin or oleandomycin.

The G + C content of the DNA ranges from 40.6–42.0 moles % (T_m).

Type species: *Actinobacillus lignieresii* Brumpt 1910, 849.

TABLE 8.29

Differences between **Actinobacillus lignieresii** *and* **Actinobacillus equuli**

	A. lignieresii	*A. equuli*
Broth deposit	−[a]	+
Methylene blue reduction	+	−
Voges-Proskauer test	d	−
Sodium hippurate hydrolysis	−	+
Gelatinase	−	+
Fermentation of		
Raffinose	−	+
Trehalose	−	+

[a] − = most (90% or more) strains negative in this test; + = most (90% or more) strains positive in this test; d = some (less than 90%) strains positive, some negative.

Genus **Cardiobacterium** *Slotnick and Dougherty 1964, 271*

Rods, 0.5–0.75 by 1–3 μm, arranged singly, in pairs, short chains and clusters. Pleomorphic. No flagella, endospores or capsules demonstrated. **Gram-negative with possible retention of stain in parts of the cells.**

Chemoorganotroph, fermentative. Acid but no gas produced from fructose, glucose, mannose, sorbitol and sucrose. Glucose fermented largely to lactate and smaller amounts of pyruvate, formate and propionate.

Cytochrome oxidase and small amounts of indole produced. Catalase not produced, nitrates not reduced. Does not liquefy gelatin or serum; does not produce acetylmethylcarbinol, urease, ornithine or lysine decarboxylase, arginine dihydrolase or phenylalanine deaminase.

Facultative anaerobe. Carbon dioxide needed by some strains on isolation. Favored by high humidity, may not grow in dry incubator. Grows at 30–37 C, no growth at 22 or at 42 C. Optimum pH 7–7.2.

The G & C content of the DNA is 61.7 moles % (T_m).

Type species: *Cardiobacterium hominis* Slotnick and Dougherty 1964, 271.

TABLE 8.30

Distinguishing characteristics of **Cardiobacterium** *and similar genera*

	Catalase	Oxidase	Indole	Nitrate reduced	Acid from glucose	Hugh and Leifson O/F test	% G + C
Cardiobacterium hominis	−[a]	+	+	−	+	F	61.7[b]
Bacteroides corrodens	−	+	−	+	−	N	56.2–58.2[b]
Moraxella kingii	−	+	−	d	+	F	44.5[c]
Other *Moraxella*	+	+	−	D	−	N	41–46[c,d]
Haemophilus aphrophilus	−	−	−	+	+	F	40[e]
Actinobacillus actinomycetemcomitans	+	−	−	+	+	F	39[f]
Other *Haemophilus*	+	+	D	+	+	F	38–42[g]
Streptobacillus moniliformis	−	−	−	−	+	NG	23.9[h]
Pasteurella sensu stricto	+	+	D	+	+	F	33–46[g,i,j]
Brucella	+	+	−	+	−	N	56–58[g,i,k,m]
Bordetella (excluding *B. bronchiseptica*)	+	D	−	−	−	N	61–66[g,l]

[a] − = most (90% or more) strains negative in this character; + = most (90% or more) strains positive in this character; F = fermentative; N = no reaction or an alkaline reaction; d = some (less than 90%) strains positive, some negative; D = different species vary in their results; NG = fails to grow in medium.

[b] Hill, *et al.*, 1970.

[c] Bøvre *et al.*, 1969.

[d] Bøvre, 1967.

[e] Lapage, unpublished.

[f] Goodman, 1972.

[g] Hill, 1966.

[h] Williams *et al.*, 1969 (L-form only).

[i] Bohaček and Mraz, 1967.

[j] Bailie, 1969.

[k] Hoyer and McCullough, 1968, 444.

[l] Bacon *et al.*, 1967.

[m] Hoyer and McCullough, 1968, 1783.

Genus **Streptobacillus** *Levaditi, Nicolau and Poincloux 1925, 1188*

Rods, 0.3–0.7 by 1–5 μm long, with rounded or pointed ends, **frequently in chains** and **filaments** 0.5–0.9 μm by 10–150 μm long. Rods may show a central thickening; filaments often have a series of oval to elongated **bulbous swellings,** 1–3 μm in diameter, giving the appearance of a string of beads. True branching does not occur.

Non-encapsulated. Non-motile. No resting stages known. **Gram-negative.** Non-acid-fast.

Spontaneous conversion to **L phase** or to **transitional phase** variant (McGee and Wittler, 1969) occurs during cultivation *in vitro* and may occur *in vivo.*

Chemoorganotrophs: **metabolism fermentative.**

Serum, ascitic fluid or blood required for growth. Neopeptone not recommended for optimal cultivation (Brown and Nunemaker, 1942). Minimal nutritional requirements unknown.

Facultative anaerobe. CO_2 and moisture required for fresh isolates and for some laboratory strains.

Temperature optimum 35–37 C; range 30–38 C; no growth at 23 C. pH optimum 7.4–7.6; range 7.0–8.0.

Parasitic to **pathogenic** for **rats** and **other mammals.**

The G + C content of the DNA of the bacillary phase of the organism (three strains) is 24–25 moles %, of the L phase variant strains L_1 Rat 30 and MLS, 24–26 moles % (T_m) (Williams *et al.*, 1969 and unpublished data of Williams and Wittler, 1970).

Type species: *Streptobacillus moniliformis* Levaditi *et al.* 1925, 1188.

Genus **Calymmatobacterium** *Aragao and Vianna 1913, 221*

Pleomorphic rods, 1–2 μm in length, **with rounded ends,** occurring singly and in clusters. Exhibit **single or bipolar condensation of chromatin,** the latter giving rise to the characteristic "safety pin" forms. Usually **encapsulated** and readily demonstrated by Wright's stain as blue bacillary bodies surrounded by well defined, dense pinkish capsules. Non-motile. **Gram-negative.**

May be **isolated and grown on fresh egg yolk medium** (Dienst, 1948). **After adaptation grows on artificial media.** Colonies on Levinthal beef heart infusion agar shiny and translucent, up to 1.5 mm diameter; gradually become gray, later brownish.

Type species: *Calymmatobacterium granulomatis* Aragao and Vianna 1913, 221

Genus. **Noguchia** *Olitsky et al., 1934, 382.**

Small, slender rods. Encapsulated. Motile by means of peritrichous flagella. Gram-negative. Produce a mucoid type of growth which, on initial isolation, occurs with some difficulty on ordinary media. Aerobic to facultatively anaerobic. Optimum temperature, between 28° and 31° C. Found in the conjunctiva of man and animals affected by a follicular type of disease.

The type species is *Noguchia granulosis* (Noguchi) Olitsky et al.

Key to the species of genus **Noguchia.**

I. Produce acid from carbohydrates.
   A. Produces acid from raffinose, maltose and salicin.
      1. *Noguchia granulosis.*
   B. Does not produce acid from raffinose, maltose or salicin.
      2. *Noguchia simiae.*
II. Does not produce acid from carbohydrates.
      3. *Noguchia cuniculi.*

ADDENDUM TO PART 8

Bacteria Symbiotic or Parasitic in Protozoa

Protozoologists have occasionally found various species of protozoa to be infected with organisms, many of which are of the size of ordinary bacteria. Some are found within the host, others on the surface; some have been reported to be parasitic, others are known to be symbionts. Some have the morphology of cocci, others are spirally shaped and still others appear to be spore-bearing rods. Unfortunately, because of their specialized ecological niche, only a few of these organisms have been cultured away from the eukaryotic cells they inhabit and then only with great difficulty. Consequently, little information has been obtained that would permit assessment of the taxonomic affinities of these organisms. In the 6th edition of the *Manual*, these bacterial inhabitants of protozoa were placed in an appendix to the Rickettsiales, but in the seventh edition, because of their greater similarities to true bacteria, they were described in an addendum to the Class Schizomycetes. Information is still too scanty to warrant any taxonomic positioning of these organisms and may in fact be quite diverse. However, as they are all Gram-negative they are now placed as an Addendum to Part 8. It is important that bacteriologists recognize their existence, for they represent an interesting group worthy of more detailed study.

The best information comes from the work of protozoologists who, until recently, have not been too concerned with the taxonomic relationships of the bacteria living in protozoa to well studied groups of bacteria. Dr. John R. Preer, an expert on the intracellular bacteria of *Paramecium* has provided a brief summary of the knowledge available about these organisms. It is presented here for the information of bacteriologists.

For information on other genera and species of bacterial parasites of protozoa, the reader is referred to pages 927–930 of the 7th edition of the *Manual*.

The Editors

* Taken from the Seventh Edition of the *Manual*, p. 421.

The Intracellular Bacteria of *Paramecium aurelia*

Many kinds of bacteria live within *Paramecium* and other ciliated protozoa; those associated with *P. aurelia* are best known. In many localities 50% or more of the individuals in natural populations are infected.

All are Gram-negative rods and show typical bacterial structure in the electron microscope. Limited growth free of *Paramecium* has been reported for two of the forms.

The bacteria are highly restricted in the strains of paramecia they can infect. Infection generally depends on the presence of one or two specific genes in the protozoon. Under certain conditions in the laboratory the infected individuals have a competitive advantage over uninfected paramecia, while under other conditions the opposite is true. Most of the bacteria convert their hosts into killers, which liberate toxic particles (generally, if not always, the bacteria themselves) into the medium and thereby kill sensitive strains of paramecia. Although each host paramecium is made resistant to the toxin by the bacterial strain it harbors, organisms free of the bacteria are generally sensitive.

The first form discovered was recognized genetically and given the designation kappa. The practice of using Greek letters has been continued for other forms (Table 8.34).

Kappas are of many kinds and found only in syngens (varieties) 2 and 4 of *P. aurelia*. They are distinguished by the presence of refractile (R) bodies in 5–30% of each population. R bodies are unique and are seen as very distinctive rolls of "ribbon," which may suddenly unroll; they arise in association with phage-like particles which infect kappa and are involved in the toxic action of kappa-bearing (killer) strains of paramecia. The cytochromes of kappa are completely different from those of the host organisms and very similar to those of many of the Enterobacteriaceae and of *Brucella*.

Mus are found in syngens 1, 2 and 8 and lead to mate-killing (death of mates at conjugation); no active toxin is liberated into the medium. Diaminopimelic acid has been demonstrated in mu.

The spiral form alpha is restricted to the macronucleus and is apparently closely related to *Drepanospira* and *Holospora*.

Recent reviews are by Beale, Jurand and Preer (1969), Preer (1971), and Preer *et al.* (*Bacteriological Reviews 38:* 113–163, 1974).

TABLE 8.31

Main characteristics of intracellular bacteria in **Paramecium aurelia**

	Morphology	Flagellation	Syngen or varieties infected	Action on sensitives
Kappa	Rods with R bodies	None	2, 4	Paralysis, spinning, vacuolization, hump formation, depending upon strain of Kappa
Mu[a]	Rods	None	1, 2, 8	Mate killing
Lambda[a]	Large straight rods	Peritrichous	4, 8	Rapid lysis
Sigma	Large curved rods	Peritrichous	2	Rapid lysis
Delta	Rods thick cell wall	Sparse	1, 2, 4, 6	Weak killer or no action
Gamma	Small rods; enclosed in extra set of membranes	None	8	Strong vacuolization
Tau	Rods	None		Weak paralysis
Nu	Rods	None	1, 2, 5	No toxic action
Alpha	Spiral	None	2	No toxic action

[a] Restricted *in vitro* culture in a complex medium has been reported for these forms (van Wagtendonk *et al.*, 1963; Williams, 1971).

PART 9

GRAM-NEGATIVE ANAEROBIC BACTERIA

FAMILY I. BACTEROIDACEAE PRIBRAM 1933, 10

Gram-negative, uniform or pleomorphic rods. Non-motile or motile with peritrichous flagella. Non-spore-forming.

Chemoorganotrophs.

Obligate anaerobes (require reduced oxygen tension).

Isolated from the natural cavities of man and other animals and the intestinal tract of insects; also isolated from infections of man and other animals. Some species are pathogenic.

Key to the genera of family **Bacteroidaceae**

I. Produce (from peptone or glucose) mixtures of acids including succinic, acetic, formic, lactic, propionic; butyric acid usually not a major product. Some species produce a mixture of butyric, isobutyric and isovaleric acids, along with major amounts of succinic acid.

Genus I. *Bacteroides*, p. 385*

II. Produce (from peptone or glucose) butyric acid (without isobutyric and isovaleric acids) as a major product.

Genus II. *Fusobacterium*, p. 404

III. Produce (from peptone or glucose) lactic acid as the only major fermentation acid.

Genus III. *Leptotrichia*, p. 416

Note: Below is a synopsis of the six genera of uncertain affiliation. They are distinguished from the family *Bacteroidaceae* mainly by flagellation and cell shape.

I. Curved rods. Motile by polar flagella. Reduce sulfates to H_2S.

Genus *Desulfovibrio*, p. 418

II. Curved rods. Motile by monotrichous polar or sub-polar flagella. Produce mainly butyrate from glucose.

Genus *Butyrivibrio*, p. 420

III. Curved rods. Motile by monotrichous polar flagella. Produce succinate, acetate and formate, but not butyrate, from glucose.

---

* All page numbers cited refer to *Bergey's Manual of Determinative Bacteriology*, 8th edition, The Williams & Wilkins Company, Baltimore, 1974.

Genus *Succinivibrio*, p. 422

IV. Straight rods. Motile by monotrichous polar flagella. Produce large amounts of succinate but not butyrate from glucose.

Genus *Succinimonas*, p. 422

V. Curved rods. Motile by monotrichous lateral flagella. Produce ethanol, lactate, formate, acetate and CO_2 from glucose.

Genus *Lachnospira*, p. 423

VI. Curved rods. Tumbling motility by a tuft of flagella from the concave side of cell.

Genus *Selenomonas*, p. 424

Recently, Moore *et al.* (*International Journal of Systematic Bacteriology 26*: 238–252, 1976) have described a new member of this family, the genus *Desulfomonas*, which contains anaerobic, straight, non-motile, non-sporeforming, sulfate-reducing rods.

Genus I. **Bacteroides** *Castellani and Chalmers 1919, 959*

Gram-negative, non-spore-forming rods. Non-motile or motile with peritrichous flagella.

Chemoorganotrophs. **Metabolize carbohydrates or peptone. Fermentation products** of saccharoclastic species **include combinations of succinic, lactic, acetic, formic or propionic acids,** sometimes with short-chained alcohols; **butyric acid is usually** not a **major product.** When *n*-butyric acid is produced, isobutyric and isovaleric acids are also present. From peptone non-saccharoclastic species produce either (a) trace to moderate amounts of succinic, formic, acetic and lactic acids or (b) major amounts of acetic and butyric acids (sometimes also succinic acid) with moderate amounts of alcohols and isovaleric, propionic and isobutyric acids. Carbon dioxide utilized or required by strains of several intestinal species and is incorporated into succinic acid (Caldwell *et al.*, 1969).

Obligately anaerobic. The maximum Eh that will allow initiation of growth varies among species and, within species, differs with the size of inoculum used. Eh sensitivity ranges from those that can initiate growth in partially oxidized media or in a candle jar to those that have not yet been cultured in media with an initial oxidation-reduction potential above minus 100 mV, even when large inocula are used. Oxidized media are inhibitory for some species. Addition of serum or ascitic fluid to such media may enable them to support growth.

Catalase usually not produced, or only in trace amounts. Hippurate usually not hydrolyzed. Lecithinase and lipase usually not produced on McClung-Toabe egg yolk agar. Meat usually not digested.

Growth usually most rapid at 37 C and pH near 7.0.

The G + C content of the DNA (in the species examined) ranges from 40–55 moles % (chromatographic separation (Sebald, 1962) and T_m (J. L. Johnson, personal communication)).

Found in cavities of man and other animals, infections of soft tissue, and sewage. Some species may be pathogenic.

Characteristics by which species in the genus can be differentiated are given in the Key and in Tables 9.1, 9.2, 9.4, 9.5, 9.6 and 9.7. Additional characteristics are given in the text concerning each species. The drawings of each species (Figs. 9A–9D) are composites of numerous cultures. Individual cultures usually show somewhat less variation in cellular morphology than is depicted. Cells may be more pleomorphic when grown in media that are not highly reduced (Doetsch *et al.*, 1957). Terminal or central swellings, vacuoles, or filaments are common.

The type species: *Bacteroides fragilis* (Veillon and Zuber) Castellani and Chalmers 1919, 959.

Key to species in the genus **Bacteroides**

Key characteristics

	Acid from:	
Production of black pigment on laked blood agar or gelatin	trehalose	Requirement for hemin
Degree of carbohydrate fermentation	rhamnose	Cell shape
Production of succinate, propionate, butyrate or lactate from glucose	mannitol	Motility
Growth in 20% bile (2% ox gall)	mannose	Gas production
Production of indole	arabinose	Nitrate reduction
	lactose	Liquefaction of gelatin
	melizitose	Hydrolysis of esculin
	cellobiose	Saccharolysis
	glucose	Proteolysis

I. Black pigment not produced.
  A. Carbohydrates strongly fermented.
    1. Succinic acid a major product.
      a. Good growth in 20% bile (2% ox gall).
        b. Indole not produced.
          c. Trehalose not acid.
            d. Rhamnose not acid.
              1a. *B. fragilis* subsp. *fragilis*
            dd. Rhamnose acid.
              1b. *B. fragilis* subsp. *vulgatus*
          cc. Trehalose acid. 1c. *B. fragilis* subsp. *distasonis*
        bb. Indole produced.
          c. Mannitol acid. 1d. *B. fragilis* subsp. *ovatus*
          cc. Mannitol not acid.
              1e. *B. fragilis* subsp. *thetaiotaomicron*
      aa. Growth inhibited by 20% bile (2% ox gall).
        b. Mannose acid.
          c. Hemin required for growth.
              2a. *B. ruminicola* subsp. *ruminicola*
          cc. Hemin not required for growth.
            d. Arabinose acid.
              2b. *B. ruminicola* subsp. *brevis*
            dd. Arabinose not acid.
              e. Long, thin cells present.
                3. *B. ochraceus*
              ee. Long, thin cells not present.
                4. *B. oralis*
        bb. Mannose not acid. 5. *B. amylophilus*
    2. Succinic acid not a major product.
      a. Propionic acid a major product.
        b. Not motile. 6. *B. hypermegas*
        bb. Motile. 7. *B. serpens*
      aa. Propionic acid not a major product.
        b. Not motile.
          c. Lactose not acid. 8. *B. termitidis*
          cc. Lactose acid.
            d. Melezitose acid.

9. *B. biacutus*

dd. Melezitose not acid.

10a. *B. clostridiiformis* subsp. *clostridiiformis*

(Also see *Clostridium ramosum*, which now includes organisms formerly identified as *Bacteroides trichoides* (*B. terebrans*) and in which spores usually are not seen.)

bb. Motile.

c. Abundant gas produced.

10b. *B. clostridiiformis* subsp. *girans*

cc. Little, if any, gas produced.

11. *B. constellatus*

B. Carbohydrates not fermented, or only weakly so.

1. Indole produced.

a. Propionic acid produced.

12. *B. putredinis*

aa. Propionic acid not produced.

13. *B coagulans*

2. Indole not produced.

a. Nitrate reduced.

b. Butyric acid produced.

14. *B. praeacutus*

bb. Butyric acid not produced.

15. *B. corrodens*

aa. Nitrate not reduced.

b. Gelatin completely liquefied.

16. *B. nodosus*

bb. Gelatin not completely liquefied.

c. Lactic acid a major product.

17. *B. furcosus*

cc. Lactic acid not a major product.

d. Esculin hydrolyzed.

18. *B. capillosus*

dd. Esculin not hydrolyzed.

e. Cellobiose weak acid.

19. *B. succinogenes*

ee. Cellobiose not acid.

20. *B. pneumosintes*

II. Black pigment on laked blood agar.

A. Glucose acid.

1. Highly saccharolytic, non-proteolytic.

21a. *B. melaninogenicus* subsp. *melaninogenicus*

2. Moderately saccharolytic, moderately proteolytic.

21b. *B. melaninogenicus* subsp. *intermedius*

B. Glucose not acid.

21c. *B. melaninogenicus* subsp. *asaccharolyticus*

III. Black pigment in gelatin culture after incubation for several days.

22. *B. niger*

TABLE 9.1

Fermentation reactions of the species of genus **Bacteroides** *species 1–5*[a]

	1a. *B. fragilis* ss. *fragilis*	1b. *B. fragilis* ss. *vulgatus*	1c. *B. fragilis* ss. *distasonis*	1d. *B. fragilis* ss. *ovatus*	1e. *B. fragilus* ss. *thetaiotaomicron*	2a. *B. ruminicola* ss. *ruminicola*	2b. *B. ruminicola* ss. *brevis*	3. *B. ochraceus*	4. *B. oralis*	5. *B. amylophilus*
Products from peptone yeast extract glucose broth[a]	SAp(l *ibiv*f)	SA(Lfp *ibiv*)	SAp(fl *ibiv*)	SA(Lfp *ibiv*)	SA(Lpf *ibiv*)	SA(flp *ib*b*iv*)	SA(lfp *ibiv*v)	SA(l)	SA(lf *ibiv*)	Sfa (pl)
Products from lactate	$-^{p}$	−	−	$-^{p}$	$-^{p}$	−	−	−	−	−
Acid produced from:										
Adonitol	−	−	$-^{w}$	−	−	$-^{w}$	$-^{w}$	−	$-^{w}$	v
Amygdalin	+	v	w^{+}	w^{+}	$+^{w}$	$+^{-}$	$+^{w}$	+	$+^{w}$	−
Arabinose	−	$+^{-}$	$+^{-}$	w^{+}	$+^{w}$	v	$+^{w}$	$-^{w}$	−	−
Cellobiose	$-^{w}$	$-^{w}$	v	w^{+}	$+^{-}$	v	$+^{w}$	w^{+}	$+^{-}$	−
Dulcitol	−	−	−	−	−	−	$-^{w}$	−	$-^{+}$	−
Erythritol	−	−	$-^{w}$	−	−	$-^{w}$	$-^{+}$	−	$-^{w}$	−
Esculin	$-^{w}$	$-^{w}$	w^{-}	w^{-}	v	d	v	$-^{w}$	$-^{+}$	−
Esculin hydrolyzed	+	$+^{-}$	+	+	+	+	+	+	$+^{-}$	−
Fructose	$+^{w}$	$+^{w}$	$+^{w}$	$+^{w}$	$+^{w}$	$+^{w}$	$+^{w}$	w^{+}	$+^{w}$	$-^{w}$
Galactose	w^{+}	$+^{w}$	$+^{w}$	$+^{w}$	$+^{w}$	w^{+}	$+^{w}$	w^{+}	$+^{w}$	−
Glucose	$+^{w}$	$+^{w}$	$+^{w}$	$+^{w}$	$+^{w}$	$+^{w}$	+	+	$+^{w}$	−
Glycerol	−	−	−	−	−	$-^{w}$	$-^{+}$	−	$-^{w}$	−
Glycogen	w^{+}	$+^{-}$	$-^{+}$	v	$+^{w}$	$+^{-}$	$+^{-}$	$+^{w}$	$+^{w}$	$+^{w}$
Inositol	−	−	−	−	−	$-^{w}$	$-^{w}$	−	$-^{w}$	−
Inulin	v	v	v	w^{+}	$+^{-}$	$-^{+}$	$+^{-}$	$+^{w}$	d	−
Lactose	w^{+}	$+^{w}$	$+^{w}$	$+^{w}$	$+^{w}$	$+^{w}$	$+^{w}$	$+^{w}$	$+^{w}$	−
Maltose	$+^{w}$	$+^{w}$	$+^{w}$	$+^{w}$	$+^{w}$	$-^{+}$	$+^{w}$	v	$+^{-}$	+
Mannitol	−	−	−	w	−	$-^{w}$	−	−	−	−
Mannose	$+^{w}$	$+^{w}$	$+^{w}$	$+^{w}$	$+^{w}$	$+^{w}$	+	$+^{w}$	$+^{w}$	−
Melezitose	−	−	v	$-^{w}$	v	−	$-^{w}$	−	$-^{w}$	−
Melibiose	w^{+}	$+^{-}$	$+^{w}$	w^{+}	$+^{w}$	$-^{+}$	$+^{-}$	$-^{w}$	v	−
Raffinose	$+^{w}$	$+^{w}$	$+^{w}$	$+^{-}$	$+^{w}$	$-^{+}$	$+^{w}$	w^{+}	$+^{w}$	−
Rhamnose	−	$+^{w}$	v	v	v	$-^{w}$	$-^{+}$	$-^{w}$	$-^{+}$	−
Ribose	$-^{w}$	d	v	v	v	$-^{w}$	v	$-^{w}$	v	$-^{+}$
Salicin	−	−	$-^{+}$	v	d	$-^{+}$	v	$-^{w}$	$-^{+}$	−
Sorbitol	−	−	−	$-^{w}$	−	$-^{+}$	$-^{w}$	−	$-^{w}$	·
Sorbose	−	−	$-^{w}$	$-^{w}$	−	$-^{w}$	$-^{w}$	$-^{w}$	$-^{w}$	−
Starch	$+^{w}$	$+^{w}$	w^{+}	w^{+}	$+^{w}$	d	$+^{w}$	$+^{w}$	+	+
Starch hydrolyzed	+	$+^{-}$	d	+	+	$+^{-}$	+	$+^{-}$	+	+
Sucrose	$+^{w}$	$+^{w}$	$+^{w}$	$+^{w}$	$+^{w}$	$+^{-}$	+	$+^{w}$	$+^{w}$	$-^{w}$
Trehalose	−	−	w	v	$+^{-}$	$-^{w}$	$-^{+}$	−	$-^{w}$	−
Xylose	$+^{w}$	$+^{w}$	$+^{w}$	$+^{w}$	$+^{w}$	$-^{+}$	$+^{-}$	$-^{w}$	$-^{w}$	−

[a] Products from PYG broth: Capital letters indicate >1 meq acid/100 ml broth; small letters <1 meq/100 ml. A, acetic; B, butyric; F, formic; iB, isobutyric; iV, isovaleric; L, lactic, P, propionic; S, succinic; v, valeric; 2, ethanol; ( ) products of occasional strains. +, strong acid, i.e. pH 5.5 or below in 90–100% of strains. w, weak acid, i.e. pH 5.5–6.0 in 90–100% of strains. −, acid not produced, i.e. pH 6.0 or above in 90–100% of strains. (Uninoculated xylose and arabinose, under CO_2 (pre-reduced media), are often pH 5.9. For cultures in these media, we consider pH 5.4 to 5.7 weak acid and pH below 5.4 strong acid.) d, reaction positive 40-60% strains. v, reaction variable within a strain. When any of the above used as subscripts—reaction of 10-40% of strains. ·, not tested. Acid products determined by procedures given in Holdeman and Moore, 1972.

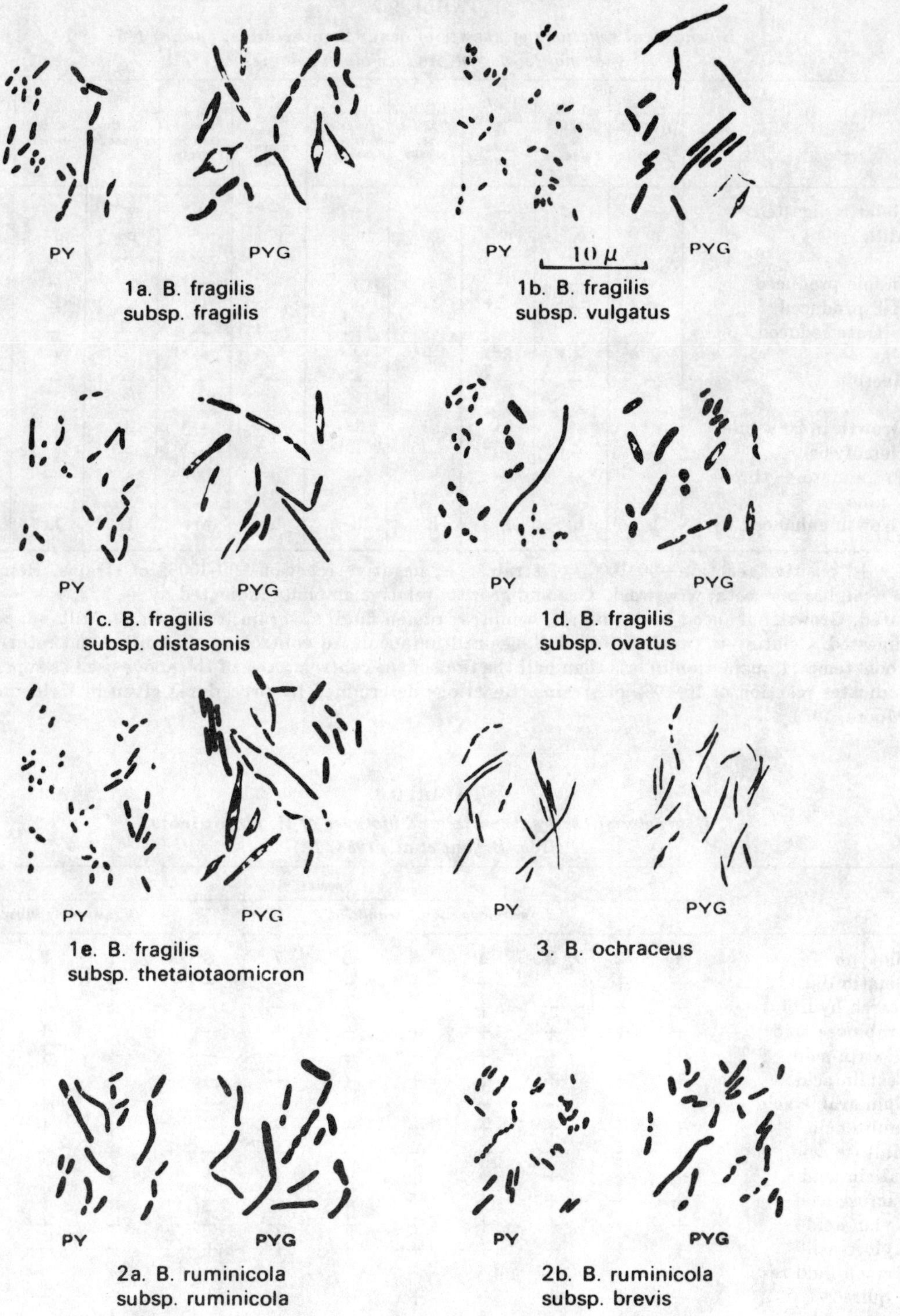

Fig. 9A. *Bacteroides*. The line drawings are composites of several cultures and individual cultures usually show somewhat less variation in cellular morphology. PY, peptone-yeast extract broth; PYG, PY-glucose broth. Scale: 9/16 inch = 10 μm.

TABLE 9.2

Biochemical reactions of species of genus **Bacteroides**, *species 1–5*
(not motile, do not produce black colonies)[a]

	1a. *B. fragilis* ss. *fragilis*	1b. *B. fragilis* ss. *vulgatus*	1c. *B. fragilis* ss. *distasonis*	1d. *B. fragilis* ss. *ovatus*	1e. *B. fragilis* ss. *thetaiotaomicron*	2a. *B. ruminicola* ss. *ruminicola*	2b. *B. ruminicola* ss. *brevis*	3. *B. ochraceus*	4. *B. oralis*	5. *B. amylophilus*
Gelatin digested	$-^{w}$	$-^{+}$	$-^{+}$	$-^{w}$	$-^{w}$	$-^{+}$	$+^{w}$	w^{-}	$+^{-}$	$+^{w}$
Milk	c	c	c	c	c	c^{-}	c^{d}	c	c^{-}	·
Indole produced	−	−	−	+	+	−	$-^{+}$	−	−	−
H_2S produced	$-^{+}$	$-^{+}$	$-^{+}$	$-^{+}$	d	−	−	−	−	−
Nitrate reduced	−	−	−	−	−	$-^{+}$	−	−	−	−
Gas	2^{-}	$+^{4}$	3^{-}	$+^{3}$	$+^{3}$	$-^{2}$	$-^{3}$	−	$-^{2}$	−
Acetoin	−	−	$-^{w}$	−	$-^{w}$	$-^{w}$	$-^{+}$	$-^{+}$	$-^{+}$	$-^{+}$
Growth in 20% bile	s^{4}	s^{4}	s^{4}	s^{4}	3^{s}	−	$-^{2}$	2^{-}	$-^{2}$	−
Hemolysis	−	$-^{b}$	$-^{w}$	$-^{b}$	$-^{ba}$	$-^{+}$	$-^{b}$	−	b^{-}	−
Propionate ← threonine	−	−	−	−	−	$+^{-}$	$-^{+}$	$-^{+}$	−	−
Growth enhanced by	h	h	h	h	h	hsr	hrt	th	hrt	CO_2

[a] +, positive reaction—90–100% of strains. −, negative reaction—90–100% of strains. Hemolysis, a = alpha; b = beta; w = weak. Gas and growth, relative amounts indicated by +, 2, 3, 4; s = stimulated. Growth enhanced by b, bile; h, hemin; r, rumen fluid; s, serum; t, Tween 80. Milk—c, curd; d, digested. Gelatin—w (weak) after chilling until uninoculated controls are solidified and returning to room temp., liquefaction in less than half the time of the controls. Any of the above used as superscript indicates reaction of 10–40% of strains. Reactions determined by procedures given in Holdeman and Moore, 1972.

TABLE 9.3

Characteristics of subspecies and biotypes of **B. ruminicola**
(from Bryant et al., 1958, 17)

	Species										
	B. ruminicola subsp. *ruminicola*								*B. ruminicola* subsp. *brevis*		
Biotype	1	2	3	4	5	6	7	8	1	2	3
Gelatin dig.	+	+	−	+	+	−	−	−	+	+	+
Starch hydrol.	+	+	+	+	−	−	−	−	+	+	+
Arabinose acid	+	+	+	+	+	−	+	−	+	+	+
Dextrin acid	+	+	+	+	−	−	−	−	+	+	+
Esculin acid	+	+	+	+	+	+	+	−	+	+	+
Gum arabic acid	−	−	−	−	−	−	−	−	+	+	−
Inulin acid	+	+	+	+	+	−	−	−	+	+	+
Maltose acid	+	+	+	+	−	−	+	−	+	+	+
Salicin acid	+	+	+	+	+	+	+	+	+	−	+
Sucrose acid	+	+	+	+	+	+	−	−	+	+	+
Xylan acid	+	+	+	+	+	+	+	+	−	−	+
Xylose acid	+	+	+	+	+	+	+	+	−	−	+
Rumen fluid required[a]	+	+	+	−	+	+	+	+	−	−	−
H_2S prod.[b]	−	±	−	±	−	−	−	−	+	−	+

[a] Hemin replaces the rumen fluid requirement of the majority of strains (Caldwell *et al.*, 1965).
[b] Medium of Bryant and Small, 1956.

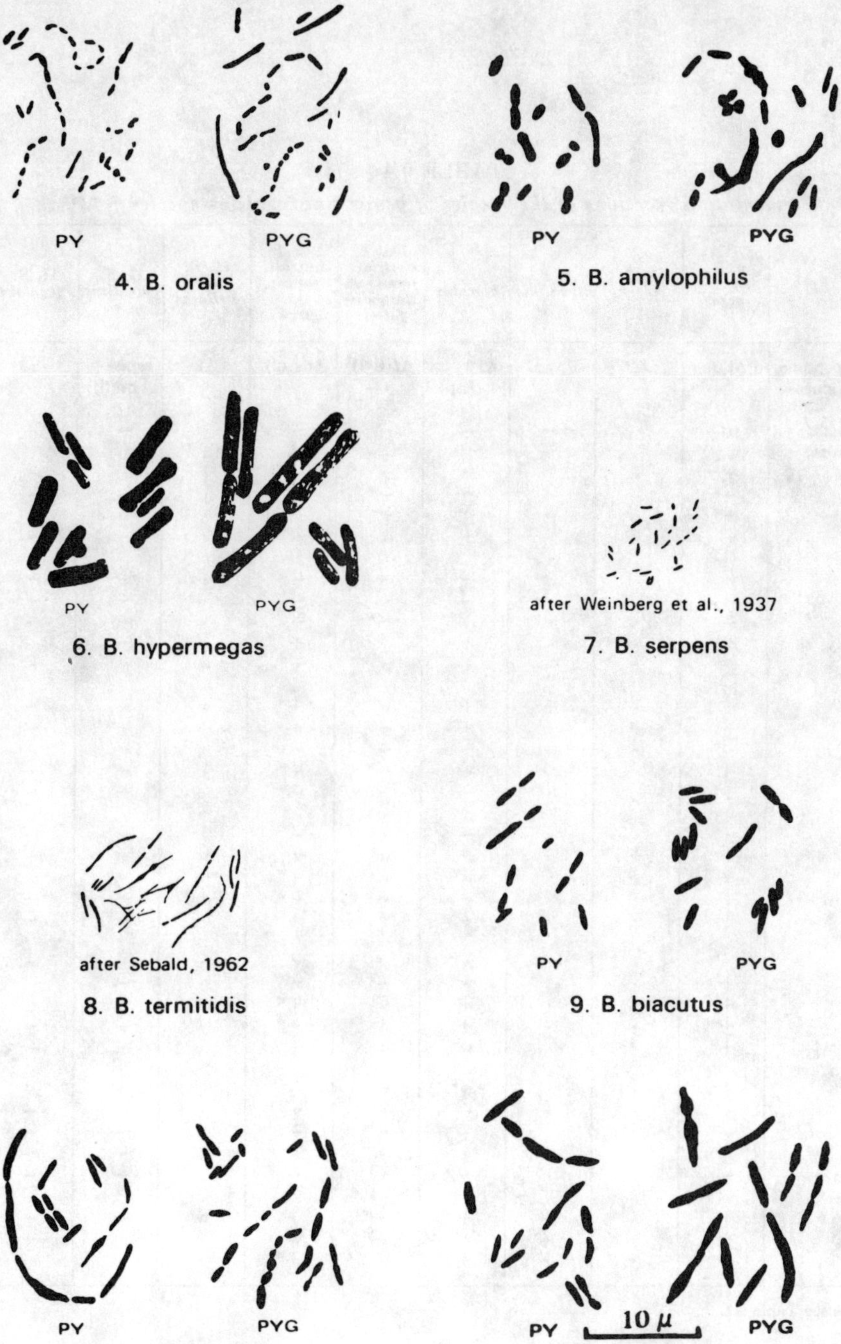

Fig. 9B. *Bacteroides*. The line drawings are composites of several cultures and individual cultures usually show somewhat less variation in cellular morphology. PY, peptone-yeast extract broth; PYG, PY-glucose broth. Scale: 9/16 inch = 10 μm.

TABLE 9.4

Fermentation reactions of the species of genus **Bacteroides** *species 6–14*[a]

	6. *B. hypermegas*	7. *B. serpens*	8. *B. termitidis*	9. *B. biacutus*	10a. *B. clostridiiformis* ss. *clostridiiformis*	10b. *B. clostridiiformis* ss. *girans*	11. *B. constellatus*	12. *B. putredinis*	13. *B. coagulans*	14. *B. praeacutus*
Products from peptone yeast extract glucose broth	PAl (fs)	AP	FALs2	AF2 (Lsp)	Af(sl 2)	Afls (2)	A2	aspiv bib(lf)	alsf	ABiViB p(slf)
Products from lactate	p^{-}	·	·	−	−	−	·	−	−	−
Acid produced from:										
Adonitol	−	·	·	−	−	$-^{w}$	·	−	−	−
Amygdalin	$-^{w}$	·	·	−	$-^{w}$	$-^{w}$	·	−	−	−
Arabinose	$+^{w}$	·	−	w^{-}	w^{-}	$+^{w}$	·	−	−	−
Cellobiose	$+^{-}$	·	·	+	d	$-^{+}$	·	−	−	−
Dulcitol	v	·	·	−	−	−	·	−	−	−
Erythritol	−	·	·	−	−	−	·	−	−	−
Esculin	−	·	·	−	$-^{w}$	−	·	−	−	−
Esculin hydrolyzed	+	·	·	+	+	+	·	−	−	−
Fructose	+	+	+	+	w^{+}	$+^{w}$	w	−	−	−
Galactose	+	+	+	+	w^{+}	$+^{w}$	w	−	−	−
Glucose	+	+	+	+	$+^{w}$	+	+	$-^{w}$	−	−
Glycerol	$-^{+}$	·	−	−	−	−	w	−	−	−
Glycogen	−	·	·	−	$-^{w}$	$-^{w}$	·	−	−	−
Inositol	$+^{-}$	·	·	−	−	$-^{w}$	·	−	−	−
Inulin	$+^{-}$	·	·	−	−	−	·	−	−	−
Lactose	+	+	−	+	w^{+}	v	w	−	−	−
Maltose	+	+	+	w^{+}	v	$+^{w}$	+	−	−	−
Mannitol	+	·	·	−	−	$-^{w}$	·	−	−	−
Mannose	+	·	·	+	v	+	·	$-^{w}$	−	−
Melezitose	$-^{w}$	·	·	w^{+}	−	v	·	−	−	−
Melibiose	+	·	·	w^{-}	$-^{+}$	v	·	−	−	−
Raffinose	+	·	·	+	v	$+^{w}$	·	−	−	−
Rhamnose	$+^{-}$	·	·	+	d	$+^{w}$	·	−	−	−
Ribose	$+^{w}$	·	·	w^{-}	d	$+^{-}$	·	−	−	−
Salicin	$+^{-}$	·	·	w^{+}	v	$-^{+}$	·	−	−	−
Sorbitol	+	·	·	−	−	v	·	−	−	−
Sorbose	$-^{w}$	·	·	−	$-^{w}$	$-^{w}$	·	−	−	−
Starch	$-^{w}$	·	·	w^{-}	$-^{w}$	w^{-}	+	−	−	−
Starch hydrolyzed	−	·	·	−	$-^{+}$	−	·	−	−	−
Sucrose	+	·	+	+	$+^{w}$	+	w	−	−	−
Trehalose	$+^{w}$	·	·	w	$-^{w}$	w^{-}	·	−	−	−
Xylose	+	·	+	+	d	$+^{w}$	·	−	−	−

[a] For symbols see Table 9.1.

TABLE 9.5

Biochemical reactions of species of genus **Bacteroides**, *species 6–14*[a]
(do not produce black colonies)

	6. *B. hypermegas*	7. *B. serpens*	8. *B. termitidis*	9. *B. biacutus*	10a. *B. clostridiiformis* ss. *clostridiiformis*	10b. *B. clostridiiformis* ss. *girans*	11. *B. constellatus*	12. *B. putredinis*	13. *B. coagulans*	14. *B. praeacutus*
Gelatin digested	−	+	−	w^-	$-^w$	−	−	+	+	+
Milk	c	c	−	c	$-^c$	c	c	d^c	c^d	−
Indole produced	−	·	−	−	$-^+$	−	·	+	+	−
H_2S produced	−	+	·	−	$-^+$	+	·	$+^-$	−	+
Nitrate reduced	−	−	−	+	d	+	+	−	−	+
Gas	2^+	$+^3$	4	4	4	4	$+^-$	4^-	$-^2$	$+^2$
Acetoin	w^-	·	−	+	$+^-$	$+^-$	·	−	$-^w$	−
Growth in 20% bile	$+^4$	·	·	3^4	4^-	4	·	3^-	$+^s$	+
Hemolysis	−	−	−	−	$-^b$	$-^a$	+	−	−	−
Motility	−	+	−	−	−	+	+	−	−	+
Propionate ← threonine	−	·	·	−	$-^+$	−	·	$-^+$	−	−
Growth enhanced by	−	·	·	−	−	−	s	−	r	−

[a] For symbols see Table 9.2.

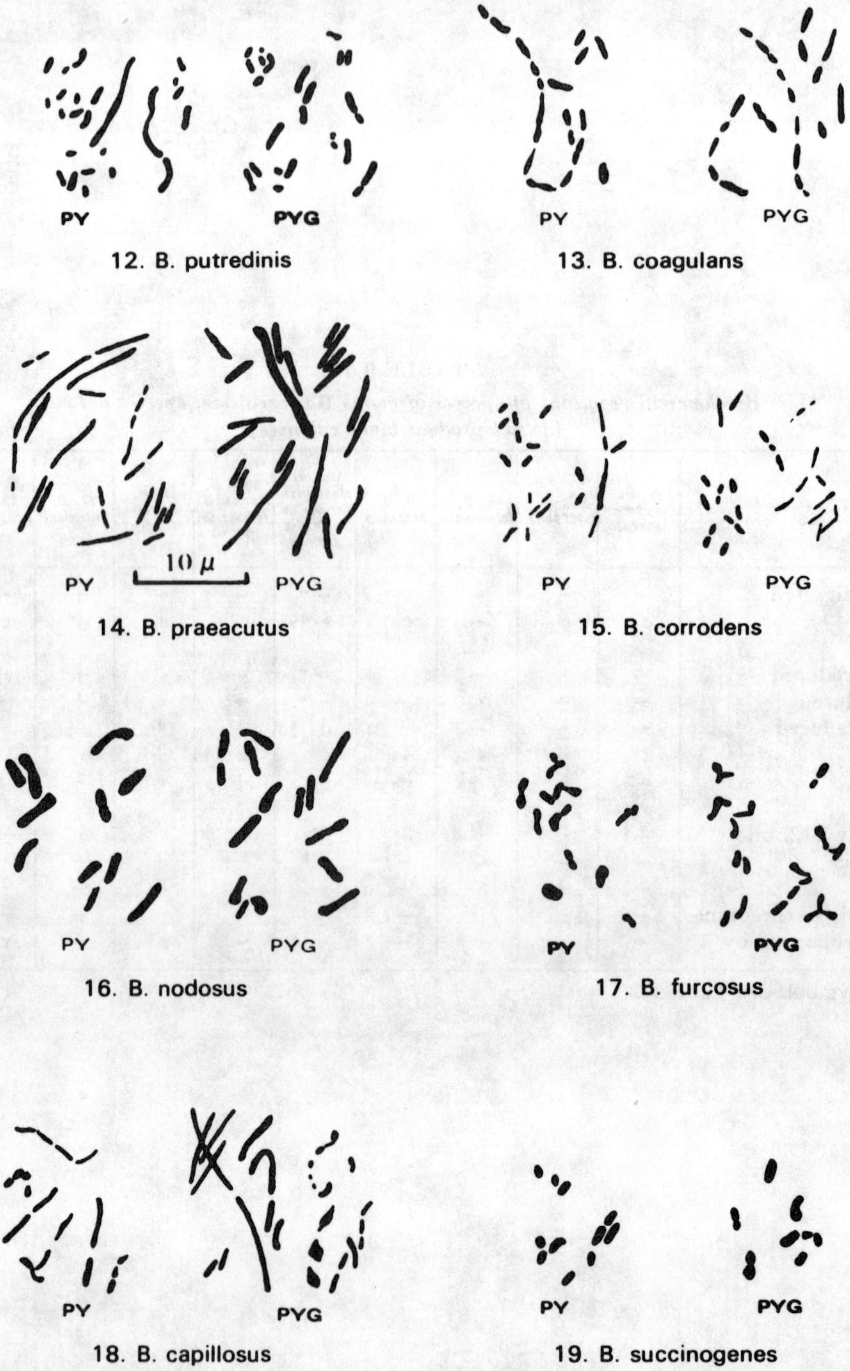

Fig. 9C. *Bacteroides*. The line drawings are composites of several cultures and individual cultures usually show somewhat less variation in cellular morphology. PY, peptone-yeast extract broth; PYG, PY-glucose broth. Scale: 9/16 inch = 10 μm.

TABLE 9.6

Fermentation reactions of species of genus **Bacteroides**, *species 15–22*[a]

	15. *B. corrodens*	16. *B. nodosus*	17. *B. furcosus*	18. *B. capillosus*	19. *B. succinogenes*	20. *B. pneumosintes*	21a. *B. melaninogenicus* ss. *melaninogenicus*	21b. *B. melaninogenicus* ss. *intermedius*	21c. *B. melaninogenicus* ss. *asaccharolyticus*	22. *B. niger*
Products from peptone yeast extract glucose broth	alsf	sa (lfp)	Lsa2 (f)	Sa(lf)	SA*piv* (f)	s(afl*iv*)	Sl (AF *ivib*)	SA*ibiv*	AB*ibiv* (splf)	AFL
Products from lactate	−	−	−	−	−	−	−	$-^{P}$	$-^{P}$	·
Acid produced from:										
Adonitol	−	−	−	−	·	−	$-^{w}$	−	−	·
Amygdalin	−	−	−	$-^{w}$	·	−	$+^{-}$	w^{-}	−	·
Arabinose	−	−	−	−	−	−	w^{-}	$-^{w}$	−	·
Cellobiose	−	−	−	$-^{w}$	w	−	$+^{-}$	−	−	·
Dulcitol	−	−	−	−	·	−	$-^{w}$	$-^{w}$	−	·
Erythritol	−	−	−	−	−	−	−	−	−	·
Esculin	−	−	−	−	−	−	−	−	−	·
Esculin hydrolyzed	−	−	$+^{-}$	+	−	−	$+^{-}$	−	−	·
Fructose	−	−	w	$-^{w}$	−	−	+	v	−	−
Galactose	−	−	w	$-^{w}$	−	−	+	d	−	·
Glucose	−	−	w^{+}	$+^{-}$	w	−	+	$+^{w}$	−	−
Glycerol	−	−	−	−	·	−	$-^{+}$	$-^{+}$	−	·
Glycogen	−	−	−	−	·	−	+	$+^{-}$	−	·
Inositol	−	−	$-^{w}$	−	·	−	−	$-^{w}$	−	·
Inulin	−	−	−	−	·	−	$+^{-}$	$-^{w}$	−	·
Lactose	−	−	−	$-^{w}$	v	−	+	$-^{+}$	−	−
Maltose	−	−	−	$-^{w}$	v	−	+	v	−	−
Mannitol	−	−	−	−	−	−	−	−	−	·
Mannose	−	−	−	$-^{w}$	−	−	+	d	−	·
Melezitose	−	−	−	−	−	−	−	−	−	·
Melibiose	−	−	−	−	−	−	w^{-}	−	−	·
Raffinose	−	−	−	−	−	−	+	$-^{+}$	−	·
Rhamnose	−	−	−	−	·	−	$-^{w}$	−	−	·
Ribose	−	−	−	$-^{w}$	·	$-^{w}$	$-^{w}$	$-^{w}$	−	·
Salicin	−	−	$-^{w}$	−	−	−	$-^{w}$	−	−	·
Sorbitol	−	−	−	−	−	−	−	−	−	·
Sorbose	−	−	−	−	·	−	$-^{w}$	−	−	·
Starch	−	−	−	$-^{w}$	v	−	+	$+^{w}$	−	·
Starch hydrolyzed	−	−	−	$-^{+}$	v	−	+	$+^{-}$	−	·
Sucrose	−	−	v	$-^{w}$	−	−	+	v	−	−
Trehalose	−	−	−	$-^{w}$	v	−	−	$-^{w}$	−	·
Xylose	−	−	−	$-^{w}$	−	−	$-^{w}$	$-^{w}$	−	·

[a] For symbols see Table 9.1.

TABLE 9.7

Biochemical reactions of species of genus **Bacteroides** *species 15–22*[a]

	15. *B. corrodens*	16. *B. nodosus*	17. *B. furcosus*	18. *B. capillosus*	19. *B. succinogenes*	20. *B. pneumosintes*	21a. *B. melaninogenicus* ss. *melaninogenicus*	21b. *B. melaninogenicus* ss. *intermedius*	21c. *B. melaninogenicus* ss. *asaccharolyticus*	22. *B. niger*
Black colonies	−	−	−	−	−	−	+	+	+	+
Gelatin digested	−	+	$-^{w}$	$-^{w}$	−	$-^{w}$	+	+	+	−
Milk	−	d^{c}	−	$-^{c}$	−	−	c	d^{c-}	d^{-}	−
Indole produced	−	−	−	−	−	−	−	$+^{-}$	$+^{-}$	−
H_2S produced	−	$+^{-}$	−	$-^{+}$	−	−	−	$-^{+}$	$+^{-}$	+
Nitrate reduced	+	−	−	−	−	−	−	−	$-^{+}$	+
Gas	−	$+^{-}$	$-^{2}$	v	−	−	$-^{+}$	$+^{-}$	2^{-}	−
Acetoin	−	−	−	−	−	−	v	−	$-^{+}$	·
Growth in 20% bile	+	−	$+^{-}$	$+^{s}$	−	$-^{+}$	$-^{3}$	2^{-}	$-^{+}$	·
Hemolysis	−	−	−	−	·	$-^{a}$	b	b^{a-}	$-^{b}$	·
Motility	−	−	−	−	−	−	−	−	−	−
Propionate ← threonine	−	$-^{+}$	−	−	−	−	−	−	−	·
Growth enhanced by	−	−	t	bth	CO_2	−	h	h	hr	s

[a] For symbols see Table 9.2.

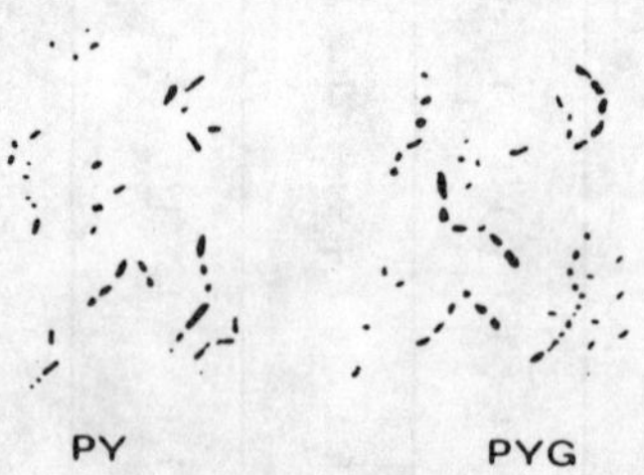

20. B. pneumosintes

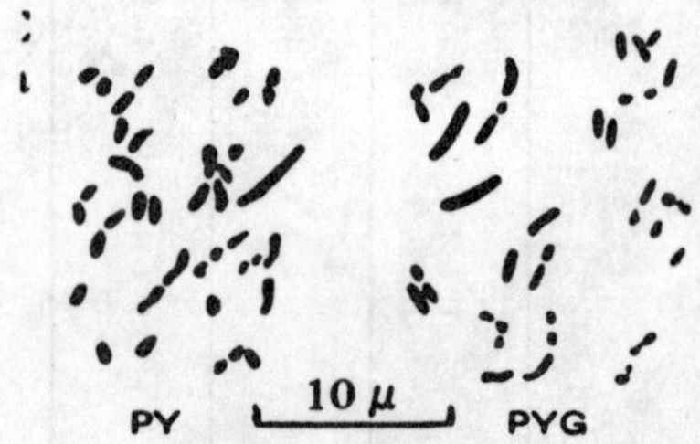

21a. B. melaninogenicus subsp. melaninogenicus

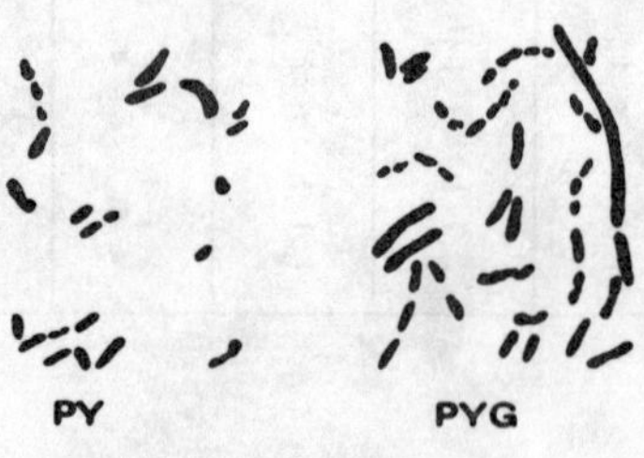

21b. B. melaninogenicus subsp. intermedius

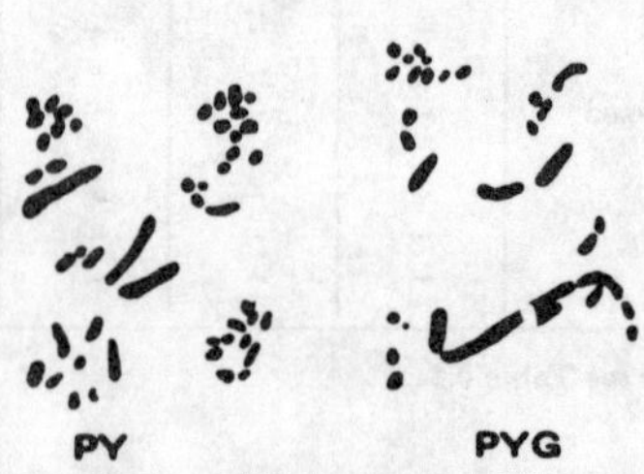

21c. B. melaninogenicus subsp. asaccharolyticus

Fig. 9D. *Bacteroides*. The line drawings are composites of several cultures and individual cultures usually show somewhat less variation in cellular morphology. PY, peptone-yeast extract broth; PYG PY-glucose broth. Scale: 9/16 inch = 10 μm.

Genus II. **Fusobacterium** *Knorr 1922, 4*

Gram-negative; non-spore-forming rods. Non-motile or motile with peritrichous flagella.

Chemoorganotrophs: Metabolize carbohydrates or peptone. **Major products from carbohydrate or peptone include butyric acid,** often with acetic and lactic acids, and lesser amounts of propionic, succinic, and formic acids and short-chained alcohols. Strains of some species convert threonine to propionate. Pyruvate converted to acetate and butyrate and sometimes also to formate, succinate and lactate.

Adonitol, dulcitol, erythritol, glycerol, inositol, melezitose, rhamnose, ribose, sorbitol and sorbose usually not attacked. Catalase usually not produced.

Obligately anaerobic. Will not grow on the surface of agar plates incubated aerobically. The maximum Eh permitting growth varies, depending upon the species, the size of inoculum, and the medium. Oxidized media are inhibitory for some species; addition of serum or ascitic fluid to such media may enable them to support growth.

Growth usually most rapid at 37 C and pH near 7.

The G + C content of the DNA in most of the species examined ranges from 26–34 moles % (chromatographic separation, Sebald, 1962 and T_m, Johnson, personal communication).

Found in cavities of man and other animals. Some species are pathogenic and occur in various purulent or gangrenous infections and in organ infarcts.

Characteristics by which species in the genus can be differentiated are given in the key and in Tables 9.8–9.11. Other characteristics of the species in the genus are given in the text.

Type species: *Fusobacterium nucleatum* Knorr 1922, 17.

Key for some species in the genus **Fusobacterium**

Key characteristics

Production of indole
Hydrolysis of esculin
Production of propionate from lactate or threonine
Acid from:
- lactose
- mannose
- maltose
- mannitol
- sucrose

Gas production in glucose agar deeps
Cell shape
Nitrate reduction
Motility

I. Indole produced.
- A. Esculin not hydrolyzed.
  - 1. Propionate not formed from lactate.
    - a. Lactose not acid.
      - b. Mannose not acid.
        - c. Propionate formed from threonine.
          - d. Little or no gas formed in glucose agar deeps.
            - 1. *F. nucleatum*
          - dd. Much gas formed in glucose agar deeps.
            - 2. *F. gonidiaformans*
        - cc. Propionate not formed from threonine.
          - 3. *F. naviforme*
      - bb. Weak acid from mannose.

4. *F. varium*

aa. Lactose acid.

5. *F. glutinosum*

2. Propionate formed from lactate.

6. *F. necrophorum*

B. Esculin hydrolyzed.

7. *F. aquatile*

II. Indole not produced.

A. Abundant gas produced in glucose agar deeps.

1. Maltose and mannitol acid.

8. *F. stabile*

2. Maltose and mannitol not acid.

a. Lactose acid.

b. Esculin hydrolyzed.

9. *F. mortiferum*

bb. Esculin not hydrolyzed.

10. *F. symbiosum*

aa. Lactose not acid.

b. Sucrose acid.

c. Elongate cells present.

11. *F. necrogenes*

cc. Short ovoid cells.

12. *F. perfoetens*

bb. Sucrose not acid.

c. Esculin hydrolyzed.

11. *F. necrogenes*

cc. Esculin not hydrolyzed.

d. Threonine converted to propionate.

4. *F. varium*

dd. Threonine not converted to propionate.

10. *F. symbiosum*

B. Little or no gas produced in glucose agar deeps.

1. Esculin hydrolyzed.

13. *F. prausnitzii*

2. Esculin not hydrolyzed.

a. Nitrate reduced.

14. *F. plauti*

aa. Nitrate not reduced.

b. Motile.

15. *F. bullosum*

bb. Not motile.

16. *F. russii*

TABLE 9.8

Fermentation reactions of species of genus **Fusobacterium**, *species 1–8*[a]

	1. *F. nucleatum*	2. *F. gonidiaformans*	3. *F. naviforme*	4. *F. varium*	5. *F. glutinosum*	6. *F. necrophorum*	7. *F. aquatile*	8. *F. stabile*
Products from peptone-yeast extract-glucose broth	Baps (Lf)	Bap (lfs)	BLap (fs)	BLA (p)	Bap (Lfs)	Bap (Lsf)	Ba(ls)	FBL
Lactate → propionate	−	−	−	−	−	P	−	·
Acid produced from:								
Amygdalin	−	−	−	−	−	−	w	·
Arabinose	−	−	−	−	−	−	w^{+}	·
Cellobiose	−	−	−	−	−	−	+	·
Dextrin	−	−	−	−	−	−	d	·
Fructose	w^{-}	−	−	w	+	$-^{w}$	+	+
Galactose	−	−	−	$-^{w}$	w	−	w	·
Glucose	$-^{w}$	w^{-}	$-^{+}$	w^{+}	+	$-^{w}$	$+^{w}$	+
Glycogen	−	−	−	−	−	−	$+^{w}$	·
Inulin	−	−	−	−	−	−	w	·
Lactose	−	−	−	−	+	−	$+^{w}$	+
Maltose	−	−	−	−	w	−	$+^{w}$	+
Mannitol	−	−	−	−	−	−	−	+
Mannose	−	−	−	w	−	−	+	·
Melibiose	−	−	−	−	−	−	$-^{w}$	·
Raffinose	−	−	−	−	−	−	w	·
Salicin	−	−	−	−	−	−	$+^{w}$	·
Starch	−	−	−	−	−	−	w	+
Starch hydrolyzed	−	$+^{-}$	−	−	−	−	+	·
Sucrose	−	−	−	−	+	−	+	+
Trehalose	−	−	−	−	−	−	−	·
Xylose	−	−	−	−	−	−	w^{+}	·

[a] Products from PYG broth: Capital letters indicate >1 meq acid/100 ml broth; small letters <1 meq/100 ml. A, acetic; B, butyric; F, formic; iV, isovaleric; L, lactic; P, propionic; S, succinic; v valeric; 2, ethanol; ( ) products of occasional strains. +, strong acid, i.e. pH 5.5 or below in 90–100% of strains. w, weak acid, i.e. pH 5.5–6.0 in 90–100% of strains. (Uninoculated xylose and arabinose, under CO_2 (pre-reduced media), are often pH 5.9. For cultures in these media, pH 5.4–5.7 are considered weak acid, below pH 5.4 strong acid.) −, acid not produced, i.e. pH 6.0 or above in 90–100% of strains. d, reaction positive 40–60% strains. v, reaction variable within a strain. When any of the above used as superscripts, reaction of 10–40% of strains. ·, not tested. Acid products determined by procedures given in Holdeman and Moore, 1972.

TABLE 9.9

Biochemical reactions of species of genus **Fusobacterium**, *species 1–8*[a]

	1. *F. nucleatum*	2. *F. gonidiaformans*	3. *F. naviforme*	4. *F. varium*	5. *F. glutinosum*	6. *F. necrophorum*	7. *F. aquatile*	8. *F. stabile*
Gelatin	$-^{w}$	w^{-}	−	$-^{w}$	−	v	−	d
Milk	−	−	−	−	c	cd^{-}	c	c
Indole produced	+	+	+	$+^{-}$	+	+	+	−
Nitrate reduced	−	−	−	−	−	−	−	−
Esculin hydrolyzed	−	−	−	−	−	−	+	·
Gas	$-^{3}$	4^{2}	$-^{2}$	4	2	4^{2}	4^{2}	4
Acetoin	$-^{+}$	−	−	−	−	−	+	·
Lecithinase	−	−	−	−	−	−	−	·
Lipase	−	−	−	−	−	$+^{-}$	−	·
Hemolysis	$-^{ab}$	a^{-}	$-^{b}$	$-^{a}$	w	b^{a-}	−	·
Motility	−	−	−	−	−	−	+	−
Hippurate hydrolyzed	$-^{+}$	v	$-^{+}$	$-^{+}$	+	−	d	·
20% Bile − growth	$-^{2}$	−	$-^{2}$	4	4	$-^{4}$	s	·
Propionate ← threonine	+	+	−	+	+	+	d	·
Gr. enhanced by	t	−	tr	−	−	−	b	·

[a] +, positive reaction—90–100% of strains. −, negative reaction—90–100% of strains. Hemolysis—a = alpha; b = beta; w = weak. Gas and growth—relative amounts indicated by +, 2, 3, 4; s = stimulated; growth enhanced by b, bile; h, hemin; r, rumen fluid; t, Tween 80. milk—c, curd, d, digested. gelatin—w (weak) after chilling until uninoculated controls are solidified and returning to room temp., liquefaction in less than half the time of the controls. Any of the above used as superscripts indicates reaction of 10–40% of strains. Reactions determined by procedures given in Holdeman and Moore (1972).

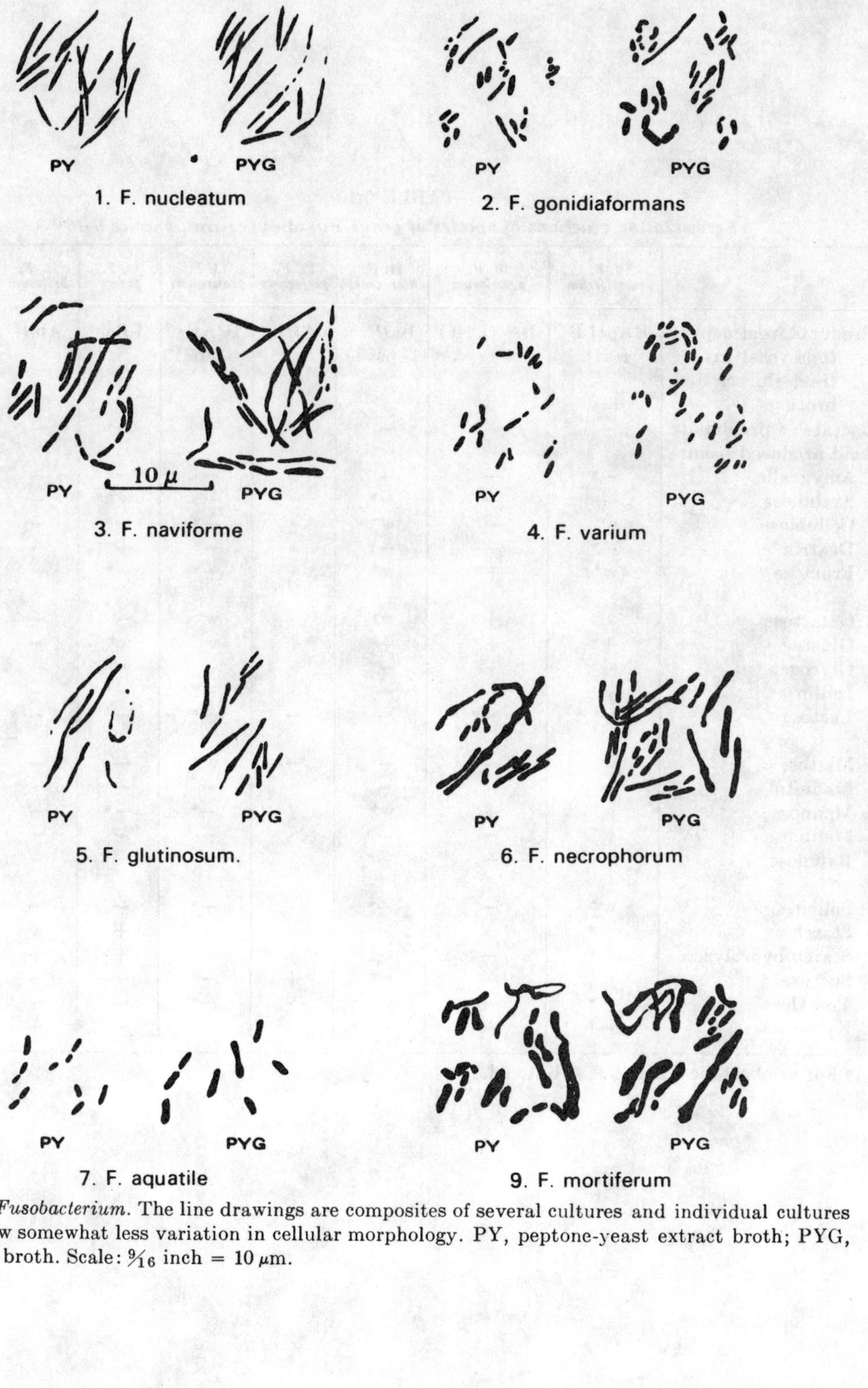

Fig. 9E. *Fusobacterium.* The line drawings are composites of several cultures and individual cultures usually show somewhat less variation in cellular morphology. PY, peptone-yeast extract broth; PYG, PY-glucose broth. Scale: 9/16 inch = 10 μm.

TABLE 9.10

Fermentation reactions of species of genus **Fusobacterium,** *species 9–16*[a]

	9. *F. mortiferum*	10. *F. symbiosum*	11. *F. necrogenes*	12. *F. perfoetens*	13. *F. prausnitzii*	14. *F. plauti*	15. *F. bullosum*	16. *F. russii*
Products from peptone-yeast extract-glucose broth	BAp(LF *iv*s4)	BA (L24f)	Ba(lf psv4)	ABL	B(AF lsp2)	LBas	Abp4	Ba(lf s24)
Lactate → propionate	−	−	−	·	−	−	−	−
Acid produced from:								
Amygdalin	$-^{w}$	−	−	·	−	−	−	−
Arabinose	−	$-^{+}$	$-^{w}$	·	−	w^{-}	−	−
Cellobiose	w^{-}	$-^{+}$	w^{-}	·	$-^{w}$	$-^{w}$	−	−
Dextrin	$-^{w}$	−	$-^{w}$	·	$-^{w}$	$+^{-}$	−	−
Fructose	w^{+}	$+^{w}$	w^{+}	·	$-^{w}$	$-^{w}$	−	−
Galactose	v	$+^{-}$	w^{+}	·	w^{-}	−	−	−
Glucose	$+^{w}$	$+^{w}$	$+^{w}$	+	v	w^{+}	−	$-^{w}$
Glycogen	−	−	−	·	−	w^{+}	−	−
Inulin	−	−	−	·	$-^{w}$	$-^{w}$	−	−
Lactose	w^{+}	$-^{w}$	−	−	$-^{w}$	−	−	−
Maltose	w^{-}	−	−	−	w^{-}	−	−	−
Mannitol	−	w^{-}	−	·	−	−	−	−
Mannose	w^{+}	$+^{-}$	w^{+}	·	$-^{w}$	$-^{w}$	−	−
Melibiose	v	−	$-^{w}$	·	−	−	−	−
Raffinose	v	−	$-^{w}$	·	−	−	−	−
Salicin	w^{-}	−	$-^{w}$	·	$-^{w}$	−	−	−
Starch	$-^{w}$	−	−	·	−	+	w	−
Starch hydrolyzed	$-^{+}$	−	−	·	−	$+^{-}$	+	−
Sucrose	v	−	w^{-}	+	$-^{w}$	−	−	−
Trehalose	$-^{w}$	$-^{w}$	w^{-}	·	$-^{w}$	−	−	−
Xylose	$-^{+}$	−	−	·	−	−	−	−

[a] For symbols see Table 9.8; 4, butanol.

TABLE 9.11

Biochemical reactions of species of genus **Fusobacterium**, *species 9–16*[a]

	9. *F. mortiferum*	10. *F. symbiosum*	11. *F. necrogenes*	12. *F. perfoetens*	13. *F. prausnitzii*	14. *F. plauti*	15. *F. bullosum*	16. *F. russii*
Gelatin	−	$-^{w}$	−	·	$-^{w}$	−	−	$-^{w}$
Milk	−	$-^{c}$	−	−	−	−	−	−
Indole produced	−	−	−	−	−	−	−	−
Nitrate reduced	−	−	−	·	−	+	−	−
Esculin hydrolyzed	+	−	+	·	+	−	−	−
Gas	4	4^{2}	4	4	$-^{2}$	$-^{+}$	2	2^{-}
Acetoin	−	v	−	·	$-^{w}$	−	−	−
Lecithinase	−	−	−	·	−	−	−	−
Lipase	−	−	−	·	−	−	−	−
Hemolysis	−	$-^{b}$	$-^{b}$	·	$-^{a}$	−	−	−
Motility	−	d	−	−	−	$+^{-}$	+	−
Hippurate hydrolyzed	−	−	−	·	−	−	+	$-^{+}$
20% Bile − growth	4^{s}	v	4	·	4^{+}	$+^{2}$	2	3^{-}
Propionate ← threonine	+	−	+	·	−	−	−	−
Gr. enhanced by	b	−	−	·	rh	−	−	−

[a] For symbols see Table 9.9.

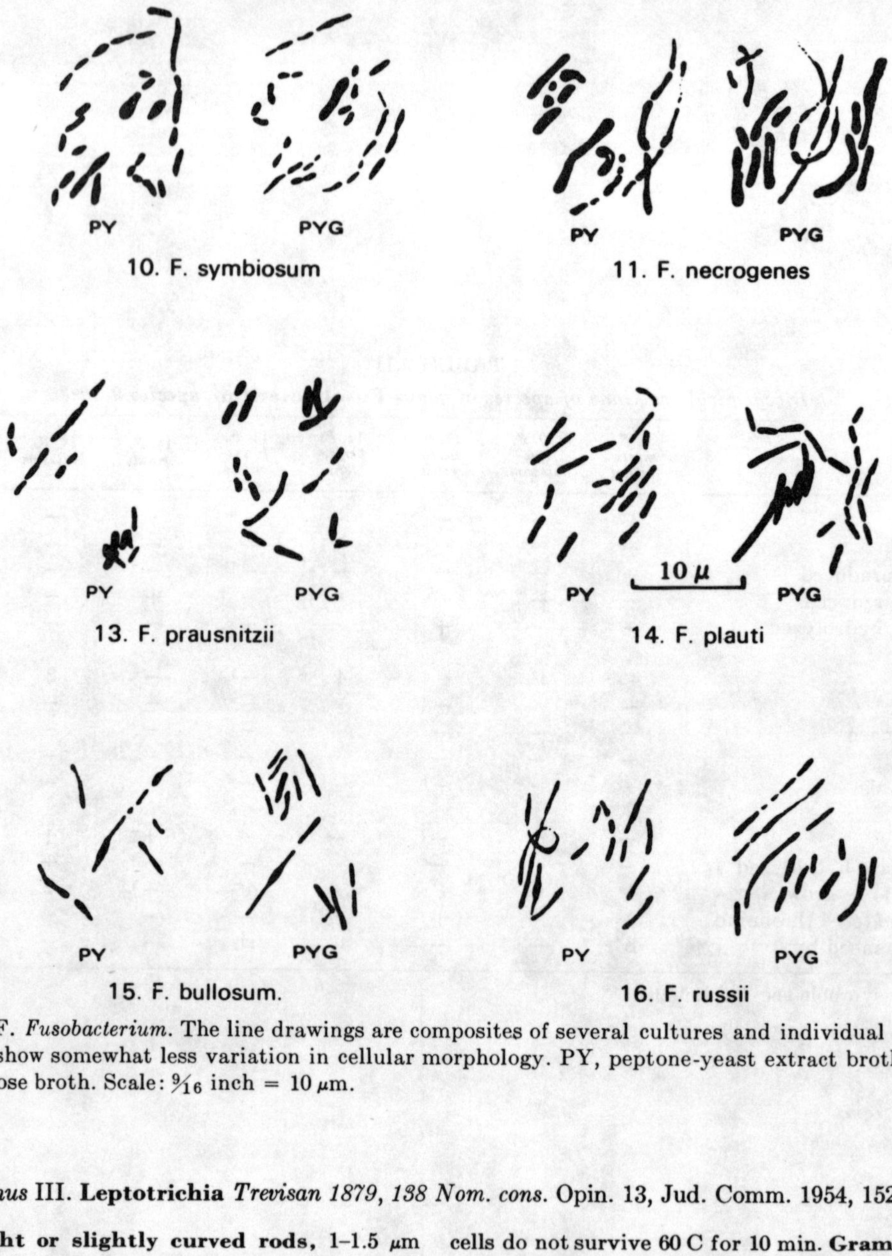

Fig. 9F. *Fusobacterium.* The line drawings are composites of several cultures and individual cultures usually show somewhat less variation in cellular morphology. PY, peptone-yeast extract broth; PYG, PY-glucose broth. Scale: 9/16 inch = 10 μm.

Genus III. **Leptotrichia** *Trevisan 1879, 138 Nom. cons.* Opin. 13, Jud. Comm. 1954, 152

Straight or slightly curved rods, 1–1.5 μm wide by 5–15 μm long, with one or both ends rounded or pointed; **many cells with pointed ends are generally found** in every culture. Two or more cells arranged in septate filaments of varying length, in older cultures filaments up to 200 μm long may twist around each other and large coccoid bodies or bulbous swellings within a filament may sometimes be found as cells lyse.

There is **no club formation or branching. Cells are non-motile.** No resting stages known; cells do not survive 60 C for 10 min. **Gram-negative.** Granules, which may be Gram-positive, are distributed rather evenly along the long axis but one or more large granules may localize near the end of the cell.

Heterotrophic with complex nutritional requirements. **Acid** without gas is **produced from glucose;** the **major products are DL-lactic acid,** about 90%, and acetic acid, about 10%; **butyric acid is not a product.**

Hydrogen sulphide and indole are not pro-

duced; **catalase, benzidine** and Voges Proskauer reactions are **negative;** nitrate usually not reduced; gelatin is not liquefied.

Anaerobic; 5% carbon dioxide is essential for successful isolation and optimal growth.

Good growth at 35–37 C; little or no growth at 25 C, no growth at 45 C.

Found in the oral cavity of man.

The **G + C content** of one strain examined was **31.5 moles %,** chromatographic analysis (Lee *et al.*, 1956); a second strain had a **G + C content of 34 moles %** (T_m) (Hofstad, 1970).

Type species: *Leptotrichia buccalis* (Robin) Trevisan 1879, 147.

TABLE 9.12

Characteristics differentiating **Leptotrichia, Fusobacterium,** *and* **Bacterionema**

	Leptotrichia buccalis	*Fusobacterium nucleatum*	*Bacterionema matruchotii*
Morphology	Unbranched rods, pointed ends present	Unbranched rods, pointed ends present	Branched filaments, bacillus-like body at one end
Gram-reaction and cell structure	Gram-negative	Gram-negative	Gram-positive
Relation to oxygen	Anaerobic	Anaerobic	Preferentially aerobic
G + C moles %	31.5–34	27–28	?
Catalase	−	−	+
H_2S	−	+	−
Indole	−	+	−
Nitrate reduction	−	±	+
Heterofermentative	−	+	+
Butyric acid produced	−	+	−
Propionic acid produced	−	−	+
Ethanol produced	−	+	?
Ammonia produced	−	+	?
CO_2 produced	−	+	+
Voges Proskauer	−	?	+
Foul odor	−	+	−

GENERA OF UNCERTAIN AFFILIATION

Genus Desulfovibrio *Kluyver and van Niel 1936, 397*

Curved rods, sometimes sigmoid or spirilloid. Morphology influenced by age and environment. Descriptions refer to freshly grown cultures in Baars's medium (1930) containing 0.02–0.1% NaCl. **Motile** by means of **polar flagella. Do not form endospores. Gram-negative.**

Chemoorganotrophs which obtain energy by anaerobic respiration reducing sulfates or other reducible sulfur compounds to H_2S. Lactate, pyruvate and usually malate, are oxidized to acetate and CO_2. Carbohydrates rarely utilized, gas never formed from carbohydrates.

Cells contain C_3 cytochromes and desulfoviridin, absorption band 630 nm; the latter is responsible for the **characteristic red fluorescence of cells** if inspected in light of 365 nm immediately after the addition of a few drops of 2.0 N NaOH; this reaction releases the chromophore of the pigment. Organic growth factors not required.

Hydrogenase usually present. Gelatin not liquefied, nitrates not reduced. Nitrogen sometimes fixed.

Strict anaerobes.

Upper temperature limit for growth 44 C. Optimum influenced by origin and previous history, usually 25–30. Some strains can grow at temperatures at or below 0 C.

Some species and subspecies moderately halophilic.

Species generally show some degree of antigenic cross-reaction. Pathogenicity not recorded.

Type species: *Desulfovibrio desulfuricans* (Beijerinck) Kluyver and van Niel 1936, 397.

TABLE 9.13
Differential characteristics of species of genus **Desulfovibrio**

	1a. *D. desulfuricans* subsp. *desulfuricans*	1b. *D. desulfuricans* subsp. *aestuarii*	2a. *D. vulgaris* subsp. *vulgaris*	2b. *D. vulgaris* subsp. *oxamicus*	3. *D. salexigens*	4. *D. africanus*	5. *D. gigas*
Dimensions μm	0.5–1 by 3–5		0.5–1 by 3–5		0.5–1 by 3–5	0.5 by 5–10	1.2–1.5 by 5–10
Flagella polar							
Single	+	+	+	+	+	−	−
Lophotrichous	−	−	−	−	−	+	+
Thickness of flagellum (nm)	20–25	20–25	20–25		20–25	12	9
Growth in[a]							
Pyruvate minus sulfate	+	+	−	+	−	−	−
Malate plus sulfate	+	+	−	−	+	+	−
Malate minus sulfate	−	−	−	−	−	−	−
Choline plus sulfate	+	+	−	+	−	−	−
Choline minus sulfate	+	+	−	+	−	−	−
NaCl requirement	−	+	−	−	+	−	−
Hibitane resistance mg/l	10–25	10–25	2.5	2.5	1000	2.5	2.5
G + C mole %	55.3 ± 1		61.2 ± 1		46.1 ± 1	61.2 ± 1	60.2
Eh requirement (mv) for growth at pH 7.2	−100		−100		−100	−100	80

[a] All species grow in lactate with sulfate and in pyruvate with sulfate.

Genus **Butyrivibrio** *Bryant and Small 1956, 18*

Curved rods, single or in chains or filaments which may or may not be helical. **Motile** with **monotrichous, polar** or sub-polar **flagella.** No resting stages known. **Gram-negative.**

Chemoorganotrophs: **Metabolism fermentative, carbohydrates** being the main fermentable substrates. **Glucose is fermented with production of butyrate** as one of the important products. Some strains in some media produce much lactic acid and little butyric acid (Gill and King, 1958; Lee and Moore, 1959).

Strict anaerobe.

Type species: *Butyrivibrio fibrisolvens* Bryant and Small 1956, 19.

Genus **Succinivibrio** *Bryant and Small 1958, 22*

Curved rods with **pointed ends. Cells** are **helical** with less than one to three or more coils per cell and may become straight to slightly curved with pointed ends after maintenance on artificial media. Rapid progressive vibrating motility with **monotrichous** and **polar flagellation.** No resting stages known. **Gram-negative.**

Chemoorganotrophs: **Metabolism fermentative, carbohydrates** being the fermentable substrates. **Major products** from glucose are **succinate, acetate,** formate and sometimes lactate. **Butyrate or hydrogen are not formed.** A large net uptake of CO_2 may occur.

Strict anaerobe.

Type species: *Succinivibrio dextrinosolvens* Bryant and Small 1958, 22.

Genus **Succinimonas** *Bryant, Small, Bouma and Chu 1958, 21*

Straight rods with rounded ends, usually short to coccoid. Motile, with monotrichous polar flagella. No spores or other resting stages known. **Gram-negative.**

Chemoorganotrophs: **Metabolism fermentative, carbohydrates** being the **fermentable substrates. Large amounts of succinate** and some acetate are the major products of glucose fermentation. **No butyrate or gas is formed.**

Strict anaerobe. Catalase is not produced.

Type species: *Succinimonas amylolytica* Bryant *et al.* 1958, 21.

Genus **Lachnospira** *Bryant and Small 1956, 24*

Curved rods, motile with lateral to subterminal monotrichous flagella. Weakly Gram-positive.

Ferments glucose producing large amounts of ethanol, lactic, formic and acetic acids, and CO_2; small amounts of hydrogen also produced. Not proteolytic.

Anaerobic.

Found in bovine rumen in significant numbers and probably present in rumen of other animals.

Type species: *Lachnospira multiparis* Bryant and Small 1956, 24.

Genus **Selenomonas** *von Prowazek 1913, 36 Nom. cons.* Opin. 21. Jud. Comm. 1958, 163

Curves **curved to helical rods,** ends usually tapered and rounded to give **kidney- to crescent-shaped short cells.** Long cells and chains of cells are helical. Strains differ markedly in size; pure cultures are often smaller than those in the natural habitat. **Motile with active tumbling;** up to 16 **flagella present as a tuft often near the center of the concave side** at or near the point of cell fission. Long cells with several tufts often seen. Spheroplasts with flagella attached are sometimes seen. Capsules not formed. Resting stages not known. Gram-negative.

Chemoorganotrophs: **Metabolism fermentative, carbohydrates and, sometimes, certain amino acids and lactic acid being the fermentable substrates. Fermentation of glucose yields chiefly acetate, propionate, CO_2 and/or lactate.** Organic growth factors required but not adequately studied. Usually grown in complex media.

Strict anaerobes, catalase not produced.

Temperature optimum 35–40 C; maximum about 45 C; minimum 20–30 C. Grow at lower pH (4.5–5.0) than most anaerobic, Gram-negative, motile rods.

Found mainly in the gastrointestinal tract of mammals but seen in dirty river water.

The **G + C content** of the DNA ranges from **53–61 moles %** (buoyant density).

Type species: *Selenomonas sputigena* (Flügge) Boskamp 1922, 70 Opin. 21, Jud. Comm. 1958, 163.

Key to the species of genus **Selenomonas**

I. Found in the human buccal cavity. Cellobiose, salicin and dulcitol are not fermented; H_2S is usually not produced. G + C content of DNA is 60.6 ± 0.8 moles %.

1. *Selenomonas sputigena*

II. Found in rumen contents of ruminants. Cellobiose, salicin and dulcitol are fermented; H_2S is usually produced. G + C content of DNA is 54.0 ± 0.8 moles %.

2. *Selenomonas ruminantium*

PART 10

GRAM-NEGATIVE COCCI AND COCCOBACILLI

FAMILY I. **NEISSERIACEAE** PRÉVOT 1933, 119

Organisms spherical in pairs or in masses with adjacent sides flattened, or rod-shaped in pairs or short chains. Not flagellated. Some species show twitching motility. Gram-negative.

Some species form xanthophyll pigment. Some species have complex growth requirements immediately after isolation, but may later grow in simple defined media. Catalase and cytochrome-oxidase produced by most species.

Aerobic.

Optimum temperature about 32–37 C.

Neisseria, *Branhamella* and *Moraxella* parasitic; *Acinetobacter* saprophytic or an opportunistic pathogen. Other characteristics differentiating the genera are given in Table 10.1.

The G + C content of the DNA ranges from 39–52 moles %.

Type genus: *Neisseria* Trevisan 1885, 105.

TABLE 10.1

Differential characteristics of the genera of family **Neisseriaceae**[a]

	I. *Neisseria*	II. *Branhamella*	III. *Moraxella*	IV. *Acinetobacter*[b]
Cells	Coccal	Coccal	Plump rods in pairs or short chains	Rod-shaped in exponential phase but nearly spherical in stationary phase
Division	Two planes	Two planes	One plane	One plane
Growth in simple defined media	d	−	d	+
Penicillin	Originally sensitive	Originally sensitive	Originally sensitive	Originally resistant
Oxidase	+	+	+	−
Reduction of nitrate	− or d	+	d	−
G + C moles %	47–52	40–45	40–46	39–47

[a] +: Most strains positive (≧90%). −: Most strains negative (≧90%). d: Some strains positive, some negative.

[b] Temporarily associated with *Neisseriaceae* (Report Subcommittee on Taxonomy of *Moraxella* and Allied Bacteria 1971. Int. J. Syst. Bacteriol., *21:* 213. Min. 13.

Note: There are two genera of uncertain affiliation recognized: *Paracoccus*, members of which reduce nitate to nitrogen gas, and *Lampropedia*, which contains cocci occurring in square, flat tablets.

Genus I. **Neisseria** *Trevisan 1885, 105. Nom. cons.* Opin. 13, Jud. Comm. 1954, 153

Cocci, 0.6–1.0 μm in diameter, occurring singly but often **in pairs** with **adjacent sides flattened.** Division in two planes at right angles to each other sometimes resulting in the formation of tetrads. Cell wall of the Gram-negative type (Plate 10.1, Fig. 1). No formation of endospores; non-motile.

Capsules and fimbriae may be present.

Gram-negative.

Yellow-greenish pigment produced by two species.

Complex growth requirements. Some species are hemolytic.

Chemoorganotrophic. **Few carbohydrates utilized** (Table 10.2).

Aerobic or facultatively anaerobic. Temperature optimum about 37 C.

Catalase and cytochrome-oxidase produced.

Parasites of mucous membranes of mammals.

The G + C content of the DNA (reported for all species) ranges **from 47.0–52.0 moles %**. Few or no genetic affinities with *Branhamella*.

Type species: *Neisseria gonorrhoeae* (Zopf) Trevisan 1885, 106.

Further Comments

Members of the genus *Neisseria* are susceptible to drying and exposure to sunlight. *N. gonorrhoeae* and *N. meningitidis* autolyse easily; they are also less resistant than other *Neisseria* to disruption by pressure in a Ribi Cell Fractionator (Martin *et al.*, 1969).

Generally, the sensitivity to various antibiotics is taxonomically not very important since this quality varies from one species to another and also within species. The tendency to *in vivo* and *in vitro* development of decreased sensitivity and complete resistance should be borne in mind. However, most *Neisseria* are sensitive to penicillin, streptomycin, tetracyclines, the macrolide group and polymyxin B sulfate and related drugs, (*N. gonorrhoeae* and *N. meningitidis* are relatively resistant to polymyxin B sulfate); they are relatively resistant to vancomycin and related drugs. *N. mucosa* is very sensitive to chloramphenicol (Véron *et al.*, 1959; Berger and Miersch, 1970). Exact inhibitory concentrations depend upon the method used.

The differential diagnosis between *N. gonorrhoeae* and *N. meningitidis* is sometimes difficult as acid production from glucose and/or maltose may be weak or absent. After 24 hr. incubation on 10% horse blood agar plates single colonies of *N. meningitidis* are generally larger than those of *N. gonorrhoeae*, and the occurrence of β-hemolysis on 5% horse blood agar after 48–72 hr. incubation points to *N. meningitidis*. Lack of acid production from glucose and maltose, and growth at 22 C points to *Branhamella catarrhalis* (see p. 432), especially if nitrates are reduced.

TABLE 10.2

Characteristics differentiating the species of genus **Neisseria**[a]

	1. *N. gonorrhoeae*	2. *N. meningitidis*	3. *N. sicca*	4. *N. subflava*	5. *N. flavescens*	6. *N. mucosa*
Capsules	−	v	v	+	−	+
Acid from:						
Glucose	+	+	+	+	−	+
Maltose	−	+	+	+	−	+
Fructose	−	−	+	v	−	+
Sucrose	−	−	+	v	−	+
Starch	−	−	v	v	−	+
Polysaccharide produced from 5% sucrose	0	0	+	d	+	+
Production of H_2S	−	−	+	+	+	+
Reduction of:						
Nitrate	−	−	−	−	−	+
Nitrite	−	d	+	+	+	+
Pigment	−	−	d	+	+	−
Extra CO_2	//	/	*	*	*	*
Growth at 22 C	−	−	d	d	+	+
G + C moles %	49.5–49.6	50.0–51.5	49.0–51.5	48.0–50.5	46.5–50.1	50.5–52.0

[a] +: Most strains positive (≧90%). −: Most strains negative (≧90%). d: Some strains positive, some negative. v: Character inconstant and in one strain may sometimes be positive, sometimes negative. 0: No growth on medium with 5% sucrose. G + C: Guanine + cytosine in the deoxyribonucleic acid. ?: Not tested. //: Very important. /: Important. *: not necessary.

Genus II. **Branhamella** *Catlin 1970, 157*

Cocci commonly arranged in pairs with adjacent sides flattened; division in two planes at right angles to each other. Cell wall of the Gram-negative type (Plate 10.2, Fig. 1). No formation of endospores; non-motile. **Gram-negative** with a tendency to resist decoloration.

Chemoorganotrophic. **Acid not produced from carbohydrates.** No xanthophyll pigment.

Growth occurs on customary media without blood. Aerobic. Temperature optimum about 37 C. **Catalase and cytochrome-oxidase produced; nitrates usually reduced.**

Parasites of mucous membranes of mammals.

The G + C content of the DNA ranges **from 40 to about 45 moles %**. Differs from *Neisseria* in DNA base content and fatty acid composition; few or no genetic affinities with *Neisseria*.

Type species: *Branhamella catarrhalis* (Frosch and Kolle) Catlin.

Genus III. **Moraxella** *Lwoff 1939, 173 Nom. cons.* Opin. 41, Jud. Comm. 1971, 106

Rods, usually very short and plump, typically 1.0–1.5 by 1.5–2.5 μm; often approaching coccus shape, predominantly in pairs and short chains. Some cultures fairly uniform, others pleomorphic showing variation in cell size and shape, filaments and long chains. Pleomorphism enhanced by lack of oxygen and above optimum temperatures. No spores formed; flagella not present. Some strains show under special conditions a "twitching" motility on solid surfaces. Capsules and fimbriae may or may not be present. Gram-negative.

Chemoorganotrophic with oxidative metabolism. Most strains more or less fastidious, but specific growth requirements not known. A limited number of organic acids, alcohols and amino acids serve as carbon and energy sources. **Carbohydrates not utilized.**

Oxidase positive, catalase usually positive. Indole, acetoin and H_2S not produced.

Strict aerobes. Optimum temperature 32–35 C. Optimum pH 7–7.5. **Highly sensitive to penicillin.**

Parasites on the mucous membranes of man and warm-blooded animals and possibly also saprophytes.

The G + C content of the DNA ranges from 40–46 moles % (buoyant density).

Type species: *Moraxella lacunata* (Eyre) Lwoff 1939, 173.

Key to the species of genus **Moraxella**

Key characteristics

Growth on mineral medium with acetate and ammonium salts
Liquefaction of coagulated serum
Hemolysis of chocolate and blood agars
Reduction of nitrate to nitrite
Production of phenylalanine deaminase

I. Do not grow in mineral medium with acetate and ammonium salts.
  A. Coagulated serum liquefied. Hemolysis on chocolate agar.
    1. Nitrite usually produced. Blood agar not hemolyzed.
      1. *M. lacunata*
    2. Nitrite not produced. Blood agar usually hemolyzed.
      2. *M. bovis*
  B. Coagulated serum not liquefied. No hemolysis on chocolate agar.
    1. Phenylalanine deaminase absent.
      3. *M. nonliquefaciens*
    2. Phenylalanine deaminase present.
      4. *M. phenylpyruvica*
II. Grow in mineral medium with acetate and ammonium salts.
      5. *M. osloensis*

Genus IV. **Acinetobacter** *Brisou and Prévot 1954, 727*

Rods, usually very short and plump, typically 1.0–1.5 by 1.5–2.5 μm in logarithmic phase, approaching coccus shape in stationary phase; predominantly in pairs and short chains. Large, irregular cells and filaments in small numbers in all cultures, but may dominate. No spores formed; flagella not present. Some strains show under special conditions a "twitching" motility on solid surfaces. Capsules and fimbriae may or may not be present. Intracellular poly-β-hydroxybutyrate inclusions do not occur. Gram-negative.

Chemoorganotrophic with oxidative metabolism. Usually versatile in the utilization of organic compounds as carbon and energy source. No specific growth requirements. Acid may or may not be formed from sugars.

Oxidase negative, catalase positive. Acetoin, indole and H_2S not produced.

Strict aerobes. Optimum temperature 30–32 C. Optimum pH about 7. **Resistant to penicillin.**

Free-living ubiquitous saprophytes.

The G + C content of the DNA ranges from 40–47 moles % (thermal melting point).

Type species: *Acinetobacter calcoaceticus* (Beijerinck) Baumann, Doudoroff and Stanier 1968, 1538 (designated by Baumann *et al.*, 1968, 1538).

GENERA OF UNCERTAIN AFFILIATION

Genus **Paracoccus** *Davis in Davis, Doudoroff, Stanier and Mandel 1969, 384*

Cells **spherical** or nearly spherical, occurring singly, in pairs or aggregates. Dimensions, 0.5–1.1 μm in diameter. (Short rod-shaped cells may occur in young cultures.) **Non-motile.** Accumulate granules of poly-β-hydroxybutyrate as intracellular carbon reserve, especially in nitrogen-deficient media. No resting stages are known. **Gram-negative.**

Chemoorganotrophs: Metabolism **respiratory,** never fermentative. One species is **facultatively chemolithotrophic,** able to use the oxidation of H_2 as energy source for autotrophic growth. **Molecular oxygen or nitrate** can serve as alternative electron acceptors. Nitrate is reduced to nitrous oxide and **molecular nitrogen** under anaerobic conditions. Oxidase and catalase reactions positive.

Aerobic, except in media containing nitrate.

The G + C content of the DNA, *ca.* 64–67 moles % (from buoyant density).

Type species: *Paracoccus denitrificans* (Beijerinck) Davis.

Key to the species of genus **Paracoccus**

1. Facultatively chemolithotrophic, not halophilic, no growth factor requirements, gelatin not hydrolyzed.
   1. *Paracoccus denitrificans*
2. Not chemolithotrophic, halophilic, complex media required, gelatin hydrolyzed.
   2. *Paracoccus halodenitrificans*

Genus **Lampropedia** *Schroeter 1886, 151*

Cells rounded or almost cubical when packed together, 1.0–1.5 by 1.0–2.5 μm (Plate 10.3, Fig. 1), occurring in **pairs, tetrads and regular, squared tablets of cells.** Division takes place alternately (and synchronously) in two planes. The cells forming a tablet are enclosed within a **complex, structured envelope** (Plate 10.3, Figs. 2–4); each cell is enclosed by a conventional wall of **Gram-negative** type. **Not motile** and non-flagellated but exhibiting a **flickering movement of groups of cells** in rapidly growing cultures. No resting stages known.

Growth forms a thin, dry, wrinkled pellicle on the surface of both solid and liquid media.

Chemoorganotroph: Respiratory energy-yielding pathway. Energy sources limited to Krebs cycle intermediates; carbohydrates, alcohols, glucosides, fatty acids and many related compounds not utilized. Limited exoenzyme production and limited amino acid utilization. **Carotenoid or photosynthetic pigments not formed.** Prominent **inclusions of poly-β-hydroxybutyrate. Do not contain gas vacuoles or sulfur inclusions.** Vitamins may be required.

Obligate aerobe. Grows over temperature range 10–35 C and at pH 6–8.5.

Ecological niche unknown but rich, organic environments likely.

The G + C content is in region of 61 moles % (buoyant density).

Type species: *Lampropedia hyalina* Schroeter 1886, 151.

PART 11

GRAM-NEGATIVE ANAEROBIC COCCI

FAMILY I. **VEILLONELLACEAE** ROGOSA 1971, 232

Cocci, varying in diameter from small (*ca.* 0.3–0.5 μm) to large (*ca.* 2.5 μm); occur characteristically in pairs; single cells, masses or chains may also occur although the chains may show gaps, illustrating the basic diplococcal arrangement. Adjacent sides of cell pairs may be flattened. No endospores. Non-motile; no flagella. Gram-negative but tend to resist decolorization.

Chemoorganotrophic. Possess complex nutritional requirements. Gas is produced, frequently abundantly. Carbohydrates may or may not be fermented. Lactic acid may not be produced and if present is not a major product; lactate is fermented by some genera with the production of CO_2, H_2 and various lower volatile fatty acids containing 2–6 C atoms.

Anaerobic. Cytochrome oxidase negative. Catalase negative, but some strains decompose peroxide by a pseudocatalase (non-heme-containing).

Parasites of homothermic animals such as man, ruminants, rodents and pigs, particularly found in the alimentary tract.

The type genus: *Veillonella* Prévot 1933, 118.

The key differentiating characteristics of the genera are shown in Table 11.1.

TABLE 11.1

Characteristics of the genera of family **Veillonellaceae**[a]

	Cell diameter	G + C	Carbohydrates fermented	Lactate fermented	Amino acids main energy source	Pyruvate utilized	Succinate decarboxylated	Products in growth media: Gases CO_2	Gases H_2	Volatile fatty acid C_2	C_3	C_4	C_5	C_6
	μm	moles %												
Veillonella	0.3–0.5	40.3–44.4	−	+	−	+	+	+	+	+	+	−	−	−
Acidaminococcus	0.6–1.0	56.6	−[b]	−	+	−	−	+	−	+	−	+	−	−
Megasphaera	1.7–1.9 2.4–2.6	53.1–54.1	+	+	−	+	−	+	w	+	+	+	+	+

[a] + = positive reaction; − = negative reaction; w = slight.

[b] Generally negative, slight or variable.

Genus I. **Veillonella** *Prévot 1933, 118, emend. mut. char. Rogosa 1965, 706*

Small cocci, 0.3–0.5 μm in diameter. By light microscopy appear as spherical **diplococci, masses** and short chains. By electron microscopy, logarithmic-phase single cells appear spherical, whereas diplococci have a flattening at the cell junctions. Non-motile. Non-sporulating. **Gram-negative.**

Lactate agar media containing 7.5 μg/ml of vancomycin favor isolation; colonies in poured plates 1–3 mm in their greatest dimension; smooth; entire; lens, diamond, or heart shaped; opaque; grayish white and butyrous.

Carbohydrates and **polyols not** fermented. D-Ribose not fermented but is incorporated into nucleic acids. **Glucokinase, fructokinase** and **glucose permease not** detectable.

Cultures produce **acetate, propionate, CO_2** and **H_2 from lactate.** Hypoxanthine may be fermented.

Resting cells respire on lactate; attack oxaloacetate aerobically producing CO_2 and H_2; and anaerobically metabolize pyruvate, oxaloacetate, malate, fumarate and succinate. **CO_2** and **propionate** produced **from succinate.** Formate, citrate, isocitrate and malonate not attacked.

H_2S produced from reduced glutathione, cysteine, cystine, thiosulfate, thiocyanate and thioglycolate. Gelatin not liquefied. Non-hemolytic. Indole negative. Nitrate reduced.

Cytochrome oxidase and **benzidine negative;** but some species produce an atypical catalase lacking porphyrin.

Chemoorganotrophic. Complex nutritional requirements. CO_2 required. No growth in amino acid media containing necessary vitamins and other co-factors unless supplemented with pyruvate, which supports growth best, or with lactate, malate, fumarate and oxaloacetate; no growth with added succinate, carbohydrates, polyols, phosphorylated hexoses or trioses. Riboflavin and folic acid not required; niacin and calcium pantothenate often stimulatory but dispensable; biotin and *p*-aminobenzoic acid frequently stimulatory and sometimes indispensable; pyridoxal and thiamine indispensable; some organisms require putrescine or cadaverine.

Growth in 1% Tween 80, variable or inhibited by 10% bile, 0.0002% crystal violet or brilliant green and 1% NaCl; no growth in 4% NaCl, 0.25% phenethyl alcohol or 0.001% potassium tellurite. A variety of **inorganic compounds, dyes** and **indicators** are **reduced;** 2,3,5-triphenyl tetrazolium chloride not reduced.

Sensitive to penicillin G (0.4–3.1 units/ml); chloramphenicol, chlortetracycline, oxytetracycline, polymyxin B (all <1.0 μg/ml), and erythromycin (1.3–5.0 μg/ml); resistant to streptomycin (>25 μg/ml), neomycin (>50 μg/ml) and vancomycin (500 μg/ml).

Anaerobic.

Growth good from 30–37 C at pH 6.5–8.0, poor at 40 C, slow at 24 C, and negative at 18 and 45 C; cells killed at 60 C for 30 min. Cells suspended in skim milk can be successfully lyophilized.

Serologically specific **endotoxins** (lipopolysaccharides) induce pyrogenicity and the Schwartzman reaction in rabbits.

Parasitic in the mouth and in the intestinal and respiratory tracts of man and other animals.

The G + C content of DNA ranges from 40–44 moles % (buoyant density). Values of 34–36.5 moles % (chromatographic) cited by Rosypal and Rosypalova (1966) for *V. parvula* are incorrect.

Type species: *Veillonella parvula* (Veillon and Zuber) Prévot 1933, 119.

Key to the species of genus Veillonella

I. Catalase-negative. Does not require putrescine or cadaverine for growth.
  1. *V. parvula*

II. Catalase-positive. Requires putrescine or cadaverine for growth.
  2. *V. alcalescens*

Genus II. Acidaminococcus *Rogosa 1969, 765*

Cocci, 0.6–1.0 μm in diameter, often as oval or kidney-shaped **diplococci. Gram-negative.** Lipopolysaccharide (endotoxin) outer cell wall layer demonstrable in thin sections by electron microscopy, and by the Schwartzman reaction in rabbits. Flagella, motility and spores not demonstrable.

Weak saccharoclastic activity; glucose is not fermented by about 60% of strains and only weakly catabolized by the remainder. **Lactate, fumarate, malate, succinate, citrate and pyruvate are not used** as energy sources; pyruvate suppresses growth completely.

Amino acids, of which glutamic acid is the most important, **can serve as sole energy source for growth;** in such media acetic and butyric acids accumulate in a molar ratio of 2:1 and CO_2 is also formed; hydrogen and propionic acid are not detectable.

Nutritional requirements multiple. Tryptophan, glutamic acid, valine and arginine are required; cysteine and histidine are required by 93% of strains, tyrosine by 79%, phenylalanine and serine by 50%. Glycine is sometimes stimulatory. Alanine, leucine, isoleucine, proline, threonine, methionine, lysine and aspartic acid are not required for growth. In amino acid media, vitamin B_{12}, pyridoxal, pantothenate and biotin are indispensable for growth; *p*-aminobenzoic acid is essential or highly stimulatory; exogenous putrescine, folic acid, folinic acid, thiamine, niacin and riboflavin are not required. No growth in lactate or pyruvate media which support growth of *Veillonella*.

Ammonia produced. Gelatin generally not liquefied although slow and partial liquefaction may sometimes occur. H_2S and indole not produced. Cytochrome oxidase, catalase, benzidine and nitrate reduction tests negative. Sulfonthalein indicators not reduced.

Anaerobic. No growth on surface of agar media incubated in air.

Good growth at 30–37 C; poor or absent at 25 and 45 C. Cells do not survive 60 C for 30 min. Growth occurs at initial pH values between 6.2–7.5, although best growth at neutral reaction; final pH values in media initially at pH 7.5 range from about 6.1–6.7.

No serological cross-reactions between strains of *Acidaminococcus* and either *Veillonella* serotypes or *Peptococcus aerogenes*.

Resistant to vancomycin (7.5 μg/ml).

The G + C content of the DNA of 15 strains was 56 ± 0.9 moles % (buoyant density).

Type species: *Acidaminococcus fermentans* Rogosa 1969, 765.

Genus III. **Megasphaera** *Rogosa 1971, 187*

Relatively large cocci, 2 μm or more in diameter, in pairs; occasionally the diplococci are arranged in chains. Non-motile; non-sporing. Gram-negative.

Chemoheterotrophic. Gas produced. Lactate and glucose are fermented with the production of lower fatty acids, CO_2 and some H_2. Succinate, fumarate and malate are not attacked.

Complex nutritional requirements.

Obligately anaerobic.

Types species: *Megasphaera elsdenii* (Gutierrez *et al.*) Rogosa 1971, 189.

PART 12

GRAM-NEGATIVE CHEMOLITHOTROPHIC BACTERIA

Key to the Gram-negative chemolithotrophic bacteria

I. Organisms oxidizing ammonia or nitrite.
Family *Nitrobacteraceae*, p. 450*

II. Organisms metabolizing sulfur and sulfur compounds.
p. 456

III. Organisms depositing iron and/or manganese oxides.
Family *Siderocapsaceae*, p. 464

FAMILY I. NITROBACTERACEAE BUCHANAN 1917, 349, *emend. mut. char.* STARKEY 1948, 69; WATSON 1971, 262

Rod-shaped, ellipsoidal, spherical, spirillar and lobular cells without endospores. Flagella subpolar or peritrichous and often absent. Gram-negative. Cells derive energy from the oxidation of ammonia or nitrite and satisfy their carbon needs by the fixation of CO_2. Only a few strains of one species, *Nitrobacter winogradskyi*, shown to be facultatively chemoorganotrophic. Not parasitic. Commonly found in soil, fresh water and sea water. All of the organisms placed in this family are obligate aerobes and none requires organic growth factors. Cells are rich in cytochromes but no other pigments have been demonstrated.

The type genus of the family is *Nitrobacter* Winogradsky 1892, 127 (*gen. cons.* Opin. 23. Jud. Comm. 1958, 169).

Key to the genera of family **Nitrobacteraceae**

Key characteristics

Oxidation of nitrite or ammonia
Cell shape and size (some of the spiral forms are tightly packed and appear as rods; see Plate 12.2, Fig. 4)
Presence, location and form of cytomembranes (requires electron microscopy but this feature is not essential to using the key)

* All page numbers cited refer to *Bergey's Manual of Determinative Bacteriology*, 8th edition, The Williams & Wilkins Company, Baltimore, 1974.

I. Nitrite oxidized to nitrate.
  A. Cells rod shaped.
    1. Short rods, often wedge or pear shaped, with a polar cap of cytomembranes.
      Genus I. *Nitrobacter*
    2. Long, slender rods with no extensive cytomembrane system.
      Genus II. *Nitrospina*
  B. Cells spherical (diameter 1.54 μm or more) with cytomembranes forming a branched, tubular network in the cytoplasm.
      Genus III. *Nitrococcus*

II. Ammonia oxidized to nitrite.
  A. Cells are straight rods with *peripheral* membranes occurring as flattened lamellae.
      Genus IV. *Nitrosomonas*
  B. Cells are not straight rods and do not contain flattened lamellae in the peripheral regions.
    1. Cells are spirals with no evident cytomembranes.
      Genus V. *Nitrosospira*
    2. Cells are not spirals but do contain cytomembranes.
      a. Cells spherical (diameter 1.5 μm or more) with cytomembranes forming flattened lamellae in center of cells.
        Genus VI. *Nitrosococcus*
      b. Cells lobular and partially compartmentalized by cytomembranes.
        Genus VII. *Nitrosolobus*

Note: Harms *et al.* (*Archives of Microbiology 108:* 105–111, 1976) have recently described a new genus and species, *Nitrosovibrio tenuis*, of ammonia-oxidizing bacteria which produce slender, curved, Gram-negative, aerobic rods with sub-polar to lateral flagella and without an extensive cytomembrane system.

Genus I. Nitrobacter *Winogradsky 1892, 127, nom. cons.* Opin. 23. Jud. Comm. 1958, 169

Cells **short rods, often wedge** or **pear shaped. Reproduce by budding.** Cells possess a **polar cap of cytomembranes. Usually non-motile.** No resting stages known. **Gram-negative.** Cells rich in cytochromes imparting a yellowish color to cell suspensions; void of other pigments.

Some strains are obligate chemolithotrophs which oxidize nitrite to nitrate and fix CO_2 to fulfill energy and carbon needs. The autotrophic medium consists of fresh water or sea water enriched with nitrite and other inorganic salts; no organic growth factor being required. **Some strains** can be grown **heterotrophically,** but the growth rate is slower than when the cells are grown autotrophically. **Strictly aerobic** using oxygen as terminal electron acceptor. **pH range for growth, 6.5–8.5. Temperature range for growth, 5–40 C.**

The **G + C content** of the DNA ranges from 60.7–61.7 moles % (buoyant density; one species, six strains).

Habitat: **Soils, fresh water and sea water.**

Type species: *N. winogradskyi* Winslow *et al.* 1917, 552, *nom. cons.* Opin. 23. Jud. Comm. 1958, 169.

Genus II. Nitrospina *Watson and Waterbury 1971, 225*

Cells are **straight, slender rods; spherical forms,** are found in senescent cultures. There is **no extensive cytomembrane system. Gram-negative. Non-motile.** Cells have cytochromes but no other pigments.

Obligately chemolithotrophic bacteria which **oxidize nitrite to nitrate** and **fix CO_2** to fulfill energy and carbon needs. **Grow in sea-water** enriched with nitrite and inorganic salts; **no organic growth factors** are **required.**

Strictly aerobic using oxygen as a terminal electron acceptor.

Temperature range for growth, 20–30 C. pH range for growth, 7.0–8.0. Optimum growth in **70–100% sea water;** no growth in distilled water mineral salts medium even if NaCl is included.

The **G + C content** of the DNA is 57.7 moles % (buoyant density; one species, one strain).

Habitat: South Atlantic Ocean.

Type species: *N. gracilis* Watson and Waterbury, 1971, 225.

Genus III. **Nitrococcus** *Watson and Waterbury 1971, 224*

Cells **spherical, 1.5 μm or larger. Gram-negative. Motile** by means of **one to two subterminally inserted flagella.** Cells are rich in cytochromes imparting a yellowish to reddish color to cell suspensions; void of other pigments.

Obligately chemolithotrophic bacteria which **oxidize nitrite to nitrate** and **fix CO_2** to fulfill energy and carbon needs. **Growth medium** is **sea water** enriched with nitrite and other inorganic salts, **no organic growth factor** being **required.**

Strictly aerobic using oxygen as a terminal electron acceptor.

Temperature range for growth, 15–30 C. pH range for growth, 6.8–8.0.

The **G + C content** of the DNA is **61.2** moles % (buoyant density; one species, one strain).

Habitat: South Pacific Ocean.

Type species: *N. mobilis* Watson and Waterbury 1971, 224.

Genus IV. **Nitrosomonas** *Winogradsky 1892, 127, nom. cons.* Opin. 23. Jud. Comm. 1958, 169

Cells **ellipsoidal** or **short rods, motile or nonmotile, occurring singly,** in **pairs** or as **short chains. Gram-negative.** Possess **cytomembranes** which **occur** as **flattened vesicles in the peripheral regions of the cytoplasm.** Cells grow free in the medium or are embedded in a slime matrix. Cells are rich in cytochromes which impart a yellowish to reddish color to cell suspensions; void of other pigments.

Obligately chemolithotrophic bacteria which **oxidize ammonia to nitrite** and **fix CO_2** to fulfill energy and carbon needs. **Grow** in **fresh water or sea water** enriched with ammonia and inorganic salts, **no organic growth factors** being **required. Strictly aerobic,** using oxygen as a terminal electron acceptor. **Temperature range for growth, 5–30 C. pH range for growth, 5.8–8.5.**

The **G + C content** of the DNA ranges from **47.4–51.0** moles % (buoyant density; 13 strains).

Habitat: **Soils, fresh water and sea water.**

Type species: *N. europaea* Winogradsky 1892, 127, *nom. cons.* Opin. 23. Jud. Comm. 1958, 169.

Genus V. **Nitrosospira** *Winogradsky and Winogradsky 1933, 406*

Cells **spiral shaped. Gram-negative. Cells lack cytomembranes** (Plate 12.2, Fig. 6). **Nonmotile or motile** by means of **peritrichous flagella.** Grow free in liquid medium. Cells are rich in cytochromes which impart a yellowish to reddish color to cell suspensions; void of other pigments.

Obligately chemolithotrophic bacteria which **oxidize ammonia to nitrite** and **fix CO_2** to fulfill energy and carbon needs. Grow in **fresh water** enriched with ammonia and inorganic salts; **no organic growth factors** are **required. Strictly aerobic** using oxygen as a terminal electron acceptor. **Temperature range for growth, 15–30 C. pH range for growth, 6.5–8.5.**

The **G + C content** of the DNA is **54.1** moles % (buoyant density; one species, one strain).

Isolated from soils from Brie, France; Island of Crete; the summit of Mt. Pilatus in Switzerland; and the Parthenon in Athens, Greece.

Type species: *N. briensis* Winogradsky and Winogradsky 1933, 407.

Genus VI. **Nitrosococcus** *Winogradsky 1892, 127, nom. cons.* Opin. 23. Jud. Comm. 1958, 169

Cells **spherical. Motile or non-motile. Gram-negative.** Grow **singly,** in **pairs** or in **tetrads** either suspended free in liquid medium or embedded in slime to form aggregates which are either attached to vessel walls or are suspended in the liquid medium.

Obligately chemolithotrophic bacteria which **oxidize ammonia to nitrite** and **fix CO_2** to fulfill energy and carbon needs. Grow in **fresh water** or **sea-water** enriched with ammonia and inorganic salts, **no organic growth factors** being **required. Strictly aerobic** using oxygen as a terminal electron acceptor. **Temperature range for growth, 2–30 C. pH range for growth, 6.0–8.0.**

The **G + C content** of the DNA ranges from **50.5–51.0** moles % (buoyant density; one species, three strains).

Habitat: **Soils and sea water.**

Type species: *N. nitrosus* (Migula) Buchanan 1925, 402. (Type species not presently in culture.)

Note: Two species are recognized. *N. nitrosus* will grow in fresh water-based medium, whereas *N. oceanus* requires sea water.

Genus VII. **Nitrosolobus** *Watson, Graham, Remsen and Valois 1971, 200*

Cells **pleomorphic** and **lobate,** 1.0–1.5 μm in diameter; **division by constriction. Gram-negative.** Cells partially **compartmentalized by the invagination of the plasma membrane** and other segments of the cell envelope into the cytoplasm forming vesicular regions. **Motile** by means of **peritrichous flagella.** Grow free in liquid medium. Cells are rich in cytochromes which impart a yellowish to reddish color to cell suspensions; void of other pigments.

Obligately chemolithotrophic bacteria which **oxidize ammonia to nitrite** and **fix CO_2** to fulfill energy and carbon needs. Grow in fresh water enriched with ammonia and inorganic salts; **no organic growth factors** are **required. Strictly aerobic** using oxygen as a terminal electron acceptor. **Temperature range for growth, 15–30 C. pH range for growth, 6.0–8.2.**

The **G + C content** of DNA ranges from **53.6–55.1** moles % (buoyant density; one species, three strains).

Habitat: Soils from Surinam, South America; Galapagos Archipelago; Southwest Africa; and Russia.

Type species: *N. multiformis* Watson *et al.* 1971, 200.

ORGANISMS METABOLIZING SULFUR AND SULFUR COMPOUNDS

Key to genera of organisms metabolizing sulfur

Key characteristics

Energy source
Presence of sulfur granules
Cell shape
Cells in gelatinous mass
Flagellation

I. Organisms metabolizing sulfur and sulfur compounds.
(Although the genus *Thiobacillus* could be placed in a monogeneric family Thiobacillaceae, with a description identical to that of the genus, this seems to serve no useful purpose and until all genera have been grown in pure culture it seems more appropriate to consider them all as genera of uncertain affiliation.)

A. Obtain energy from oxidization of reduced or partially reduced sulfur compounds.

1. Rods, motile.
   1. *Thiobacillus*, p. 456

2. Cells spherical with lobes.
   2. *Sulfolobus*, p. 461

B. Have not been grown in pure culture, cells contain sulfur granules.

1. Rods, embedded in gelatinous mass, non-motile.
   3. *Thiobacterium*, p. 462

2. Cells not embedded in gelatinous mass, motile.

a. Cylindrical cells, flagella polar.
   4. *Macromonas*, p. 462

b. Ovoid cells, flagella peritrichous.
   5. *Thiovulum*, p. 463

c. Spiral cells, flagella polar.
   6. *Thiospira*, p. 464

PLATE SECTION

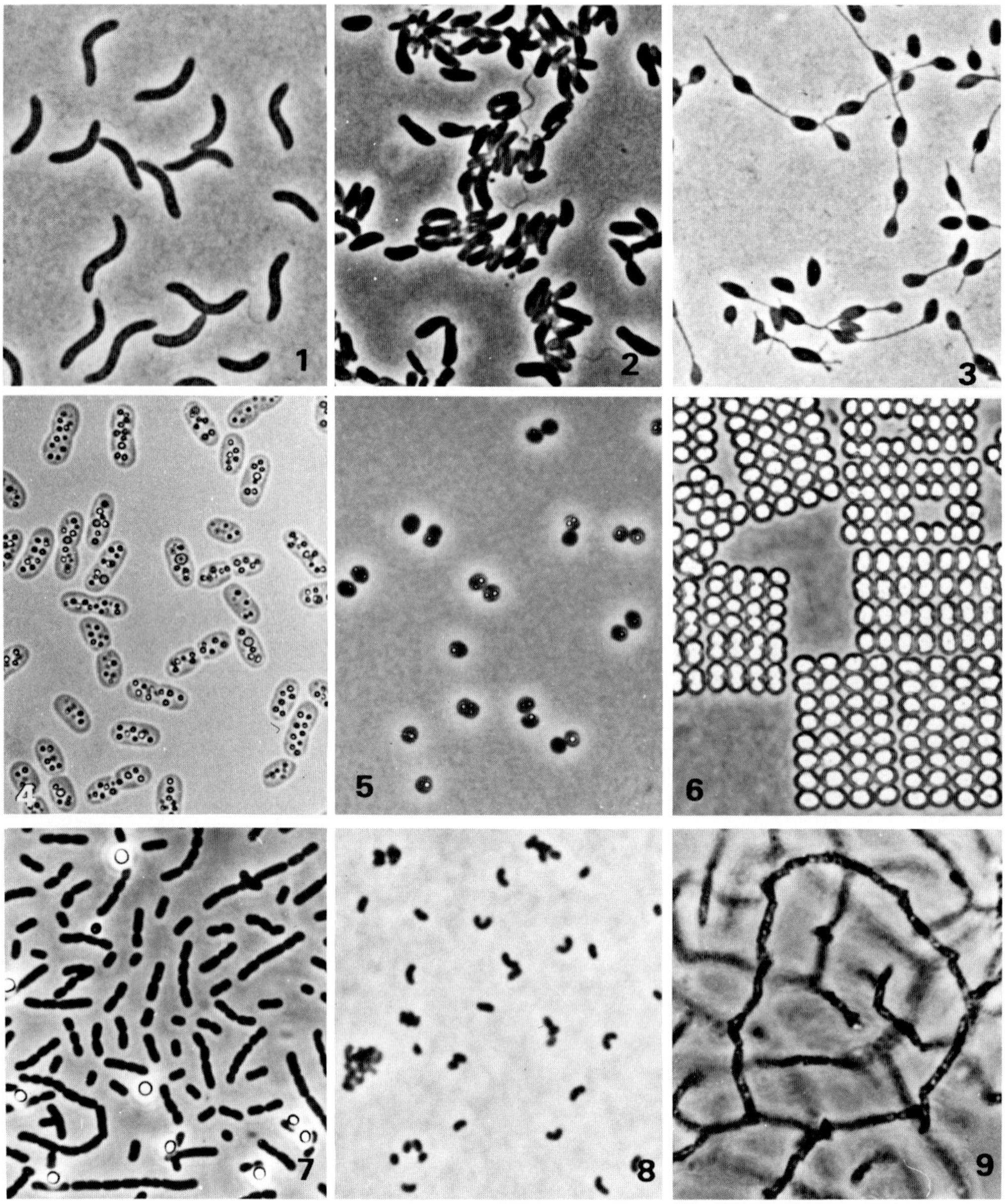

Plate 1.1. Morphology of typical representatives of the phototrophic bacteria during the exponential growth phase. (All organisms were grown anaerobically under optimum conditions.)

Fig. 1. *Rhodospirillum molischianum* strain NTHC 131. Phase contrast: × 2000.

Fig. 2. *Rhodopseudomonas acidophila* strain SMG 137. Phase contrast: × 2000.

Fig. 3. *Rhodomicrobium vannielii* strain SMG 163. Phase contrast: × 2000.

Fig. 4. *Chromatium vinosum* strain SMG 186. Light field: × 2000. Note the intracellular sulfur globules.

Fig. 5. *Thiocapsa roseopersicina* strain SMG 219. Phase contrast: × 2000. Note the intracellular sulfur globules.

Fig. 6. *Thiopedia rosea* from nature. Phase contrast: × 2000. Note the gas vacuoles.

Fig. 7. *Chlorobium limicola* strain SMG 249. Phase contrast: × 2000. Note the extracellular sulfur globules.

Fig. 8. *Chlorobium phaeovibrioides* strain SMG 269. Phase contrast: × 3000.

Fig. 9. *Pelodictyon clathratiforme* strain 1831. Phase contrast: × 2000. Note the gas vacuoles.

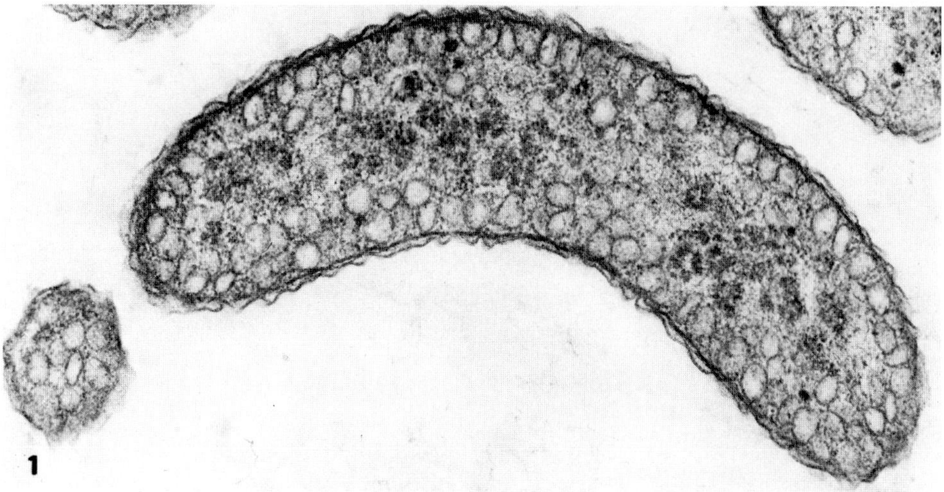

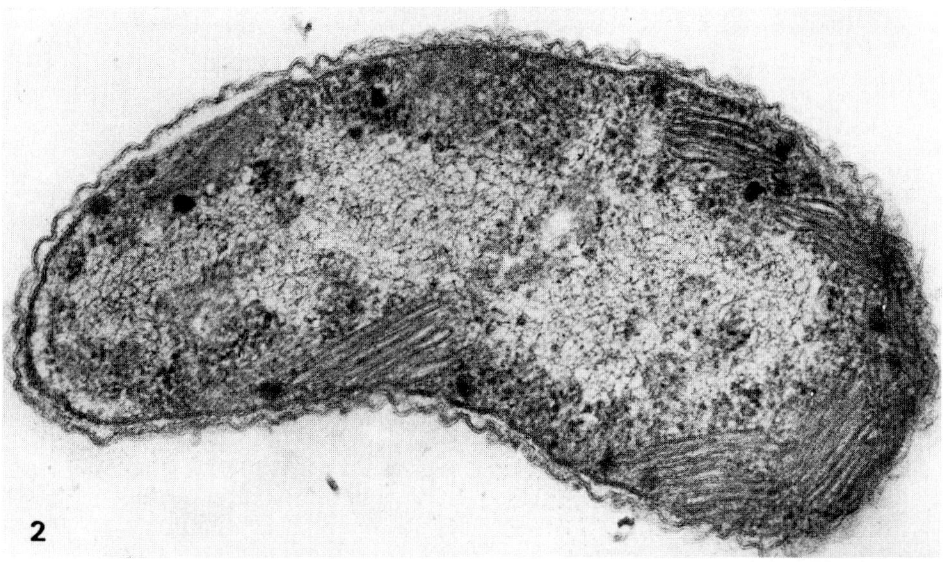

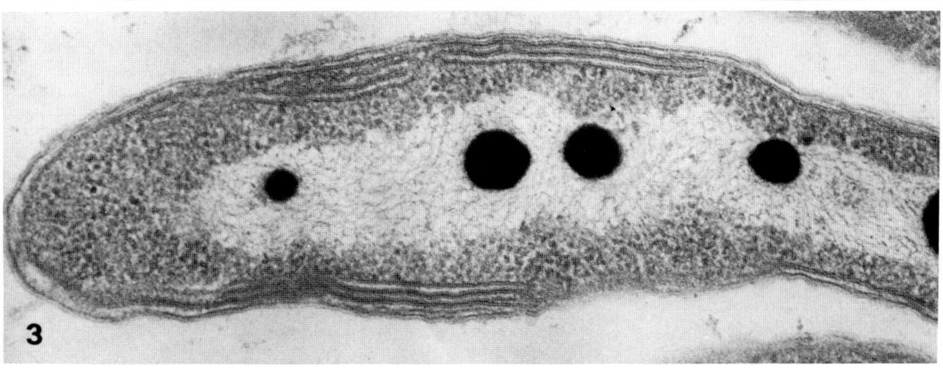

Plate 1.2. Fine structure of some Rhodospirillaceae

Fig. 1. *Rhodospirillum rubrum* strain FR1 grown anaerobically in the light. Note the vesicular structure of the intracytoplasmic membrane system. × 51,000. Courtesy of G. Drews and R. Ladwig.

Fig. 2. *Rhodospirillum molischianum* grown anaerobically in the light. Note the position and the lamellar stack type of the intracytoplasmic membrane system. × 90,000. Courtesy of G. Drews and R. Ladwig.

Fig. 3. *Rhodopseudomonas palustris* strain 11/1 grown semi-aerobically in the dark. Note position and type of intracytoplasmic membrane system. × 60, 000. Courtesy of H. D. Tauschel and R. Ladwig.

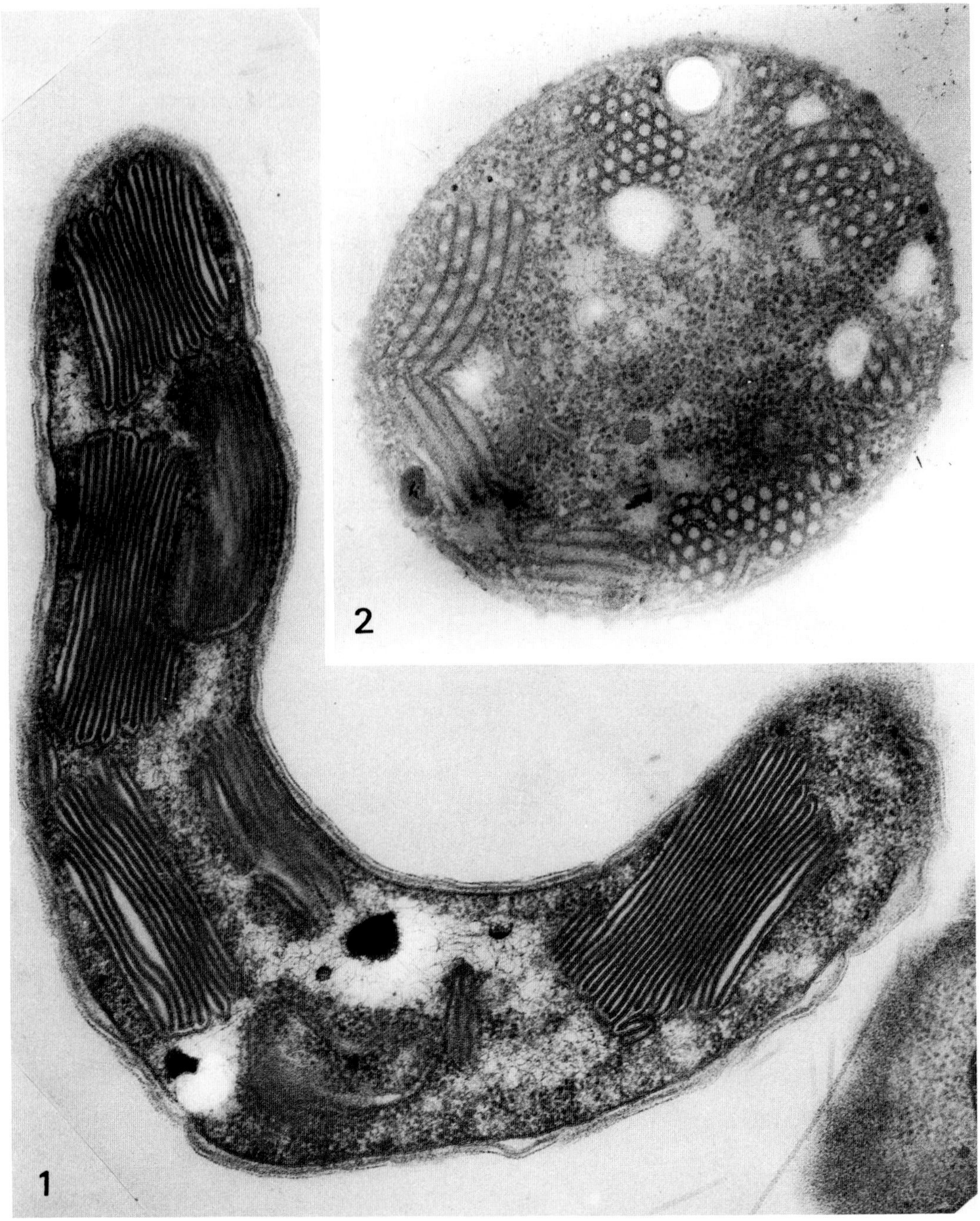

Plate 1.3. Fine structure of some Chromatiaceae

FIG. 1. *Ectothiorhodospira mobilis* strain SMG 237. Note the position and lamellar stack type of the intracytoplasmic membrane system. × 60,000. Courtesy of C. C. Remsen, S. W. Watson and J. B Waterbury.

Fig. 2. *Thiocapsa pfennigii* strain SMG 228. Note the bundled tube type of the intracytoplasmic membrane system. × 60,000. Courtesy of S. W. Watson, J. B. Waterbury and C. C. Remsen.

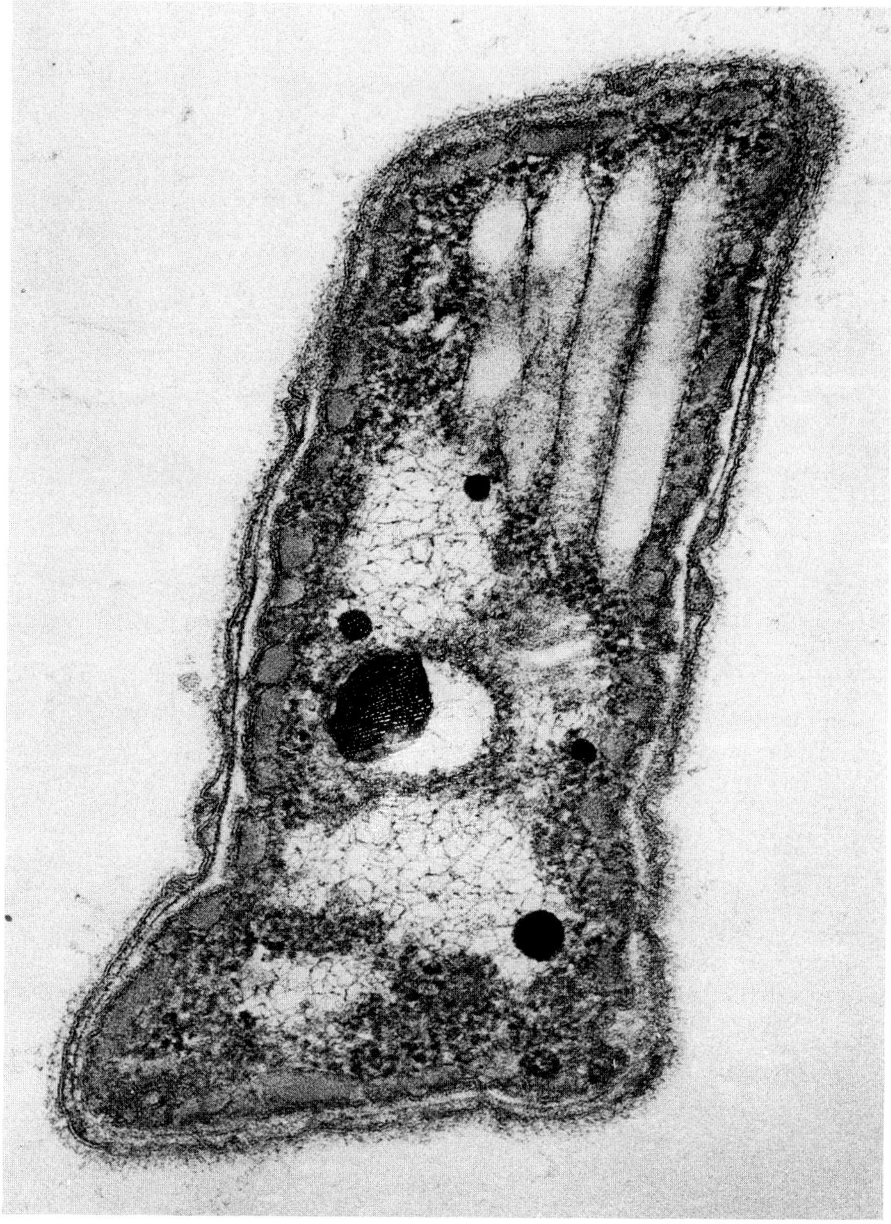

Plate 1.4. Fine structure of a green sulfur bacterium

Pelodictyon clathratiforme strain 1831. Note the Chlorobium vesicles (dark grey) underlying and attached to the cytoplasmic membrane, the bundle of gas vacuoles (light grey with pointed ends) in the upper part of the section, and the pointed ends of the cell indicating ternary fission. × 105,000. Courtesy of G. Cohen-Bazire.

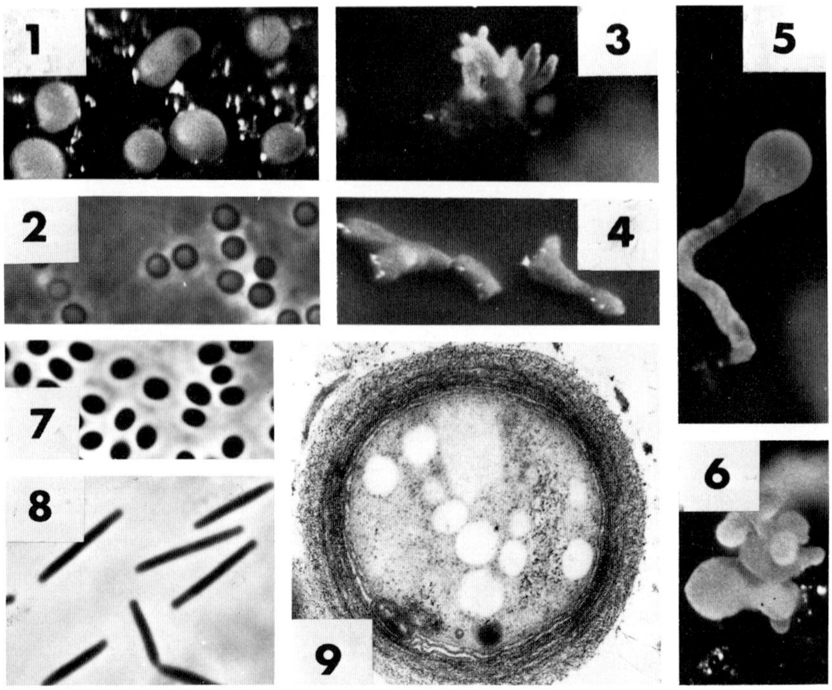

Plate 2.1. *Myxococcus*

Fig. 1. *Myxococcus fulvus* fruiting body on dung (× 30).
Fig. 2. *Myxococcus coralloides* microcysts (× 2000).
Fig. 3. *Myxococcus coralloides* fruiting body on bark (× 100).
Fig. 4. *Myxococcus coralloides* fruiting body on agar (× 100).
Fig. 5. *Myxococcus stipitatus* fruiting body on dung (× 100).
Fig. 6. *Myxococcus stipitatus* fruiting on dung (× 30).
Fig. 7. *Myxococcus stipitatus* microcysts (× 2000).
Fig. 8. *Myxococcus fulvus* vegetative cells (× 2000).
Fig. 9. *Myxococcus xanthus* electron micrograph of microcyst (ca. × 40,000). Courtesy of Herbert Voelz.

Figs. 3, 4 and 5. Reproduced by permission of the National Research Council of Canada from McCurdy, *Canadian Journal of Microbiology 15:* 1453–1461, 1969.

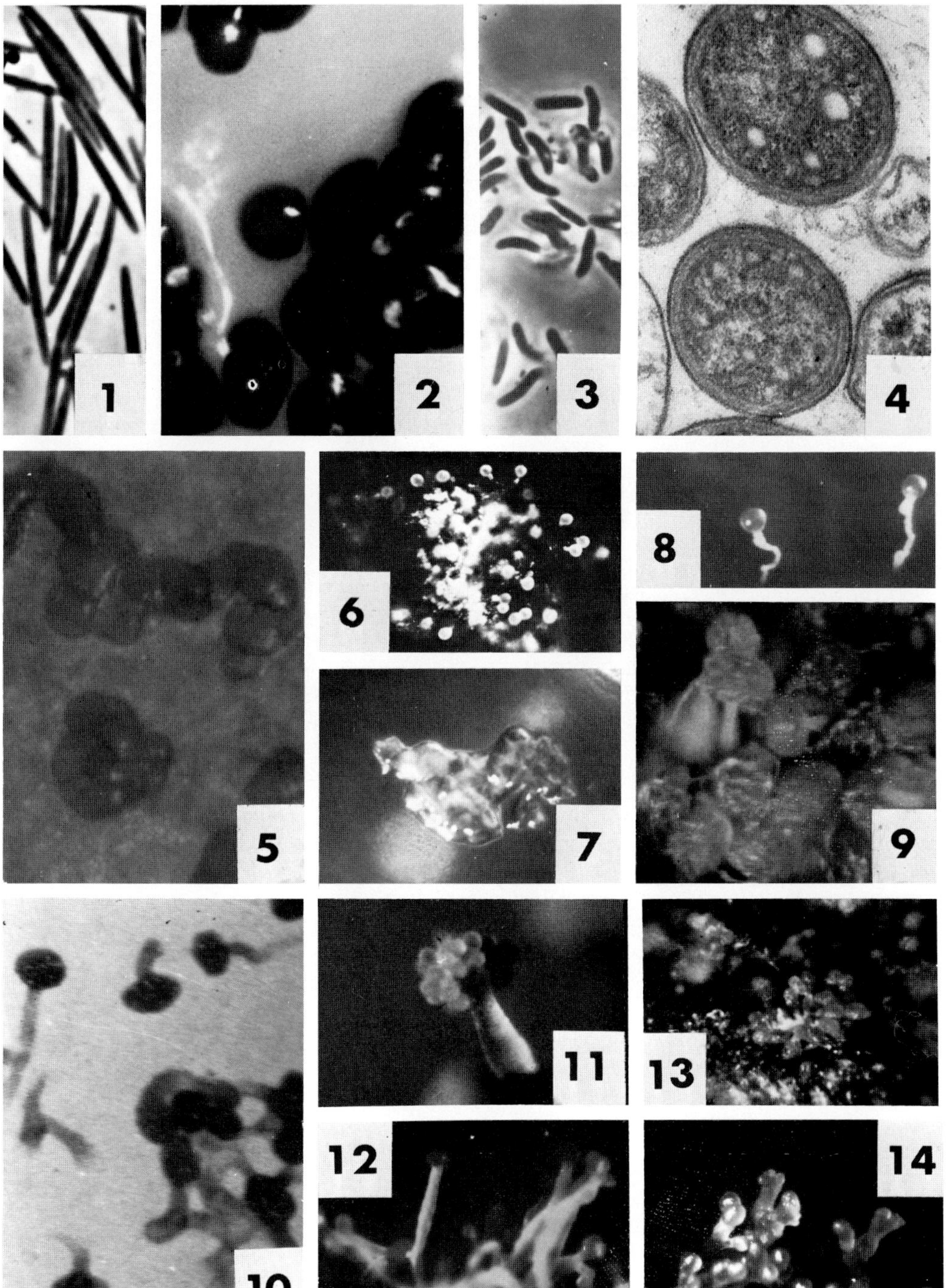

Plate 2.2.

Plate 2.2. *Cystobacter, Melittangium, Stigmatella*

Fig. 1. *Cystobacter fuscus* vegetative cells (× 2000).
Fig. 2. *Cystobacter fuscus* sporangia (× 100).
Fig. 3. *Cystobacter fuscus* microcysts (× 2000).
Fig. 4. *Cystobacter fuscus* microcyst, electron micrograph of cross-section (× 50,000).
Fig. 5. *Cystobacter ferrugineum* fruiting bodies (× 120).
Fig. 6. *Melittangium lichenicola* fruiting bodies on bark (× 100).
Fig. 7. *Melittangium lichenicola* archangium-like fruiting body on agar (× 25).
Fig. 8. *Melittangium alboraceum* fruiting bodies (ca. × 130).
Fig. 9. *Melittangium boletus* fruiting bodies (× 120).
Fig. 10. *Melittangium boletus* fruiting bodies on agar (ca. × 100).
Fig. 11. *Stigmatella aurantiaca* fruiting body on bark (× 90).
Fig. 12. *Stigmatella aurantiaca* fruiting bodies on agar (× 100).
Fig. 13. *Stigmatella erecta* fruiting body on bark (× 60).
Fig. 14. *Stigmatella erecta* fruiting body on agar (× 100).
Figs. 6 and 7. From McCurdy, *International Journal of Systematic Bacteriology 21:* 50-54, 1971.
Fig. 8. Courtesy of J. E. Peterson.
Figs. 9 and 10 From Krzemieniewska and Krzemieniewski, *Acta Soc Bot Polon 4:* 1–54, 1926.
Fig. 14. Reproduced by permission of the National Research Council of Canada from McCurdy and Khouw, *Canadian Journal of Microbiology 15:* 731–738, 1969.

Fig. 1. *Polyangium vitellinum* vegetative cells (× 2000).
Fig. 2. *Polyangium vitellinum* sporangia (× 100).
Fig. 3. *Polyangium aureum* sporangia (× 1200).
Fig. 4. *Polyangium fumosum* sori (× 120).
Fig. 5. *Polyangium fumosum* migrating mass and grooved trail (× 50).
Fig. 6. *Polyangium sorediatum* fruiting body on dung (ca. × 80).
Fig. 7. *Polyangium cellulosum* (× 300).
Fig. 8. *Polyangium cellulosum* (× 300).
Figs. 2 and 3. From McCurdy, *International Journal of Systematic Bacteriology 20:* 283–296, 1970.

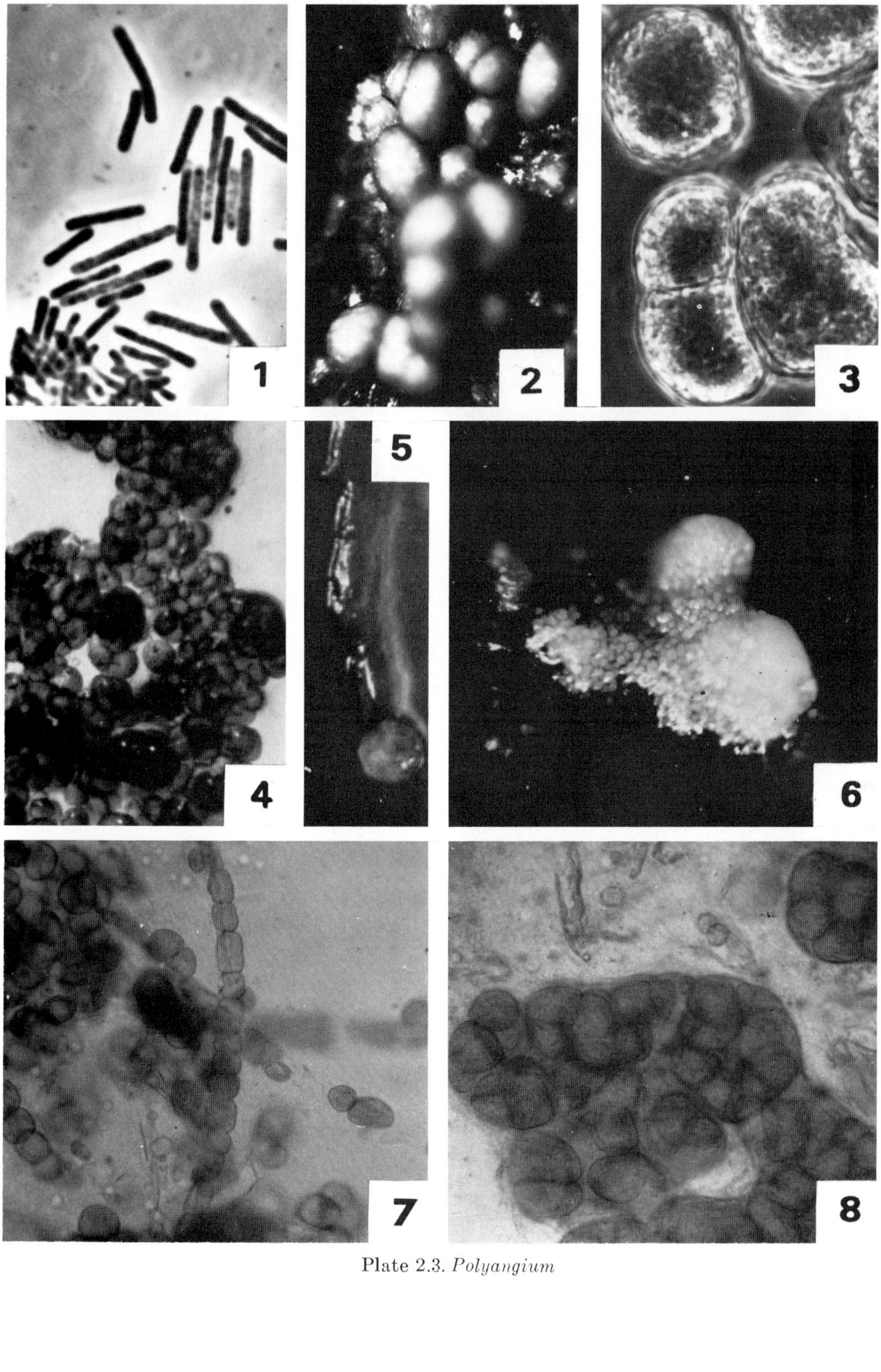

Plate 2.3. *Polyangium*

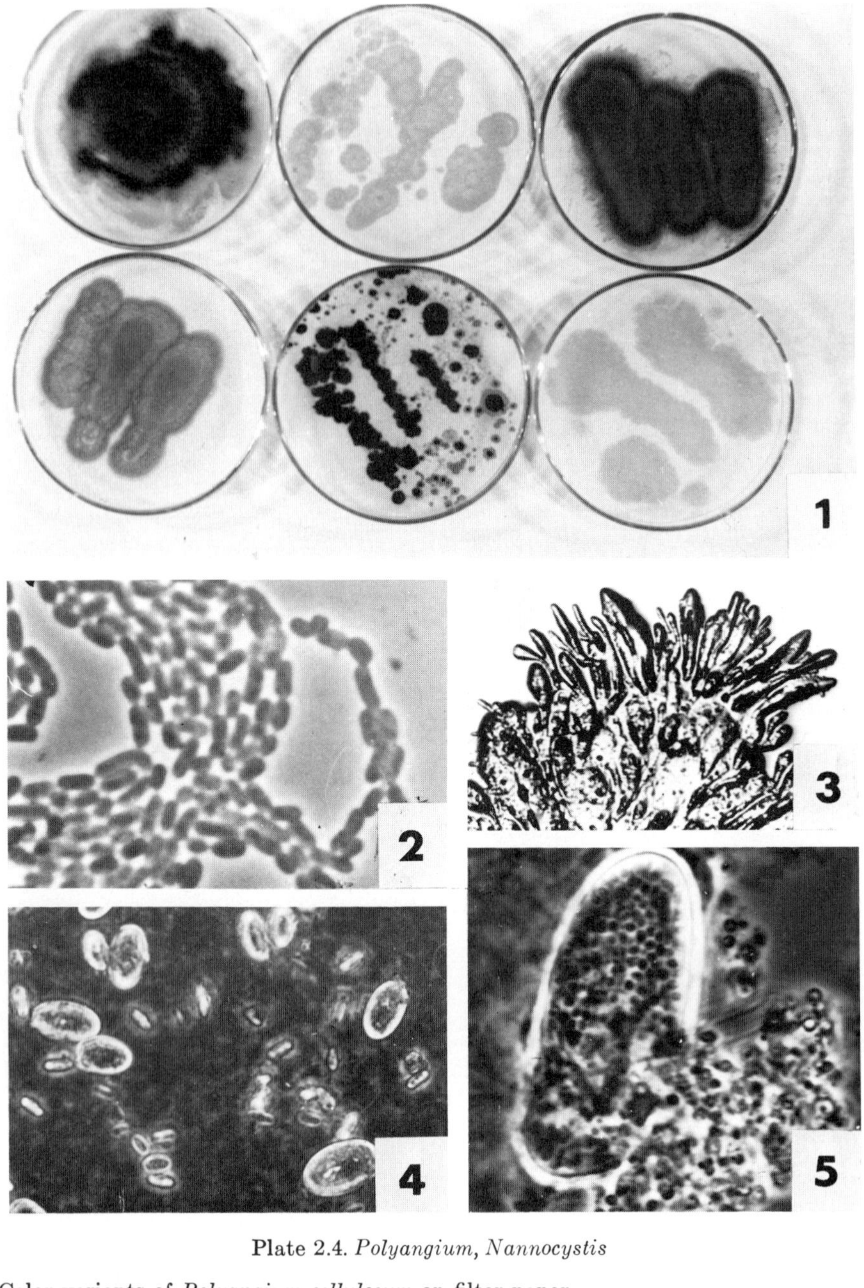

Plate 2.4. *Polyangium, Nannocystis*

Fig. 1. Color variants of *Polyangium cellulosum* on filter paper.
Fig. 2. *Nannocystis exedens* vegetative cells (× 2000).
Fig. 3. *Nannocystis exedens* swarm pattern etched into agar (× 4.7).
Fig. 4. *Nannocystis exedens* sporangia of various sizes embedded in agar (× 500).
Fig. 5. *Nannocystis exedens* squashed sporangium with myxospores (× 1500).
Fig. 1. Courtesy of J. E. Peterson.
Figs. 2 through 5. Courtesy of H. Reichenbach.

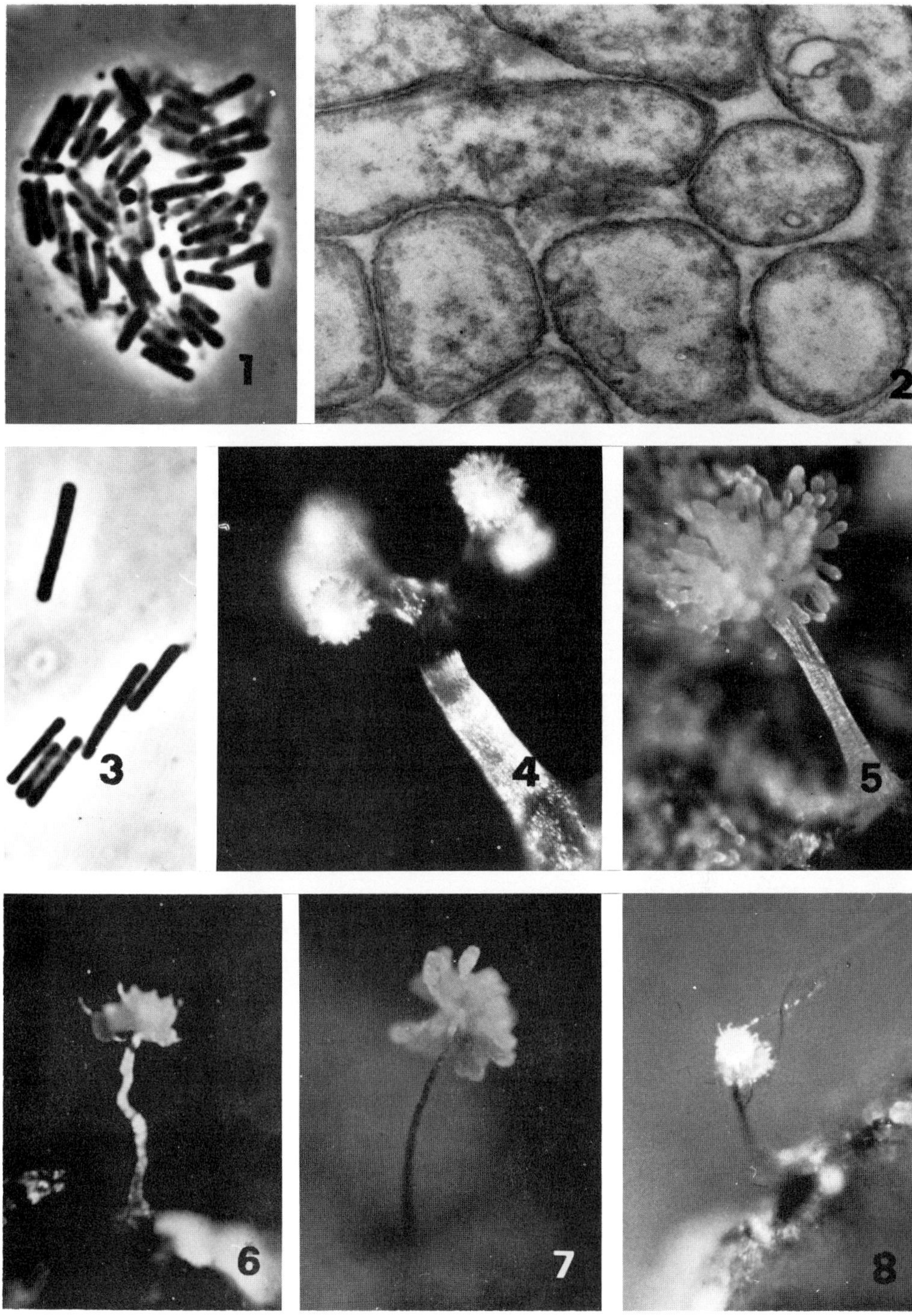

Plate 2.5.

Plate 2.5. *Chondromyces*

Fig. 1. *Chondromyces crocatus* myxospores (× 2000).
Fig. 2. *Chondromyces crocatus* myxospores in sporangium (electron micrograph, × 30,000).
Fig. 3. *Chondromyces crocatus* vegetative cells (× 2000).
Fig. 4. *Chondromyces crocatus* fruiting body (× 130).
Fig. 5. *Chondromyces catenulatus* fruiting body (× 120).
Fig. 6. *Chondromyces apiculatus* fruiting body (× 100).
Fig. 7. *Chondromyces pediculatus* fruiting body (× 100).
Fig. 8. *Chondromyces lanuginosus* fruiting body (× 90).
Figs. 1, 3, and 5. From McCurdy, *International Journal of Systematic Bacteriology 21:* 40–49, 1971.
Fig. 2. From McCurdy, *Arch Mikrobiol 65:* 380–390, 1968.
Figs. 6, 7, and 8. Reproduced by permission of the National Research Council of Canada from McCurdy, *Canadian Journal of Microbiology 15:* 1453–1461, 1969.

Plate 2.6. *Thioploca ingrica* from Lake Erie

Fig. A. Portion of mass of filaments washed free of mud. Scale 200 μm. *b*, area enlarged in Fig. B.

Fig. B. Fascicle of filaments (*f*) enclosed by the sheath (*s*); free filaments (*t*) emerged from a break in the sheath. Phase-contrast. Scale 50 μm.

Fig. C. Note sulfur inclusions and the obtuse (*top*) and tapered (*bottom*) terminal segments. Phase-contrast. Scale 10 μm.

Plate 2.7. *Simonsiella, Alysiella*

Fig. 1. *Simonsiella crassa* short filament.

Fig. 2. Microcolony of *S. crassa* showing filaments coiled on edge and commencement of migration on flat side of filaments at left.

Fig. 3. *Alysiella filiformis* microcolony showing very short filaments.

Fig. 4. *A. filiformis* microcolony showing long filaments some on edge.

All young cultures on serum agar, phase-contrast microscopy. Lines = 10 μm. From *Journal of General Microbiology 29:* 624, 1962.

Plate 2.8. *Toxothrix trichogenes*

Figs. 1 and 2. Living trichomes during 1st min of laboratory observation; phase-contrast microscopy, wet mount.

Figs. 3 to 6. Fate of filaments during laboratory microscopic observation; Fig. 3 at beginning and Fig. 4 breakage after 5 min of observation; Fig. 5 after breakage and Fig. 6 disintegration 3 min after breakage.

Figs. 7 and 8. The excreted mucoid material partially encrusted with iron; Fig. 7 fan-shaped structure; Fig. 8 "twisted rope sheaths," fans and double tracks.

Figs. 9 and 10. Double tracks as produced by the ends of U-shaped filament moving over glass slide; dark-field. All μm lines = 10 μm, except as noted.

All figures from Krul *et al.*, *Antonie van Leeuwenhoek Journal of Microbiology and Serology 36:* 409–420, 1970.

Plate 2.9. Cells of *Achromatium oxaliferum*

Fig. 1. Untreated.

Figs. 2 and 3. After immersion in 0.001 M acetic acid for 10 and 30 min, respectively. Scale line represents 20 μm.

Fig. 4. Slime and filaments, electron micrograph, osmic acid fixation, Pt shadowed, × 31,160. From de Boer, la Rivière and Schmidt, *Antonie van Leeuwenhoek Journal of Microbiology and Serology 37:* 553–563, 1971.

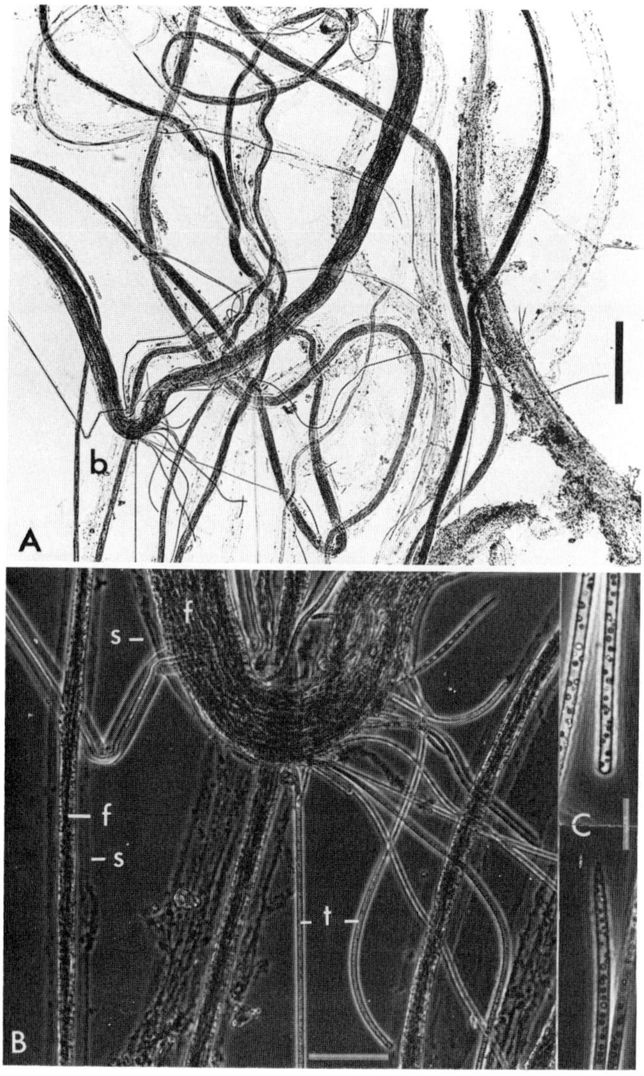

Plate 2.6.

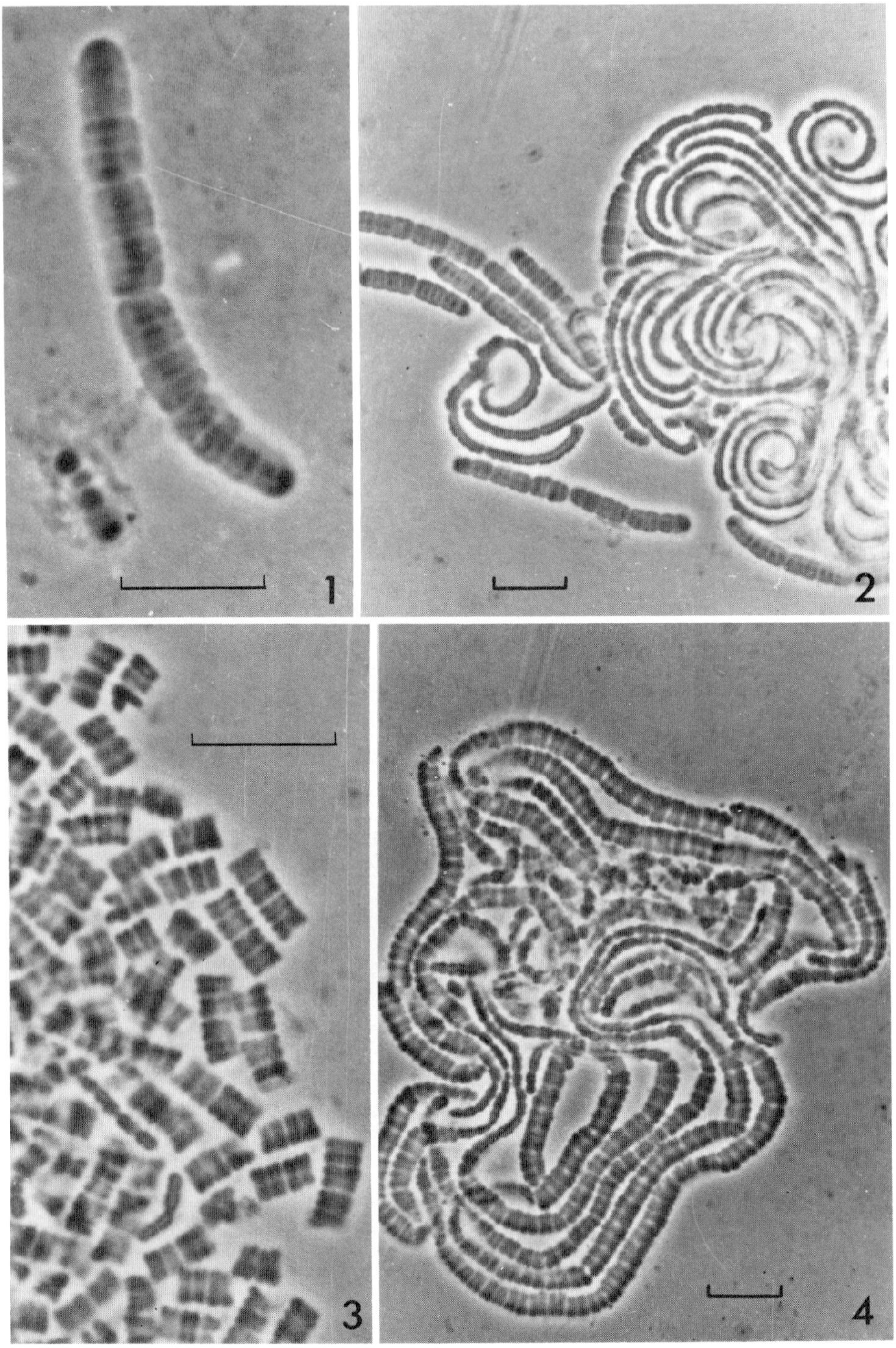

Plate 2.7.

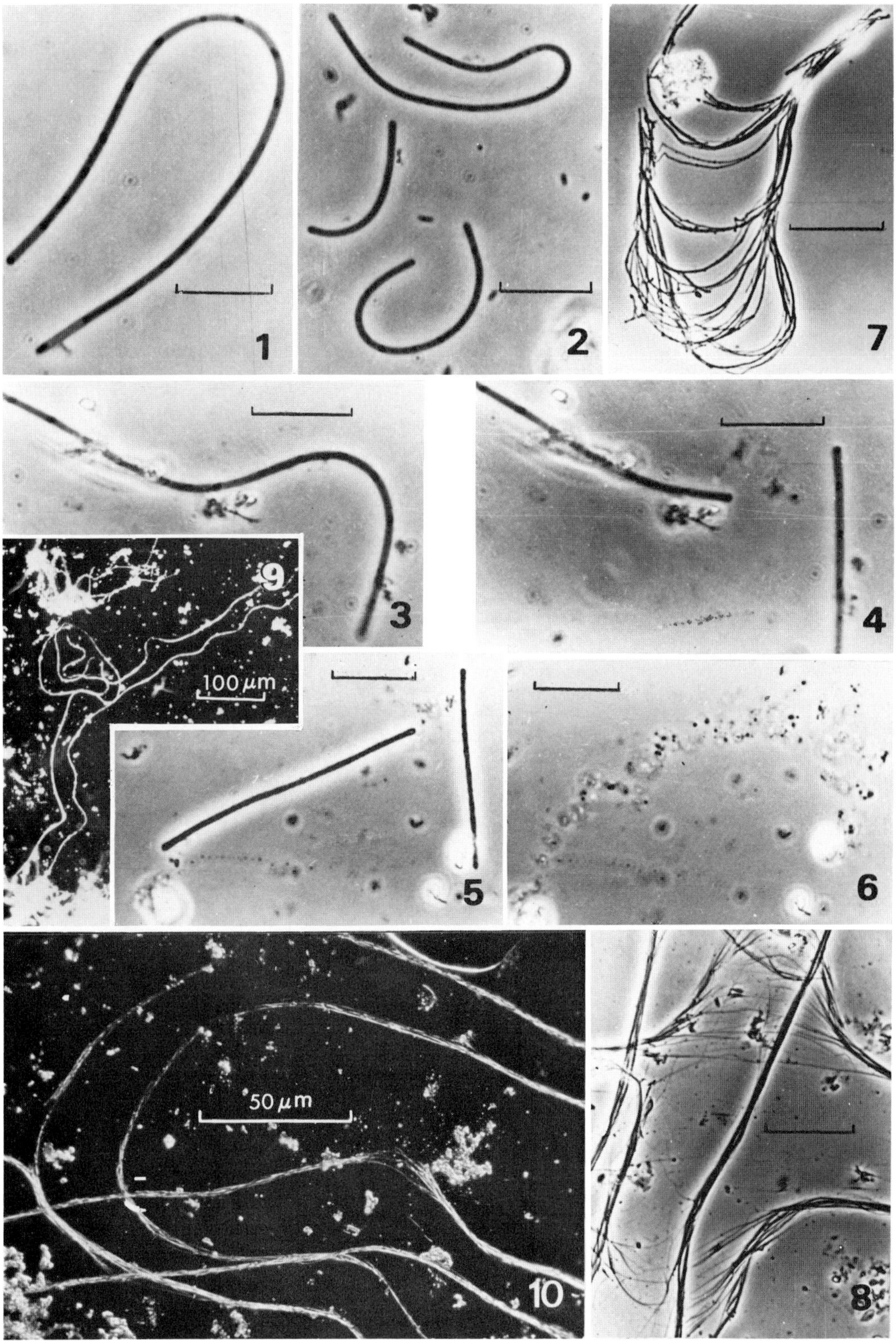

Plate 2.8.

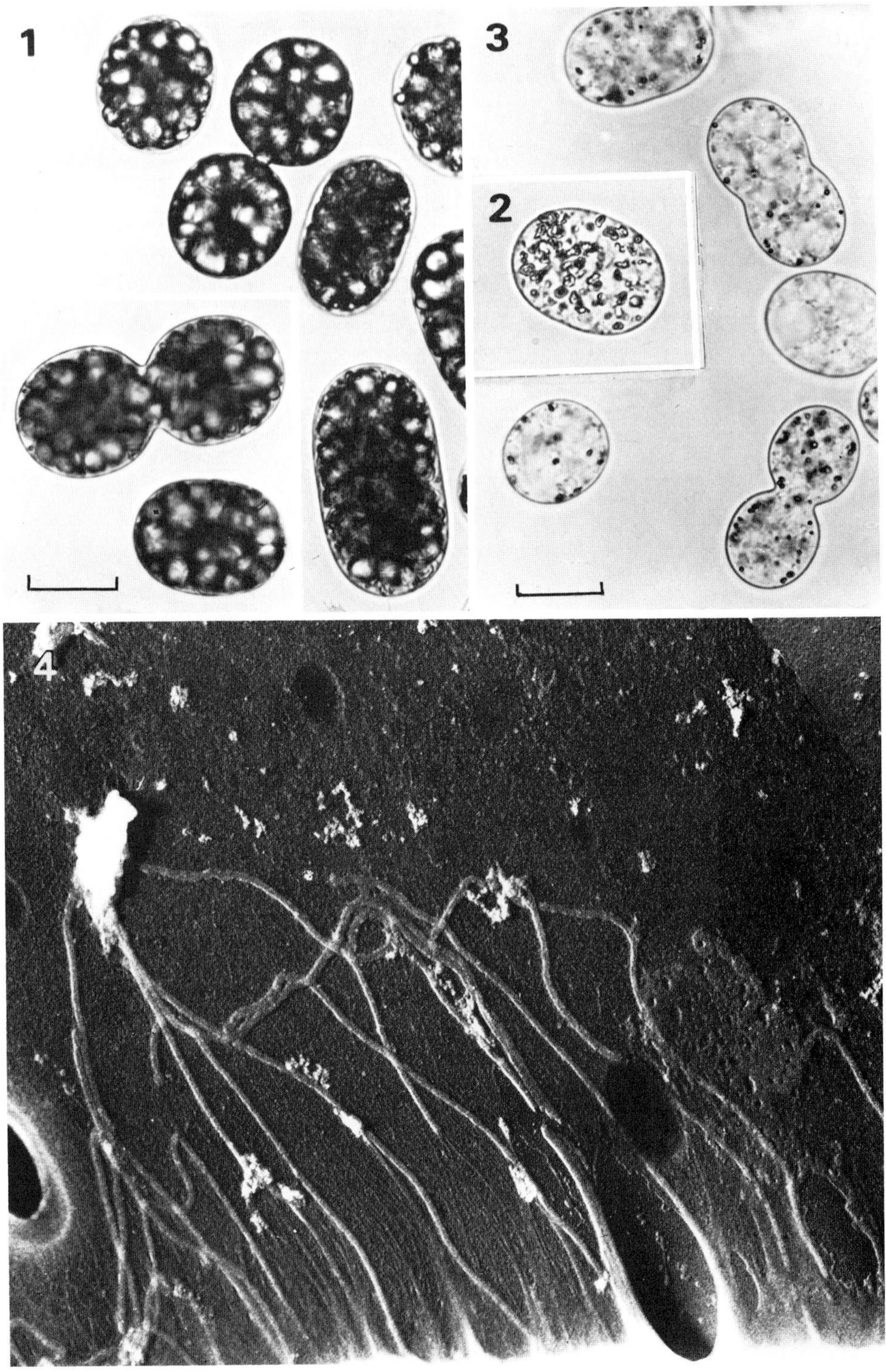

Plate 2.9.

Plate 3.1. *Sphaerotilus* and *Leptothrix*

Fig. 1. *Sphaerotilus natans* in nutrient medium containing ferric iron as the quinic acid complex; sheaths and holdfasts. × 812.

FIG. 2. *Sphaerotilus natans* in basal culture solution containing 2.5 g of glucose and 1 g of peptone/liter. Cells within sheaths, as in rough colonies (Fig. 4). Globular inclusions are poly-β-hydroxybutyrate. × 812. From Poindexter, J. S. 1971. Microbiology, An Introduction to Protists. The Macmillan Company, New York.

Fig. 3. Flagella of *Sphaerotilus natans*. × 10,360. From Sykes, G. and Skinner, F. A. (editors). 1971. Microbiol aspects of pollution. The Society for Applied Bacteriology: Symposium Series 1. Academic Press, London.

Fig. 4. *Sphaerotilus natans*. Rough colony on basal agar containing 1 g of glucose and 1 g of peptone/liter. × 14.5. From Antonie van Leeuwenhoek, 29, 121–153 (1963).

Fig. 5. *Sphaerotilus natans*. Smooth colony on basal agar containing 5 g of glucose and 5 g of peptone/liter. × 14.5. From Antonie van Leeuwenhoek, 29, 121–153 (1963).

Fig. 6. *Sphaerotilus natans*. Cells without sheaths from a smooth colony (Fig. 5). × 812. From Antonie van Leeuwenhoek, 29, 121–153 (1963).

Fig. 7. *Leptothrix pseudo-ochracea* in slowly running sterilized iron-containing soil extract. × 812. From Antonie van Leeuwenhoek, 29, 121–153 (1963).

Fig. 8. *Leptothrix pseudo-ochracea* on basal agar medium containing 1 g of glucose and 1 g of peptone/liter, showing circular growth. × 812. From Antonie van Leeuwenhoek, 29, 121–153 (1963).

Fig. 9. Empty sheaths of an enrichment culture of *Leptothrix ochracea* in slowly running iron-containing soil extract (artificial ditch water). × 812. From Journal of Applied Bacteriology 27, 151–173 (1964).

Plate 3.2. *Leptothrix and Streptothrix*

Fig. 1. Flagellum of *Leptothrix discophora*. × 27,100. From Revue d'écologie et de biologie du sol, 9, 1972.

Fig. 2. Formation of empty sheaths by an enrichment culture of *Leptothrix ochracea* in iron-containing soil extract. Cells leave old sheaths and form new sheaths with a rapidity of approximately 1 μm/min. × 812. From Anreicherungskultur und Mutantenauslese. 1965. Supplementheft 1 zum Zentralblatt für Bakteriologie, 1. Abt.

Fig. 3. *Leptothrix cholodnii* containing globules of poly-β-hydroxybutyrate. × 1620.

Fig. 4. Granular deposition of MnO_2 on sheaths of *Leptothrix cholodnii* grown in a basal medium containing 0.2 g of glucose, 0.2 g of peptone and 25 mg of $MnSO_4 \cdot H_2O$/liter. × 812. From Antonie van Leeuwenhoek, 29, 121–153 (1963).

Fig. 5. *Leptothrix discophora* in a basal medium containing 0.2 g of glucose, 0.2 g of peptone and 25 mg of $MnSO_4 \cdot H_2O$/liter. × 812. From Journal of Applied Bacteriology, 27, 151–173 (1964).

Fig. 6. *Leptothrix discophora* in slowly running sterilized iron-containing soil extract (artificial ditch water). × 812. From Antonie van Leeuwenhoek, 29, 121–153 (1963).

Fig. 7. *Leptothrix lopholea* in a basal culture solution supplied with 2.5 g of glucose and 2.5 g of peptone/liter. Chains of cells arising from a cluster of holdfasts. × 812. From Antonie van Leeuwenhoek, 29, 121–153 (1963).

Fig. 8. Iron-impregnated holdfasts of *Leptothrix lopholea* in a culture medium with ferric iron as the quinic acid complex. × 812.

Fig. 9. *Streptothrix hyalina* grown in soil extract enriched with trypticase soy broth, glucose and vitamins. × 325.

Fig. 10. *Streptothrix hyalina*, sheaths and cells. × 24,600.

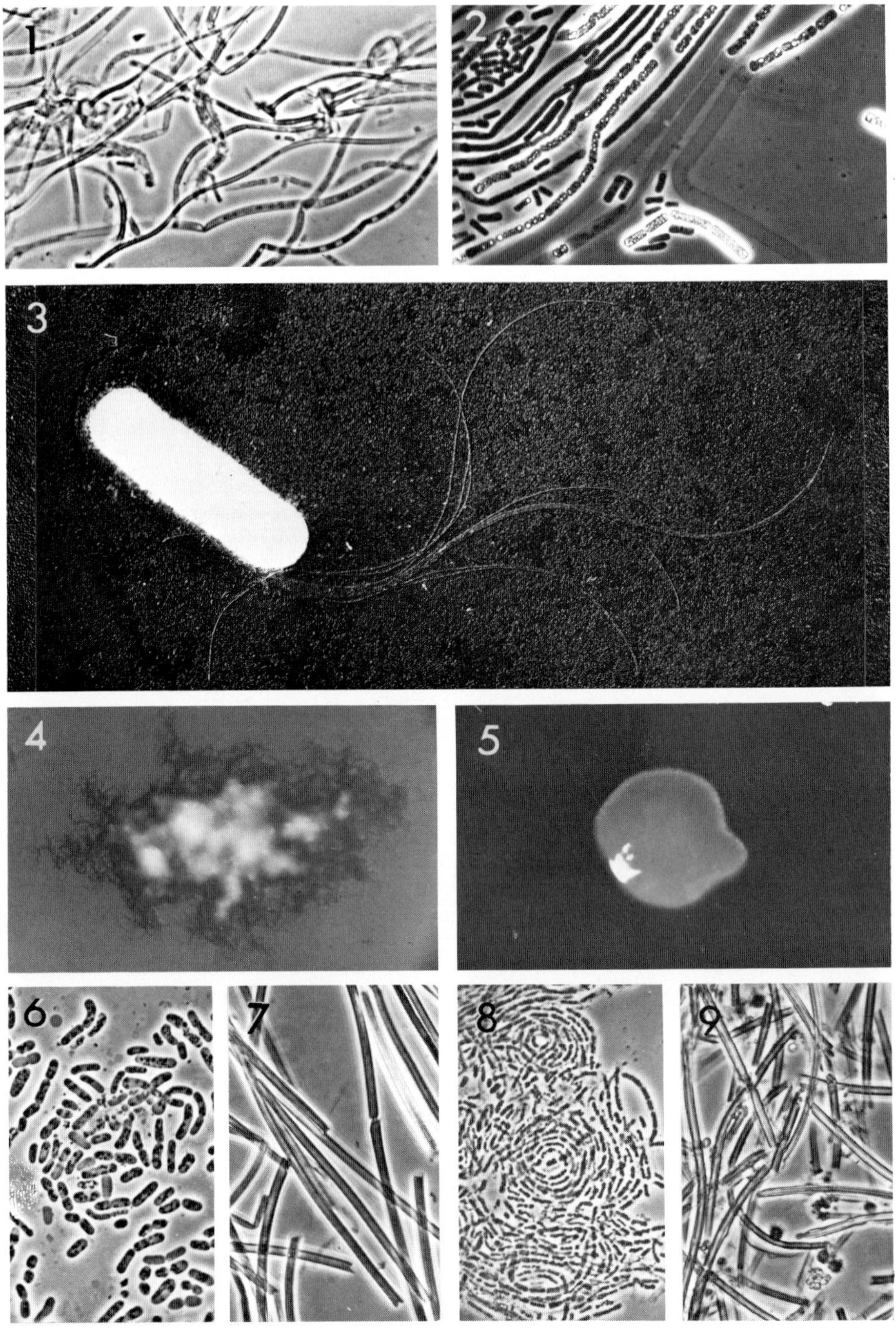

Plate 3.1

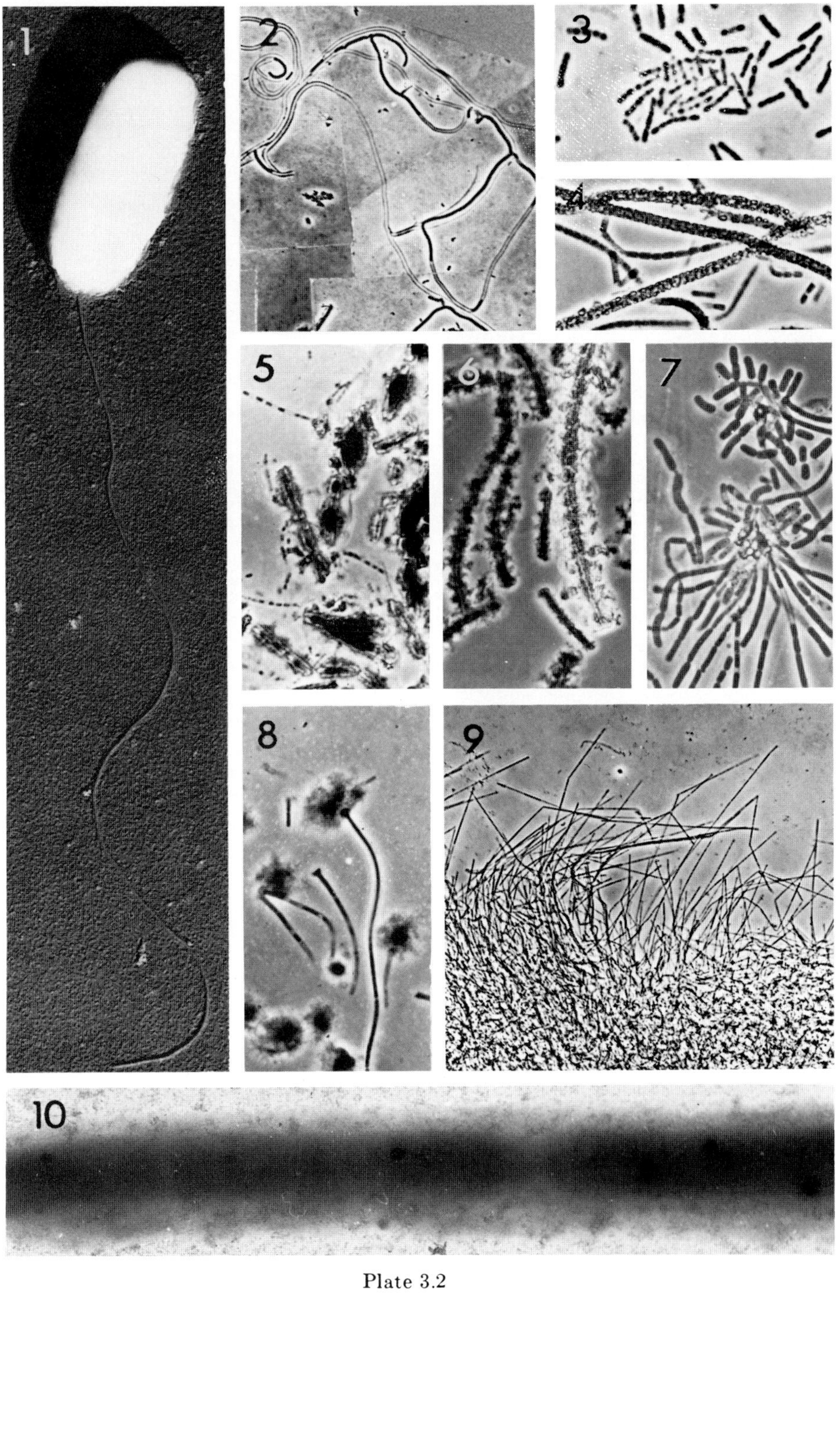

Plate 3.2

Plate 3.3*A*. *Phragmidiothrix multiseptata*. Original line drawings of Engler

Fig. 24. Filament showing arrangement and septation of disc-shaped cells.

Fig. 25. Part of filament showing enlarged end and individual disc-shaped cells of uneven diameter.

Fig. 26. Part of a filament with much cross- and longitudinal septation.

Fig. 26*a* shows a part at higher magnification.

Fig. 27. A bristle of *Gammarus locusta* with two attached filaments of *Phragmidiothrix* (wide filaments) and several narrow filaments of young "*Beggiatoa alba*"?.

Magnification: approximately × 330 (original magnification × 400). Reproduced from Engler (1883).

Plate 3.3*B*. *Clonothrix fusca*

Fig. 1. End of adult filament with false lateral branches.

Fig. 2. End of young filament with false lateral branches.

Fig. 3. Cell emerging from a broken filament.

Fig. 4. Young filament growing from a broken adult filament.

Fig. 5. Mode of insertion of two adult filaments, that on right forming a false lateral branch.

Fig. 6. A filament carrying an ascending series of ampoule-like swellings (buds).

Fig. 7. End of a filament terminating in an ampoule-like swelling.

Fig. 8. A filament terminating in swelling but also carrying a series of lateral swellings.

Figs. 9–11. Buds showing vacuoles, and granules.

Magnification: Figs. 1–4, 6–8 approximately × 365. Fig. 5, 9–11 approximately × 650. Reproduced from Roze, 1896; original magnification 450 and 800, respectively.

Plate 3.4. *Crenothrix polyspora*

Fig. 1. Filaments showing variation in size and shape of individual cells (Wolfe, 1960).

Figs. 2–4. Reproductive cells in terminal swellings. The thin sheath clearly visible in Fig. 4. All living filaments in water; phase contrast photomicrographs by R. S. Wolfe.

PHRAGMIDIOTHRIX A

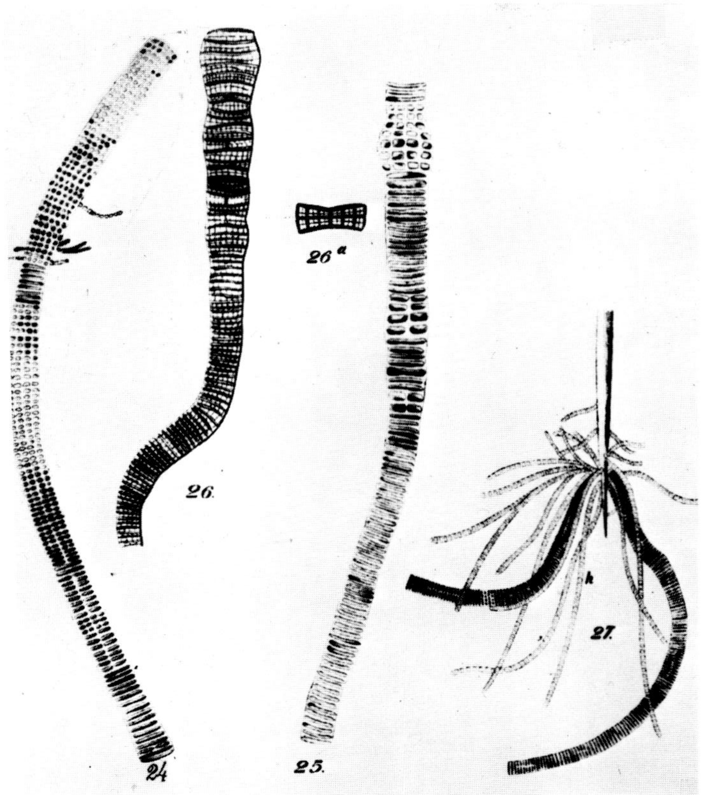

CLONOTHRIX B

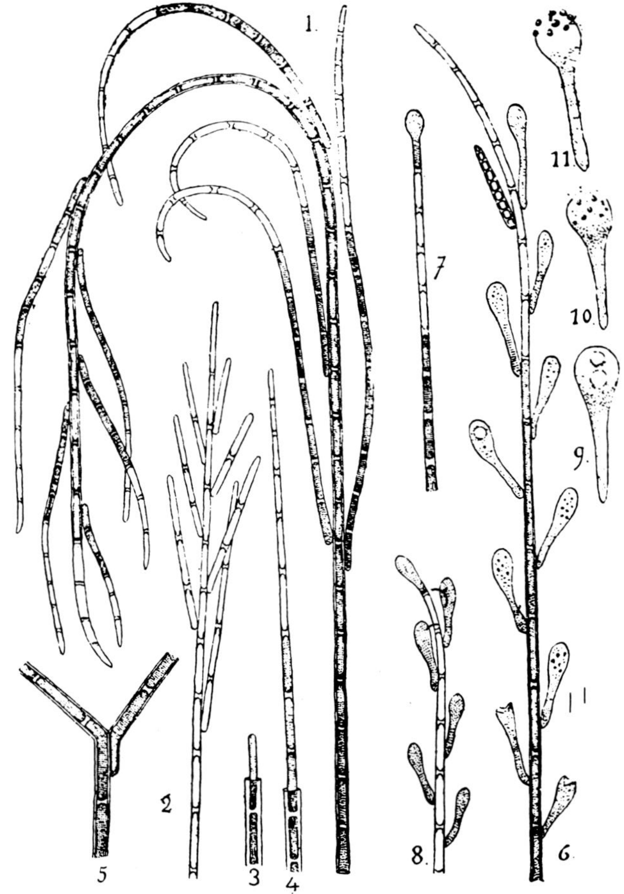

Plate 3.3

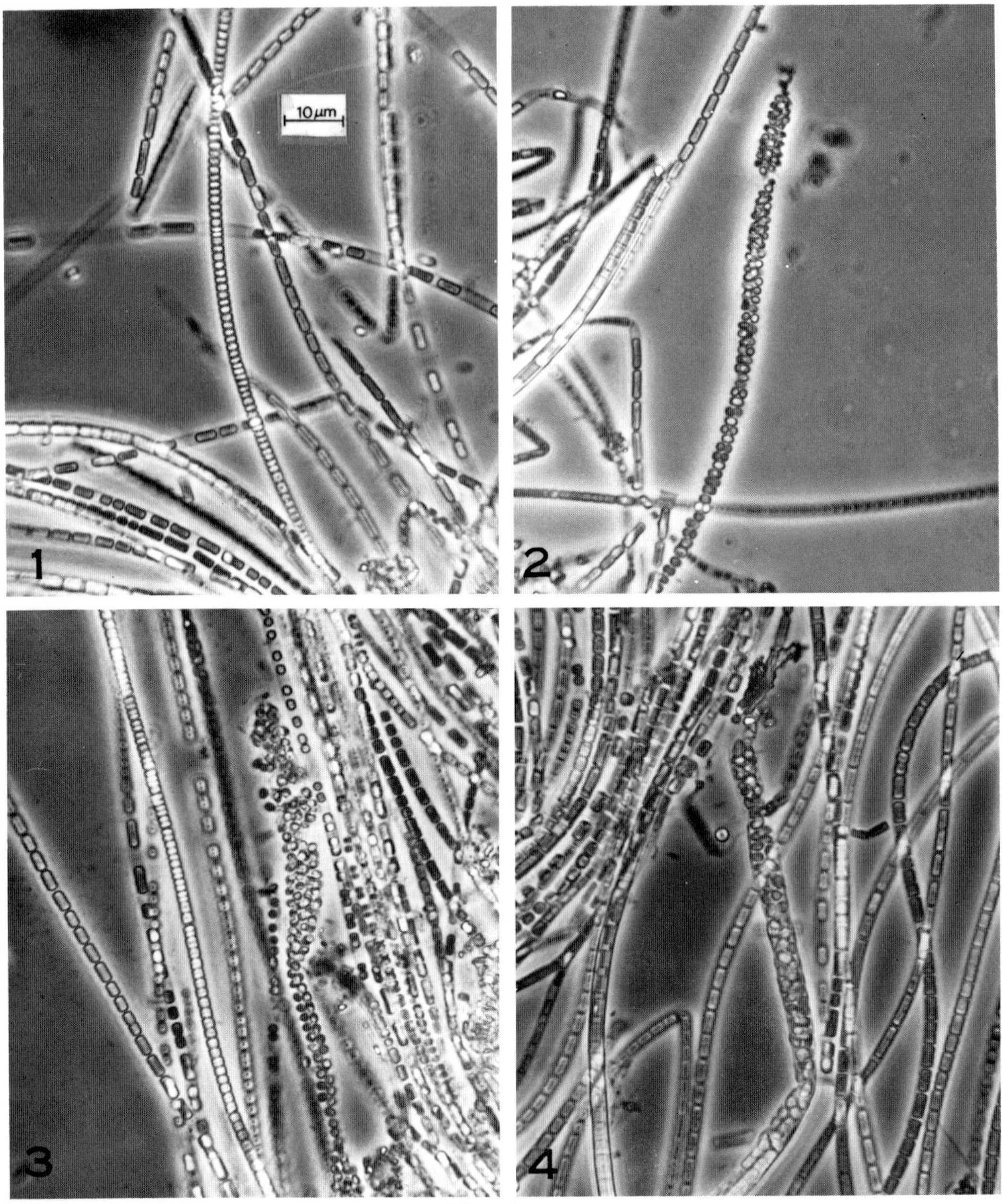

Plate 3.4

Plate 4.1. *Hyphomicrobium* and *Pasteuria*

Fig. 1. *Hyphomicrobium* sp. (strain P-546). Mature cell showing hyphae and terminal bud cells. Isolated from soil. Electron micrograph, negatively stained with phosphotungstate. × 20,500.

Figs. 2–4. *Pasteuria ramosa*. Spherical bodies multiplying by budding.

Fig. 4. On antennae of *Daphnia* sp. Wet mounts on agar slides. × 1250. From Hirsch, P. (1972) Int. J. Syst. Bacteriol. *22:* 112–116.

Plate 4.2. *Ancalomicrobium*, *Prosthecomicrobium*, *Thiodendron*

Fig. 1. *Ancalomicrobium adetum*. Electron micrograph, negatively stained. From Staley, J. T. (1968) J. Bacteriol. *95:* 1921–1924.

Fig. 2. *Prosthecomicrobium pneumaticum*. Electron transparent inclusions are vesicles of gas vacuoles. Electron micrograph, negatively stained. From Staley, J. T. (1968) J. Bacteriol. *95:* 1921–1924.

Figs. 3 and 4. *Thiodendron* sp. observed in mud and water samples from a pond. Stalks ca. 0.2 μm in diameter. Courtesy Dr. Hans Hippe.

Plate 4.3. *Seliberia*

Figs. 1 and 2. Star-shaped aggregates; stained with erythrosin. × 2000. Courtesy of T. V. Aristovskaya.

Fig. 3. Aggregates showing "buds" at ends and sides of rod-shaped cells. Phase contrast. × 2000. Courtesy of T. V. Aristovskaya.

Fig. 4. Electron micrograph showing twisted and aggregated cells. × 6000. Courtesy of T. V. Aristovskaya.

Fig. 5. *Seliberia* sp. from iron spring in Michigan showing spirally twisted cell and flagellum. × 50,000. From Hirsch, P. and Pankratz, H. (1970) Z. Allg. Mikrobiol. *10:* 589.

Plate 4.4. *Metallogenium*

Fig. 1. *Metallogenium symbioticum*. Free cells from 48-hr liquid culture. From Zavarzin, G. A. (1964) Z. Allg. Mikrobiol. *4:* 390–395.

Fig. 2. *M. symbioticum*. Part of young microcolony; old filaments are covered with manganese oxide; younger parts are less encrusted. An arai is visible at a rupture of the coating (*b*). From Zavarzin, G. A. (1964) Z. Allg. Mikrobiol. *4:* 390–395.

Figs. 3–5. *M. personatum*. Young colonies forming by germination of reproductive (round) cells. × 1000.

Fig. 6. *M. personatum*. Formation of reproductive cells on ends of young branches. × 2000.

Fig. 7. *M. personatum*. Young colonies. Show development of secondary microcolonies. × 1000.

Fig. 8. *M. personatum*. Young colonies. × 1000.

Fig. 9. *M. personatum*. Young colonies. Show development of secondary microcolonies. The beginning of mineralization. × 1000.

Figs. 3–9. From Perfil'ev, B. W., Gabe, D. R., Gal'perina, A. M., Rabinovich, V. A., Sapotnitskii, A. A., Sherman, É. É. and Troshanov, É. P. 1964 *Applied Capillary Microscopy*. Translated by F. L. Sinclair. 1965. Consultants Bureau, New York.

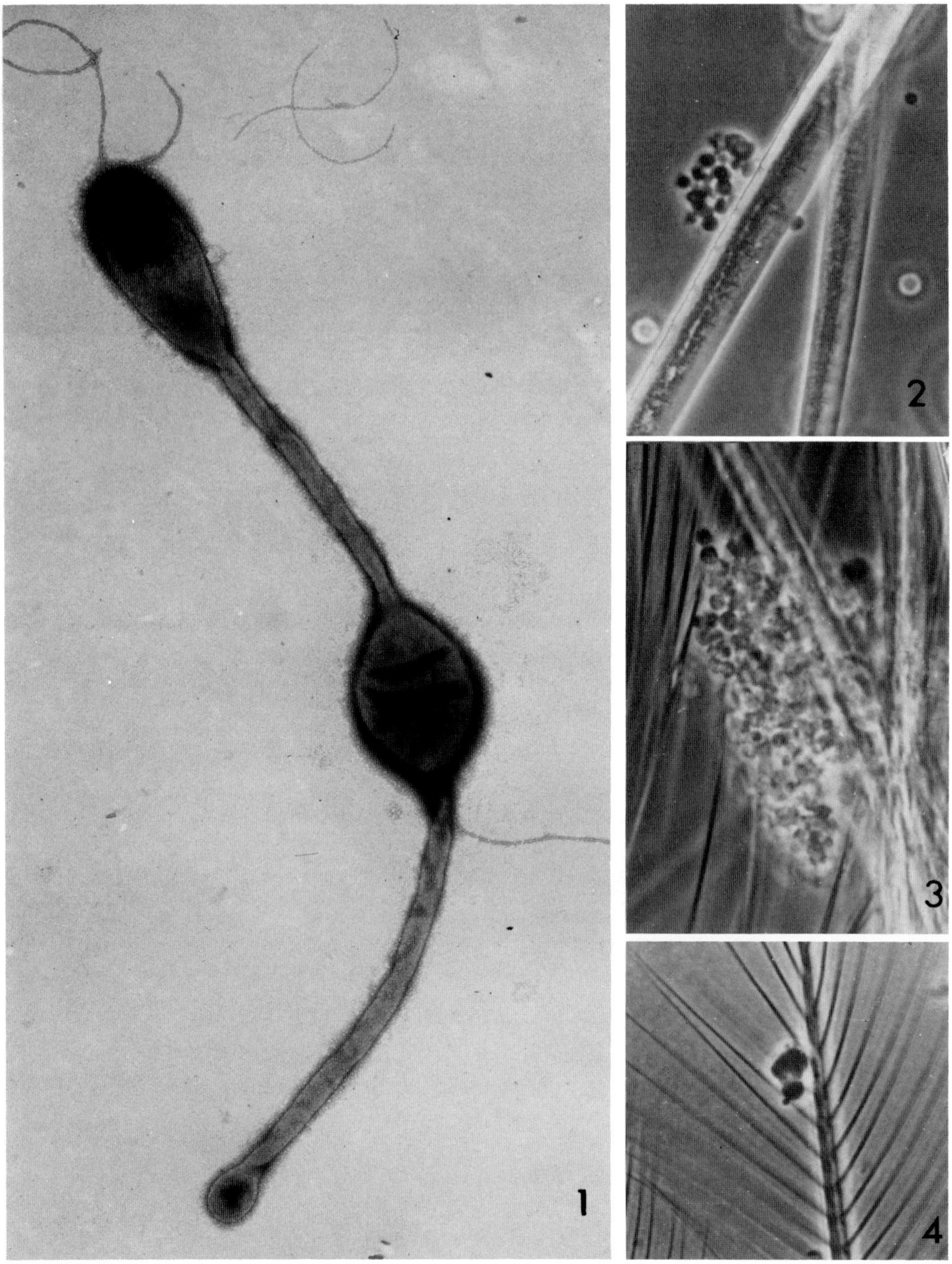

Plate 4.1

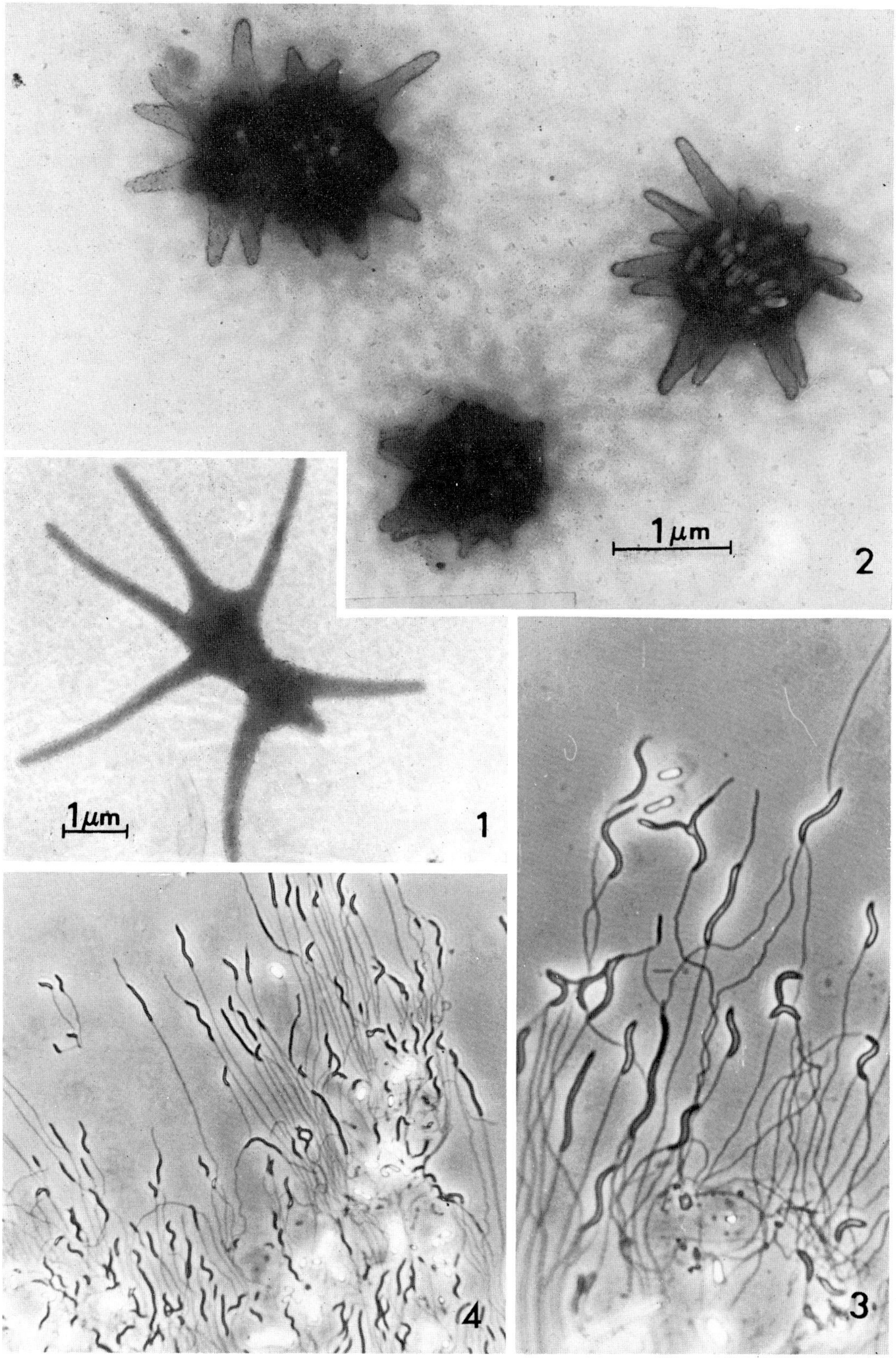

Plate 4.2

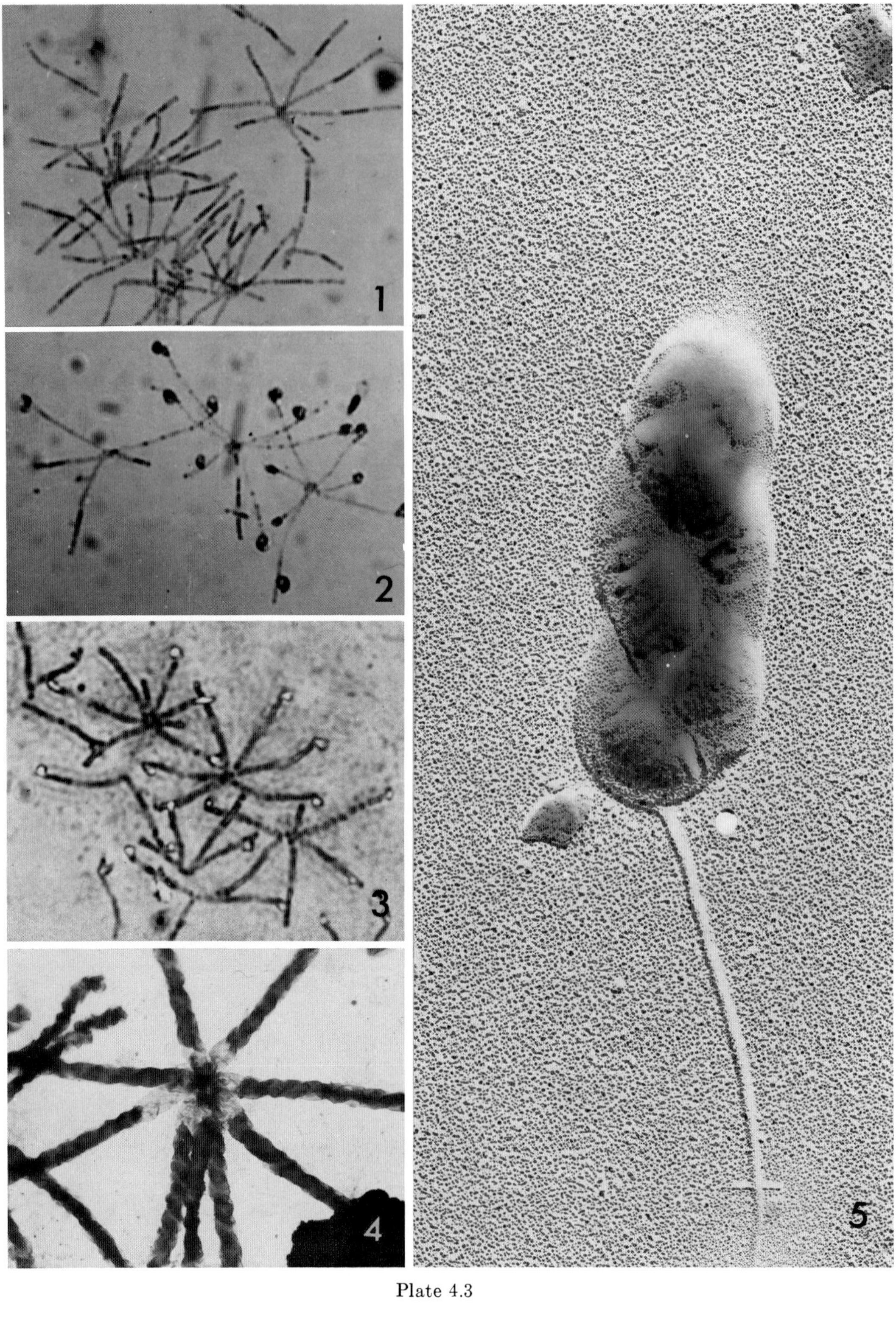

Plate 4.3

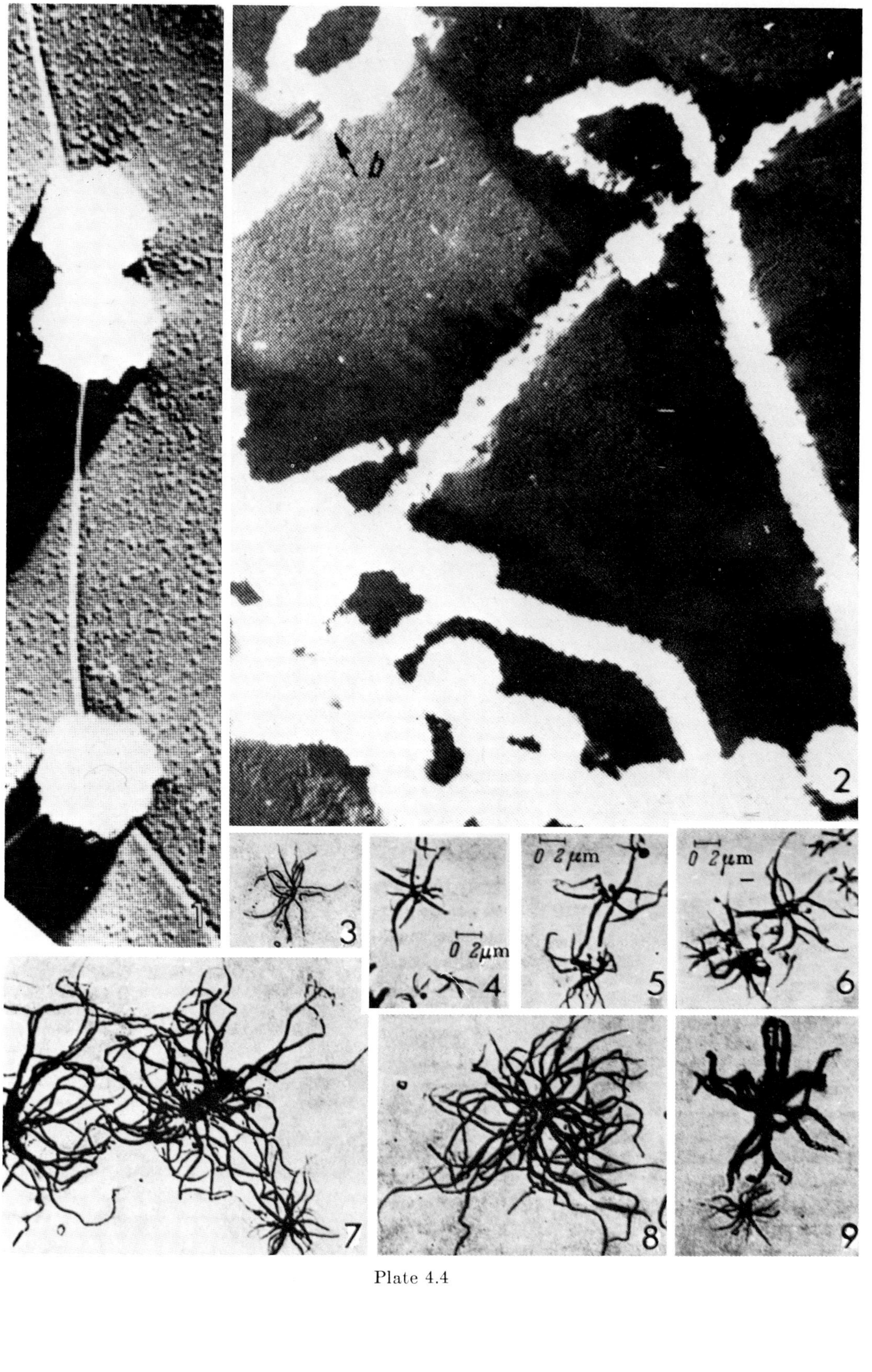

Plate 4.4

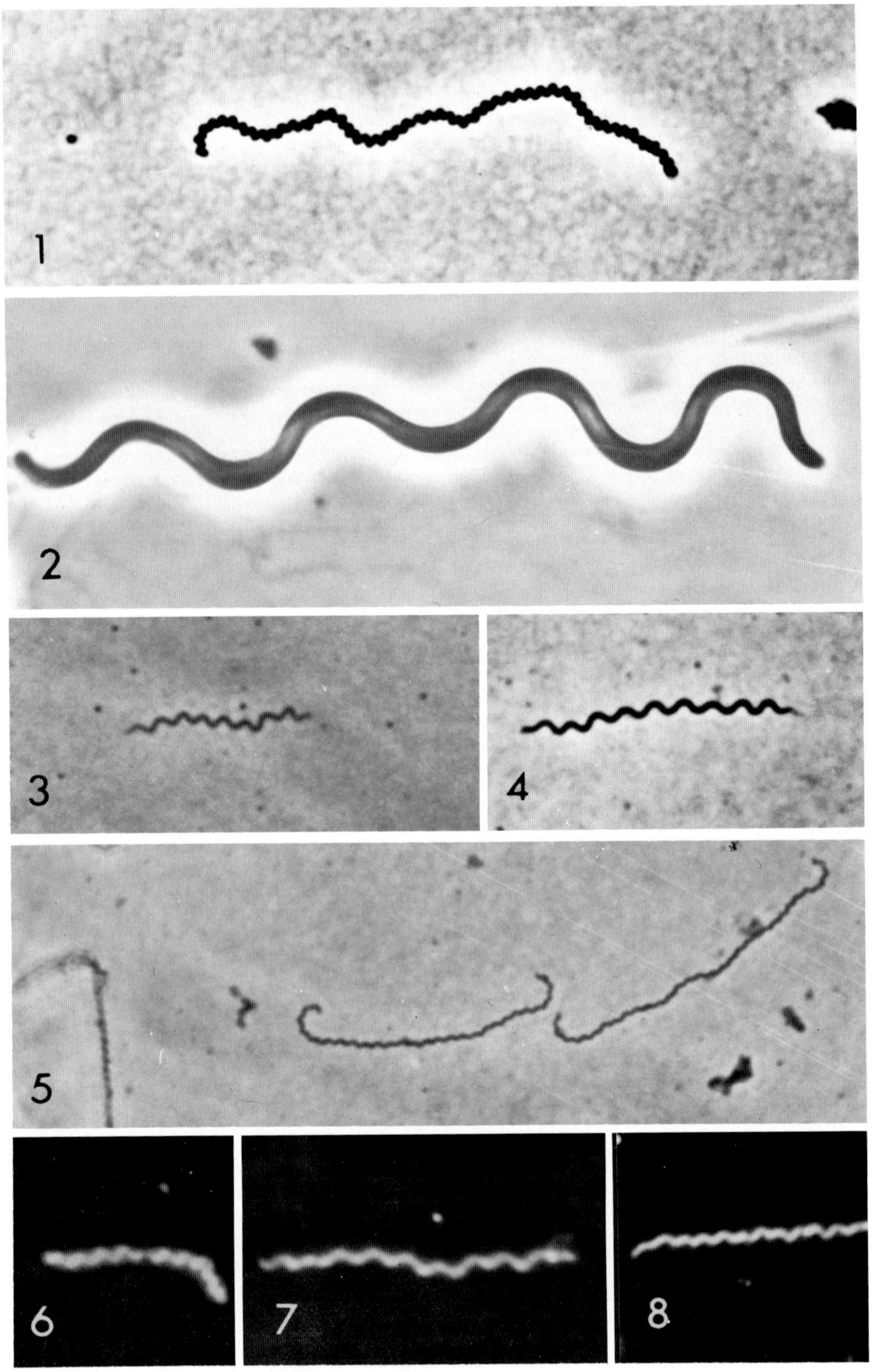

Plate 5.1

Plate 5.1

Plate 5.1. Spirochetes

Fig. 1. *Spirochaeta plicatilis* from activated sludge.
Fig. 2. *Cristispira* sp. from crystalline style of clam, *Cryptomya californica*.
Fig. 3. *Treponema pallidum* from rabbit testis.
Fig. 4. *Borrelia anserina* from chicken blood.
Fig. 5. *Leptospira interrogans* serotype icterohaemorrhagiae from Fletcher's medium.
Fig. 6. *Treponema phagedenis* biotype reiter.
Fig. 7. *Treponema refringens*.
Fig. 8. *Treponema denticola*.
Figs. 1–5. Phase-contrast photomicrographs by Dr. Daisy A. Kuhn, magnification × 2200. Material for *Treponema* supplied by Dr. James N. Miller, for *Leptospira* and *Borrelia* by Dr. Ernst L. Biberstein; Figs. 6–8. Dark-field microscopy, magnification × 2300, by Dr. R. M. Smibert.

Plate 6.1. *Bdellovibrio* and *Pelosigma*

Fig. 1. *Bdellovibrio bacteriovorus* ATCC 15143. Negatively stained preparation showing the vibrio-shaped cell with its thick ensheathed polar flagellum. From Burnham, J. C., Hashimoto, T. and Conti, S. F. 1968. J. Bacteriol. *96:* 1369.

Fig. 2. *Bdellovibrio bacteriovorus* ATCC 15143. Negatively stained flagellum showing blebs formed by sheath pulling away from the flagellar core. ×34,300.

Fig. 3. *Bdellovibrio bacteriovorus* ATCC 15143. A bdellovibrio penetrating an *Escherichia coli* (ATCC 15144) host cell (*p*) and several *Escherichia coli* cells already infected (*i*). Phase-contrast. ×1960.

Fig. 4. *Bdellovibrio bacteriovorus* ATCC 15143. Two *Escherichia coli* B/r cells being attacked by a number of *B. bacteriovorus* cells. Phase-contrast. ×1764.

Fig. 5. *Bdellovibrio bacteriovorus* ATCC 15143. Schematic representation of the life cycle of parasitic, host-independent and facultative strains of the genus *Bdellovibrio*. Modified from Burnham, J. C., Hashimoto, T. and Conti, S. F. 1970. J. Bacteriol. *101:* 1004.

Fig. 6. *Pelosigma* sp. a coenobium from the mud of a fresh water lake (Plussee, Germany) growing attached to a submerged glass slide. ×1254.

Plate 7.1. *Zoogloea* (page 287)

Fig. 1. *A*, finger-like outgrowth of floc, showing cells of *Z. ramigera*. Bright phase-contrast. × 833. *B*, Flocs of *Z. ramigera* at low magnification, showing star-like masses. × 25.

Fig. 2. *A*, tapelike growth in broth of *Z. ramigera*, strain I-16-M. *B* and *C*, lacy hanging growth in broth of *Z. filipendula*, strain P-8-4 and P-77-4 respectively.

Fig. 3. Fibrils interlacing cells of *Z. ramigera*, strain I-16-M. Electron micrograph, shadow cast with carbon platinum. From Friedman, B. A., Dugan, P. R., Pfister, R. M. and Remsen, C. C. 1969 J. Bacteriol. *98:* 1330.

Plate 7.2. *Azotobacter, Azomonas and Derxia* (page 288) (Photographs by J. H. Becking)

Fig. 1. *Azotobacter chroococcum*. Strain with typical rod-shaped cells; 3-day-old cells on nitrogen-free glucose agar.

Fig. 2. *Azotobacter chroococcum*. Cysts and germinating cysts; 14-day-old culture on nitrogen-free glucose agar.

Fig. 3. *Azomonas agilis*. Large, almost spherical yeast-like cells; 3-day-old culture on nitrogen-free glucose agar.

Fig. 4. Seven-day-old cells of *Derxia gummosa* on nitrogen-free agar containing 2% glucose.

Fig. 5. Ten-day-old cells of *Derxia gummosa* on peptone agar with 2% glucose.

Fig. 6. Three-week-old cells of *Derxia gummosa* on nitrogen-free glucose agar. The cells show shrinkage.

Fig. 7. Three-month-old cells of *Derxia gummosa* on nitrogen-free glucose agar. The cells show shrinkage and are enclosed by a thick slime envelope.

All phase-contrast photographs of living cells. Figures 1–3: × 1500, Figures 4–7: × 950.

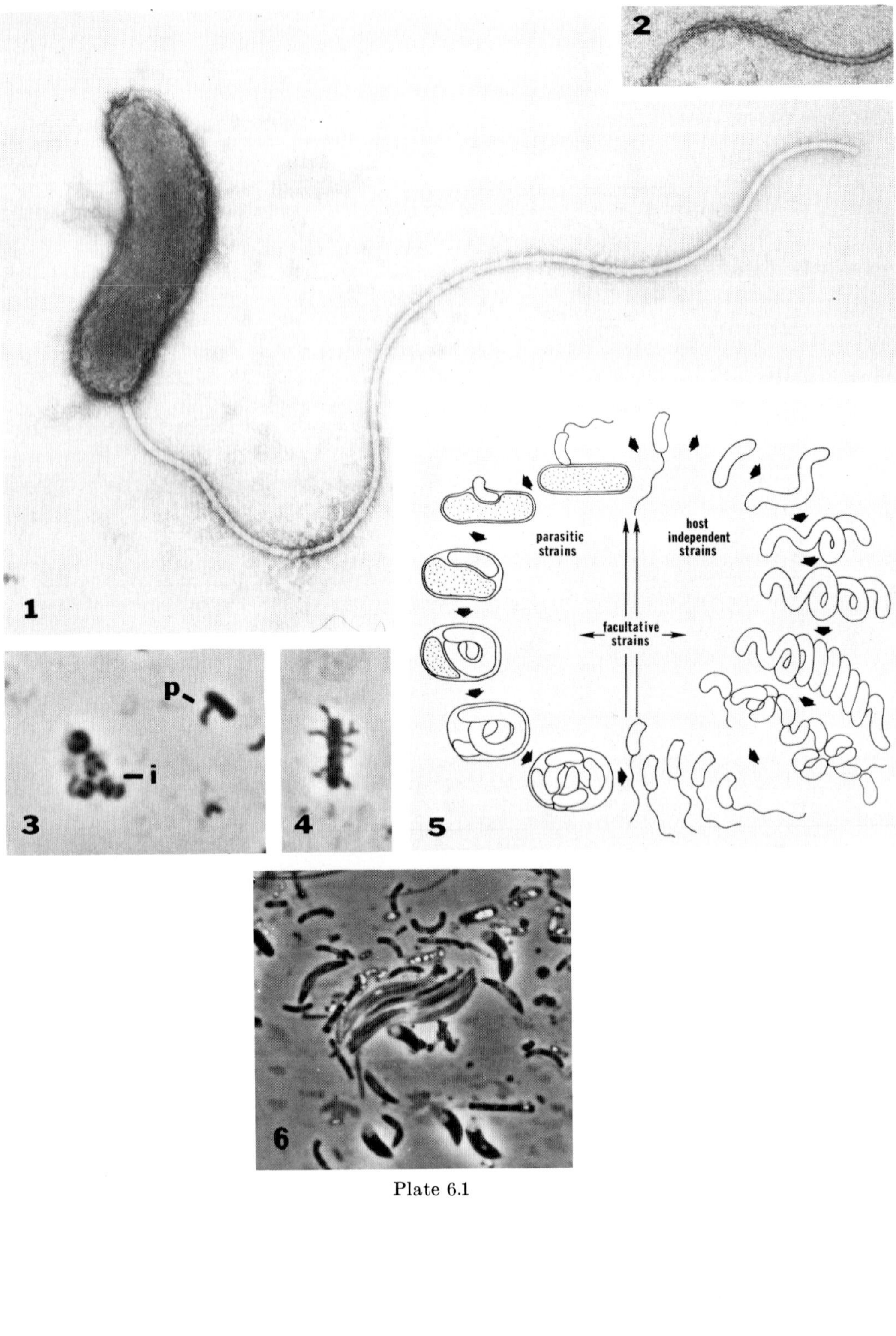
1
2
3
p
i
4
5
parasitic strains
host independent strains
facultative strains
6

Plate 6.1

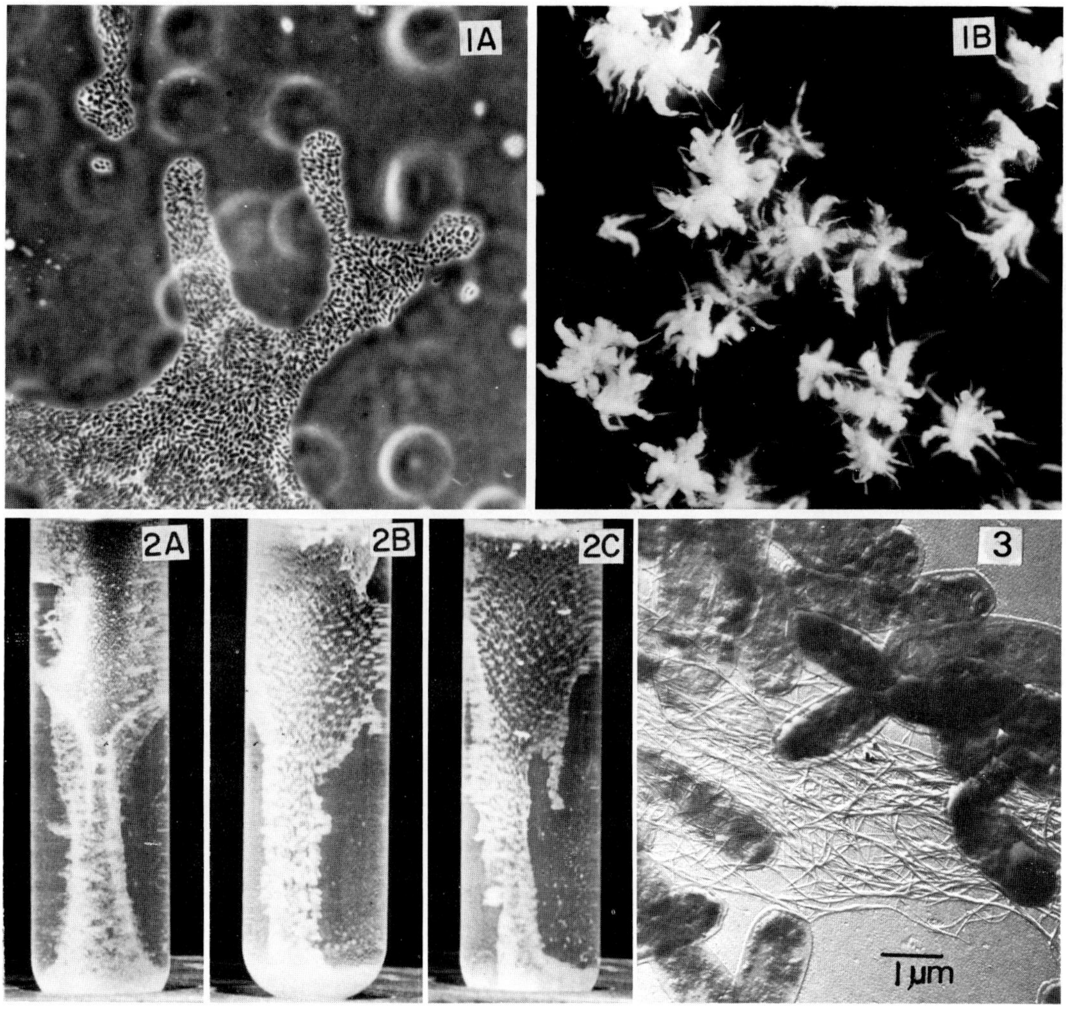

Plate 7.1

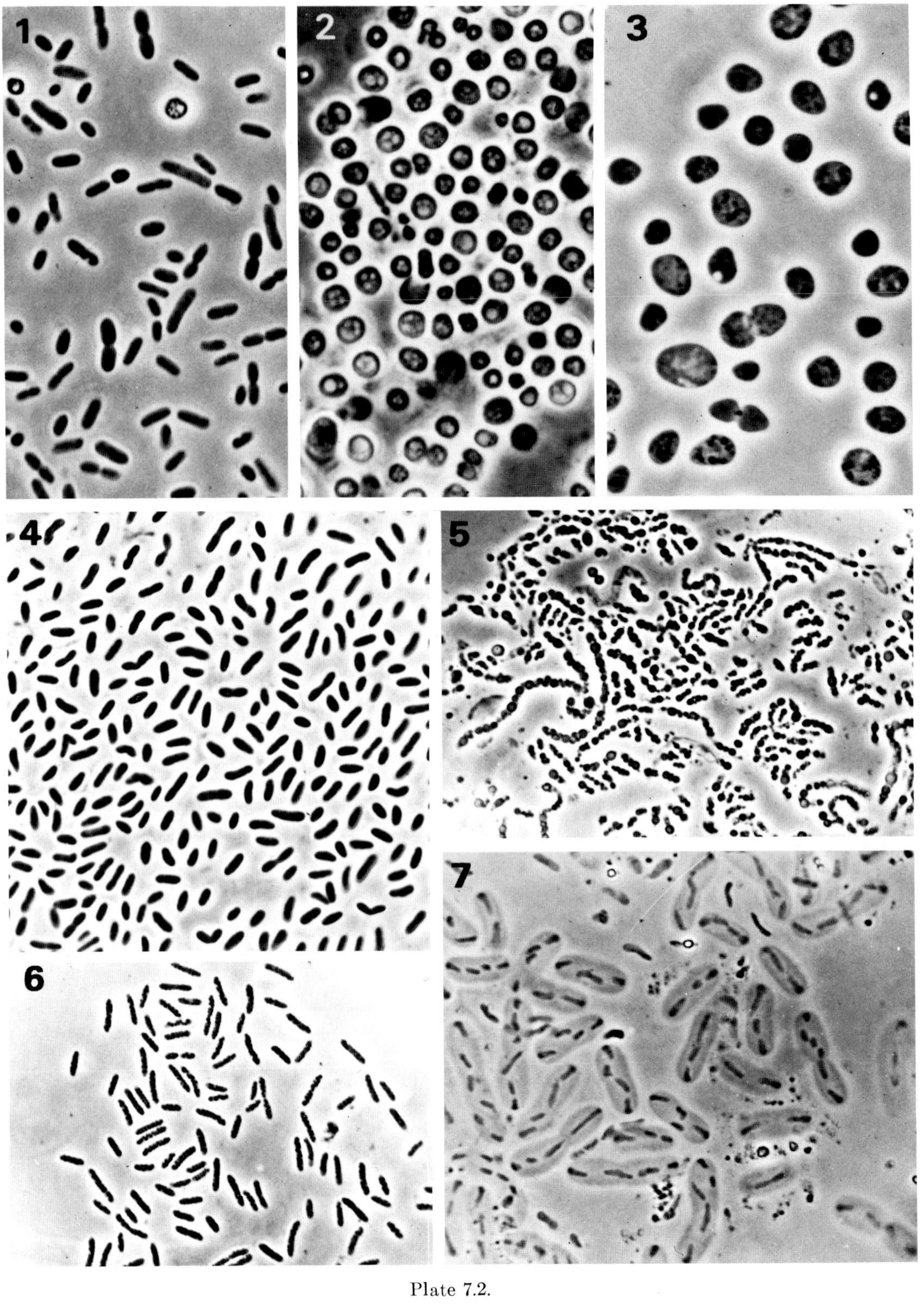

Plate 7.2.

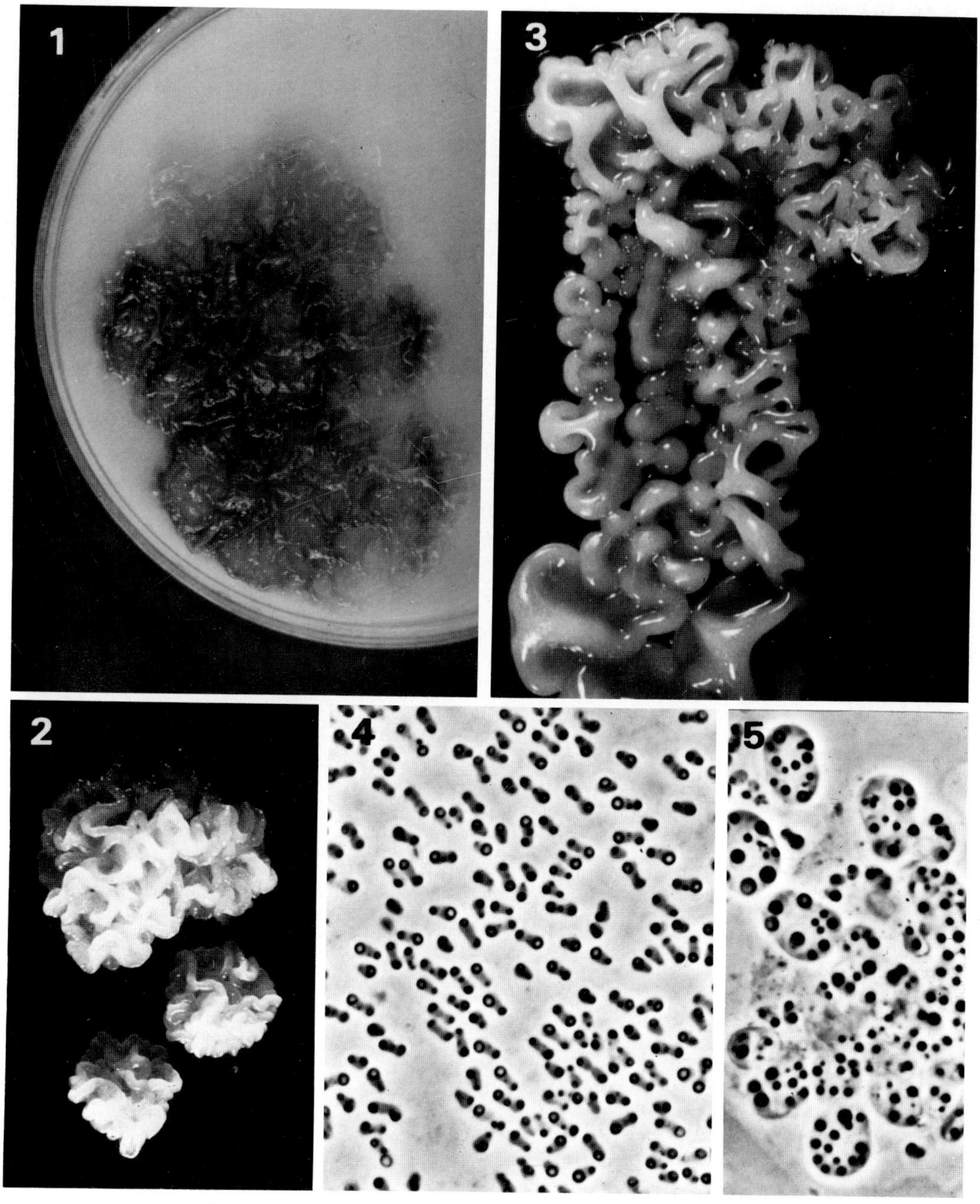

Plate 7.3. *Derxia, Beijerinckia* (Photographs by J. H. Becking)

Fig. 1. Colony-type of *Derxia gummosa* on nitrogen-free glucose agar with calcium carbonate. × 1.0.

Fig. 2. Typical colony of *Beijerinckia fluminensis* showing very plicated surface, while the slime is of more granular consistency. × 4.0.

Fig. 3. *Beijerinckia indica* colony on nitrogen-free glucose agar without calcium carbonate. × 2.0.

Fig. 4. Typical *Beijerinckia* cells (*B. indica*) showing characteristic polar lipoid bodies. × 1500.

Fig. 5. Capsule formation in *Beijerinckia fluminensis* on nitrogen-free glucose agar. × 1500.

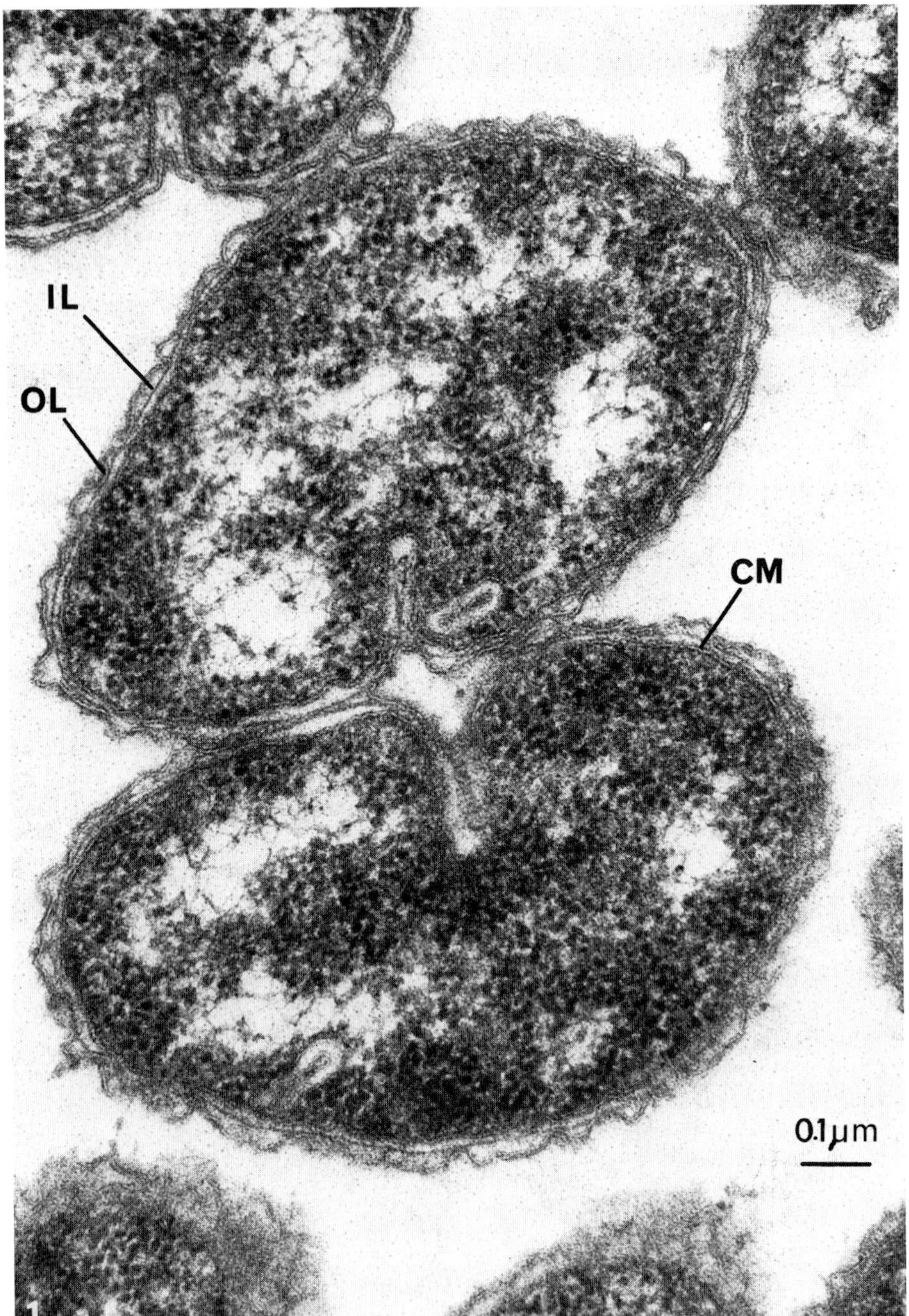

Plate 10.1. *Neisseria*

Fig. 1. Section of dividing cells of *Neisseria gonorrhoeae* (Statens Seruminstitut, Copenhagen, 87115/1961). Note outer layer (*OL*) and intermediate layer (*IL*) of cell wall and cytoplasmic membrane (*CM*). Stained with magnesium uranyl acetate and lead citrate. × 90,000. From unpublished work of A. Reyn and A. Birch-Andersen.

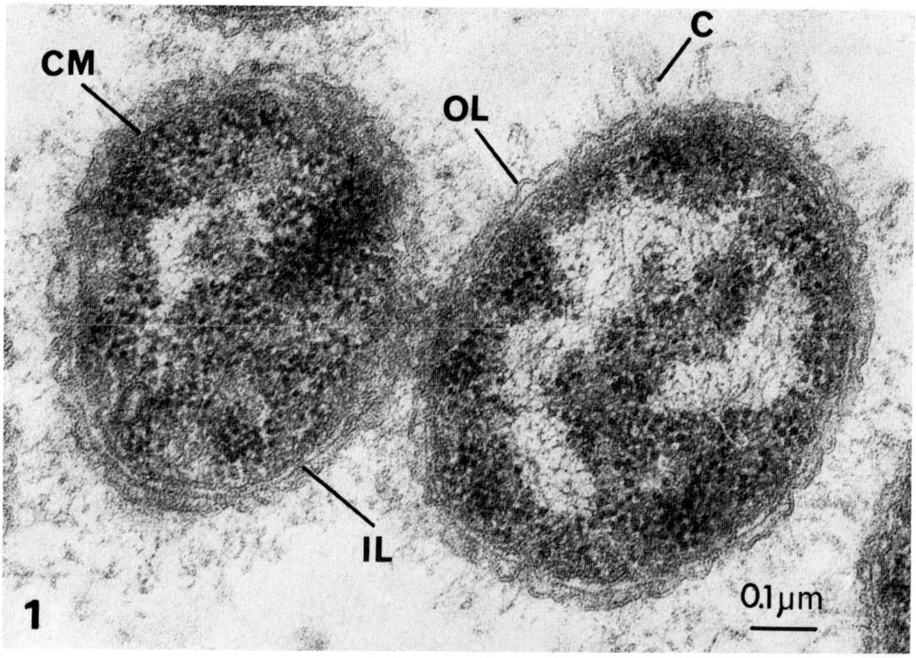

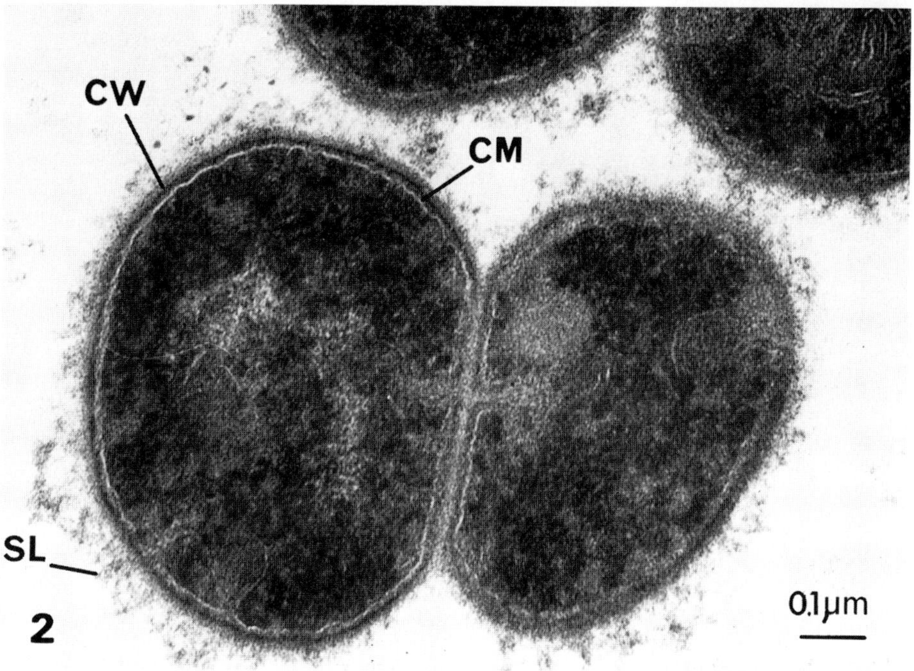

Plate 10.2. *Branhamella* and *Gemella*

Fig. 1. Section of dividing cell of *Branhamella catarrhalis* (Statens Seruminstitut, Copenhagen, 87895/1959). Note outer layer (*OL*) and intermediate layer (*IL*) of cell wall, cytoplasmic membrane (*CM*) and capsule (*C*). Stained with magnesium uranyl acetate and lead citrate. × 90,000. From unpublished work of A. Reyn and A. Birch-Andersen.

Fig. 2. Section of dividing cells of *Gemella haemolysans* (ATCC 10379). Note slime layer (*SL*), cell wall (*CW*) and cytoplasmic membrane (*CM*). Stained with magnesium uranyl acetate and lead citrate. × 90,000. From unpublished work of A. Reyn and A. Birch-Andersen. (See genus *Gemella*, Part 14.).

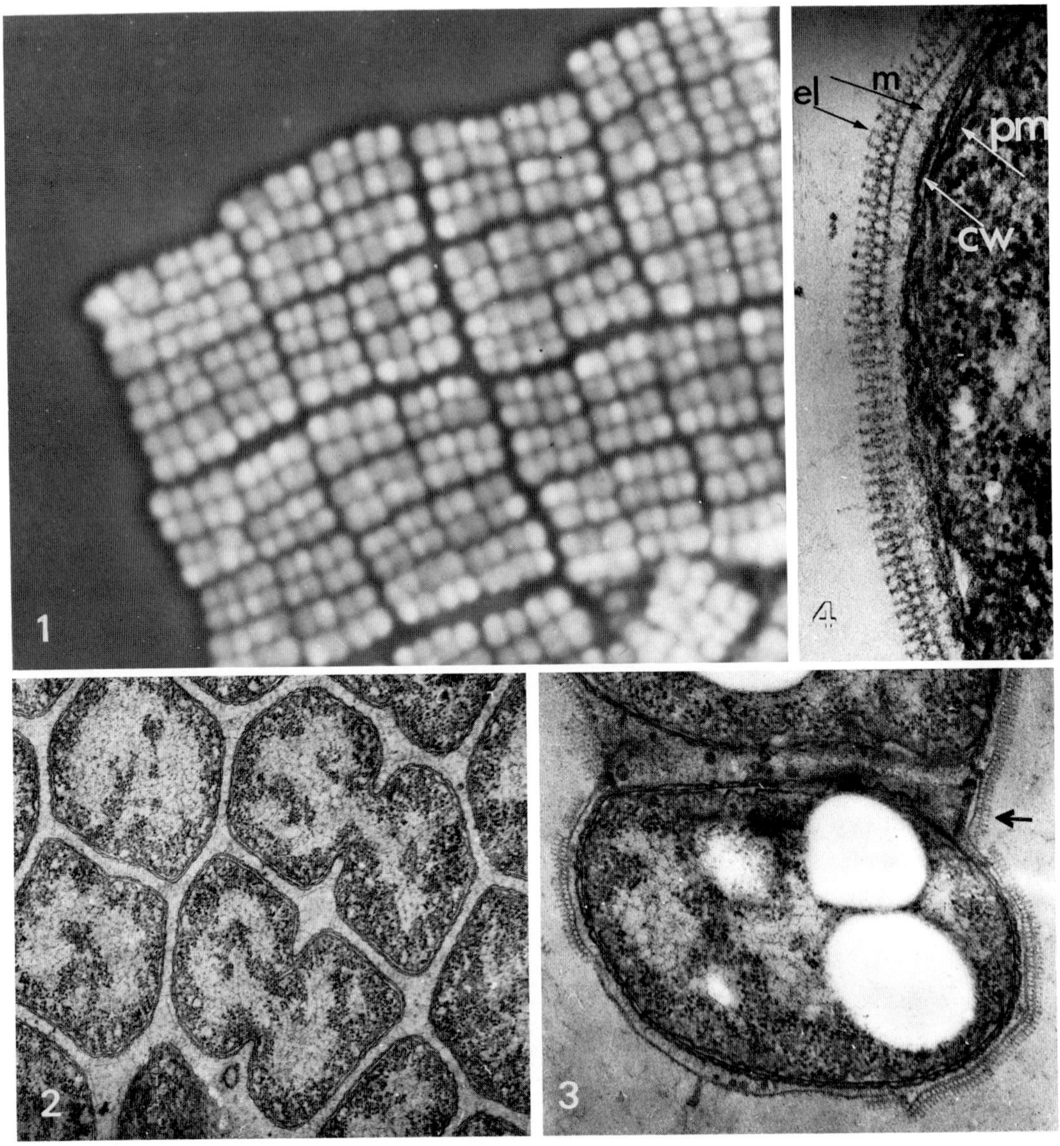

Plate 10.3. *Lampropedia hyalina*

Fig. 1. Light micrograph of a nigrosin preparation showing a corner of a sheet of actively growing tablets of cells. Adjacent tablets are almost synchronized in division. × 2880.

Fig. 2. Electron micrograph of a section in the plane of a tablet to show division and the relatively large volume of nucleoplasm. A matrix substance separates the cell walls (of usual Gram-negative character, see Fig. 4) of adjacent cells. Divisions always show a "constrictive" form. × 25,200.

Fig. 3. Section transverse to the edge of a tablet of cells showing a bipartite external structured envelope. The inner layer is obvious in an area exposed by stripping. The envelope encloses the tablet and bridges over the matrix separating adjacent cells (*arrow*). The low density vesicles represent poly-β-hydroxybutyrate granules. × 42,300.

Fig. 4. A high magnification of a section showing the complex structured envelope (*el*), the matrix (*m*), the cell wall (*cw*) and plasma membrane (*pm*). × 99,000. (Micrographs by R. G. E. Murray and S. Lanys.)

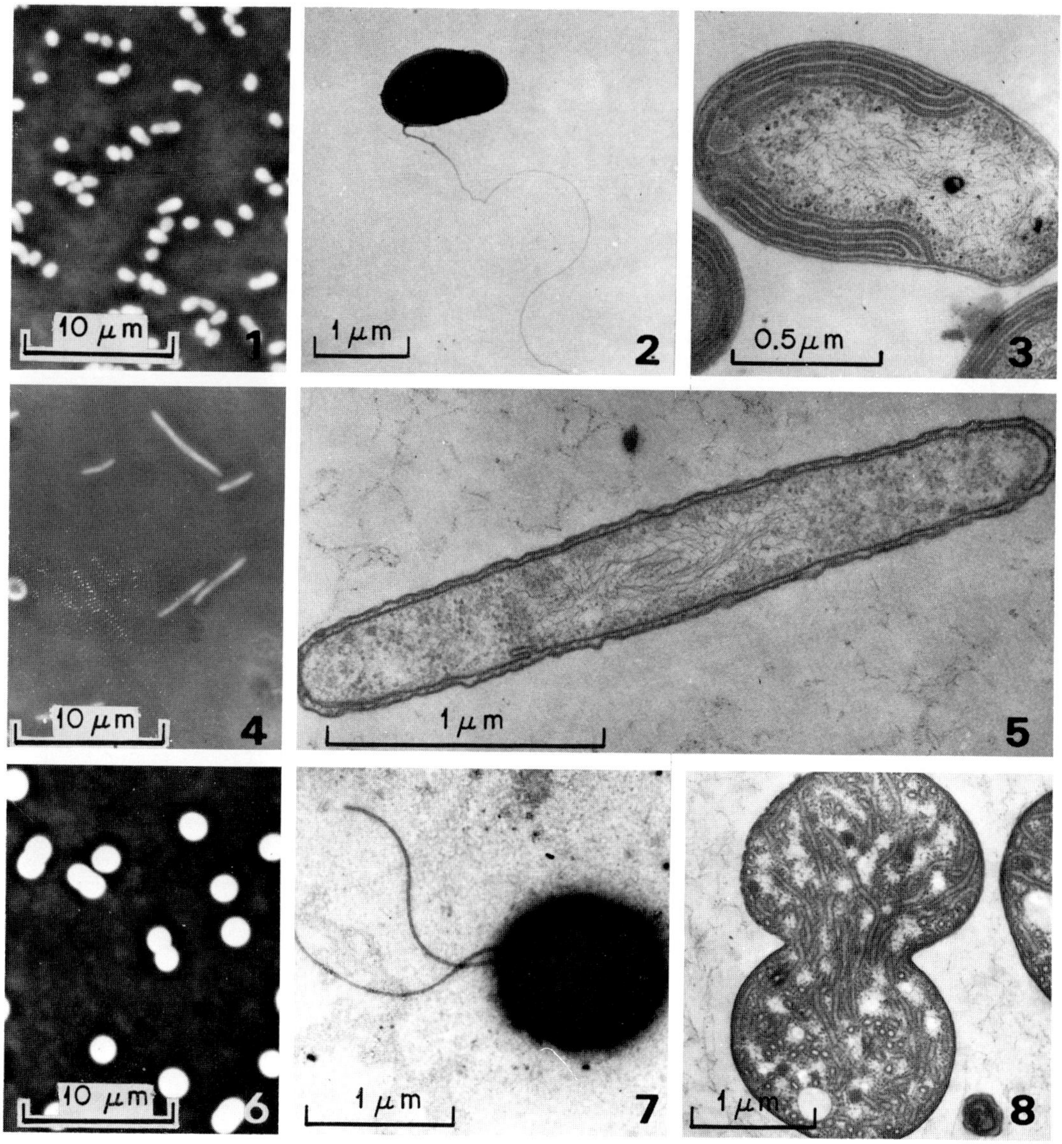

Plate 12.1.

Plate 12.1. *Nitrobacteraceae*

Fig. 1. Phase-contrast photomicrograph of *Nitrobacter winogradskyi* showing pear-and wedge-shaped cells. × 2000.

Fig. 2. Electron micrograph of a negatively stained cell of *Nitrobacter winogradskyi* showing single subpolar flagellum. × 12,800.

Fig. 3. Electron micrograph of a section of *Nitrobacter winogradskyi* showing polar cap of peripheral cytomembranes. × 41,000.

Fig. 4. Phase-contrast photomicrograph of *Nitrospina gracilis* showing the shape and size of cells. × 2000.

Fig. 5. Electron micrograph of a section of *Nitrospina gracilis* showing the lack of extensive cytomembranes found in most other nitrifying bacteria but shows small bleb-like intrusion of plasma membrane. × 37,500.

Fig. 6. Phase-contrast photomicrograph of *Nitrococcus mobilis* showing size and shape of cells. × 2000.

Fig. 7. Electron micrograph of a negatively stained cell of *Nitrococcus mobilis* showing pair of flagella. × 19,300. Figures 5 and 7 from Watson, S. W., Waterbury, J. B.: Characteristics of Two Marine Nitrite-Oxidizing Bacteria, *Nitrospina gracilis* nov. gen. nov. sp. and *Nitrococcus mobilis* nov. gen. nov. sp. Arch Mikrobiol. *77:* 203–230, 1971. Berlin-Heidelberg-New York: Springer.

Figure 8. Electron micrograph of a section of *Nitrococcus mobilis* showing tubular type cytomembranes extending throughout the cytoplasm. × 16,000. With permission of the American Society for Microbiology. Journal of Bacteriology, *107;* 563–569, 1971.

Plate 12.2. *Nitrobacteraceae*

Fig. 1. Phase-contrast photomicrograph of *Nitrosomonas europaea* showing shape and size of cells. × 2000.

Fig. 2. Electron micrograph of a negatively stained cell of *Nitrosomonas europaea* showing pair of flagella with subpolar insertion. × 13,600.

Fig. 3. Electron micrograph of a section of *Nitrosomonas europae* showing peripheral cytomembranes. × 35,200.

Fig. 4. Phase-contrast photomicrograph of *Nitrosospira briensis* showing tightly coiled spirals. × 2000.

Fig. 5. Electron micrograph of a negatively stained cell of *Nitrosospira briensis* showing peritrichous flagella. × 6,000.

Fig. 6. Electron micrograph of a section passing through a spiral of a single cell of *Nitrosospira briensis*, note the lack of cytomembranes. × 32,500. Figures 5 and 6 from Watson, S. W.: Reisolation of *Nitrosospira briensis* S. Winogradsky and H. Winogradsky, 1933. Arch Mikrobiol *75:* 179–188, 1971. Berlin-Heidelberg-New York: Springer.

Fig. 7. Phase-contrast photomicrograph of *Nitrosococcus oceanus* showing shape and size of cells. × 2000.

Fig. 8. Electron micrograph of a shadowed cell of *Nitrosococcus oceanus* showing peritrichous flagella. × 10,000.

Fig. 9. Electron micrograph of a section of *Nitrosococcus oceanus* showing cytomembranes arranged as flattened lamellae in the central region of the cell. × 22,500.

Fig. 10. Phase-contrast photomicrograph of *Nitrosolobus multiformis* showing the lobular shape of cells. × 2000.

Fig. 11. Electron micrograph of a negatively stained cell of *Nitrosolobus multiformis* showing peritrichous flagella. × 8900.

Fig. 12. Electron micrograph of a section of *Nitrosolobus multiformis* showing its lobular shape and cytomembranes which partially compartmentalize the cell. × 22,500. Figures 11 and 12 from Watson, S. W., Graham, L. B., Remsen, C. C., Valois, F. W.: A Lobular, Ammonia-Oxidizing Bacterium, *Nitrosolobus multiformis* nov. gen. nov. sp. Arch Mikrobiol *76:* 183–203, 1971. Berlin-Heidelberg-New York: Springer.

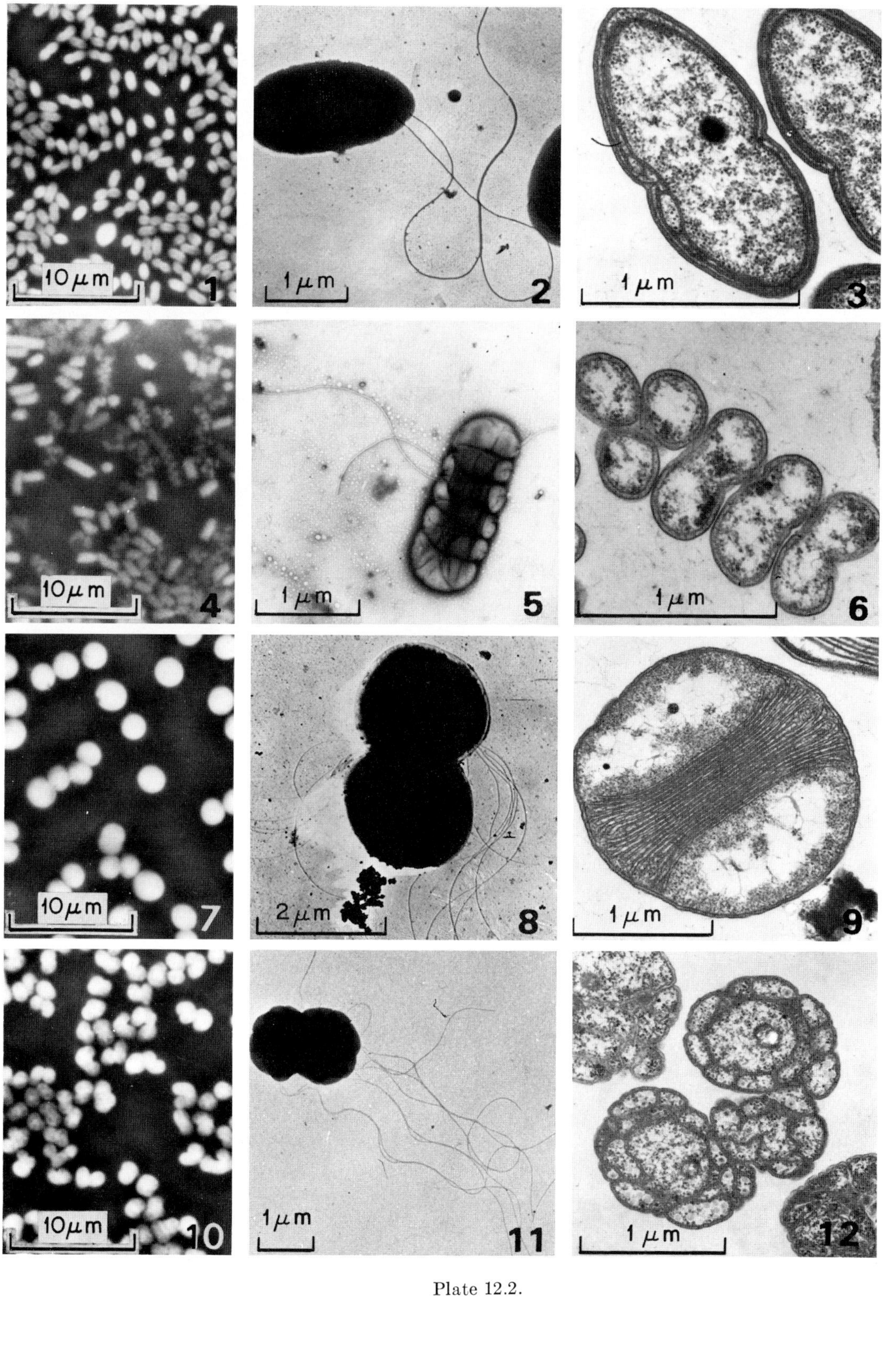
10 μm
1
1 μm
2
1 μm
3
10 μm
4
1 μm
5
1 μm
6
10 μm
7
2 μm
8
1 μm
9
10 μm
10
1 μm
11
1 μm
12

Plate 12.2.

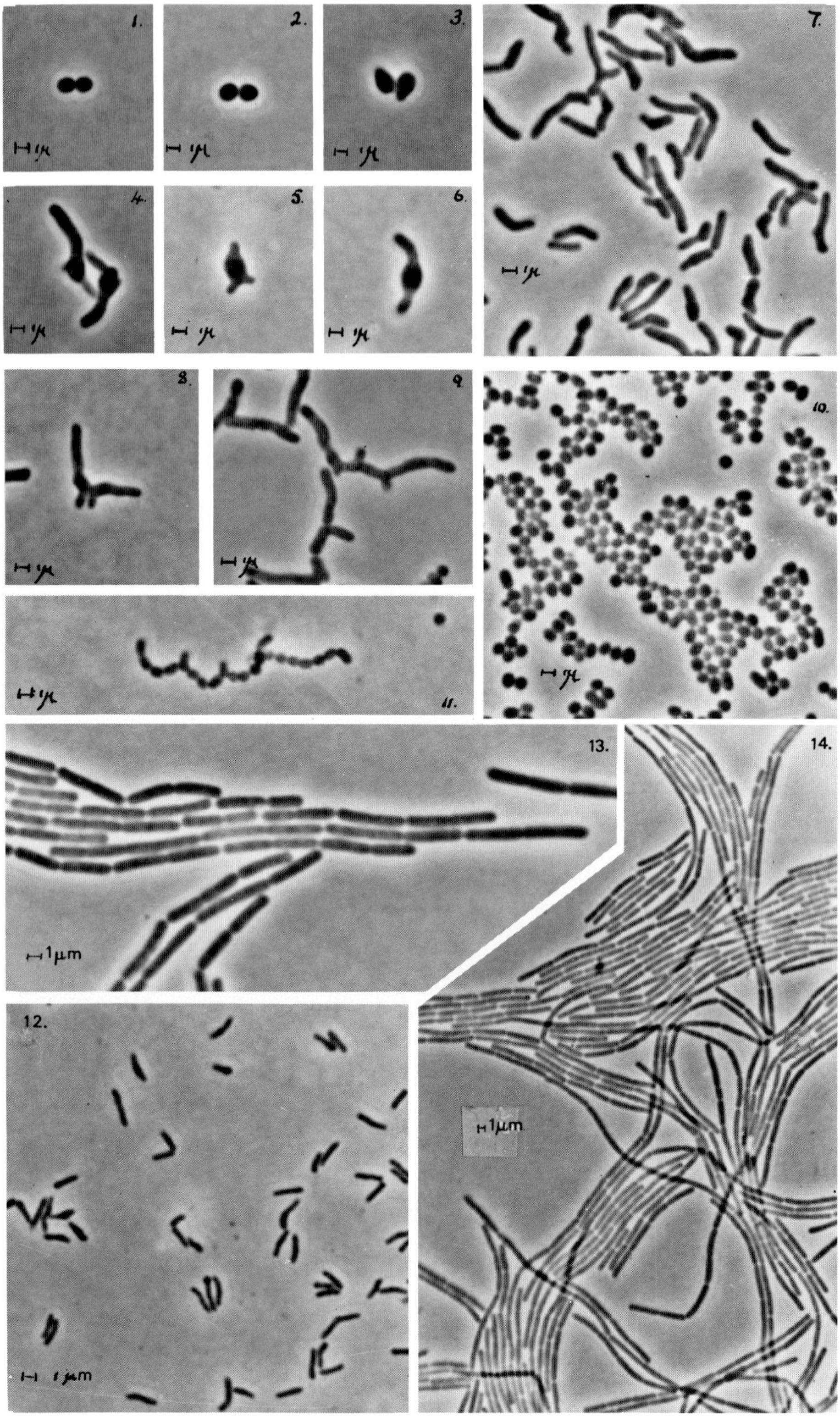

Plate 17.1.

Plate 17.1. Morphology of Species of *Arthrobacter*, *Cellulomonas* and *Kurthia*.

A. globiformis (ATCC 8010, Figs. 1–7, 10; ATCC 4336, Figs. 8, 9, 11). Cultures were at 25 C.

On YSXA medium (0.1% w/v yeast extract, soil extract agar): Figures 1–4 slide cultures after 1.5, 4.25, 6 and 9.75 hr, respectively; Figures 7–8 after 24 hr.

On MYA medium (0.1% w/v yeast extract, mineral salts agar): Figures 5–6 after 5 hr; Figure 11 after 3 days.

On TSXA medium (0.25% w/v yeast extract and peptone, 25% v/v soil extract agar): Figure 9 after 11.5 hr; Figure 10 after 3 days.

Figure 12. *Cellulomonas flavigena* (ATCC 482) on TSXA medium after 45 hr at 30 C.

Figures 13–14. *Kurthia zopfii*. edge of colony on nutrient agar plate incubated at 30 C for 24 hr. All phase contrast by R. M. Keddie. Figure 14 × 940; all others × 1875.

Plate 17.2. *Actinomyces*

Fig. 1. *Actinomyces israelii*, Gram-stain showing diphtheroid morphology.

Fig. 2. *A. israelii*, Gram-stain showing branching morphology.

Fig. 3. *A. bovis*, microcolony, 24 hour growth on brain heart infusion (BHI).

Fig. 4. *A. israelii*, microcolony, 24 hour growth on BHI.

Plate 17.3. *Actinomyces*

Fig. 1. *A. israelii*, mature colony, 14 days growth on BHI.

Fig. 2. *A. naeslandii*, microcolony, 48 hour growth on BHI.

Plate 17.4. *Bacterionema* and *Rothia*

Figs. 1–2. *Bacterionema matruchotii*: Fig. 1. Stained by Gram-reaction from BHI plus yeast extract; Fig. 2. Typical microcolony.

Figs. 3–4. *Rothia dentocariosa*: Fig. 3. Stained by Gram-reaction showing coccoid and filamentous forms; Fig. 4. Rough and smooth colonies of a single isolate; both figures from 7-day-old culture on trypticase soy agar.

Plate 17.5. *Frankia*

Fig. 1. *Frankia alni*. Section through nodule of *Alnus glutinosa* showing hyphae in living host cell: NE, nuclear material; P, plasmalemmosome; CH, cytoplasmic membrane of host. Electron photomicrograph. × 47,000. From Becking, J. H., de Boer, W. E. and Houwink, A. L. 1964 Electron microscopy of the endophyte of *Alnus glutinoas*. Antonie van Leeuwenhoek *30:* 343–376.

Plate 17.6. *Frankia*

Fig. 1. *Frankia alni*. Longitudinal section of vesicle at end of hypha in *Alnus glutinosa* host cell, showing complicated internal structure with many incomplete cell walls and the associated membranes. Electron photomicrograph. × 34,500. J. H. Becking *et al.*, 1964.

Fig. 2. Bacteria-like cells or "bacteroids" in dead host cell (*Alnus glutinosa*), arranged in groups in different stages of development. The irregular mature cells contain reserve material, probably lipid, resistant to sectioning. W, host-cell wall. Electron photomicrograph. × 8,500. J. H. Becking *et al.*, 1964.

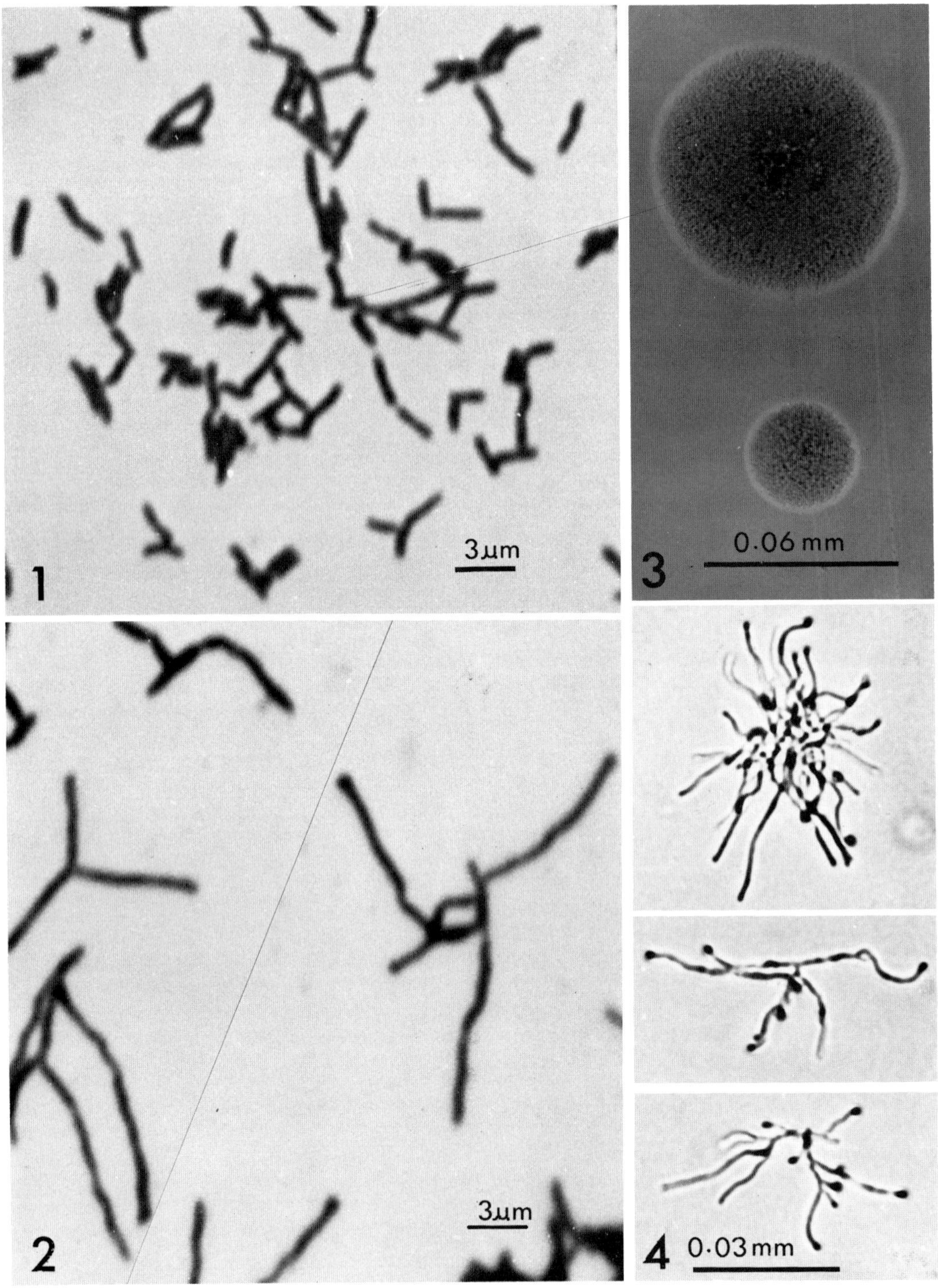

Plate 17.2.

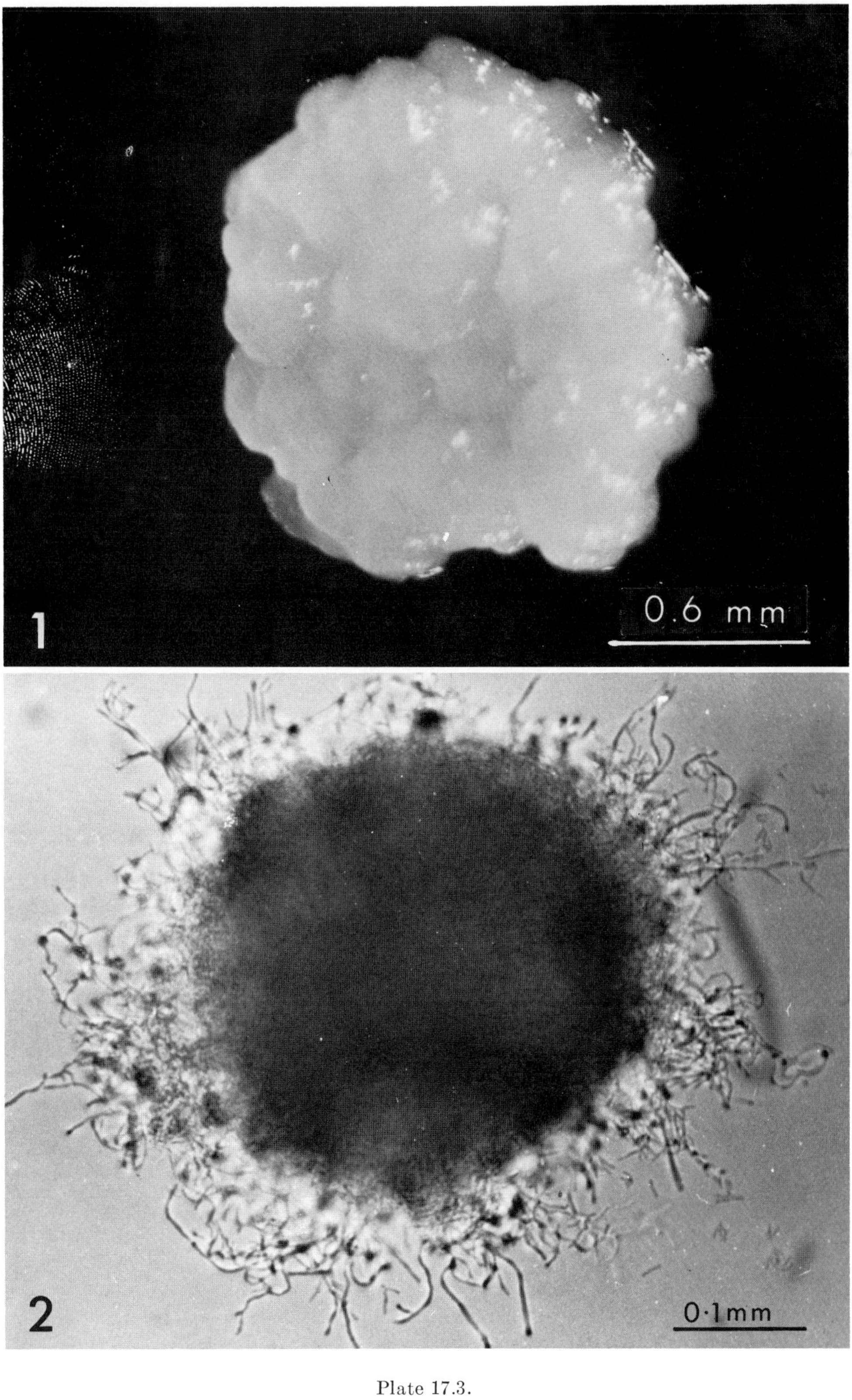

Plate 17.3.

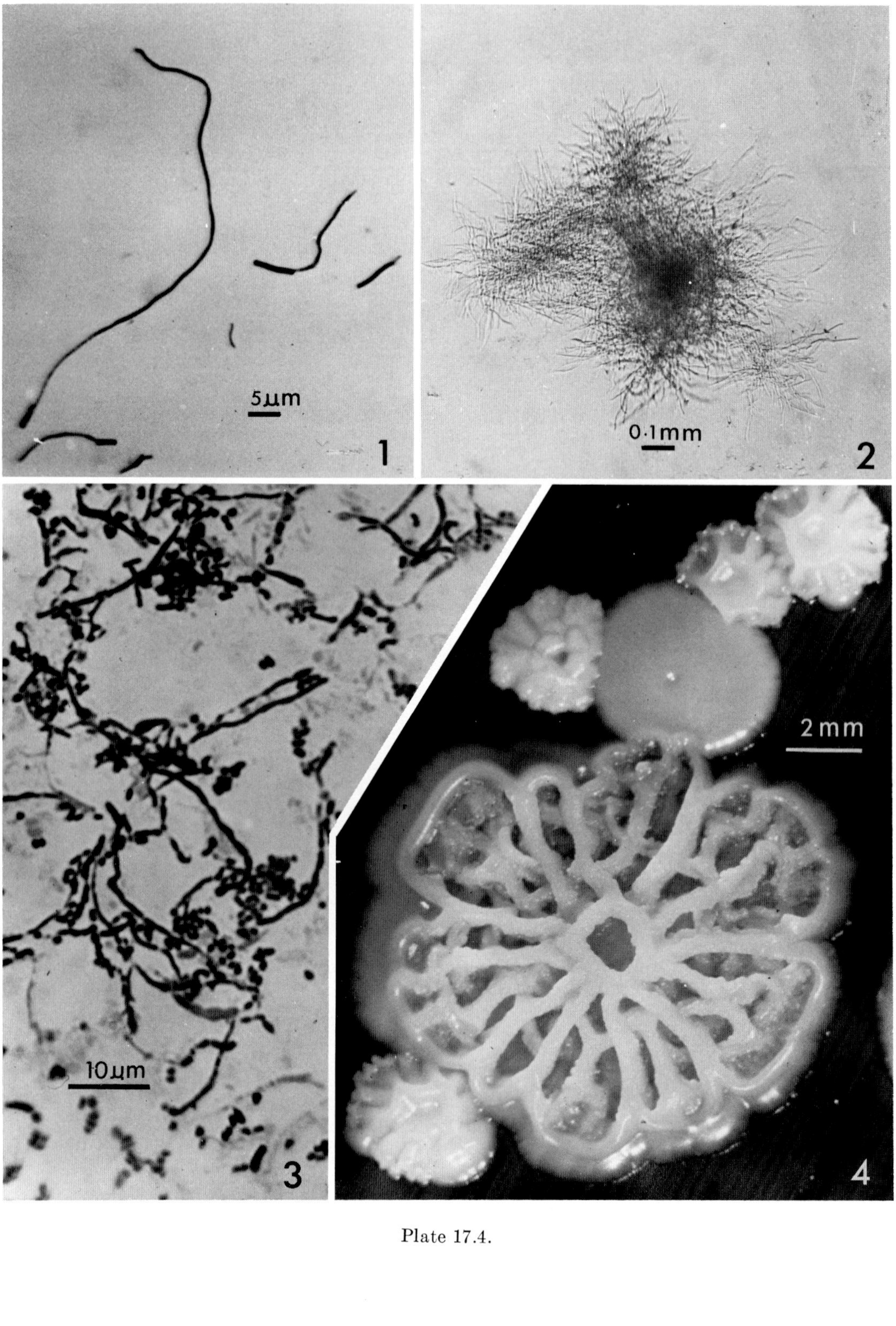

Plate 17.4.

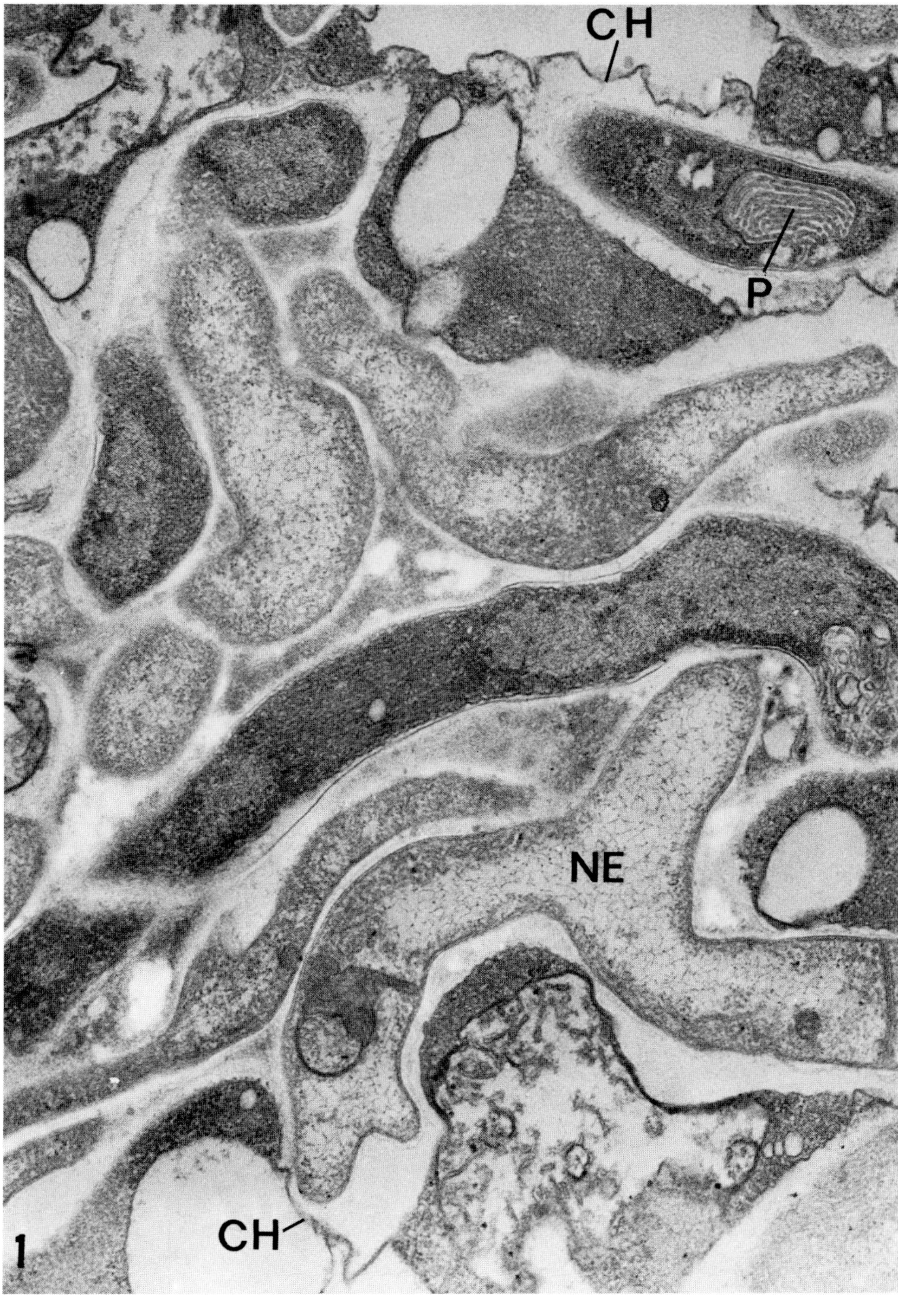

Plate 17.5.

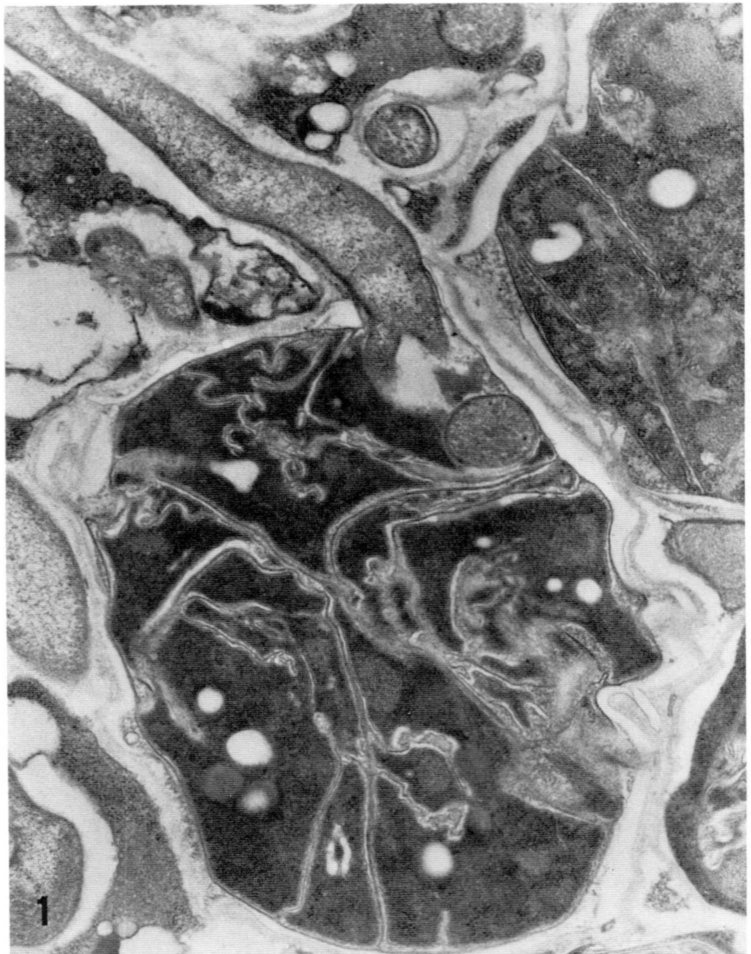

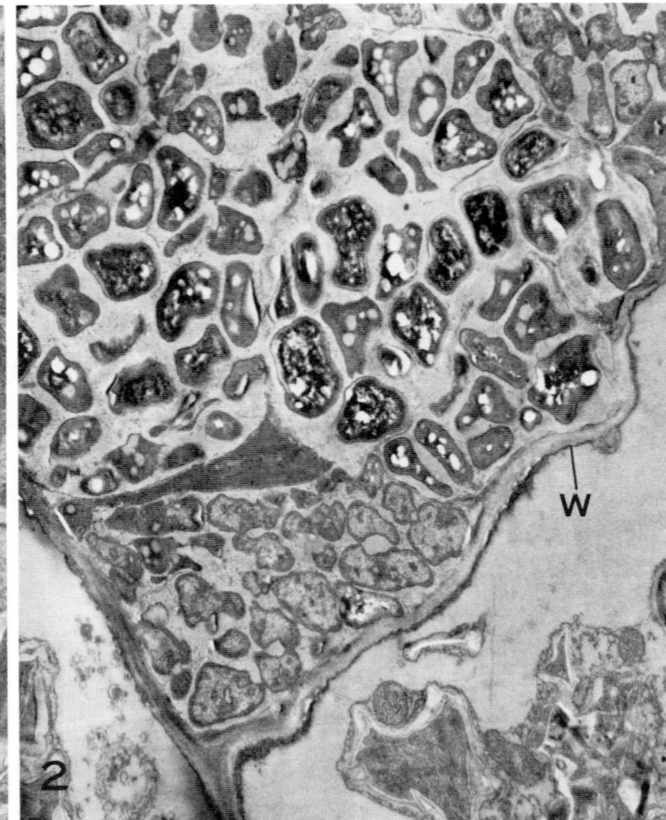

Plate 17.6.

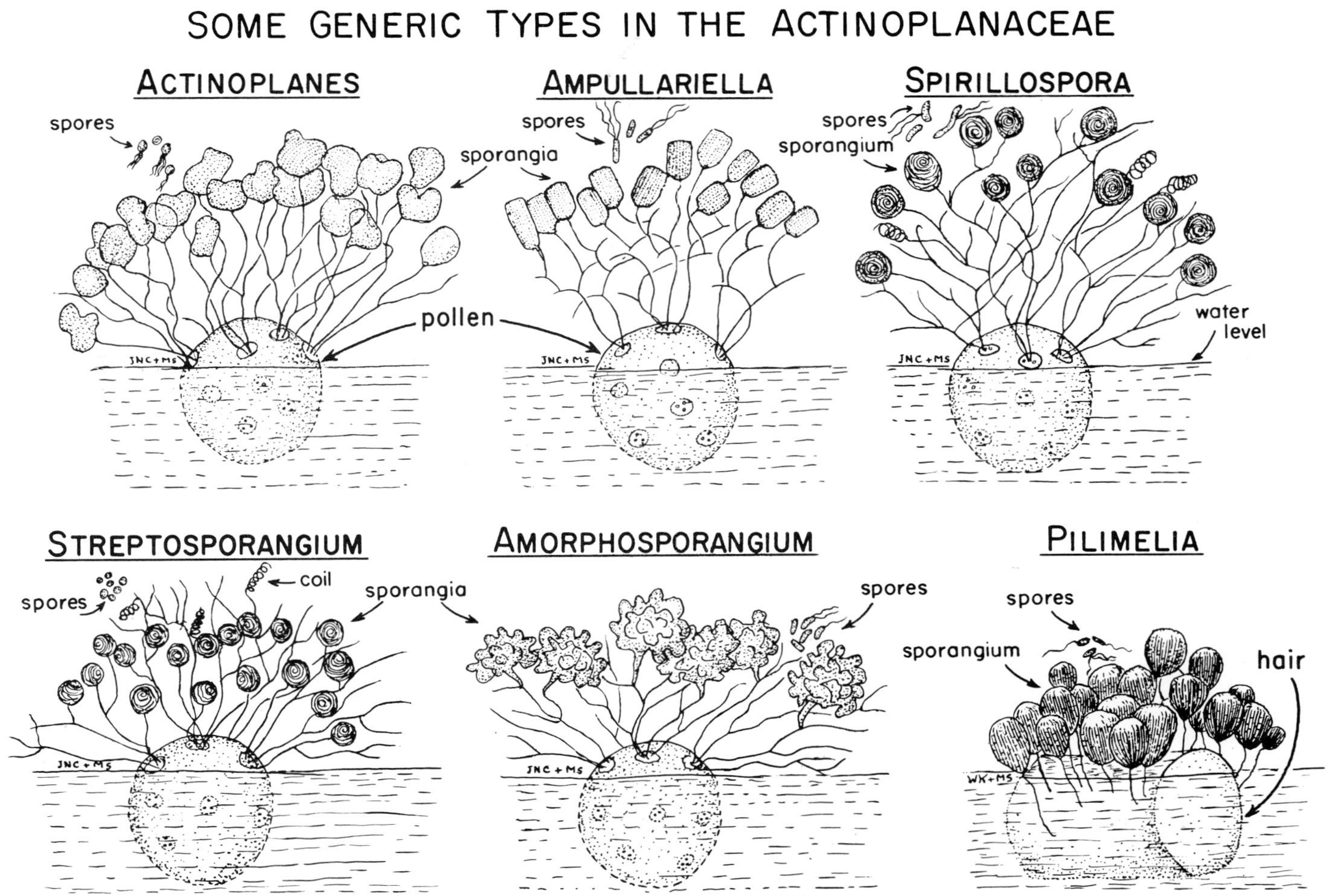

Plate 17.7. *Actinoplanaceae*

Semi-diagrammatic drawings of six of the commoner genera of the *Actinoplanaceae*, five growing on the pollen of *Liquidambar styraciflua* and one on a piece of human hair. The pollen floats at the surface of the water, the hyphae and sporangiophores emerging into the air through the small circular pits in the wall of the pollen grains. × 500.

From *McGraw-Hill Encyclopedia of Science and Technology*, volume 1. Copyright 1968. Used with permission of McGraw-Hill Book Company.

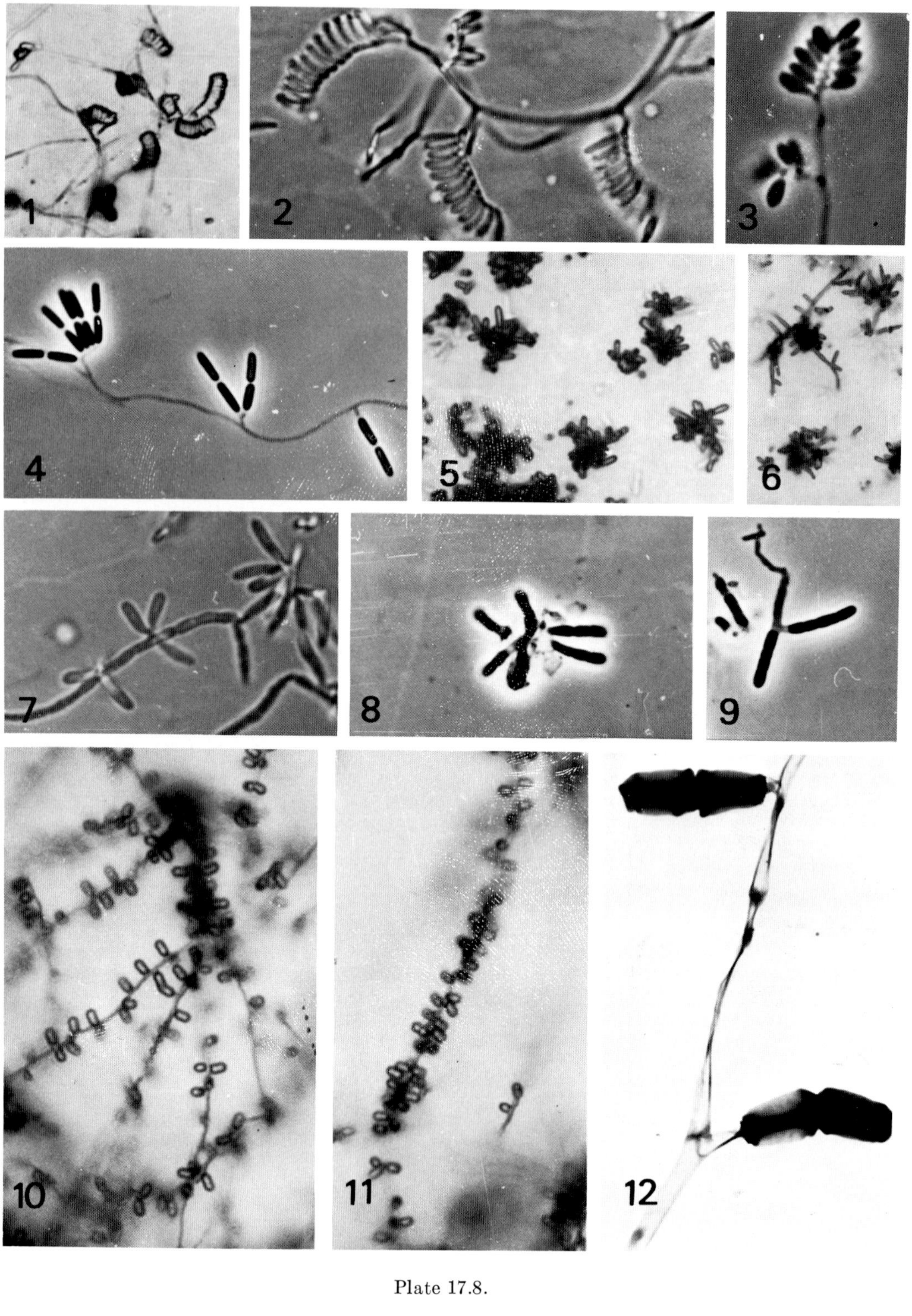

Plate 17.8.

Plate 17.8. *Planomonospora*, *Dactylosporangium* and *Microbispora*

Figs. 1, 2. *Planomonospora parontospora* sporogenous hyphae. Fig. 1. Series of sporangia (× 534). Fig. 2. Developing on agar media (× 1335).

Fig. 3. *Pl. venezuelensis* wet mount showing "palm-leaf" pattern of the sporangia (× 1335).

Fig. 4. *Planobispora longispora* typical sporangia with longitudinal pair of sporangiospores attached to aerial mycelium (× 1335).

Figs. 5–9. *Dactylosporangium.*

Fig. 5. *D. aurantiacum* typical unbranched, finger-shaped sporangia formed at surface of culture media (× 534).

Fig. 6. *D. sp.* branched sporangia developing on surface of culture media (× 356).

Fig. 7. as Fig. 6 (× 1669).

Fig. 8. *D. aurantiacum* wet mount of sporangia (× 1335).

Fig. 9 as Fig. 8. D. *sp.* (× 1335).

Figs. 10–12. *Microbispora.*

Figs. 10–11. *M. rosea*, aerial mycelium with spores (× 712).

Fig. 12. Electron micrograph showing paired spores of *M. bispora* (× 1068).

Figs. 1–9 by J. E. Thiemann. Figs. 10–11 by T. Cross. Fig. 12 by A. Henssen.

Plate 17.9. *Dermatophilus congolensis*

Fig. 1. *Dermatophilus congolensis*, wet mount of beef infusion peptone broth culture, showing various stages in development, from fine hyphae to multiseptate coarse filaments transforming into chains of coccal packets. Dark-field microscopy. × 945.

Fig. 2. Same preparation as in Fig. 1; final transformation of mature hypha into an agglomeration of motile spores, in process of dispersion. Dark field microscopy. × 2100.

Figs. 1 and 2 from Streptothricosis: a new zoonotic disease. New York State Journal of Medicine *61:* 1283–1287, 1961; Fig. 3, A and B by permission of The Medical Society of the State of New York.

Fig. 3 Fully segmented hyphae forming cubical packets of coccoid spores; wet mount of broth culture. Dark-phase contrast. × 1750.

Fig. 4. Encapsulated motile spore of *D. congolensis* with tuft of flagella; negative stain (phosphotungstic acid). × 17,500.

Figs. 3 and 4 from The Genus Dermatophilus, J. Bacteriol., *88:* 509–522, 1964, Figs. 33 and 25, respectively. By permission of the American Society for Microbiology.

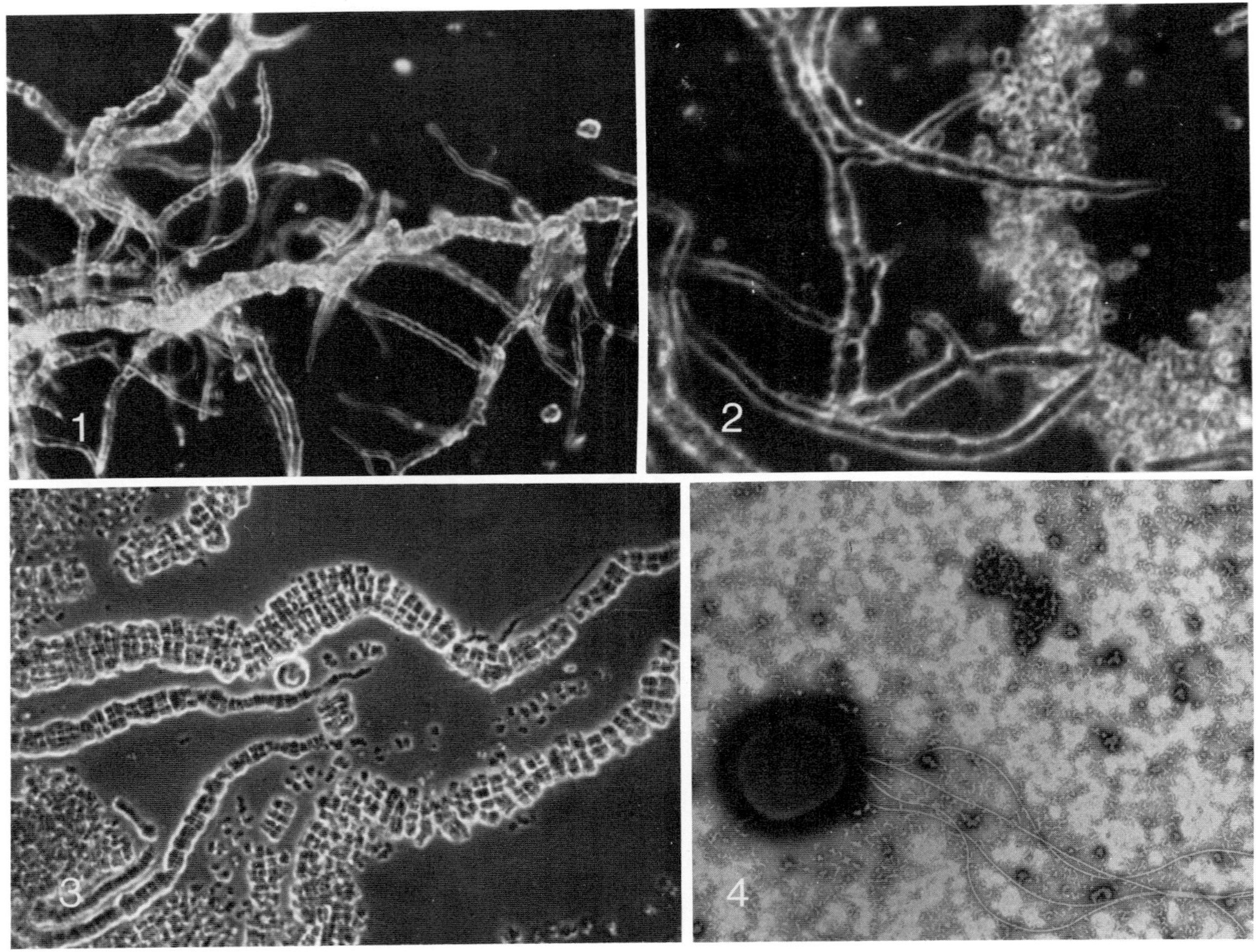

Plate 17.9.

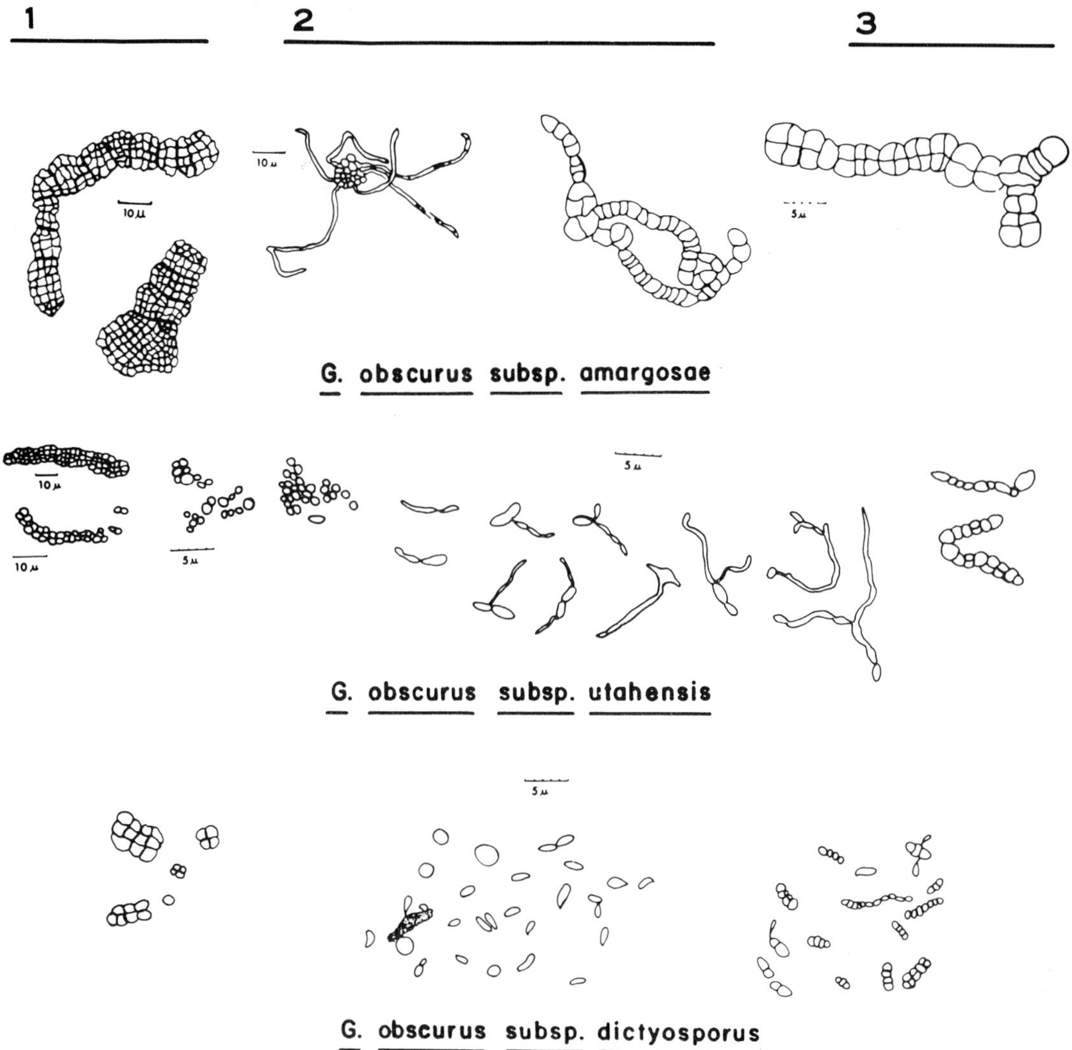

Plate 17.10

Line drawings illustrating the variation in the morphological development of three subspecies of *Geodermatophilus obscurus* and the complexity of the life cycle.

Left to right: (*1*) variation in size and cell number of multiloculate thalli; (*2*) breakdown of thalli into component cells which may germinate (*top*) directly into filaments, (*center*) to form motile zoospores which then come to rest and form filaments, or (*bottom*) to form motile or non-motile cells from which small thalli develop directly rather than first developing a filamentous stage; (*3*) early stages in the formation of thalli illustrating muriform septation.

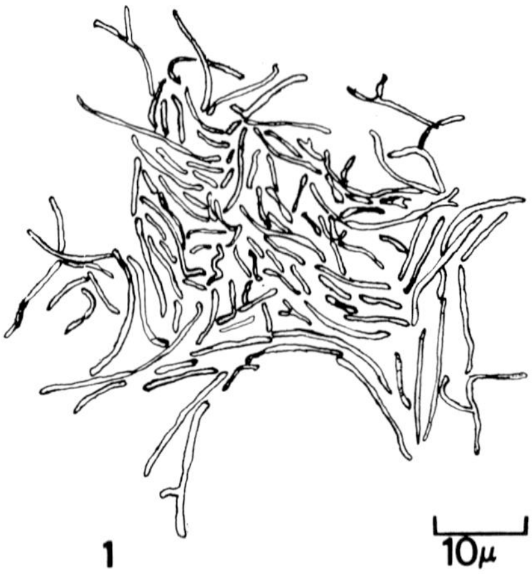

1

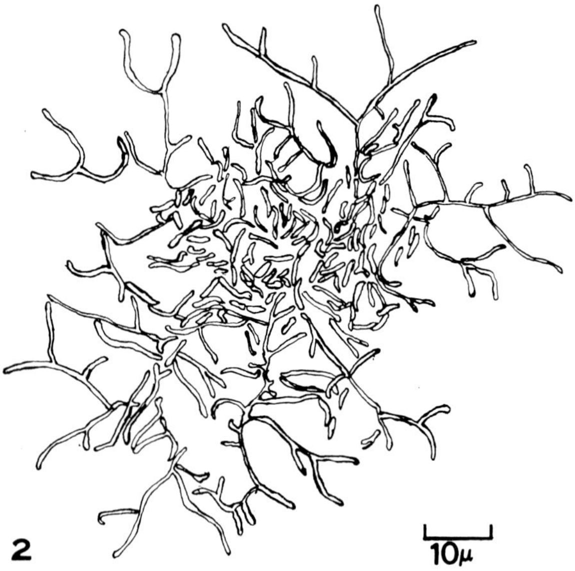

2

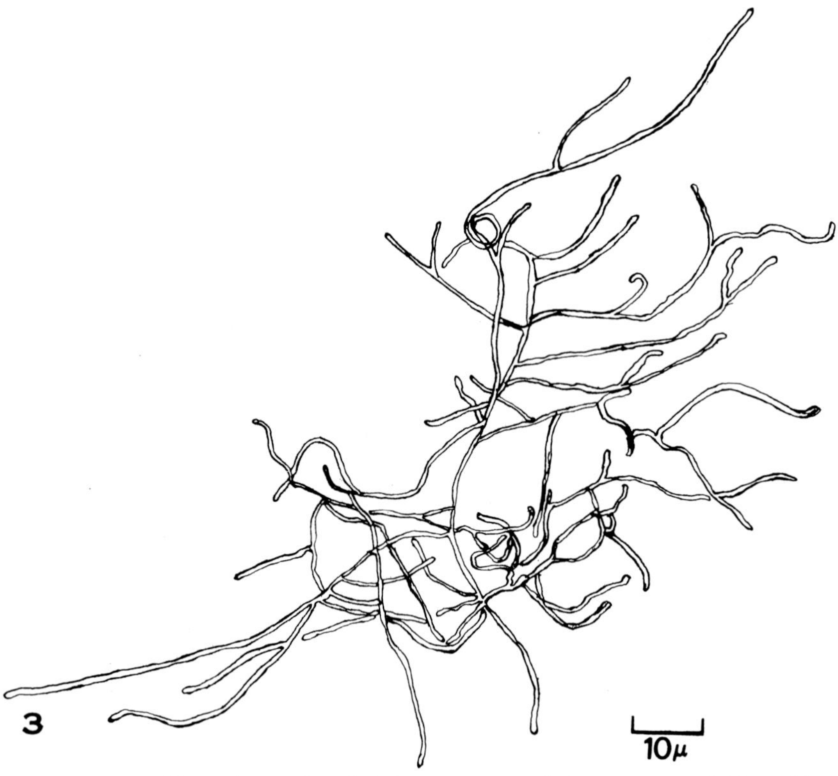

3

Plate 17.11. Mycelial development of three groups of *Nocardia*

Fig. 1. *Nocardia erythropolis* 27 hr at 28 C. Group I. Note sparse branching.

Fig. 2. *Nocardia rubra* 27 hr at 28 C. Group II.

Fig. 3. *Nocardia* sp. 27 hr at 28 C. Group III. Note extensive mycelial development and lack of fragmentation.

From *Lloydia* 12: 137–177. By permission of the editor.

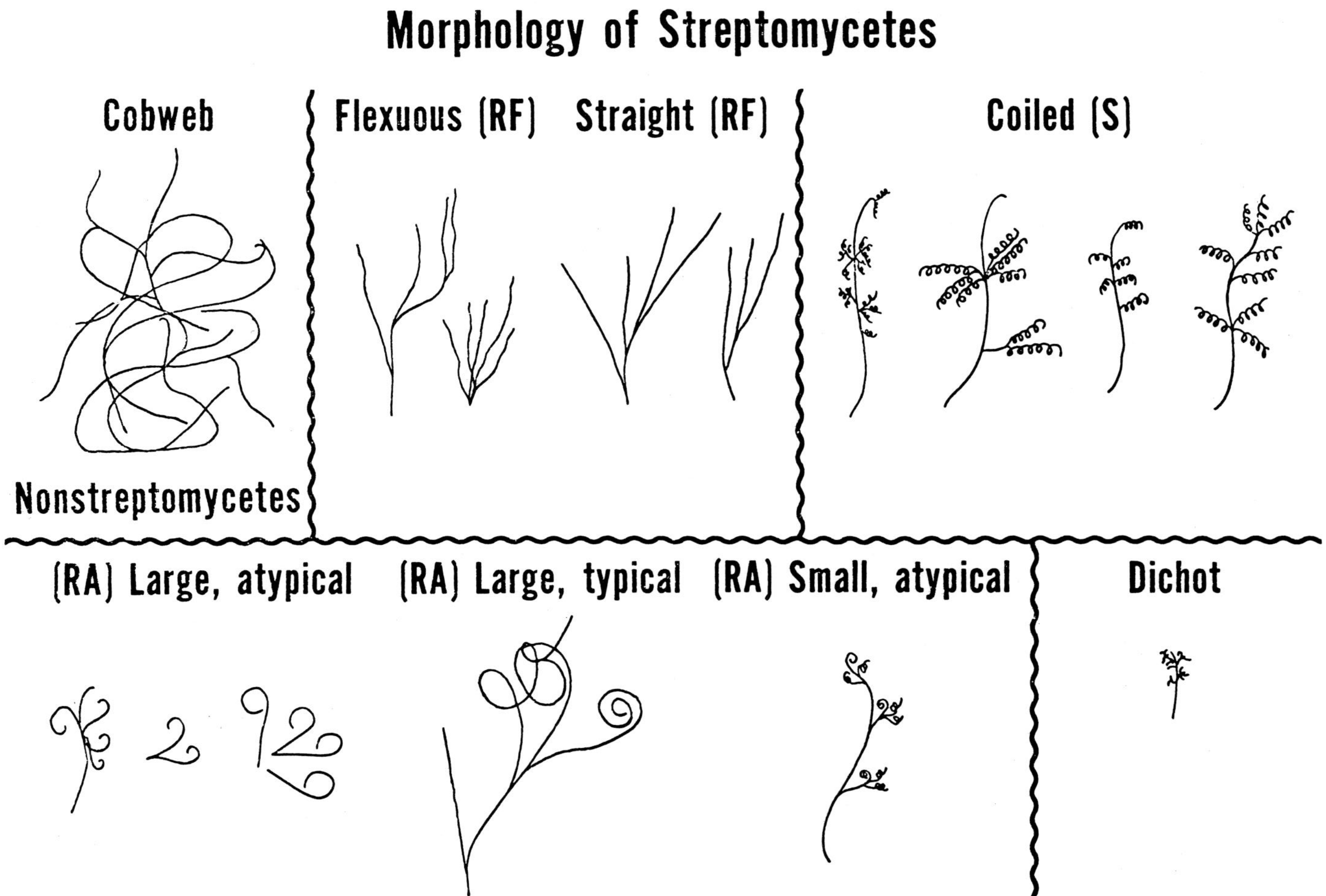

Plate 17.12
Morphology of spore chains of streptomycetes.

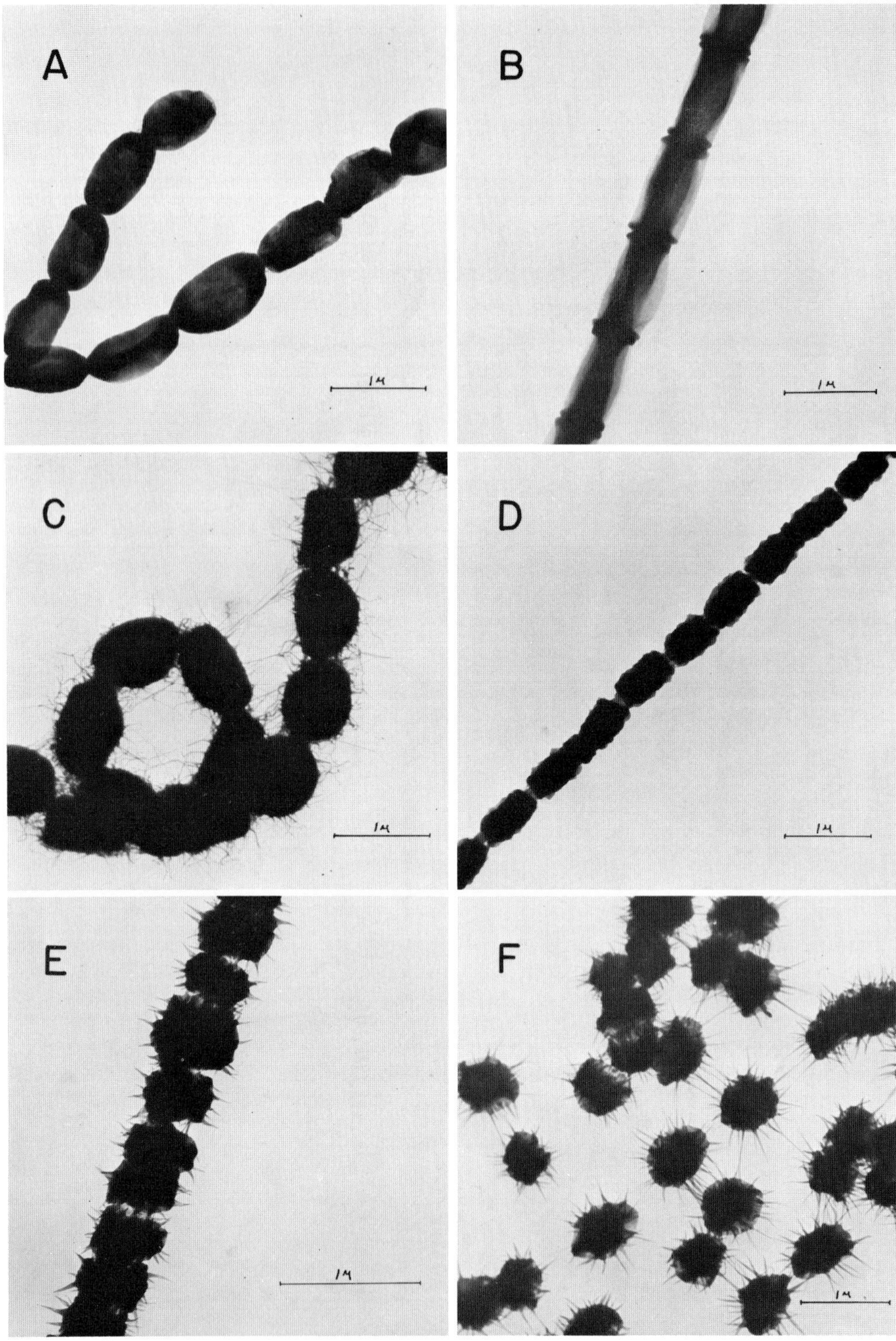

Plate 17.13. Spore surface morphology of streptomycetes

A, spore surfaces smooth. *B*, spore surfaces smooth; spores phalangiform. *C*, spore surfaces hairy. *D*, spore surfaces warty. *E*, spore surfaces with short stout spines. *F*, spore surfaces with long slender spines.

Plate 18.1. *Ehrlichia, Neorickettsia*

Fig. 1. *Ehrlichia phagocytophila* in cytoplasmic vesicle in eosinophil granulocyte of infected calf. The inclusion, ready to rupture, contains large particles of the organism. The structural units of the particles, cell wall, plasma membrane, nucleoid and cytoplasm may be seen. Ribosome-like granules occur in a ground substance of intermediate density. Almost complete division of particles is demonstrated in an oblique section *(arrow)*. The vacuole is lined by a membrane. Cytoplasmic vesicles appear inside the vacuole and especially in the space between vacuole membrane and host cell membrane which are detached from each other. Budding of still smaller vesicles into the vacuole is also visible. × 45,000. (From Tuomi and von Bonsdorff (1965) J Bacteriol *92:* 1482.)

Fig. 2. *Neorickettsia helminthoeca* as seen in imprints of mesenteric lymph node of fatally infected dog. *A*, clumps and scattered organisms from ruptured cells. *B*, morula-like colonies in the cytoplasm of a disintegrating macrophage surrounded by lymphocytes. Giemsa stain. × 1500. (From Philip, Hadlow and Hughes (1954) Exp Parasitol *3:* 345.)

Plate 18.2. *Bartonella*

Fig. 1. *Bartonella bacilliformis*, in blood films from an Oroya fever patient. An intense infection involving the erythrocytes and, a rarer finding, the mononuclear leucocytes also.

Fig. 2. Human Oroya fever blood, showing rod and ring forms of *B. bacilliformis*.

Fig. 3. Section of an Oroya fever lymph node. Development of *B. bacilliformis* in distended endothelial cells lining the vein. Intracytoplasmic distribution in rounded clumps is distinct in some heavily infected cells (redrawn from C. Uribe).

Fig. 4. Section of a human verruga. *B. bacilliformis* is distinct and usually is clearly situated in the cytoplasm. It is not observed on or in erythrocytes. When stained by Giemsa stain the organism is bright red; for a color plate see Weinman (1968).

Fig. 5. *B. bacilliformis* from an 8-day culture, hemolyzed, palladium-shadowed (From Peters and Wigand (1952) Z Tropenmed Parasitol *3:* 313–326.)

Fig. 6. A 4-day culture of *B. bacilliformis* grown with an explant of guinea pig tunica vaginalis in Maitland medium. Intracytoplasmic clusters and discrete organisms. (From Pinkerton and Weinman (1937) Proc Soc Exp Biol Med *37:* 587–590.)

Plate 18.3. *Anaplasma, Paranaplasma*

Fig. 1. *Anaplasma marginale. A*, ultrathin section of an infected erythrocyte; an inclusion body containing five initial bodies is depicted. × 65,000. *B*, intraerythrocytic inclusion bodies stained by Giemsa. × 959. (From Ristic and Watrach (1961) Amer J Vet Res *22:* 109–116.)

Fig. 2. *Paranaplasma caudata. A*, ultrathin section of an infected erythrocyte. × 40,000. (From Simpson *et al.* (1965) J Cell Biol *27:* 225.) *B*, stained by fluorescent antibody technique. × 1800. (From Kreier and Ristic (1963) Amer J Vet Res *24:* 676–687.)

Plate 18.4. *Aegyptionella, Haemobartonella*

Fig. 1. *Aegyptionella pullorum. A*, ultrathin section of an infected erythrocyte, depicting an inclusion body with 12 initial bodies. × 100,000. *B*, intraerythrocytic inclusion bodies stained by Giemsa method. × 800. (From Gothe (1967) Z Parasitenk *29:* 119–129.)

Fig. 2. *Haemobartonella muris. A*, ultrathin section showing parasite on rat erythrocyte. There is some indentation of the erythrocyte membrane at point of attachment. × 63,920. (From Tanaka *et al.* (1965) J Bacteriol *90:* 1735–1749.) *B*, *Haemobartonella felis* on erythrocytes of an infected cat. Polychrome stain. × 950. (From Flint and McKelvie (1955) Proc 92 Ann Mtg Amer Vet Med Ass 240–242.)

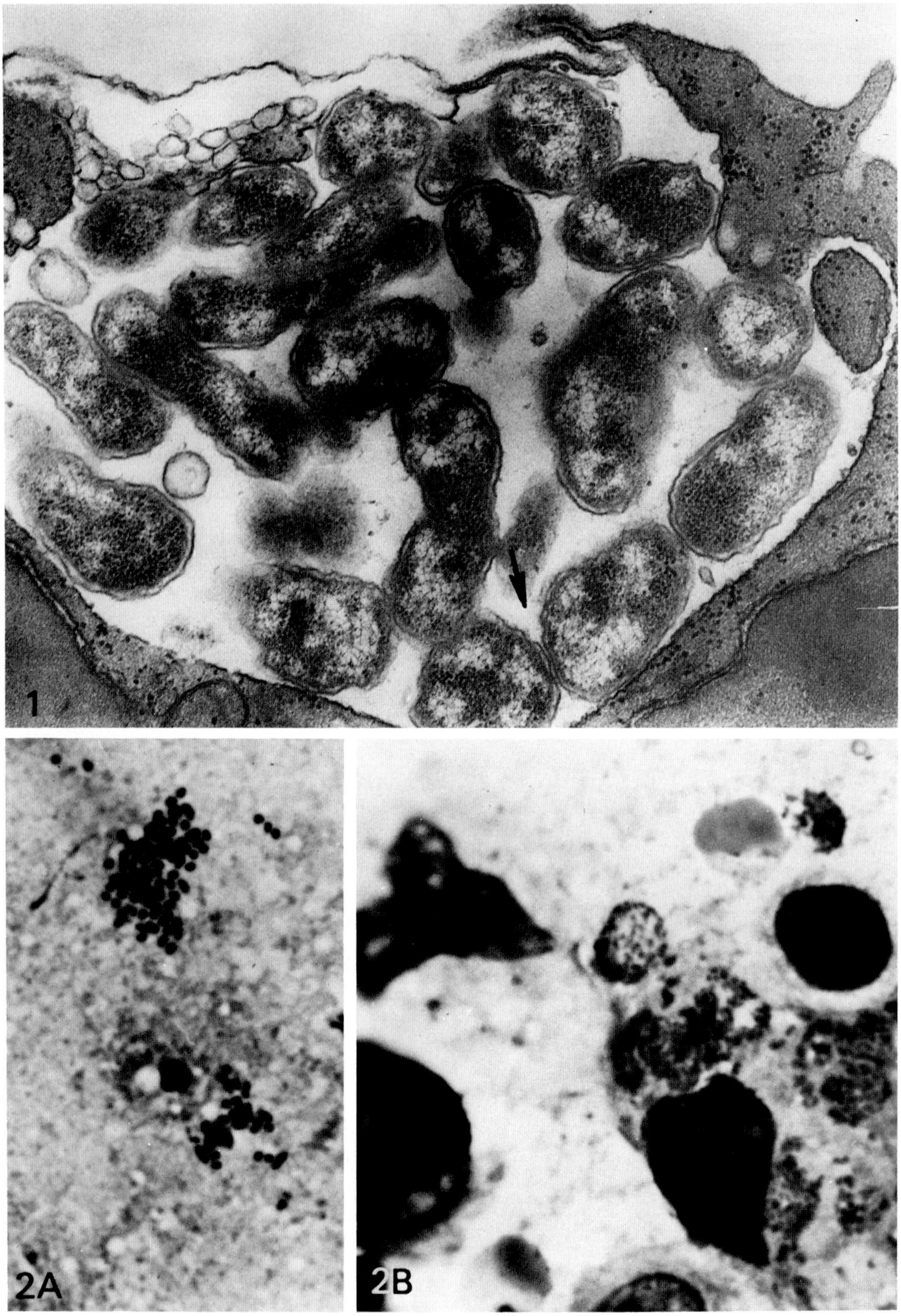

Plate 18.1.

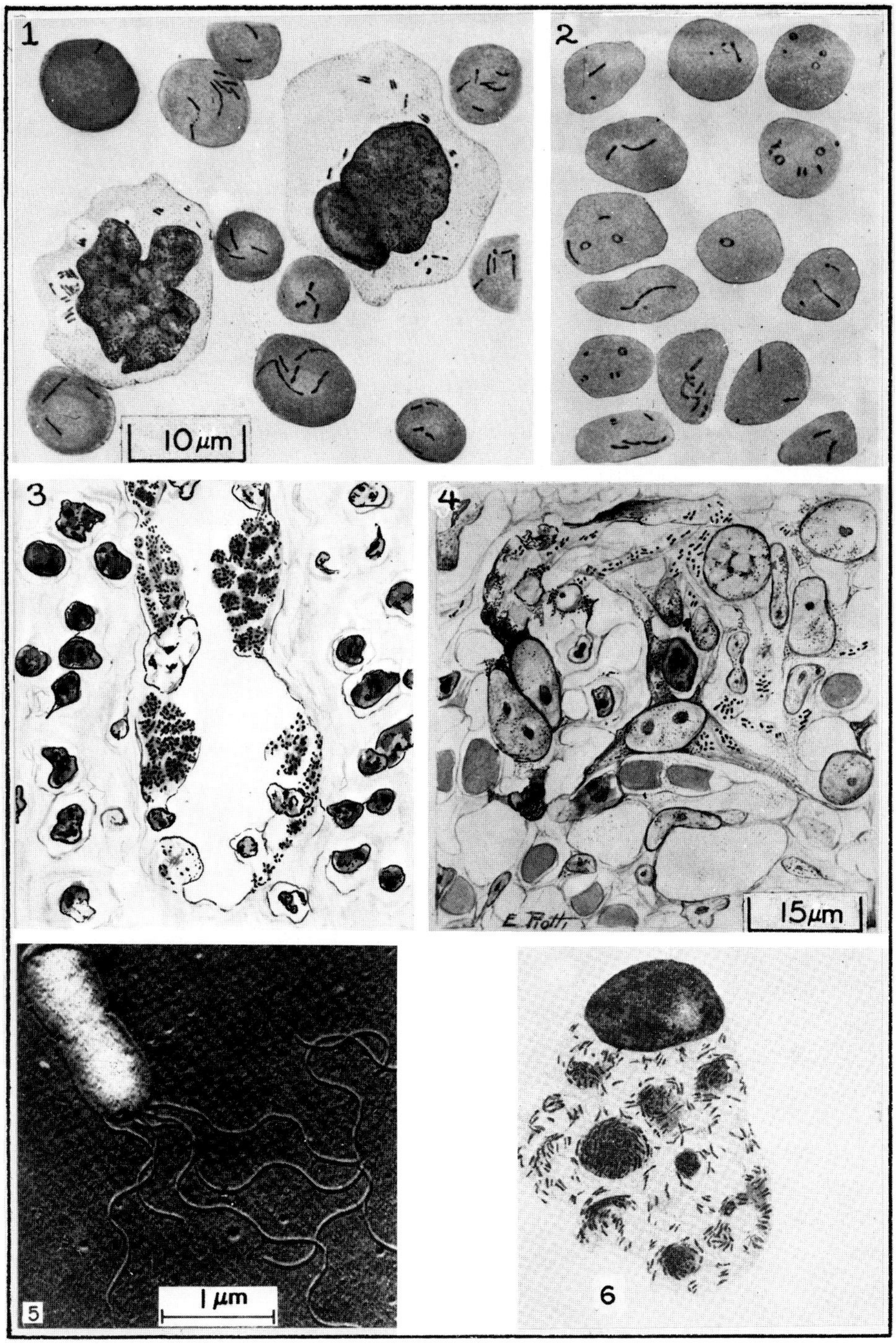

Plate 18.2.

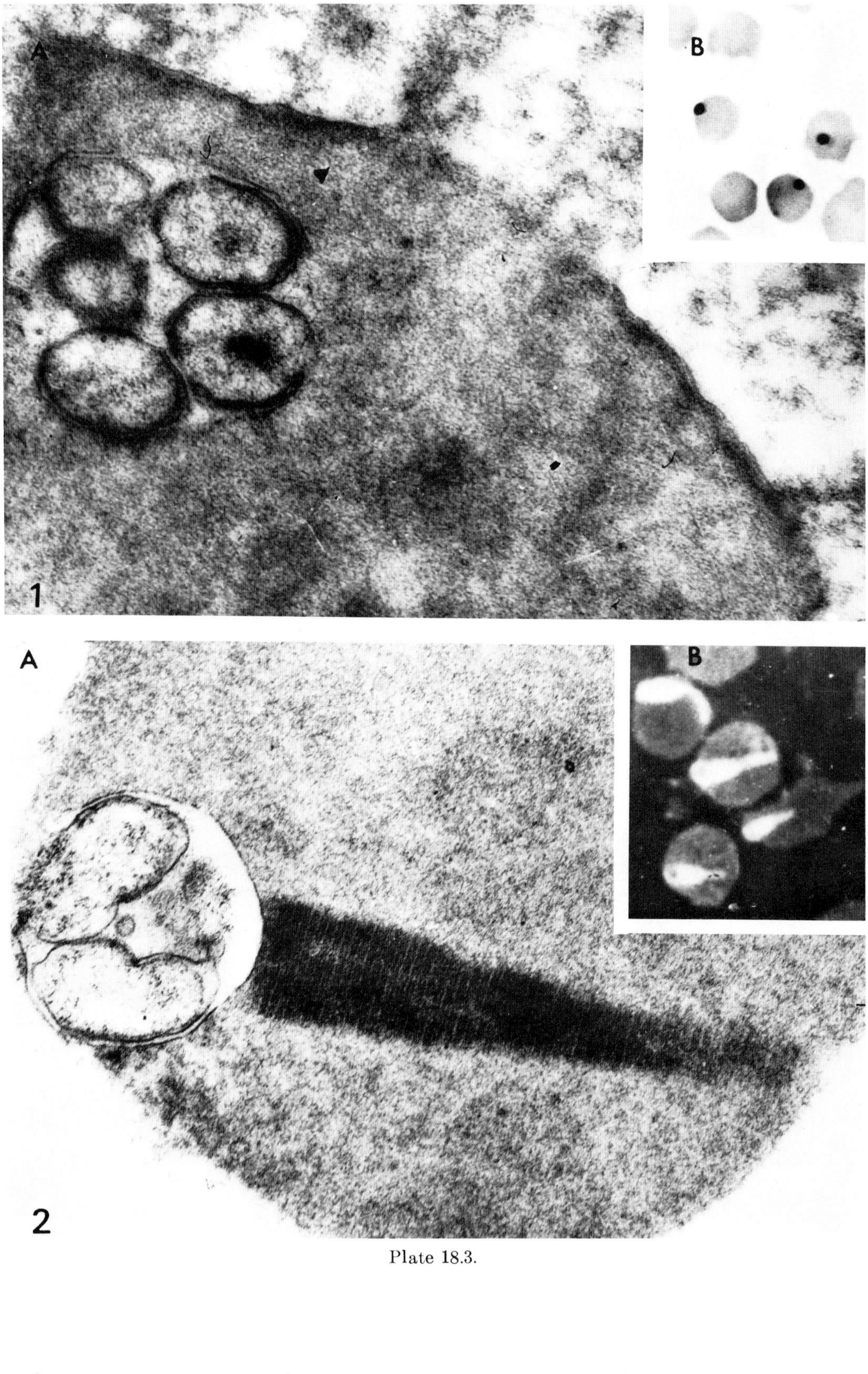

Plate 18.3.

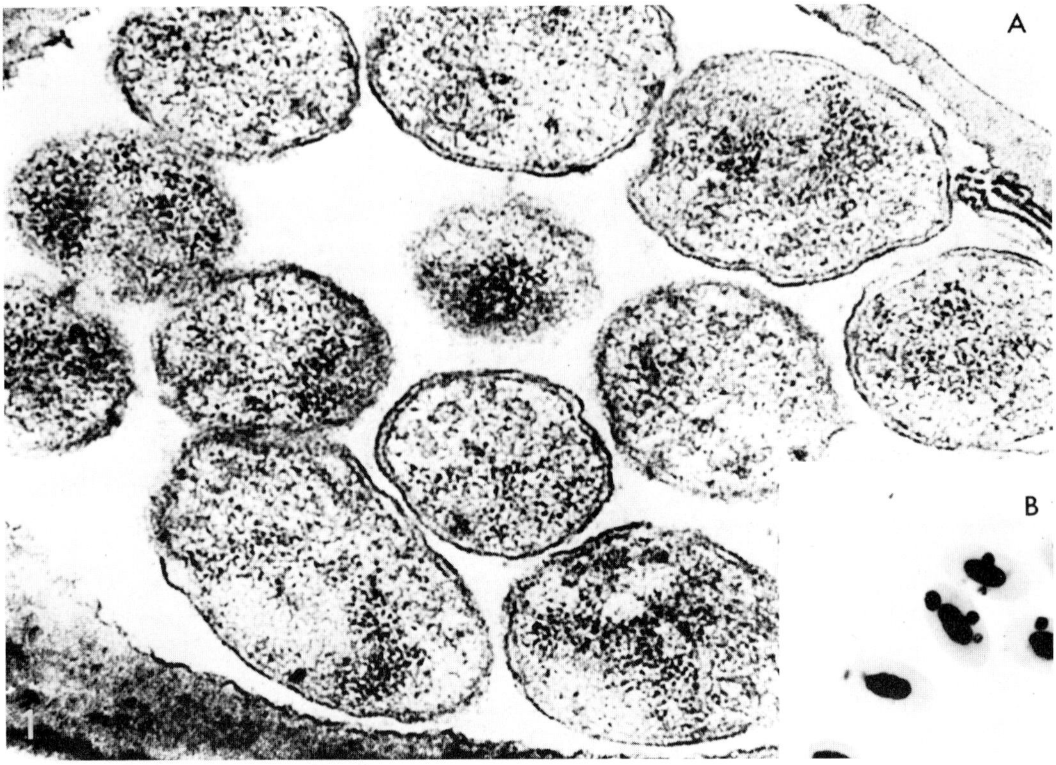

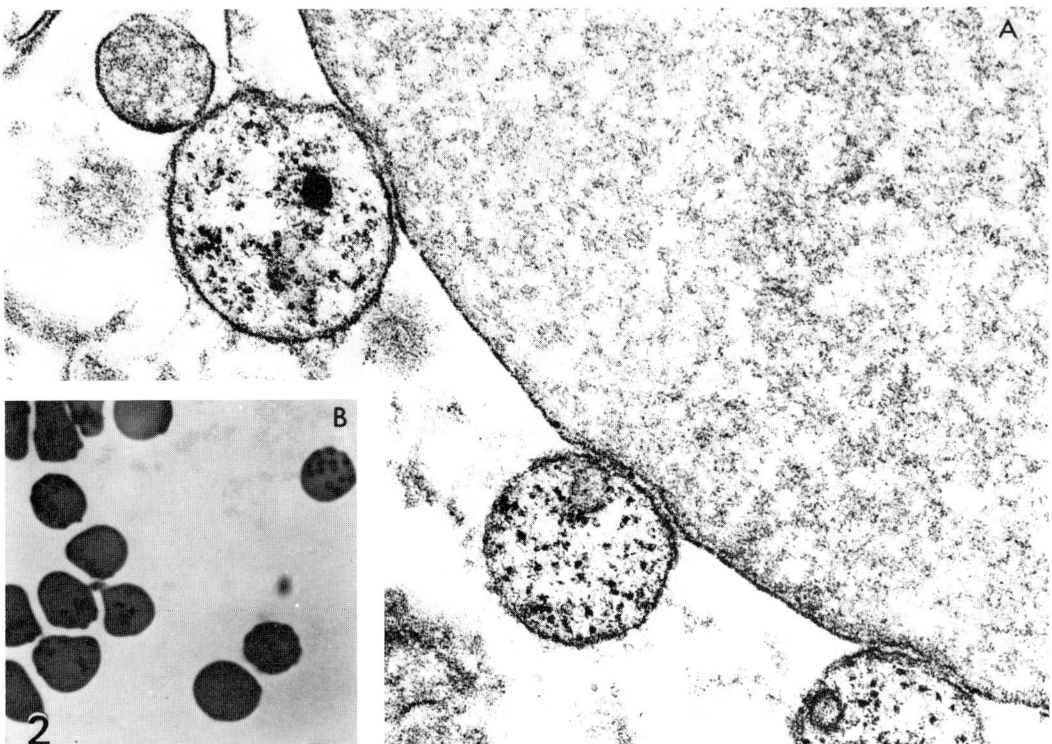

Plate 18.4.

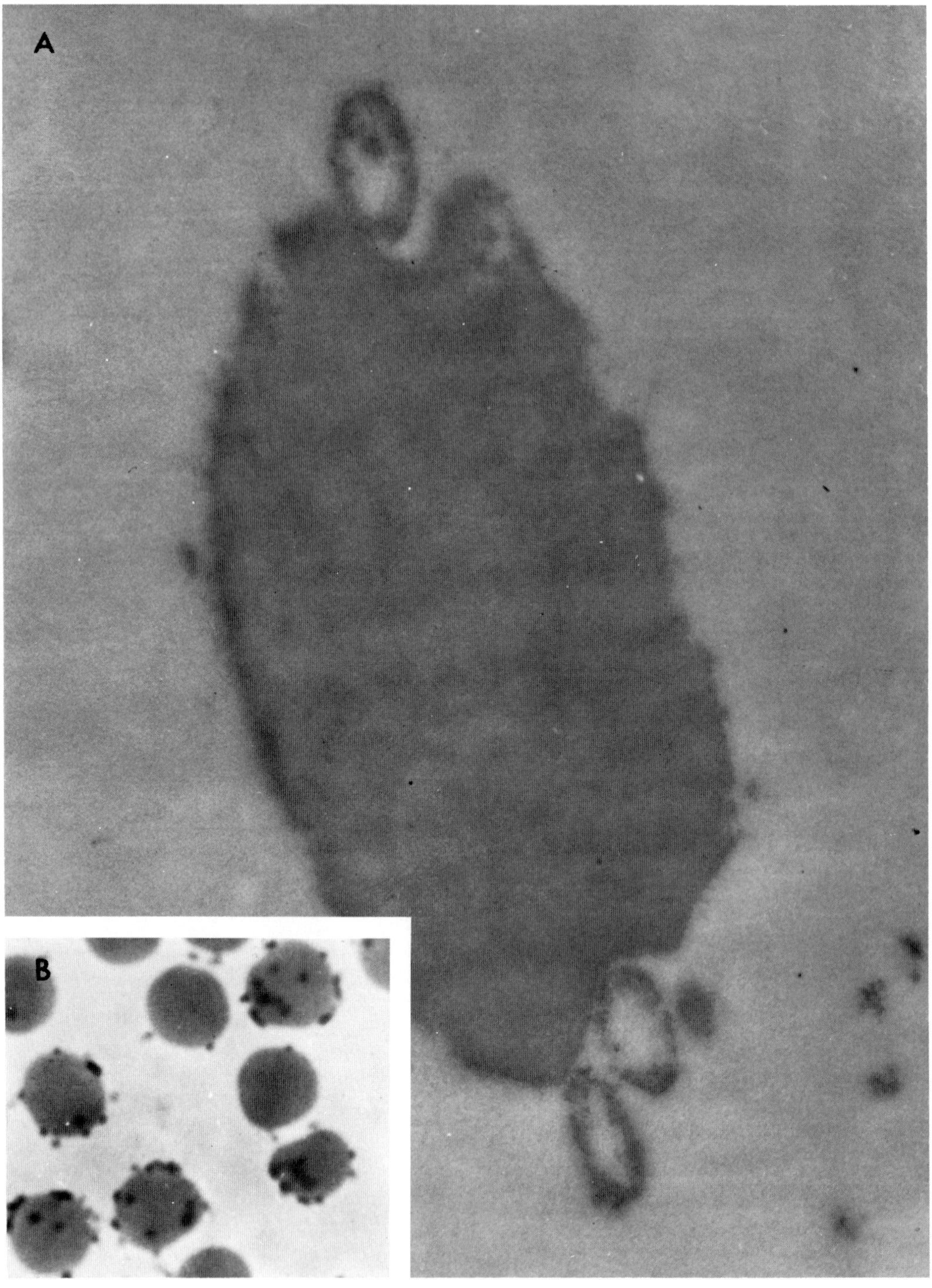

Plate 18.5. *Eperythrozoon*

Fig. 1. *Eperythrozoon ovis*. *A*, ultrathin section showing parasite on sheep erythrocytes; note erosion of the erythrocyte at point of attachment. × 21,500. *B*, Parasites on infected sheep blood. Giemsa stain. × 2400. (From Kreier and Ristic (1963) Amer J Vet Res *24:* 488–500.)

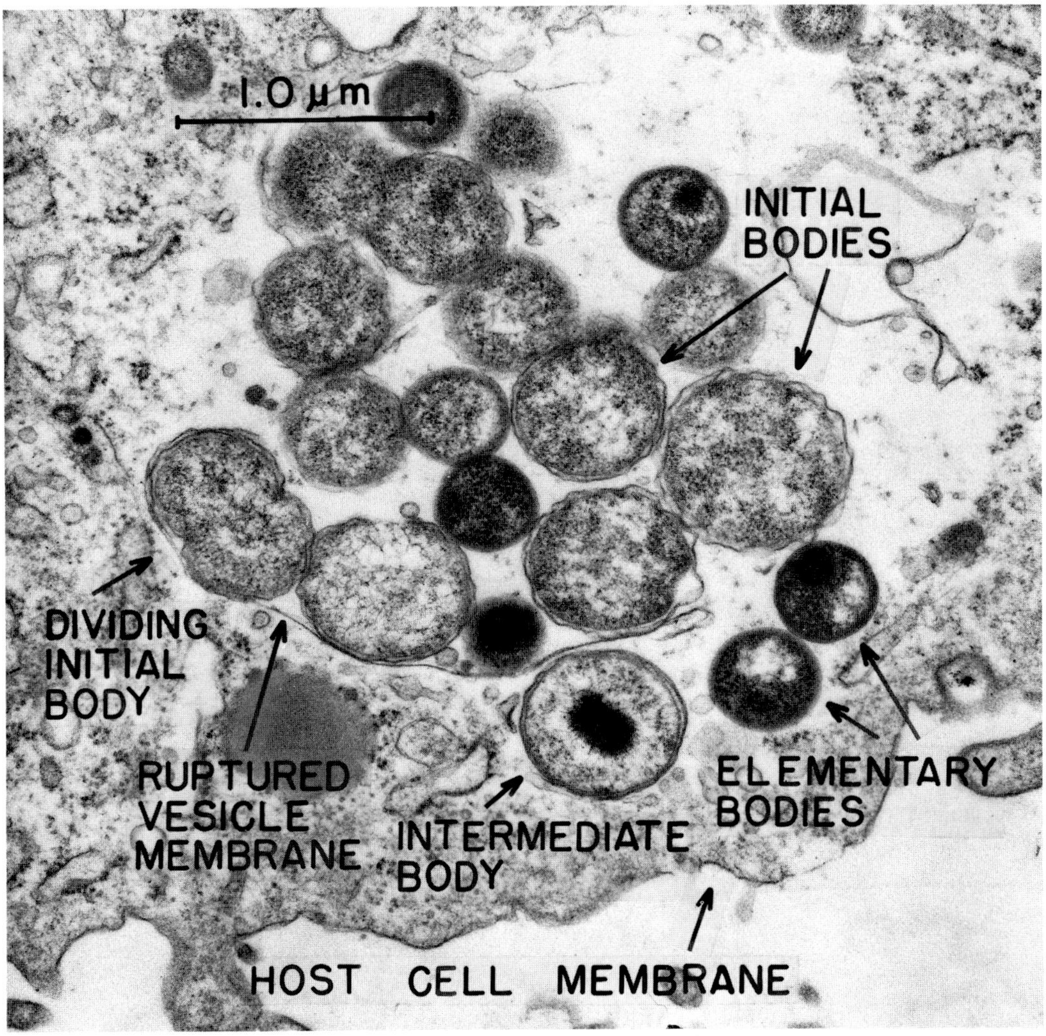

Plate 18.6. *Chlamydia*

Fig. 1. Electron micrograph of ultrathin section through a microcolony of *Chlamydia psittaci* in cytoplasm of a McCoy cell after 48-hrs incubation. The various developmental forms are labeled. The vesicular membrane has been ruptured and chlamydiae are being released into the cytoplasm. The multilaminated nature of the wall of the elementary bodies, the double unit membrane surrounding the intermediate and initial bodies are visible. × 28,700. (From Cutlip (1970) IAI *1:* 500.)

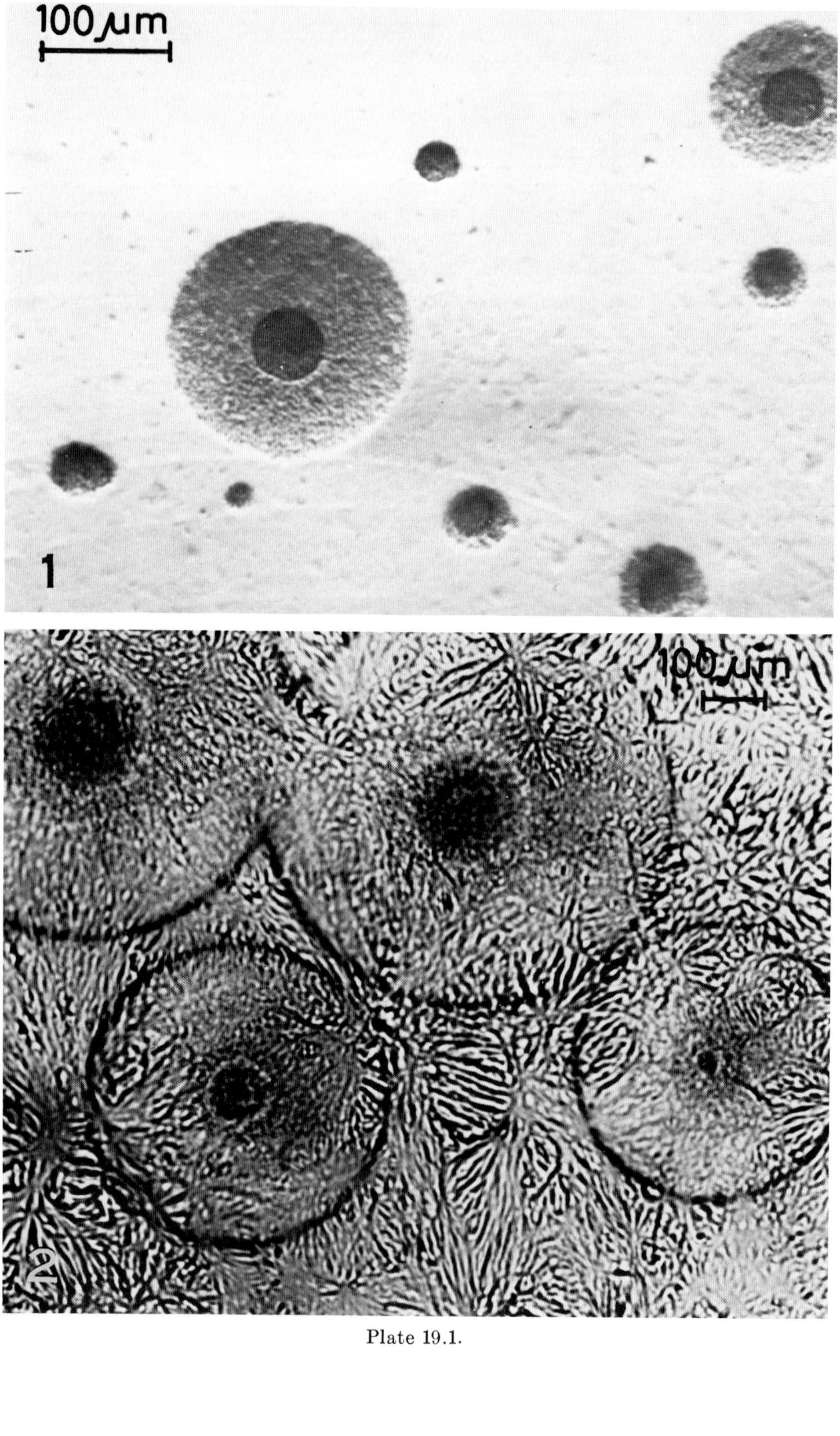

Plate 19.1.

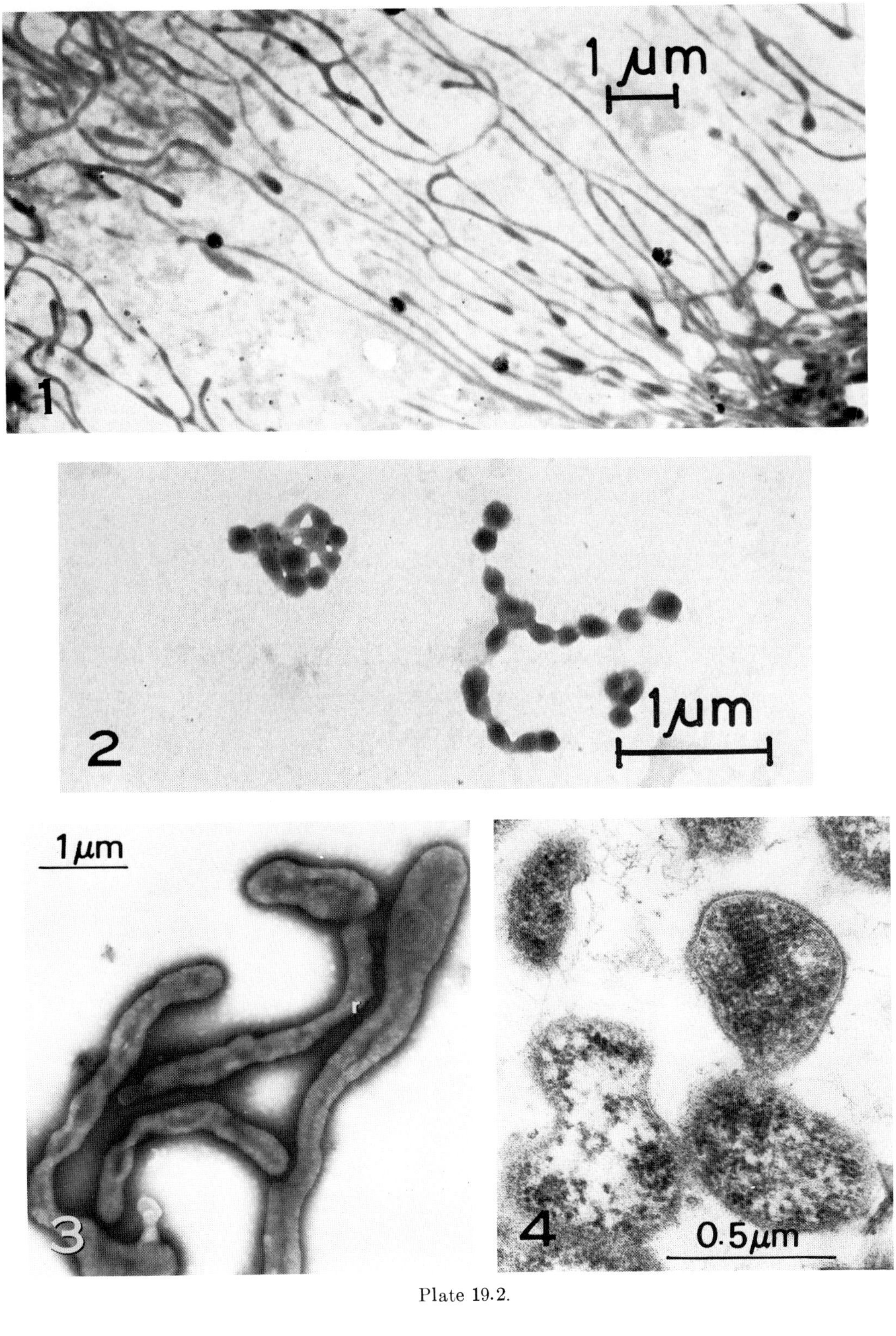

Plate 19.2.

Plate 19.1. *Mycoplasma*

Fig. 1. Biphasic colonies on horse serum agar showing typical "fried egg" appearance (*M. orale* type 1).

Fig. 2. "Film and spots phenomenon." Culture of *M. pulmonis* on horse serum agar showing biphasic colonies and pronounced development of a wrinkled, pearly film on the medium surface.

Figs. 1 and 2 from Freundt (1971) Medicinsk årbog XIV.

Plate 19.2. *Mycoplasma*

Fig. 1. Long branched filaments of *M. mycoides* subsp. *mycoides;* early exponential growth phase. Electron micrograph of cells on a collodion film floating upon serum broth.

Fig. 2. Chains of coccoid structures in liquid cultures of *M. mycoides* subsp. *mycoides;* late exponential growth phase.

Fig. 3. Short filaments of T-mycoplasma; exponential growth phase. Electron micrograph of a negatively stained specimen.

Fig. 4. Section of round to pleomorphic cells of T-mycoplasma showing triple-layered limiting membrane and arrangement of ribosomes in compact geometrical pattern; exponential growth phase.

Figs. 1 and 2 from Freundt (1958); Figs. 3 and 4 from Freundt (1971) Medicinsk årbog XIV.

Genus 1. **Thiobacillus** *Beijerinck 1904, 597*

Small rod-shaped cells. Motile by means of a single polar flagellum: two species non-motile. No resting stages known. Gram-negative.

Energy derived from the oxidation of one or more reduced or partially **reduced sulfur compounds,** including sulfides, elemental sulfur, thiosulfate, polythionates and sulfite. The final oxidation product is sulfate, but sulfur and polythionates accumulate, sometimes transiently, under certain conditions. *T. ferrooxidans* also utilizes ferrous compounds as electron donors.

The genus includes strictly autotrophic species which derive their carbon from carbon dioxide, facultative autotrophs and at least one species which requires both a partially reduced sulfur compound and organic matter for optimal growth.

Obligate aerobes, except *T. denitrificans* which grows anaerobically with nitrate as electron acceptor.

Optimum temperature about 28–30 C. pH range wide; some strains prefer strongly acid conditions, others mildly alkaline conditions.

Found in sea water, marine mud, soil, fresh water, acid mine waters, sewage, sulfur springs and in or near sulfur deposits. Especially frequent in environments in which hydrogen sulfide is produced or sulfur deposited.

The G + C content of the DNA, for the species tested, ranges from 50–68 moles %.

Type species: *Thiobacillus thioparus* Beijerinck 1904, 597.

TABLE 12.1

Types of fatty acid methyl ester (FAME) profiles found in genus **Thiobacillus**[a]

Fatty acid[b]	Total extracted fatty acids in:		
	Type I	Type II	Type III
	%	%	%
C_6			7
C_8	6	4	6
C_9			7
C_{10}	6		3
C_{11}	2		10
C_{12}	4		14
C_{13}			7
$C_{14:1}$	30	2	16
C_{14}	12	3	21
C_{15}	2	32	
$C_{16:1}$	15	2	
C_{16}	9	7	4
$C_{17:1}$		15	
C_{17}		20	

[a] Levin (1971) obtained a fatty acid profile for *T. thiooxidans* (ATCC 8085) different from Type III above (obtained from two strains maintained by D. G. Lundgren, one originally ATCC 8085) and different profiles from Types I and II. Its main feature was the predominance of a C_{19}-cyclopropane acid.

[b] Subscript gives length of carbon chain; number to right of colon indicates number of double bonds.

TABLE 12.2

Classification of genus **Thiobacillus** *based on lipid and nucleotide content*

FAME profile	G + C fraction of DNA			
	50–52% moles %	56–57 moles %	62–68 moles %	Not determined
Type I		*T. neapolitanus* *T. ferrooxidans*		
Type II			*T. thioparus* *T. novellus*	*T. denitrificans* *T. intermedius*
Type III	*T. thiooxidans*			
Not determined				*T. perometabolis*

Key to the species of genus **Thiobacillus**

Key characteristics

Strictness of autotrophism
Oxidation of sulfur or ferrous compounds
Relationship of organic compounds to above
Relationship to oxygen
FAME profiles (see Table 12.1)
Optimum pH requirements
DNA base ratio (not essential to key)

I. Strictly autotrophic
  A. Oxidize sulfur compounds only.
    1. Strictly aerobic.
      a. Grows optimally in neutral medium (pH 6.0–8.0).
        aa. FAME profile II (Table 12.1), G + C fraction 62–68%.
          1. *Thiobacillus thioparus*
        aaa. FAME profile I (Table 12.1), G + C fraction 56–57%.
          2. *Thiobacillus neapolitanus*
      b. Grows optimally in acid medium (pH 1.0–3.5).
          3. *Thiobacillus thiooxidans*
    2. Facultatively anaerobic, uses nitrate in absence of oxygen.
          4. *Thiobacillus denitrificans*
  B. Oxidizes ferrous compounds or sulfur compounds.
          5. *Thiobacillus ferrooxidans*

II. Facultatively autotrophic.
  A. Oxidation of thiosulfate repressed by organic substrates.
          6. *Thiobacillus novellus*
  B. Simultaneous oxidation of thiosulfate and organic substrates occurs.
          7. *Thiobacillus intermedius*

III. Heterotrophic, but requires simultaneous utilization of sulfur compounds and organic substrates for optimal growth.
          8. *Thiobacillus perometabolis*

Genus 2. **Sulfolobus** *Brock, Brock, Belly and Weiss 1972, 54*

Spherical cells with lobes, 0.8–1 μm in diameter. Resemble mycoplasmas but are more refractile and more regular in diameter. Not motile; **flagella and endospores absent. Gram-negative. Cell wall devoid of peptidoglycan.** Subunit structure of cell wall visible upon negative staining and electron microscopy.

Colonies smooth, glistening, non-pigmented.

Facultative autotrophs. **Use elemental sulfur as energy source.** May also use yeast extract, glutamate or ribose as a carbon and energy source.

Aerobic.

Temperature optimum 70–75 C; maximum 85 C; minimum 55 C.

pH optimum 2–3; maximum 5.8; minimum 0.9.

The G + C content of the DNA ranges from 60–68 moles % amongst five strains tested (buoyant density).

Found in solfatara areas containing hot acid environments, both soil and water.

Type species: *Sulfolobus acidocaldarius* Brock, Brock, Belly and Weiss 1972, 66.

Genus 3. Thiobacterium *Janke 1924, 68*

Rod-shaped cells, each containing one or more sulfur droplets. Cells embedded in gelatinous masses, which are spherical when free floating **or dendroid** when attached to a solid substrate. **Non-motile.** No resting stages known.

Has not been grown in pure culture.

Type species: *Thiobacterium bovista* (Molisch) Janke 1924, 68.

Genus 4. Macromonas *Utermöhl and Koppe in Koppe 1924, 632*

Cylindrical to bean-shaped cells, motile by a polar flagellum. Inclusions of calcium carbonate sometimes accompanied by sulfur globules. Multiplication by constriction followed by fission. No resting stages known. **Microaerophilic;** found in fresh water environments with low oxygen concentrations, e.g. hypolimnia and upper layers of bottom muds. Has not been grown in pure culture.

Type species: *Macromonas mobilis* (Lauterborn) Utermöhl and Koppe in Koppe 1924, 632.

Genus 5. Thiovulum *Hinze 1913, 195*

Cells **round** to **ovoid,** 5–25 μm in diameter. Cytoplasm often concentrated at one end of cell, the remaining space being occupied by a large vacuole. Cytoplasm normally **contains orthorhombic sulfur inclusions,** sometimes concentrated at one end, sometimes filling cells almost completely. **Strongly motile** by **peritrichous flagella,** forward movement accompanied by **rotation** around the long axis. Cells characterized by presence of **one polar fibrillar organelle,** visible in thin sections by electron microscopy; its function is not known. No resting stages observed.

Multiplication by constriction followed by fission.

Gram-negative.

Microaerophilic. Catalase negative.

Type species: *Thiovulum majus* Hinze **1913, 195.**

Genus 6. Thiospira *Visloukh 1914, 48*

Spirilla, usually with pointed ends, with **sulfur inclusions. Motile** by mono- or polytrichous **polar flagella. No resting stages known.** Has not been cultivated in pure culture.

Type species: *Thiospira winogradskyi* (Omelianski) Visloukh 1914, 48.

Further Comments

Pure culture studies will have to determine whether further differentiation into more than two species is warranted.

FAMILY SIDEROCAPSACEAE* PRIBRAM 1929, 377

Cells spherical, ellipsoidal or rod-shaped. May have thick or thin capsules.

Able to deposit iron and/or manganese oxides on or in capsules when present, or on extracellular material.

Found in iron-bearing waters.

---

* *Editorial Note.* The author has questioned the validity of the family and to the editors, its placement has presented a problem.

The four genera included have only one feature in common—the ability to deposit or to cause the deposition of iron and/or manganese oxides. Most are organotrophs. *Siderococcus* reproduces by budding and could be placed in Part 4. Dubinina and Zhdanov (*International Journal of Systematic Bacteriology 25*: 340–350, 1975) have studied pure cultures of *Siderocapsa eusphaera* and found them indistinguishable from members of the genus *Arthrobacter*. They placed them in a new species, *Arthrobacter siderocapsulatus*, and also strongly suggested that the family *Siderocapsaceae* be abolished. *Ochrobium* may be an alga.

As it is impossible to place the family, as here constituted, in a more logical place it is included with the chemolithotrophs, even though it is known that in many instances the deposition of iron or manganese is the result of the metabolism of the organic ion of the compound.

Key to the genera of family **Siderocapsaceae**

I. Iron or manganese oxides deposited.
  A. Oxides deposited on or in capsular material which surrounds groups of spherical or ovoid cells.
    Genus I. *Siderocapsa*
  B. Oxides form a delicate sheath (torus) which emphasizes the margin of the cell; cells usually rod-shaped.
    1. Torus completely surrounds the cell.
      Genus II. *Naumanniella*
    2. Torus is open at one end, resembling a horseshoe.
      Genus III. *Ochrobium*
II. Iron but not manganese oxide deposited in the medium. Cells coccoid.
    Genus IV. *Siderococcus*

Genus I. **Siderocapsa** *Molisch 1910, 30*

One to many **spherical to ovoid cells embedded in a common capsule,** partially **encrusted with iron and/or manganese compounds.** Cells colorless but outer portion of capsule yellowish to brownish from metal compounds, especially when old. Motility not known.

Common in fresh water; may be attached to surface of water plants or other substrates and in plankton or free floating. Generally aerobic but can grow under reduced oxygen tensions. Develop under neutral or slightly alkaline conditions.

As they have not been grown in pure culture, differentiation is based on morphological characteristics.

Type species: *Siderocapsa treubii* Molisch 1910, 30.

Key to the species of genus **Siderocapsa**

I. Grow attached to substrates, especially leaves of fresh water plants.
  A. Many cells embedded in a relatively large mass of capsular material.
    1. Cocci, 0.4–0.6 μm in diameter.
      1. *Siderocapsa treubii*
    2. Cocci, 0.7–1.8 μm in diameter.
      2. *Siderocapsa major*
  B. Single cocci in thin, closely fitting capsule, 0.5–0.75 μm.
      3. *Siderocapsa monoica*
II. Grow in plankton or free floating in surface layers of fresh water.
  A. Capsules thin, closely fitting and more or less homogeneous.
    1. Single coccus, 0.2–0.5 μm in diameter, per capsule.
      4. *Siderocapsa anulata*
  B. Capsular material more or less abundant, gelatinous and may have stratified, indistinctly radial or granular structure.
    1. More or less ovoid cells, 0.5–0.6 by 0.8 μm, usually in pairs in capsule.
      5. *Siderocapsa geminata*
    2. Generally many spherical cells in each capsule (colony).
      a. Generally neustic forms, up to 90 colonies aggregate in larger zoogloeal masses. Cells 1.0–1.2 μm in diameter.
        6. *Siderocapsa coronata*
      b. Generally planktonic forms, especially in hypolimnion; colonies do not aggregate.
        bb. Cells 0.4–1.0 μm in diameter, one to four per colony.
          7. *Siderocapsa arlbergensis*
        bbb. Cells 1–2 μm in diameter, usually up to 60 cells per colony, rarely one or two.
          8. *Siderocapsa eusphaera*

Genus II. **Naumanniella** *Dorff 1934, 19*

Rod-shaped cells, each surrounded by a delicate, very regular capsule. Capsule becomes golden yellow from deposits of iron compounds, which **emphasizes the margin and gives the cell the appearance of a minute diatom** or link of a chain ("torus"). Manganese oxide stains the capsule black and renders it opaque.

Complex organic compounds of iron and manganese are decomposed.

Widely distributed in iron-bearing waters.

Type species: *Naumanniella neustonica* Dorff 1934, 20.

Note: There are six species recognized, which are separated on the basis of cell shape and size.

Genus III. **Ochrobium** *Perfil'ev in Visloukh 1921, 88*

Ellipsoidal to rod-shaped cells, 0.5–3.0 by 1.5–5.0 μm, partially surrounded by a marginal thickening (torus) that is heavily impregnated with iron. The torus remains open at one end, resembling a horseshoe; pairs of cells may unite with the open ends together. The cells are surrounded by a delicate transparent capsule that contains a small amount of iron; clumps of several cells may form. May be motile by means of two unequal polar flagella.

Widely distributed in iron-bearing fresh water.

Type species: *Ochrobium tectum* Perfil'ev in Visloukh 1921, 88.

Genus IV. **Siderococcus** *Dorff 1934, 9*

Cells spherical 0.2–0.5 μm in diameter; occur singly or in small motile colonies of 6–15 cells resembling an ear of corn (*Zea mays*). **Motile** but type of flagellation is not known. Cells have filamentous appendages **but no capsules. Multiplication is apparently by budding,** the cells becoming pear shaped.

The organisms usually develop as a homogeneous, continuous accretion although sometimes chains of coccoid cells are arranged in a net-like structure. **Cells are not encrusted with iron compounds** although older colonies may become covered with iron oxides and the cells themselves are hardly recognizable. The orange-yellow deposits contain only ferric hydroxide.

So far growth has been observed only in the natural habitat and in isolated mud samples; *Siderococcus* microzones are found in mud horizons with low concentrations of oxygen and a neutral pH (Perfil'ev and Gabe, 1964). Widely distributed in fresh waters and bottom deposits.

Type species: *Siderococcus limoniticus* Dorff 1934, 9.

PART 13

METHANE-PRODUCING BACTERIA

FAMILY I. METHANOBACTERIACEAE BARKER 1956, 15

Cells are **rods or cocci, motile** or **non-motile,** Gram-positive or Gram-negative. Spores have not been found in any pure culture. Very strict **anaerobes** which obtain energy for growth via **formation of methane** by reduction of CO_2, utilizing electrons generated in the oxidation of compounds such as hydrogen and formate, or via fermentation of compounds such as acetate and methanol with formation of methane and CO_2 as the products. A **highly specialized physiological group** which does not utilize carbohydrate or proteinaceous materials or organic compounds, other than those listed above, as energy sources.

Key to the genera of family **Methanobacteriaceae**

I. Rods or chain-forming lancet-shaped coccoids.

Genus I. *Methanobacterium*, p. 473*

II. Cocci other than chain-forming lancets.

A. Large cocci in packets.

Genus II. *Methanosarcina*, 476

B. Cocci occurring singly, in pairs or clumps.

Genus III. *Methanococcus*, p. 477

Note: A genus of spiral-shaped methanogenic bacteria has recently been described by Ferry *et al.* (*International Journal of Systematic Bacteriology 24:* 465–469, 1974). The genus is named *Methanospirillum* and is monotypic. The reader is referred to the above paper for a complete description.

Genus I. Methanobacterium *Kluyver and Van Niel 1936, 399*

Curved, crooked, to straight **non-sporing rods,** long and filamentous to **coccoid,** about 0.5–1.0 μm in width, Gram-positive to Gram-negative, non-motile or motile with monotrichous polar flagellation.

Chemolithotrophs: **Energy for growth** obtained via **reduction of CO_2 to methane utilizing H_2** and sometimes formate as the electron donor. The type species supposedly utilized acetate and butyrate as energy sources but it was never obtained in pure culture (see *Further Comments* under the description of the family).

Autotrophic to heterotrophic in nutrition.

Very strict **anaerobes.**

Mesophiles to extreme thermophiles.

Widely distributed in nature being found in anaerobic habitats such as sediments of natural waters, soil, anaerobic sewage digestors and the gastrointestinal tract of animals.

Type species: *Methanobacterium soehngenii* Barker 1936, 437.

* All page numbers cited refer to *Bergey's Manual of Determinative Bacteriology*, 8th edition, The Williams & Wilkins Company, Baltimore, 1974.

Key to the species of the genus **Methanobacterium**

Key characteristics

Cell shape
Production of methane from acetate
Temperature limits of growth
Gram reaction
Motility

I. Rods which produce methane from acetate.
 1. *Methanobacterium soehngenii*
II. Rods or chained coccoids which do not produce methane from acetate.
 A. Slender, cylindrical, straight to irregularly crooked, non-motile rods.
  1. Grow at mesophilic temperatures.
   2. *Methanobacterium formicicum*
  2. Grow at thermophilic temperature, 40–75 C.
   3. *Methanobacterium thermoautotrophicum*
 B. Gram-positive coccoids to short, lancet-shaped rods usually in chains.
   4. *Methanobacterium ruminantium*
 C. Gram-negative, motile, short, straight or slightly curved rods.
   5. *Methanobacterium mobile*

Genus II. **Methanosarcina** *Kluyver and van Niel 1936, 400, emend. mut. char. Barker 1956, 14*

Large spherical cells, 1.5–2.5 μm in diameter, generally occurring in regular **packets.** Gram-variable to Gram-positive. Non-motile.

The energy metabolism involves **formation of methane from acetate** and sometimes from **methanol,** CO and possibly, from butyrate. H_2 may also be utilized in reduction of methanol.

Strict anaerobes.

Good growth at 30–37 C.

Found in muddy sediments of natural waters and in anaerobic sewage digestors.

Type species: *Methanosarcina methanica* (Smit) Kluyver and van Niel 1936, 400.

Note: There are two species included in the *Manual*. The type species, *M. methanica*, has not been studied in pure culture and is believed by the author (M. P. Bryant) to be identical with the more adequately studied species *M. barkeri*.

Genus III. **Methanococcus** *Kluyver and van Niel 1936, 400, emend. mut. char. Barker 1936, 430*

Spherical cells, occurring **singly,** in **pairs** or in **masses.** Motile or non-motile. Gram-positive or -negative. (It should be noted that the genus *Methanobacterium* includes Gram-positive, chain-forming coccoids.)

The **energy metabolism involves the formation of methane** from H_2 and CO_2, formate, acetate or butyrate. Authentic pure cultures have not been shown to utilize acetate or butyrate in methane formation and it is quite possible that butyrate is not fermented by methanogenic bacteria *per se* (see *Further Comments* on the family).

Strict anaerobes.

The type species: *Methanococcus mazei* Barker 1936, 433.

Descriptions of this genus and two species are only slightly modified from those of Barker (1957).

PART 14

GRAM-POSITIVE COCCI

Key to the families of Part 14

I. Aerobic, facultatively anaerobic or microaerophilic.
   A. Divide in more than one plane to form irregular clusters, tetrads, or packets.
      Family I. *Micrococcaceae*, p. 478*
   B. Divide in one or two planes to form pairs, chains or occasionally tetrads.
      Family II. *Streptococcaceae*, p. 490

II. Anaerobic.
      Family III. *Peptococcaceae*, p. 517

FAMILY I. MICROCOCCACEAE PRIBRAM 1929, 385

Cells spherical, 0.5–3.5 μm in diameter, characteristically dividing in more than one plane to form regular or irregular clusters or packets. Motile or non-motile. Resting stages not produced. Gram-positive.

Chemoorganotrophs; metabolism respiratory or fermentative. Acid without gas produced from glucose, when attacked.

Nutritional requirements variable. All strains grow in presence of 5% NaCl; many grow in 10–15%.

Catalase positive.

Aerobic or facultatively anaerobic.

The G + C content of the DNA ranges from about 30–75 moles %.

Type genus: *Micrococcus* Cohn 1872, 151.

TABLE 14.1

Differential characteristics of the genera of family **Micrococcaceae** *and of genus* **Aerococcus** (*Streptococcaceae*)[a]

	Micrococcus	*Staphylococcus*	*Planococcus*	*Aerococcus*
Cells, spherical, Gram-positive	+	+	+	+
Arrangement: Irregular clusters	+	+	−	−
Tetrads	v	+	+	+
Glucose fermentation[b]	−	+	−	d
Cytochromes	+	+	+	−
Catalases: Heme	+	+	+	−
Non-heme	−	−	−	v
Hydrogen peroxide formation	−	−	−	+
Motility	−	−	+	−
Yellow-brown pigment	−	−	+	−
G + C content of DNA (moles %)	66–75	30–40	39–52	37–41

[a] + = most (90% or more) strains positive; − = most (90% or more) strains negative; v = characters inconstant and in one strain may sometimes be positive, sometimes negative; d = some (less than 90%) strains positive.

[b] Growth and acid anaerobically from glucose in the Standard Medium proposed by the I.C.S.B. Subcommittee on the Taxonomy of Staphylococci and Micrococci (1965), with exception of strains of *S. saprophyticus* which only weakly ferment glucose (see species description).

* All page numbers cited refer to *Bergey's Manual of Determinative Bacteriology*, 8th edition, The Williams & Wilkins Company, Baltimore, 1974.

Genus I. **Micrococcus** *Cohn 1872, 151*

Cells **spherical,** 0.5–3.5 μm in diameter, occurring singly, in pairs and characteristically dividing in more than one plane to form irregular clusters, tetrads or cubical packets. Usually nonmotile. No resting stages known. Gram-positive.

Chemoorganotrophs: Metabolism strictly respiratory. Menaquinones and cytochromes form the electron transport system; carotenoid pigments may be present. **Oxygen is the universal electron acceptor.** Carbon-containing compounds such as pyruvate, acetate, lactate, succinate, glutamate and carbohydrate are usually utilized. **Glucose oxidized** to mainly acetate or completely oxidized to carbon dioxide and water. Glucose is metabolized by hexose monophosphate pathway and citric acid cycle enzymes; a functional glycolysis cycle may also be present.

Indole not produced. Teichoic acids are uncommon.

Catalase produced.

Nutritional requirements are variable. Members of some species can grow with ammonium phosphate as sole nitrogen source or with glutamic acid as carbon, nitrogen and energy source and thiamine and/or biotin as growth factors.

Aerobes. Growth optimum 25–30 C. All grow in the presence of up to 5% sodium chloride. Cells not lysed by lysostaphin endopeptidase (1 unit/ml).

Common inhabitant of soils and fresh water. Frequently found on the skin of man and other animals.

The G + C content of the DNA ranges from 66–75 moles % (T_m; buoyant density; chromatography; E_{260}/E_{280} at pH 3.0).

Type species: *Micrococcus luteus* (Schroeter) Cohn 1872, 153.

TABLE 14.2

Differential characteristics of the species of genus **Micrococcus**[a]

	1. *M. luteus*	2. *M. roseus*	3. *M. varians*
Pigment: Yellow	+	−	+
Red	−	+	−
Novobiocin sensitivity[b]	S/R	S	R
Glucose: acid[c]	−	d	+
Xylose: acid[d]	−	d	+
Fat: fatty acids[d]	+	−	d
Arginine: ammonia[d]	−	−	d
Cell wall: L-lys peptide	+	−	−
G + C content of DNA (moles %)	71–75	66–75	66–72

[a] + = most (90% or more) strains positive; − = most (90% or more) strains negative; d = some (less than 90%) strains positive.

[b] R = M.I.C. > 2.0 μg/ml; S = M.I.C. < 0.6 μg/ml (Mitchell and Baird-Parker, 1967; Jeffries, 1969).

[c] Standard method of I.C.S.B. Subcommittee on the Taxonomy of Staphylococci and Micrococci (1965).

[d] Baird-Parker (1963).

Genus II. **Staphylococcus** *Rosenbach 1884, 18 nom. cons.* Opin. 17 Jud. Comm. 1958, 153

Cells spherical, 0.5–1.5 μm in diameter occurring singly, in pairs and characteristically dividing in more than one plane to form irregular clusters. Non-motile. No resting stages known. Gram-positive. **Cell wall contains two main components; a peptidoglycan and its associated teichoic acids.** The peptidoglycan consists of glycan composed of repeating units of β-1,4 linked *N*-acetyl-glucosamine and *N*-acetyl-muramic acid residues that are linked through *N*-acetyl-muramyl-L-alanine linkages to peptide subunits consisting of N^{α}-(L-alanyl-D-isoglutamyl)-L-lysyl-D-alanine. The peptide subunits are cross-linked by penta-peptide bridges containing solely or mainly glycine. These bridges extend from the N^{ϵ}-lysine residue of one peptide subunit to the C-terminal D-alanine of a neighboring peptide subunit. Attached to or in close proximity to the peptidoglycan are teichoic acids which, depending on the species, contain either ribitol or glycerol linked to a sugar or an aminosugar.

Chemoorganotrophs: **Metabolism respiratory and fermentative.** Catalase produced. Menaquinones and cytochromes a, b_1 and O form the electron transport system; carotenoid pigments may be present. Oxygen is the universal terminal electron acceptor.

A wide range of **carbohydrates may be utilized,** particularly in the presence of air, **with the production of acid** but gas not detectable by standard procedures. Under anaerobic conditions the main product of glucose fermentation is lactic acid; in the presence of air the main product is acetic acid with small amounts of CO_2. Acetoin is usually formed as an end-product of glucose metabolism. Terminal pH in unbuffered glucose broth is 4.2–5.4. Acid usually not produced from arabinose, cellobiose, inositol, inulin or raffinose. Starch and esculin usually not hydrolyzed.

Extracellular enzymes and toxins are formed. A variety of protein and fat-containing substrates are hydrolyzed. Hippurate and arginine may be broken down. Indole not formed.

Amino acids and vitamins required for aerobic growth. In addition, uracil and a fermentable carbon source are required for anaerobic growth.

Facultative anaerobes, growth more rapid and abundant under aerobic conditions. Temperature optimum 35–40 C; growth range 6.5–46 C. pH optimum 7.0–7.5; growth range pH 4.2–9.3.

Most strains grow in the presence of 15% sodium chloride or 40% bile. **Sensitive to lysis by lysostaphin endopeptidase** (1 unit/ml). **Resistant to lysis by lysozyme** (100 μg/ml).

Usually **sensitive to antibiotics** such as the β-lactam and macrolide antibiotics, tetracylines, novobiocin and chloramphenicol but **resistant** to polymyxin and polyenes. Sensitive to antibacterials such as phenols and their derivatives, surface active compounds, salicylanilides, carbanilides, halogens (chlorine and iodine) and their derivatives such as chloramines and iodophors. Usually sensitive to heat ($D_{60\,C}$ value 3 min or less in buffer, pH 7.0). Moderately resistant to γ-radiation (D value *ca.* 20 Krads in buffer, pH 7.0).

Host for a wide range of bacteriophages which may have a narrow or wide host range. Transfer of characters by transduction has been shown for *S. aureus*.

Mainly associated with skin, skin glands and mucous membranes of warm-blooded animals. Their host range is wide and many strains are potential pathogens.

The G + C content of DNA ranges from 30–40 moles % (T_m, buoyant density, chromatography).

Type species: *Staphylococcus aureus* Rosenbach 1884, 18.

TABLE 14.3

Characteristics differentiating species of genus **Staphylococcus**[a]

	1. *S. aureus*	2. *S. epidermidis*[b]	3. *S. saprophyticus*
Coagulases	+	−	−
Mannitol:			
Acid aerobically	+	d	d
Acid anaerobically	+	−	−
α-Toxin	+	−	−
Heat-resistant endonucleases	+	−	−
Biotin for growth	−	+	NT
Cell wall:			
Ribitol	+	−	+
Glycerol	−	+	d
Protein A	+	−	−
Novobiocin sensitivity[c]	S	S	R

[a] + = most (90% or more) strains positive; − = most (90% or more) strains negative; d = some (less than 90%) strains positive, some negative; NT, not tested.

[b] Four biotypes recognized: see Table 14.4.

[c] R = MIC > 2.0 μg/ml; S = MIC < 0.6 μg/ml (Mitchell and Baird-Parker, 1967; Jeffries, 1969).

TABLE 14.4

Characteristics of biotypes of **Staphylococcus epidermidis**[a]

	Biotype 1	2	3	4
Acetoin	+	−	+	+
Phosphatase	+[b]	+	−	−
Acid aerobically from:				
Lactose	+	d	−	d
Maltose	+	−	d	d
Mannitol	−	−	−	+

[a] + = most (90% or more) strains positive; − = most (90% or more) strains negative; d = some (less than 90%) strains positive some negative. For methods see Baird-Parker (1963).

[b] On the plate test (Baird-Parker, 1963) some of these organisms are phosphatase negative but by the more sensitive tube test of Pennock and Huddy (1967) they are phosphatase positive.

Genus III. **Planococcus** *Migula 1894, 236*

Spheres, 1.0–1.2 μm in diameter, **occurring singly, in pairs, in threes or in tetrads. Motile,** each cell usually possessing one or two flagella, occasionally three to four. Non-spore-forming. **Gram-positive.**

Chemoorganotrophs: **Metabolism respiratory, never fermentative.** Do not produce acid or gas from glucose, maltose, lactose or sucrose in standard medium (Subcommittee, 1965). Produce acid but no gas from glucose, maltose and mannitol but not from lactose or sucrose in MOF medium (Leifson, 1963). Final pH in glucose broth 6.6–7.0.

Do not produce acetylmethylcarbinol, coagulase, phosphatase; do not reduce nitrate or nitrite, or hydrolyze arginine. Gelatin is hydrolyzed.

Strict aerobes.

Catalase positive; benzidine test for porphyrins is positive.

Good growth between 20 and 37 C.

Found in sea water, but grows equally well in sea water medium and in media without added salt.

The G + C content of the DNA ranges from 48–52 moles % (T_m) (Bohacek *et al.*, 1967, 1968).

Type species: *Planococcus citreus* Migula 1894, 236.

FAMILY II. STREPTOCOCCACEAE DEIBEL AND SEELEY 1974, 490

Cells spherical or ovoid, in pairs or chains of varying length or in tetrads. Non-motile or rarely motile. Endospores not formed. Gram-positive.

Chemoorganotrophs. Metabolism fermentative; lactic, acetic and formic acids, and ethanol and CO_2 formed from carbohydrates.

Nutritional requirements complex and variable.

Catalase test is variable, benzidine negative.

Facultatively anaerobic.

The G + C content of DNA ranges from 33–44 moles %.

Type genus: *Streptococcus* Rosenbach 1884, 22.

Key to the genera of family Streptococcaceae

I. Glucose is fermented by the hexose diphosphate pathway yielding chiefly dextrorotatory lactic acid (homofermentative). Cell division in one plane resulting in pairs and chains. Catalase negative.

Genus I. *Streptococcus*, p. 490

II. Glucose is fermented by the hexose monophosphate pathway yielding levorotatory lactic acid, CO_2, ethanol and/or acetic acid (heterofermentative). Cell division in one plane resulting in pairs and chains. Catalase negative.

Genus II. *Leuconostoc*, p. 510

III. Glucose is fermented yielding inactive lactic acid (homofermentative). Cell division in two planes resulting in pairs and tetrads. Catalase variable.

Genus III. *Pediococcus*, p. 513

IV. Glucose is fermented yielding dextrorotatory lactic acid (homofermentative). Cell division in two planes resulting in pairs and tetrads. Catalase variable.

Genus IV. *Aerococcus*, p. 515

V. Glucose is fermented but the products of fermentation have not yet been reported. Gram-indeterminate, but cell wall structure is of the Gram-positive type. Catalase negative.

Genus V. *Gemella*, p. 516

Genus I. Streptococcus *Rosenbach 1884, 22*

Cells spherical to ovoid, less than 2 μm in diameter (varies with species), occurring in **pairs or chains** when grown in liquid media. Occasional motile strains in serological group D. **Gram-positive.**

Chemoorganotrophs: **Metabolism fermentative**, predominant end product of glucose fermentation is **dextrorotatory** lactic acid (**homofermentative**), some species ferment organic acids (malic and citric) and amino acids (serine and arginine). Pigments, found only in groups B and D, usually red or yellow.

Pellicles are never formed. Do not contain heme compounds; benzidine negative and **catalase negative.**

Facultative anaerobes. Oxygen, or other hydrogen acceptors, may alter end products of carbohydrate metabolism. Hydrogen peroxide may accumulate in presence of oxygen.

Minimal nutritional requirements are generally complex (but variable) and may involve amino acids, purines, pyrimidines, peptides, vitamins and occasionally fatty acids and elevated CO_2 tension. In contrast, strains of one species (*S. bovis*) may only require glucose and ammonium and inorganic salts for growth.

Temperature optimum, about 37 C unless noted otherwise; maximum and minimum temperatures vary with the species.

The G + C content of the DNA (reported for 15 species) ranges from 33–42 moles % (most used method was chemical analysis).

Type species: *Streptococcus pyogenes* Rosenbach 1884, 23.

Key to the species of genus **Streptococcus**

Key characteristics

Growth at 10, 45 and 50 C
Growth in 2.0 and 6.5% NaCl broths
Growth at pH 9.6
Growth in 0.1% methylene blue milk
Hydrolysis of sodium hippurate
Requirement for high CO_2 tension for rapid growth on blood agar
Size of colonies
Production of ammonia from arginine
Fermentation of:
 inulin
 trehalose
 sorbitol
 glycerol
 raffinose
 lactose
 arabinose
Fibrinolysis
Type of hemolysis
Lancefield serological group (not essential)
Bile solubility
Production of viscous polysaccharide in 5% sucrose broth
Final pH in glucose broth
Hydrolyosis of starch
Growth in 40% bile blood agar
Production of mucoid colonies on 5% sucrose or raffinose agar
Utilization of arginine as source of energy
Requirement for either folic or folinic acid for growth in simple media

I. Does not grow at 10 or 45 C (some exceptions in *S. sanguis*). Does not grow in 6.5% NaCl broth, at pH 9.6, or in 0.1% methylene blue milk.
 A. Does not hydrolyze sodium hippurate (some exceptions in *S. dysgalactiae*).
  1. Does not require a high CO_2 tension for rapid growth on blood agar. Cells and colonies not "minute." Ammonia produced from arginine.
   a. Does not ferment inulin.
    b. Fibrinolytic.
     c. Beta-hemolytic. Ferments trehalose but not sorbitol.
      d. Does not ferment glycerol. Lancefield Group A.
       1. *Streptococcus pyogenes*

dd. Ferments glycerol aerobically. Lancefield Group C.

2. *Streptococcus equisimilis*

bb. Not fibrinolytic.

c. Beta-hemolytic.

d. Ferments sorbitol but not trehalose. Ferments glycerol aerobically. Lancefield Group C.

3. *Streptococcus zooepidemicus*

dd. Does not ferment trehalose, sorbitol or glycerol. Lancefield Group C.

4. *Streptococcus equi*

cc. Not beta-hemolytic. Ferments trehalose. Usually ferments sorbitol. Does not ferment glycerol. Lancefield Group C.

5. *Streptococcus dysgalactiae*

aa. Inulin usually fermented.

b. May or may not be beta-hemolytic. Not bile soluble. Viscous polysaccharide may be produced in 5% sucrose broth. Lancefield Group H.

6. *Streptococcus sanguis*

bb. (Viscous polysaccharide not produced in 5% sucrose broth.) Alpha-reaction on blood agar. Bile soluble. No group antigen demonstrated.

7. *Streptococcus pneumoniae*

2. High CO_2 tension required for rapid growth on blood agar. "Minute" cells and colonies. Ammonia produced from arginine. Lancefield Group F and Type 1. Group G.

8. *Streptococcus anginosus*

B. Hydrolyzes sodium hippurate.

1. Produces ammonia from arginine. Ferments glycerol aerobically. May be beta-hemolytic. Final pH in glucose broth 4.2–4.8. Lancefield Group B.

9. *Streptococcus agalactiae*

2. Does not produce ammonia from arginine. Does not ferment glycerol. Not hemolytic. Final pH in glucose broth 5.6–6.5. No group antigen demonstrated.

10. *Streptococcus acidominimus*

II. Does not grow at 10 C. Grows at 45 C (exceptions in *S. mitis*). Does not grow in 6.5% NaCl broth, at pH 9.6 or in 0.1% methylene blue milk. Does not produce ammonia from arginine. (Exceptions in *S. mitis.*) Not beta-hemolytic.

A. Does not grow at 50 C. Grows in 2% NaCl broth.

1. Does not hydrolyze starch. Does not grow on 40% bile blood agar.

a. Indifferent gamma-reaction on blood agar. Ferments raffinose and inulin. Produces large mucoid colonies on 5% sucrose and raffinose agar. Lancefield Group K.

11. *Streptococcus salivarius*

aa. Alpha-reaction on blood agar. Does not ferment inulin. May or may not ferment raffinose. Does not produce mucoid colonies on 5% sucrose agar. No group antigen demonstrated.

12. *Streptococcus mitis*

2. Hydrolyzes starch. Grows on 40% bile blood agar.

a. Ferments lactose. Lancefield Group D.

13. *Streptococcus bovis*

aa. Does not ferment lactose. Lancefield Group D.

14. *Streptococcus equinus*

B. Grows at 50C. Does not grow in 2% NaCl broth. No group antigen demonstrated.

15. *Streptococcus thermophilus*

III. Grows at 10 and 45 C.

A. Grows in 6.5% NaCl, at pH 9.6 and in 0.1% methylene blue milk.

1. Uses arginine as a source of energy; ferments sorbitol, does not ferment arabinose. Does not require folic acid for growth in simplified media. Lancefield Group D.

16. *Streptocccus faecalis*

2. Does not utilize arginine as a source of energy but hydrolyzes it to produce ammonia. Ferments arabinose, does not ferment sorbitol. Requires folic acid for growth in simplified media. Lancefield Group D.

17. *Streptococcus faecium*

B. Grows in 6.5 % NaCl broth and at pH 9.6. Does not grow in 0.1% methylene blue milk. Ferments arabinose and sorbitol. Requires folinic acid for growth in simplified media. Does not produce ammonia from arginine. Contains Lancefield Group Q and Group D antigens.

18. *Streptococcus avium*

C. Does not grow in 6.5% NaCl broth, at pH 9.6 or in 0.1% methylene blue milk. Ferments sorbitol. Does not ferment arabinose. Usually requires folic acid for growth, in simplified media. Produces ammonia from arginine.

19. *Streptococcus uberis*

IV. Does not grow at 45 C; grows at 10 C. Does not grow in 6.5% NaCl broth, or at pH 9.6. Grows in 0.1% methylene blue milk. Lancefield Group N.

A. Produces ammonia from arginine.

20. *Streptococcus lactis*

B. Does not produce ammonia from arginine.

21. *Streptococcus cremoris*

TABLE 14.5

Some chemical characteristics and cellular location of streptococcal antigens

Group antigen (hapten)			Type antigen (hapten)			Species
Designation	Chemical nature	Cellular location	Designation	Chemical nature	Cellular location	
A	Rhamnose-*N*-acetyl-glucosamine polysaccharide	Wall	M R T	Protein Protein Protein	Envelope Envelope Envelope	*S. pyogenes*
B	Rhamnose-glucosamine polysaccharide	Wall	S (5 types)	Glucose-galactose-*N*-acetyl-glucosamine polysaccharide	Envelope	*S. agalactiae*
C	Rhamnose-*N*-acetyl-galactosamine polysaccharide	Wall	(8 types described)	Protein		*S. equisimilis*
			(8 types)	Protein		*S. zooepidemicus*
			(Only 1 type described)	Protein	Envelope	*S. equi*
			(3 types)	Protein		*S. dysgalactiae*
D	Glycerol teichoic acid containing D-alanine and glucose	"Intracellular" between wall and membrane	(11 types established)	Rhamnose-glucosamine-glucose polysaccharide	Cell wall	*S. faecalis*
D			(19 types described; many more exist)			*S. faecium*
D, Q						*S. avium*
D			(Many types exist)	Carbohydrate?	Capsule	*S. bovis*
D						*S. equinus*
E	Rhamnose polysaccharide	Wall	I to V	Polysaccharide		*Streptococcus* sp.
F	Rhamnose and a glucopyranosyl-*N*-acetyl-galactosamine tetrasaccharide	Wall	I to V	Carbohydrates—some types contain glucose, galactose and rhamnose		*S. anginosus*
G	Rhamnose-galactosamine polysaccharide	Wall	(3 types described)			*Streptococcus* sp. (large colony)
H	Rhamnose polysaccharide	Wall	(5 types described)			*S. sanguis*
K	Rhamnose polysaccharide	Wall	I and II, and I-II	Galactose, glucose rhamnose Type I—*O*-β-D galactopyranosyl-(1->6)-D-galactose	Cell wall	*Streptococcus* sp., *S. salivarius*
N	Glycerol teichoic acid containing D-alanine and galactose phosphate	"Intracellular" between wall and membrane	Many types exist			*S. lactis*
						S. cremoris
D			(Only 1 type)	Contains glucose, glucosamine, galactosamine	Capsule	*S. suis*
None	Ribitol teichoic acid with choline phosphate	Wall	(80 types)	Carbohydrate	Capsule	*S. pneumoniae*
			M	Protein	Envelope	

TABLE 14.6

Tolerance tests used to differentiate streptococcal species[a]

Streptococcal species	Growth at:		Growth in media containing:			Growth initiation at pH 9.6	Heat tolerance (60 C for 30 min)
	10 C	45 C	Methylene blue (0.1% in milk)	Sodium chloride (6.5%)	Bile (40%)		
1. *S. pyogenes*	−	−	−	−	−	−	−
2. *S. equisimilis*	−	−	−	−	−	−	−
3. *S. zooepidemicus*	−	−	−	−	−	−	−
4. *S. equi*	−	−	−	−	−	−	−
5. *S. dysgalactiae*	−	−	−	−	−	−	−
6. *S. sanguis*	−	d	−	−	+	ND	d
7. *S. pneumoniae*	−	−	ND	−	−	−	−
8. *S. anginosus*	−	d	−	−	d	−	−
9. *S. agalactiae*	−	−	−	−	+	−	−
10. *S. acidominimus*	−	−	−	−	ND	−	−
11. *S. salivarius*	−	+	−	−	−	−	−
12. *S. mitis*	−	d	−	−	−	−	−
13. *S. bovis*	−	+	−	−	+	−	+
14. *S. equinus*	−	+	−	−	+	−	−
15. *S. thermophilus*	−	+	−	−	−	−	+
16. *S. faecalis*	+	+	+	+	+	+	+
17. *S. faecium*	+	+	+	+	+	+	+
18. *S. avium*	+	+	−	+	ND	+	ND
19. *S. uberis*	+	+	−	−	−	−	+
20. *S. lactis*	+	−	+	−	+	−	d
21. *S. cremoris*	+	−	d	−	+	−	d
a. *S.* sp., group G	−	−	−	−	−	−	−
b. *S.* sp., group E	−	−	−	−	−	−	−
c. *S.* sp., group L	ND	ND	ND	ND	ND	ND	ND
d. *S.* sp., group M	ND	ND	ND	−	−	ND	ND
e. *S.* sp., group O	ND	ND	ND	ND	−	ND	−
f. *S.* sp., group R	ND	−	−	−	+	−	−
g. *S. suis*, group D	ND	ND	ND	ND	d	ND	−

[a] − = most (90–100%) strains negative; + = most (90–100%) strains positive; ND = no data or insufficient data; d = some strains positive, some negative.

TABLE 14.7

Hemolysin and fibrinolysin production and key hydrolytic reactions of streptococci[a]

Streptococcal species	Soluble hemolysin production			Fibrinolysin production	Hydrolysis of:				
	O	S	Other		Gelatin	Starch	Hippurate	Esculin	Arginine
1. *S. pyogenes*	+	+	−	+	d	−	−	+	+
2. *S. equisimilis*	+	−	+	+[b]	−	d	−	d	+
3. *S. zooepidemicus*	−	−	+	−	−	+	−	+	+
4. *S. equi*	−	−	+	−	−	ND	−	−	+
5. *S. dysgalactiae*	−	−	−	d[c]	−	ND	d	d	+
6. *S. sanguis*	ND	ND	−	−	−	−	−	+	+
7. *S. pneumoniae*	+	−	−	−	−	ND	ND	ND	ND
8. *S. anginosus*	−	−	d	−	−	−	−	+	+
9. *S. agalactiae*	−	−	+	−	−	−	+	−	+
10. *S. acidominimus*	ND	ND	−	−	−	−	(+)[d]	−	−
11. *S. salivarius*	−	−	−	−	−	−	−	+	−
12. *S. mitis*	-	−	ND	−	−	−	−	−	d
13. *S. bovis*	−	−	−	−	−	+	d	+	−
14. *S. equinus*	−	−	−	−	−	+[e]	−	+	−
15. *S. thermophilus*	−	−	−	−	−	+	−	−	−
16. *S. faecalis*	−	−	ND	−	−[e]	−	+	+	+
17. *S. faecium*	−	−	ND	−	−	−	ND	+	+
18. *S. avium*	ND	ND	ND	ND	−	−	−	+	−
19. *S. uberis*	−	−	−	ND	−	d	(+)[d]	+	+
20. *S. lactis*	−	−	−	−	−	−	d	d	+
21. *S. cremoris*	−	−	−	−	−	−	−	d	−
a. *S.* sp., group G	+	ND	−	d	−	+	−	+	+
b. *S.* sp., group E	ND	ND	ND	−	d	−	−	+	+
c. *S.* sp., group L	−	−	−	ND	ND	ND	−	ND	ND
d. *S.* sp., group M	ND	ND	d	ND	ND	ND	−	−	d[b]
e. *S.* sp., group O[b]	−	−	−	−	ND	ND	−	ND	ND
f. *S.* sp., group R[b]	ND	ND	d	ND	ND	ND	−	+	+
g. *S. suis*, group D	ND	ND	ND	ND	ND	+	ND	+	ND

[a] See Table 14.6 for key to symbols.
[b] See species description.
[c] For animal fibrin only.
[d] ( ) Delayed reaction.
[e] See description of varieties.

Genus II. **Leuconostoc** *van Tieghem 1878, 198, emend. mut. char. Hucker and Pederson 1930, 66*

Cells may be **spherical but often lenticular** particularly on agar, **usually in pairs and chains.** Non-motile. **Gram-positive.** Spores not formed.

Colonies are small, usually less than 1 mm diameter, smooth, round, grayish white.

Chemoorganotrophs requiring rich media; **often having complex growth factor and**

TABLE 14.8

Differentiating characteristics of the species of genus **Leuconostoc**[a]

	1. *L. mesenteroides*	2. *L. dextranicum*	3. *L. paramesenteroides*	4. *L. lactis*	5. *L. cremoris*	6. *L. oenos*
Acid from:						
Amygdalin	d	d	(d)	−	−	.
Arabinose	+	−	d	−	−	d
Arbutin	d	−	−	−	−	.
Cellobiose	d	d	(d)	−	−	d
Fructose	+	+	+	+	−	+
Galactose	+	d	+	+	d	d
Glucose	+	+	+	+	+	+
Lactose	(d)	+	(d)	+	+	−
Maltose	+	+	+	+	d	−
Mannitol	d	d	(d)	−	−	−
Mannose	+	d	+	d	−	d
Melibiose	d	d	+	d	−	d
Raffinose	d	d	d	d	−	−
Ribose	+	.	.	.	.	.
Salicin	d	d	−	d	−	d
Sucrose	+	+	+	+	−	−
Trehalose	+	+	+	−	−	+
Xylose	d	d	d	−	−	d
Hydrolysis of esculin	d	d	d	−	−	+
Required for growth:						
Uracil	−	−	−	−	+	−
Guanine + adenine + xanthine + uracil	−	d	d	−	+	+
Riboflavin	d	d	+	+	+	+
Pyridoxal	d	d	+	−	+	+
Folic acid	d	d	+	−	+	+
Tomato juice factor	−	−	−	−	−	d
Destruction of tomato juice factor	−	−	−	−	−	d
Dextran formation	+	+	−	−	−	−
Utilize citrate (carbohydrate present)	d	d	d	d	+	d
Utilize malate:						
No carbohydrate	d	−	d	−	−	.
+ carbohydrate	d	−	d	−	−	+
Yeast glucose litmus milk acid	+	+	+	+	+	d
Clot	d	d	d	d	d	−
Reduction	d	d	d	(d)	−	d
Gas	d	d	−	−	−	−
Growth in: 3.0% NaCl	+	d	d	d	−	.
6.5% NaCl	d	−	d	−	−	.
Growth at pH:						
4.8 (initial)	−	−	d	−	−	+
6.5 (initial)	+	+	+	+	+	d
Growth at 37 C	d	+	d	+	−	d
Final pH in glucose broth	4.5	4.5	4.4	4.7	5.0	.

[a] + = > 90% strains positive; d = 10–90% strains positive; − = > 90% strains negative; ( ) = delayed reaction; . = not known.

amino acid requirements. Nicotinic acid, thiamine, pantothenic acid and biotin are required by all species but cobalamin and *p*-amino benzoic acid by none.

Growth is dependent on the presence of a fermentable carbohydrate; glucose is fermented with the production of D(−) lactic acid, ethanol and CO_2. Some strains have an oxidative mechanism and acetic acid is formed in place of ethanol. Rhamnose, melezitose, inulin, starch, glycerol, sorbitol and inositol are not fermented. Dextrin is seldom fermented. Sorbose, erythritol, adonitol and dulcitol are not fermented by species 1–5 and not recorded for 6. Other reactions on carbohydrates and growth requirements are given in Table 14.8.

Some species produce a characteristic slime from sucrose. **Mannitol is formed from fructose.**

The amino acid composition of the cross-linking peptide of the cell wall murein consists of alanine, serine and lysine. Peptides from different species are not however identical (Kandler *et al.*, 1967).

Catalase negative although some strains will destroy H_2O_2 when grown in the presence of heme and others have peroxidase activity. Orthotoluidene is not attacked. Cytochromes are absent.

Arginine is not hydrolyzed. Milk is rarely acidified and curdled unless supplemented with a fermentable carbohydrate and yeast extract.

Non-proteolytic. Indole is not formed. Do not reduce nitrates. Non-hemolytic.

Facultative anaerobes.

Optimum temperature between 20 and 30 C.

Non-pathogenic to animals and man.

The G + C content of the DNA for *L. lactis* is 43–44 moles % and for all other species from 38–42 moles % (T_m, buoyant density).

Type species: *Leuconostoc mesenteroides* (Tsenkovskii) van Tieghem 1878, 191.

Genus III. Pediococcus *Balcke 1884, 257*

Cocci occurring **in pairs or in tetrads** as the result of alternate division along the two perpendicular planes; single cells or chains are rare. Non-motile, do not form endospores. Gram-positive.

Chemoorganotrophs: **metabolism fermentative.** The fermentation is **homolactic, producing DL-lactic acid**; the L(+) enantiomorph generally predominates. Acid but no gas produced from glucose, fructose and mannose; sorbitol and starch not fermented.

Gelatin not liquefied. Nitrates not reduced to nitrites.

Nutritional requirements complex.

Microaerophilic; surface growth is poor. Generally **catalase negative** although may possess some non-heme catalase activity.

Saprophytes found in fermenting plant materials, especially in spoiled beer; rare in milk and dairy products.

The G + C content of the DNA ranges from 34–44 moles %.

Type species: *Pediococcus cerevisiae* Balcke 1884, 257.

Key to the species of genus Pediococcus

Key characteristics

Growth at pH 5.0, 7.0 and 9.0
Growth at 35 and 50 C
Relationship to oxygen
Tolerates 15% NaCl (halophilic)

I. Grow at pH 5.0 but not at pH 9.0.
  A. No growth at pH 7.0 or at 35 C; prefers anaerobic conditions.
    1. *Pediococcus cerevisiae*
  B. Growth at pH 7.0 and at 35 C; microaerophilic.
    1. Can grow at 50 C.
      2. *Pediococcus acidilactici*
    2. Can not grow at 50 C.
      3. *Pediococcus pentosaceus*

II. Do not grow at pH 5.0 but grow at pH 9.0; microaerophilic.
   A. Halophilic.
      4. *Pediococcus halophilus*
   B. Not halophilic
      5. *Pediococcus urinae-equi*

Genus IV. **Aerococcus** *Williams, Hirch and Cowan 1953, 475*

Spheres, 1.0–2.0 μm in diameter, with a **strong tendency toward tetrad formation** when grown in suitable liquid media (Deibel and Niven, 1960). Non-motile. **Gram-positive.**

Growth on solid media is generally sparse and beaded (small discrete colonies).

Chemoorganotrophic: Acid but no gas is produced from glucose, fructose, galactose, mannose, maltose and sucrose; acid is usually produced from lactose, trehalose and mannitol. Final pH in glucose broth 5.0–5.5.

Microaerophilic. In shake cultures or soft sugar agar a heavy band of discrete colonies is produced just beneath the surface. Anaerobic growth frequently absent and when it does occur is delayed and often consists of only a few discrete colonies.

Catalase activity is absent or weak and when present is a non-heme pseudocatalase. Porphyrin respiratory enzymes are absent (method of Deibel and Evans, 1960). Hydrogen peroxide is produced during aerobic growth.

The cell wall murein contains no interpeptide bridge. The carboxyl group of D-alanine is directly bound to the ε-amino group of L-lysine of an adjacent peptide subunit (Kandler *et al.*, 1970).

Widely distributed in air, in meat brines and on raw and processed vegetables.

The G + C content of DNA of the type species ranges from 36–40 moles % (T_m) (Schultes and Evans, 1971).

Type species: *Aerococcus viridans* Williams, Hirch and Cowan 1953, 477.

Genus V. **Gemella** *Berger 1960, 253*

Cocci singly or in pairs with adjacent sides flattened. Gram-indeterminate but cell wall is of the Gram-positive type (Plate 10.2, Fig. 2). Endospores not formed; non-motile.

On blood agar small smooth colonies resemble those of beta-hemolytic streptococci; the hemolysin is produced on rabbit or horse blood agar. Pigment is not produced.

Chemoorganotroph: Fermentative, acid is produced from several carbohydrates. Growth requirements complex.

Nitrates not reduced; nitrites reduced by some strains.

Aerobic or facultatively anaerobic. Respiratory system resistant to KCN.

Optimal temperature 37 C, grows at 22 C.

Parasites of mammals.

Mean G + C content of DNA 33 ± 1.6 moles %.

Type species: *Gemella haemolysans* (Thjøtta and Bøe) Berger 1960, 253.

FAMILY III. **PEPTOCOCCACEAE** ROGOSA 1971, 235

Cocci varying in diameter (0.5–2.5 μm). Occur singly, in pairs, tetrads and irregular masses, occasionally in three-dimensional cubic packets. Some organisms may form short or long chains with some elongation of individual elements. Flagella, motility and endospores are absent. Gram-positive but may stain equivocally. Anaerobic. Chemoorganotrophic with complex nutritional requirements. Gas consisting principally of CO_2 and usually H_2 is produced in cultures from amino acids or carbohydrates or both; H_2S may also be detectable. Carbohydrates may or may not be fermented. In complex media the major non-gaseous products are lower fatty acids; in some cases succinate or ethanol may also be major products. Lactate, if produced, is generally not a major product and is always associated with a considerable heterofermentation. Found in the mouth and intestinal and respiratory tracts of man and other animals; and frequently in the normal and pathological human female urogenital tracts. Packet formers are found in soil, on the surface of cereal grains and in the alimentary tract of man and animals.

The characteristics differentiating the genera of the family are shown in Table 14.9.

TABLE 14.9

Characteristics differentiating the genera of family **Peptococcaceae**[a]

	Growth at pH 2–2.5	Arrangement		G + C moles %	Carbohydrates fermented	Cellulose digested	Major products			Peptones or amino acids main N and energy source
		Chains	Cubic packets				Succinate	Ethanol	Volatile acids with >3C atoms	
Peptococcus	−	−	−	35.7–36.7	d	−	−	−	+	+
Peptostreptococcus	−	+	−	33.5	d	−	−	d	+	+
Ruminococcus	−	+ or −	−	39.8–45.4	+	+	+ or −	+ or −	−	−
Sarcina	+	−	+	28.6–30.6	+	−	−	+ or −	+ or −	−

[a] + = positive; − = negative; d = usually (90–100%) positive.

Note: Holdeman and Moore (*International Journal of Systemic Bacteriology 24:* 260–277, 1974) have recently described a new genus *Coprococcus* with three species which they feel should be included in the family *Peptococcaceae.* The organisms produce diplococci or chains; have a G + C content of 39–42 moles %; ferment carbohydrates with the production of butyric and acetic acids; and do not use peptones as a main carbon source.

Genus I. **Peptococcus** *Kluyver and van Niel 1936, 400*

Spherical bacteria about 0.5–1.0 μm or exceptionally 1.6 μm in diameter. Occur **singly,** in **pairs, tetrads** or **irregular masses** or rarely in short chains but not in three-dimensional cubical packets; long chains never formed. Non-flagellated and non-motile. No spores. **Gram-positive.**

Chemoorganotrophic. Can use protein decomposition products (peptones, amino acids) as sole energy source. Carbohydrates not required and saccharoclastic activity often limited or negative but carbohydrates may enhance growth or required fatty acids may stimulate carbohydrate fermentation of some species. Where tested, such nitrogenous compounds as **amino acids,** purines or pyrimidines are fermented. Gas produced in cultures by some species with or without carbohydrates. Ammonia produced. Depending on the substrate, acetate, propionate, butyrate, formate, succinate, lactate, CO_2, ammonia and occasionally hydrogen are produced. In complex media, various combinations of the **C_1 to C_4 lower volatile fatty acids, CO_2, H_2** and **ammonia** tend to be the **major products from amino acids;** however, lactate is also produced from histidine. Generally **lactate is not a major** product in glucose complex media and it is **not formed homofermentatively** as in *Streptococcus* but is always associated with a very considerable heterofermentation. Lactate and malate are not fermented; a number of species produce gas from pyruvate, but most species do not produce gas from citrate or tartrate.

Weak or variable catalase reaction. Rarely hemolytic. Coagulase negative. Indole sometimes produced.

Temperature range generally 25–37 C; optimal temperature 35–37 C.

Anaerobic. pH range generally 6–8; optimal pH 7–8.

Sensitive to penicillin G (1–10 units/ml; often to less than 1 unit/ml).

Isolated from the human female urogenital tract, the human intestinal and respiratory tract, inflamed appendices, cystitis, pleurisy, the gums, postpartum septicemia, tonsils, plasma, skin, caseous lymphadenitis in sheep, and tidal bay mud. Pathogenicity is uncertain because isolations are very frequently from sites where other potential pathogens are present.

Type species: *Peptococcus niger* (Hall) Kluyver and van Niel 1936, 400.

TABLE 14.10

Characteristics of the species of genus **Peptococcus**[a]

	1. *P. niger*	2. *P. asaccharolyticus*	3. *P. aerogenes*	4. *P. activus*	6. *P. anaerobius*
Sugars fermented	−	−	w	+	w
Visible gas	+	+	+	+	−
Amino acids fermented					
Glycine	NT	−	−	−	+
Serine	NT	+	+	+	−
Threonine	NT	+	+	+	−
Glutamate	NT	+	+	+	−
Histidine	NT	+	+	+	−
Purines fermented	NT	+	+	+	−
Nitrate reduction	NT	+	+	+	−
Indole	NT	+	+	+	−
Gelatin	−	−	−	+	−
H_2S	+	+	+	+	+

[a] NT = not tested; + = positive; − = negative; w = negative or feebly positive. No other amino acids than those listed are fermented. Products from glycine are CO_2, NH_3 and acetate; from serine CO_2, NH_3, H_2 and acetate; from threonine CO_2, NH_3, H_2 and propionate; from glutamate CO_2, NH_3, H_2, acetate and butyrate.

Genus II. **Peptostreptococcus** *Kluyver and van Niel 1936, 401*

Spherical to **ovoid** cells, usually 0.7–1.0 μm, but occasionally 0.3–0.5 μm in diameter, occurring in **pairs, short or long chains.** Non-flagellated and non-motile. No spores. **Gram-positive.**

Chemoorganotrophic. With one exception, **carbohydrates** are **fermented** with production of **acid, gas** or **both. Gas production in carbohydrate-free peptone media by some species. Pyruvate frequently fermented producing acid or gas or both**; malate, citrate, and tartrate **not fermented. Ammonia is produced** but an alkaline reaction may not result when acid is produced in carbohydrate medium. **Acidic products** may be any one or combinations of the following: **acetic, formic, propionic, butyric, isobutyric, valeric, isovaleric, isocaproic, caproic** and **succinic.** Amines and various alcohols may also be produced. **Lactate not produced. Catalase negative.** Nitrates not reduced. Indole generally not produced. Gelatin generally not liquefied and coagulated proteins not attacked. Very rarely hemolytic.

Anaerobic.

Initial pH growth range about 6.0–8.0; optimum 7.0–7.5. Growth generally from 25–38 C; optimum about 35–37 C. Complex media required.

Isolated from the normal and pathological female genital tract and blood in puerperal fever, from respiratory and intestinal tracts in normal humans and animals, from the oral cavity, from pyogenic infections, septic war wounds and appendicitis. May be pathogenic. Generally sensitive to penicillins (10 unit/ml) and often <1 unit/ml.

Type species: *Peptostreptococcus anaerobius* (Krönig) Kluyver and van Niel 1936, 401.

TABLE 14.11

Characteristics of the species of genus **Peptostreptococcus**[a]

	1. *P. anaerobius*	2. *P. productus*	3. *P. lanceolatus*	4. *P. micros*	5. *P. parvulus*
Gas production from complex media and from pyruvate	+	+	+	−	−
Fetid odor	+	+	+	−	−
Propionate from threonine	+	−	−	−	−
Major acids produced					
Lactate	−	− or w	−	+	+
Acetate	+	+	+	+	+
C_3–C_6	+	+	+	−	−
Carbohydrates fermented					
Arabinose	−	w	−	−	d
Fructose	w	+	−	−	−
Galactose	−	+	−	−	−
Glucose	w	+	+	−	+
Lactose	−	+	−	−	+
Maltose	w	+	−	−	−
Sucrose	−	+	+	−	−
Xylose	−	+	−	−	−
Litmus milk acid and coagulated	−	+	−	−	+

[a] + = 90% or more strains positive; − = 90% or more strains negative; d = variable; w = weak positive or trace. All species produce chains of cells. None ferments adonitol, cellulose, erythritol or glycogen. None produces lecithinase, lipase or acetylmethylcarbinol.

Genus III. **Ruminococcus** *Sijpesteijn 1948, 152*

Spherical to **elongated coccoid** cells. When elongated, ends of cells rounded or flat rather than pointed or lancet shaped. Sides of cells may be flattened when in contact with other cells, particularly with diplococci which resemble neisseria and pediococci. Occur as single cells, pairs or short chains of 3–4 cells (*R. albus*) or chains of 8–50 or more cells (*R. flavefaciens*). Average diameter varies from 0.7–1.2 μm with a modal range of 0.8–1.0 μm.

Gram-positive anatomy but may decolorize relatively easily. Some cells of all strains in young culture contain an **iodophilic** reserve **polysaccharide** (a glucose polymer) which may rapidly disappear in the stationary phase of growth. There are no endospores and the cells are non-motile.

Roll tube surface colonies after 3 days in rumen fluid glucose cellobiose agar are entire, smooth and slightly convex. In transmitted light, they are translucent to opaque but occasionally have a fluorescent to "frosted glass" appearance. Colonies may be white, light tan or yellow to orange and are generally 2–4 mm in diameter. Deep colonies are lenticular.

Cellulose is **usually digested** and reducing sugar is produced in media containing excess cellulose. **Cellobiose** is always **fermented.** Xylan is fermented by about 85% of strains. A given strain may not ferment any other sugars or may attack one or more of the following: glucose, D-xylose, esculin, fructose, sucrose, lactose and L-arabinose. A glucose permease is lacking where whole cells do not ferment glucose; a glucokinase is present in cell extracts (Ayers, 1958). No acid is produced from a wide variety of other carbohydrates and polyols.

In media **containing bicarbonate** and **gassed** with CO_2, **cellobiose is fermented; major products are acetic** and **formic** acids; **ethanol** and/or **succinic** acid may also be major products; but **when ethanol** is a **major product,** there is **more hydrogen** and **little** or **no succinic acid** (*R. albus*); and **when succinate** is a **major product** there is **little hydrogen** and **little or no ethanol** (*R. flavefaciens*). Lactic acid is usually a minor product, if produced; methane and fatty acids containing three or more carbon atoms are not formed. Succinate production involves CO_2 fixation. Organisms intermediate between *R. albus* and *R. flavefaciens* in colony color, morphology and fermentation products have been isolated (Hungate, 1957).

Chemoorganotrophic and **nutrient requirements** may be **complex** (Ayers, 1958). Rumen fluid or fecal extracts may be required for growth in usual media (Hungate, 1957); however, more than half of strains can be grown in defined media containing B vitamins, cellobiose, minerals, ammonia, sulfide, carbon dioxide, acetate and one or more of the branch-chained acids, isovalerate, isobutyrate and 2-methyl-butyrate. **Ammonia is** essential as the **main nitrogen source** for all strains adequately studied and exogenous amino acid carbon and nitrogen are very poorly utilized although a few strains require methionine or vitamin B_{12} (Allison *et al.*, 1962; Bryant and Robinson, 1961; Dehority *et al.*, 1967).

Growth in rumen fluid cellobiose broth is evenly turbid within 1 day. There is good growth from 30–37 C and no growth at 15 or 50 C. Occasional strains grow at 22 or 45 C.

Catalase, indole and **H_2S** are **not produced.** Starch is not hydrolyzed and nitrate is not reduced. Ammonia is not produced from amino acids or peptides. Voges-Proskauer reaction is variable and a small minority of strains liquefy gelatin.

Strictly anaerobic, requiring technique of Hungate (1950) or similar stringent anaerobic method for successful isolation. An initial pH range of 6.0–7.0 favors growth; a pH of about 6.5 is optimal.

There is a multiplicity of serological types (Jarvis, 1967).

Isolated from bovine and ovine rumen and from ceca of rabbits and guinea pigs; probably widely distributed in the rumen of *Ruminantia* and in the cecum and colon of herbivores and thought to play an important role in the fermentation of cellulose in these organs.

The G + C content of the DNA ranges from 39.8–45.4 moles % (T_m).

Type species: *Ruminococcus flavefaciens* **Sijpesteijn 1948, 152.**

Note: There are two species recognized, *R. flavefaciens* and *R. albus*, separated mainly on pigmentation. See The *Manual*, pp. 526–527.

Genus IV. Sarcina *Goodsir 1842, 434*

Nearly spherical cells, 1.8–3 μm in diameter, **occurring in packets** of eight or more. Some of the cells in cultures may be present singly, or as groups of fewer than eight cells. Generally the cells are flattened in the areas of contact with adjacent cells. **Division occurs in three perpendicular planes. Non-motile.** A report of spore formation by these organisms has been published (Knöll, 1965). **Gram-positive.**

Chemoorganotrophs: **Strictly fermentative metabolism, carbohydrates are the fermentable substrates.** The main products of glucose fermentation are CO_2, H_2, acetic acid as well as ethanol for one species and butyric acid for the other. Not pigmented. Catalase not present. **The minimal growth requirements include numerous amino acids and few vitamins.**

Strict anaerobes. Grow from pH values near 1 to 9.8.

The **G + C content** of the DNA ranges from **28–31 moles %** (buoyant density).

Type species: *Sarcina ventriculi* Goodsir 1842, 437.

Note: there are two species recognized. *S. ventriculi* possesses a cellulose outer layer and ferments glucose to form ethanol, acetate, lactate, CO_2 and H_2. *S. maxima* has no cellulose outer layer and ferments glucose to form butyrate, acetate, CO_2 and H_2.

PART 15

ENDOSPORE-FORMING RODS AND COCCI

FAMILY I. BACILLACEAE FISCHER 1895, 139

Cells rod-shaped, in one genus spherical. Mycelium not produced. Endospores formed. Endospores differ from vegetative cells in being more refractive and less susceptible to staining, in having greater resistance to heat and other destructive agents, and in containing dipicolinic acid (5–15% of dry weight). The spore contains a central cell (the core) which is enclosed by a cortex of peptidoglycan and an outer spore coat.

Majority Gram-positive. Motile by lateral or peritrichous flagella, or non-motile.

Aerobic, facultative or anaerobic.

Key to the genera of family **Bacillaceae**

I. Cells rod-shaped
  A. Aerobic or facultative, catalase usually produced.
    Genus I. *Bacillus*, p. 529*
  B. Microaerophilic, catalase not produced.
    Genus II. *Sporolactobacillus*, p. 550
  C. Anaerobic.
    1. Sulfate not reduced to sulfide.
      Genus III. *Clostridium*, p. 551
    2. Sulfate reduced to sulfide.
      Genus IV. *Desulfotomaculum*, p. 572

II. Cells spherical, in packets.
    Genus V. *Sporosarcina*, p. 573

Note: There is one genus of uncertain affiliation, *Oscillospira*, which grows as large rods or filaments divided into disc-shaped cells.

Genus I. **Bacillus** *Cohn 1872, 174; nom.gen.cons.* Nomencl. Comm. Intern. Soc. Microbiol. 1937, 28; Opin. A. Jud. Comm. 1955, 39

Cells rod-shaped, straight or nearly so, 0.3–2.2 by 1.2–7.0 μm. **Majority motile;** flagella typically lateral. **Heat-resistant endospores formed; not more than one in a sporangial cell.** Sporulation not repressed by exposure to air. Gram reaction: positive, or positive only in early stages of growth, or negative.

Chemoorganotrophs; metabolism strictly respiratory, strictly fermentative or both respiratory and fermentative, using various substrates. The terminal electron acceptor in respiratory metabolism is molecular oxygen, replaceable in some species by nitrate. Catalase formed by most species.

Strict aerobes or facultative anaerobes.

The G + C content of the DNA of those strains examined ranges from 32–62 moles % (T_m and buoyant density).

Type species: *Bacillus subtilis* (Ehrenberg) Cohn 1872, 174.

*All page numbers cited refer to *Bergey's Manual of Determinative Bacteriology*, 8th edition, The Williams & Wilkins Company, Baltimore, 1974.

TABLE 15.1

Differential characteristics of the species of genus* Bacillus *(Group I)

The symbols used are: E, elliptical or cylindrical; S, spherical or nearly so; C, central; T, terminal or subterminal; CT, central to terminal, variation within or between strains; W, weakly positive.

	Spore			Products of action on glucose		
	Shape	Distends sporangium distinctly	Dominant position	Acid[a]	Gas[b]	Acetoin[c]
1. *B. subtilis*	E	–	C	+	–	+
2. *B. pumilus*	E	–	C	+	–	+
3. *B. licheniformis*	E	–	C	+	W or –	+
4. *B. cereus*	E	–	C	+	–	+
5. *B. anthracis*	E	–	C	+	–	+
6. *B. thuringiensis*	E	–	C	+	–	+
7. *B. megaterium*	E	–	C	+	–	–
8. *B. polymyxa*	E	+	CT	+	+	+
9. *B. macerans*	E	+	T	+	+	–
10. *B. circulans*	E	+	CT	+	–	–
11. *B. stearothermophilus*	E	+ or –	T	+	–	–
12. *B. coagulans*	E	+ or –	CT	+	–	+ or –
13. *B. alvei*	E	+	CT	+	–	+
14. *B. firmus*	E	–	C	W	–	–
15. *B. laterosporus*	E	+	C	+	–	–
16. *B. brevis*	E	+	CT	+ or –	–	–
17. *B. sphaericus*	S	+	T	–	–	–
18. *B. pasteurii*[d]	S	+	T	–	–	–
19. *B. fastidiosus*[d]	E	–	C	–	–	–
20. *B. larvae*[d]	E	+	CT	+	–	–
21. *B. popilliae*[d]	E	+	C	+	–	–
22. *B. lentimorbus*[d]	E	+	C	+	–	–

[a] Medium 1 (grams per liter): $(NH_4)_2HPO_4$, 1.0; KCl, 0.2; $MgSO_4 \cdot 7H_2O$, 0.2; yeast extract, 0.2; agar, 15; bromcresol purple, 0.008; glucose (added after sterilization), 5; used as slants.

[b] Medium 2 (grams per liter): peptone, 5; yeast extract, 3; NaCl, 5; agar, 3; bromcresol purple, 0.008; glucose (added after sterilization), 10; pH 7.0; used for stab cultures; may show acid formation if growth fails on medium 1.

[c] Medium 3 (grams per liter): proteose peptone, 7; glucose, 5; NaCl, 5.

[d] Little or no growth in medium 1, 2 or 3; in media that support growth the results are as shown.

TABLE 15.2

Differential characteristics of **Bacillus** *species 1–3*

Common characteristics: Gram-positive; motile; catalase positive; intracellular globules unstainable by fuchsin in cells grown on glucose agar not observed; acid produced from arabinose, xylose and mannitol in Medium 1 (Table 15.1) with glucose replaced; egg yolk reaction negative; no growth in 0.02% azide by strains growing at 55 C; growth in 7% NaCl and in Sabouraud dextrose broth and/or agar; variable growth in 0.001% lysozyme; alkali on citrate-salts agar; casein hydrolyzed; tyrosine not decomposed. Data are from Gordon *et al.* (1973).

	1. *B. subtilis*	2. *B. pumilus*	3. *B. licheniformis*
Rods			
Width, μm	0.7–0.8	0.6–0.7	0.6–0.8
Length, μm	2–3	2–3	1.5–3
Hydrolysis of			
Starch	+[a]	−	+
Hippurate	−	+	−
NO_3^- to NO_2^-	+	−	+
Growth in anaerobic agar	−	−	+
Temperature for growth, C			
Maximum	45–55	45–50	50–55
Minimum[b]	5–20	5–15	15

[a] + = positive for 90–100% of strains; − = negative for 90–100% of strains.

[b] Lowest temperature tested 3 C.

TABLE 15.3

Differential characteristics of **Bacillus** *species 4–7*

Common characteristics: Gram-positive, catalase positive, cells grown on glucose agar contain intracellular globules unstainable by fuchsin, grow in 7% NaCl and in Sabouraud dextrose broth and/or agar, hydrolyze starch and casein.

	4. *B. cereus*	5. *B. anthracis*	6. *B. thuringiensis*	7. *B. megaterium*
Rods				
Width, μm	1.0–1.2	1.0–1.2	1.0–1.2	1.2–1.5
Length, μm	3–5	3–5	3–5	2–5
Intracellular protein crystals	−[a]	−	+	−
Motility	d	−	d	d
Acid from[b] arabinose, xylose and mannitol	−	−	−	d
NO_3^- to NO_2^-	+	+	+	d
Egg yolk reaction	+	+	+	−
Growth in				
Anaerobic agar	+	+	+	−
0.001% lysozyme	+	+	+	−
Alkali on citrate-salts agar	+	d	+	+
Temperature for growth, C				
Maximum	35–45	40	40–45	35–45
Minimum[c]	10–20	15–20	10–15	3–20
Tyrosine decomposed	+	d	d	d

[a] + = positive for 90–100% of strains; − = negative for 90–100% of strains; d = reactions differ, positive for 11–89% of strains.

[b] Medium 1 (Table 15.1) with glucose replaced.

[c] Lowest temperature tested 3 C.

TABLE 15.4

Differential characteristics of **Bacillus** *species 8–10*

Common characteristics: Gram-positive, -variable or -negative; catalase positive; acid produced from arabinose, xylose and mannitol in Medium 1 (Table 15.1) with glucose replaced; starch hydrolyzed; indole not produced; phenylalanine not deaminated; tyrosine not decomposed.

	8. *B. polymyxa*	9. *B. macerans*	10. *B. circulans*
Rods			
Width, μm	0.6–0.8	0.5–0.7	0.5–0.7
Length, μm	2–5	2.5–5	2–5
Motility	+[a]	+	d
Temperature for growth, C			
Maximum	35–45	40–50	35–50
Minimum	5–10	5–20	5–20
Gas from arabinose, xylose and mannitol	+	+	−
Growth in			
Anaerobic agar	+	+	d
Sabouraud dextrose broth and/or agar	+	+	d
5% NaCl	−	−	d
7% NaCl	−	−	d
0.001% lysozyme	d	−	d
Production of			
Crystalline dextrins	−	d	−
Dihydroxyacetone	+	−	−
NO_3^- to NO_2^-	+	+	d
Decomposition of casein	+	−	d

[a] + = positive for 90–100% of strains; − = negative for 90–100% of strains; d = reactions differ, positive for 11–89% of strains.

TABLE 15.5

Differential characteristics of **Bacillus** *species 11–13*

Common characteristics: Starch hydrolyzed, no growth in 7% NaCl, phenylalanine not deaminated.

	11. *B. stearothermophilus*	12. *B. coagulans*	13. *B. alvei*
Rods			
Width, μm	0.6–1	0.6–1	0.5–0.8
Length, μm	2–3.5	2.5–5	2–5
Gram reaction	d and v[a]	+[a]	d and v
Motility	+	+	d
Catalase	d	+	+
Temperature for growth, C			
Maximum	65–75	55–60	35–45
Minimum	30–45	15–25	15–20
Acid from[b] arabinose, xylose and mannitol	d	d	−
Growth in			
Anaerobic agar	−	+	+
Sabouraud dextrose broth and/or agar	−	+	−
5% NaCl	d		d
0.02% azide	−	+	
0.001% lysozyme	−	−	+
Production of			
Dihydroxyacetone	−	d	+
Indole	−	−	+
NO_3^- to NO_2^-	d	d	−
Decomposition of			
Casein	d	d	+
Tyrosine	−	−	d

[a] + = positive for 90–100% of strains; − = negative for 90–100% of strains; d = reactions differ, positive for 11–89% of strains; v = character inconstant in one strain.

[b] Medium 1 (Table 15.1) with glucose replaced.

TABLE 15.6

Differential characteristics of **Bacillus** *species 14–17*

Common characteristics: Catalase positive; dihydroxyacetone not produced.

	14. *B. firmus*	15. *B. laterosporus*	16. *B. brevis*	17. *B. sphaericus*
Rods				
Width, μm	0.6–0.9	0.5–0.8	0.6–0.9	0.6–1
Length, μm	1.2–4	2–5	1.5–4	1.5–5
Gram reaction	+[a]	d and v	d and v	d and v
Easily stainable body attached to one side of spore	−	+	−	−
Motility	d	+	+	+
Temperature for growth, C				
Maximum	40–45	35–50	40–60	30–45
Minimum	5–20	15–20	10–35	5–15
Acid from[b]				
Arabinose and xylose	d	−	−	−
Mannitol	+	+	d	−
Hydrolysis of starch	+	−	−	−
Growth in				
Anaerobic agar	−	+	−	−
0.02% azide[c]			−	
0.001% lysozyme	−	+	d	d
7% NaCl	+	−	−[d]	d
Sabouraud dextrose broth and/or agar	−	−	d	d
Production of				
Alkaline reaction in V-P broth	−	−	+	+
Indole	−	d	−	−
NO_3^- to NO_2^-	+	+	d	−
Decomposition of				
Casein	+	+	+	d
Tyrosine	d	+	+	−
Deamination of phenylalanine	d	−	−	+

[a] + = positive for 90–100% of strains; − = negative for 90–100% of strains; d = reactions differ, positive for 11–89% of strains; v = character inconstant in one strain.

[b] Medium 1 (Table 15.1) with glucose replaced.

[c] Only strains that grow at 55 C were tested.

[d] Cultures of *B. brevis* did not grow in 5% NaCl.

TABLE 15.7

Differential characteristics of **Bacillus** *species 20–22*[a]

Common characteristics: Acid produced from trehalose, acid not produced from arabinose and xylose, starch not hydrolyzed, alkali not produced in citrate-salts medium, catalase negative, dihydroxyacetone and indole not produced, phenylalanine not deaminated, tyrosine not decomposed, growth in anaerobic agar and in 0.001% lysozyme, no growth on serial transfer in nutrient broth or in 5% NaCl.

	20. *B. larvae*	21. *B. popilliae*	22. *B. lentimorbus*
Rods			
Width, μm	0.5–0.6	0.5–0.8	0.5–0.7
Length, μm	1.5–6	1.3–5.2	1.8–7
Parasporal body in sporangium	–[b]	+	–
Gram reaction	+	–[c]	–[c]
Motility	d	d	–
Temperature for growth, C			
Maximum	40	35	35
Minimum	25	20	20
Acid from mannitol	d	–	–
NO_3^- to NO_2^-	d	–	–
Growth in 2% NaCl	+	+	–
Decomposition of			
Casein	+	–	–
Gelatin	+	–	–

[a] All cultures grown in J-medium or modifications of J-medium (Gordon *et al.*, 1973).

[b] + = positive for 90–100% of strains; – = negative for 90–100% of strains; d = reactions differ, positive for 11–89% of strains.

[c] Sporangia and presporal forms are Gram-positive.

Group II

The 26 species in this group are placed together because it might be considered that they need further investigation to show how they differ from other species or from closely similar but unnamed organisms. Several have the severe limitation that the original description was based on very few strains, while not more than a single culture may now be available for study.

Some differential properties of the species in Group II are given in Table 15.8.

Editorial Note: The status of the species in this section has been the subject of considerable correspondence between the authors and editors. In many genera, most, if not all of the species in this group would have been listed as *species incertae sedis*, and one author agrees. It is evident that more work is needed to establish their status as distinct species. *B. acidocaldarius* seems quite distinct and is placed at the end because its description appeared after the manuscript and tables had been written.

TABLE 15.8

Differential characteristics of the species of genus **Bacillus** *Group II*

	Spore			Acid from glucose	Starch hydrolysis	Anaerobic growth	Growth at 3 C
	Shape	Distends sporangium distinctly	Dominant position				
23. *B. amyloliquefaciens*	E[a]	−	C	+	+	−	−
24. *B. medusa*	E	−	C	+	+	+	−
25. *B. maroccanus*	E	−	CT	+	+	−	−
26. *B. pacificus*	E	+	C	+	+	−	−
27. *B. lentus*	E	−	C	+	+	−	−
28. *B. epiphytus*	E and S	−	C	+	+	−	−
29. *B. apiarius*	E	+	C	+	+	+	−
30. *B. psychrosaccharolyticus*	E	+	C	+	+	+	+
31. *B. macquariensis*	E	+	T	+	+	+	+
32. *B. laevolacticus*	E	+	T	+	+	+	−
33. *B. racemilacticus*	E	+	T	+	+	+	−
34. *B. filicolonicus*	E	+	T	+	+	+	−
35. *B. pantothenticus*	E and S	+	T	+	+	+	−
36. *B. thiaminolyticus*	E	+	T	+	+	+	−
37. *B. pulvifaciens*	E	+	CT	+	−	+	−
38. *B. cirroflagellosus*	E	+	C	−	+	−	−
39. *B. freudenreichii*	E	−	C	−	−	−	−
40. *B. alcalophilus*	E	−	T	−	+	−	−
41. *B. badius*	E	−	CT	−	−	−	−
42. *B. aneurinolyticus*	E	v	T	−	−	−	−
43. *B. macroides*	E	v	T	−	−	−	−
44. *B. aminovorans*	S	−	C	+	+	−	−
45. *B. insolitus*	S	−	CT	−	−	−	+
46. *B. globisporus*	S	+	T	+	d	−	+
47. *B. psychrophilus*	S	v	T	+	+	−	+
48. *B. acidocaldarius*	E	+	T	−	+	−	−[b]

[a] E = elliptical or cylindrical; S = spherical or nearly so; C = central; T = terminal or subterminal; CT = central to terminal; variation within or between strains; + = positive for 90–100% of strains; − = negative for 90–100% of strains; d = reactions differ, positive for 11–89% of strains; v = character inconstant in one strain.

[b] Minimum temperature for growth 45 C.

Genus II. **Sporolactobacillus** *Kitahara and Suzuki 1963, 69*

Cells straight rods, 0.7–0.8 by 3–5 μm, occurring singly, in pairs, rarely in short chains. Motile by means of a small number of long peritrichous flagella. **Endospores formed.** Gram-positive.

Chemoorganotrophs; metabolism fermentative, hexose sugars are decomposed through the typical homo-fermentative pathway producing only lactic acid. **Do not contain heme compounds such as catalase or cytochromes.**

Microaerophilic.

Does not grow below 10 C or above 45 C. Grows well between 15 and 40 C. Optimum for growth 35 C, for fermentation 30 C.

Type species: *Sporolactobacillus inulinus* (Kitahara and Suzuki) Kitahara and Lai 1967, 197.

Genus III. **Clostridium** *Prazmowski 1880, 23*

Rods, usually motile by means of peritrichous flagella; occasionally non-motile. **Form ovoid to spherical spores that usually distend the bacilli.** Generally Gram-positive, at least in the early stages of growth.

Chemoorganotrophs. Some species are saccharolytic, some proteolytic, some both, some neither. Ferment sugars, polyalcohols, amino acids, organic acids, purines and other organic compounds. Some species fix nitrogen. **Do not reduce sulfate.**

Most strains are strictly anaerobic, although some may grow in the presence of air at atmospheric pressure. Catalase is usually not produced, but when it is, in small amounts.

Commonly found in soil, marine and fresh water sediments, and in the intestinal tract of man and other animals.

The G + C content of the DNA ranges from 23–43 moles %.

Type species: *Clostridium butyricum* Prazmowski 1880, 24.

Key to the groups of genus **Clostridium**

I. Spores subterminal.
    A. Gelatin not hydrolyzed.
        Group I. Species 1–11
    B. Gelatin hydrolyzed.
        Group II. Species 12–31
II. Spores terminal.
    A. Gelatin not hydrolyzed.
        Group III. Species 32–50
    B. Gelatin hydrolyzed.
        Group IV. Species 51–56
III. Species with special growth requirements.
        Group V. Species 57–61

TABLE 15.9

Distinguishing characteristics of species of **Clostridium** *Group I*

Spores subterminal, gelatin not hydrolyzed

	Maltose	Mannose	Raffinose	Lactose	Ribose	Starch	Dulcitol	Sorbitol	Nitrate
1. *C. butyricum*	+	+	+	+	+	+	−	−	d
2. *C. beijerinckii*	+	+	+	+	−	−	−	−	−
3. *C. oroticum*	+	+	+	+	+	−	+	−	+
4. *C. rectum*	+	+	+	+	−	−	−	−	+
5. *C. paraperfringens*	+	+	−	+	V[a]	+	−	−	V
6. *C. rubrum*	+	+	+	+	V	+	−	V	−
7. *C. fallax*	+	+	−	−	d	+	−	d	−
8. *C. pasteurianum*	+	+	+	−	−	−	−	d	−
9. *C. sticklandii*	+	−	−	−	+	−	−	−	−
10. *C. tyrobutyricum*	−	+	−	−	−	−	−	−	−
11. *C. propionicum*	−	−	−	−	−	−	−	−	−

[a] V = variable.

TABLE 15.10

Carbohydrates fermented by species of genus **Clostridium** *Group I*

Adonitol, cellulose and erythritol not fermented.

	1. *C. butyricum*	2. *C. beijerinckii*	3. *C. oroticum*	4. *C. rectum*	5. *C. paraperfringens*	6. *C. rubrum*	7. *C. fallax*	8. *C. pasteurianum*	9. *C. sticklandii*	10. *C. tyrobutyricum*	11. *C. propionicum*
Amygdalin	d	−	−	−	−	+	−	−	−	−	−
Arabinose	d	+	+	−	−	+	−	+	−	−	−
Cellobiose	+	+	+	+	+	+	−	−	−	−	−
Dulcitol	−	−	+	−	−	−	−	−	−	−	−
Esculin	d	−	−	−	−	+	+	−	−	−	−
Fructose	+	+	+	+	+	+	+	+	−	+	−
Galactose	+	+	+	+	+	+	+	+	−	−	−
Glucose	+	+	+	+	+	+	+	+	w	+	−
Glycerol	d	−	−	−	−	−	−	+	−	−	−
Glycogen	+	−	−	−	+	+	−	−	−	−	−
Inositol	d	+	+	−	−	−	−	+	−	−	−
Inulin	d	+	+	−	−	+	−	+	−	−	−
Lactose	+	+	+	+	+	+	−	−	−	−	−
Maltose	+	+	+	+	+	+	+	+	w	−	−
Mannitol	d	−	−	−	−	d	−	+	−	+	−
Mannose	+	+	+	+	+	+	+	+	−	+	−
Melezitose	−	+	+	−	−	+	−	+	−	−	−
Melibiose	+	+	−	−	−	−	−	−	−	−	−
Raffinose	+	+	+	+	−	+	−	+	−	−	−
Rhamnose	d	−	+	−	−	d	−	−	−	−	−
Ribose	+	−	+	−	d	d	d	d	w	−	−
Salicin	+	−	+		+	+	−	d	−	−	−
Sorbitol	−	−	−	−	−	d	d	+	−	−	−
Sorbose	−	−	−	−	−	−	d	+	−	−	−
Starch	+	−	−	−	+	d	+	−	−	−	−
Sucrose	+	+	+	+	+	+	−	+	−	−	−
Trehalose	+	+	+	−	d	d	−	+	−	−	−
Xylose	+	+	+	−	−	+	+	−	−	+	−

TABLE 15.11

Biochemical reactions of species of genus **Clostridium** *Group I*

Gelatin and casein are not hydrolyzed, indole or acetylmethylcarbinol not produced, urease or lipase not formed, toxin not produced, not pathogenic for laboratory animals.

	1. *C. butyricum*	2. *C. beijerinckii*	3. *C. oroticum*	4. *C. rectum*	5. *C. paraperfringens*	6. *C. rubrum*	7. *C. fallax*	8. *C. pasteurianum*	9. *C. sticklandii*	10. *C. tyrobutyricum*	11. *C. propionicum*
H_2S produced	−	−	−	+	+	−	+	−	−	−	+
Lecithinase on egg-yolk agar	−	−	−	−	+	−	−	−	−	−	−
Blood agar hemolyzed	−	+	−	−	d	−	−	−	−	−	−
Nitrate reduced	d	−	+	+	d	−	−	−	−	−	−

TABLE 15.12

Distinguishing characteristics of species of genus **Clostridium** *Group II*
Spores subterminal, gelatin hydrolyzed

	Casein digested	Lecithinase	Indole	Lipase	Glucose	Mannose	Maltose	Lactose	Salicin
12. *C. ghoni*	+	+	+	+	−	−	−	−	−
13. *C. bifermentans*[a]	+	+	+	−	+	V[g]	+	−	−
14. *C. sordellii*[a]	+	+	+	−	+	−	+	−	−
15. *C. lituseburense*	+	+	−	−	+	−	+	−	−
16. *C. limosum*[b]	+	+	−	−	−	−	−	−	−
17. *C. subterminale*[b, c]	+	V	−	−	−	−	−	−	−
18. *C. mangenotii*	+	−	+	−	−	−	−	−	−
19. *C. sporogenes*[d]	+	−	−	+	+	−	V	−	d
20. *C. botulinum*									
Types A, B, C, D, F[d]	+	−	−	+	+	−	V	−	V
Types B, C, D, E, F[d]	−	V	−	+	+	+	V	−	−
Type G[d]	+	−	−	−	−	−	−	−	−
21. *C. plagarum*	+	−	−	−	+	+	+	+	−
22. *C. acetobutylicum*	d	−	−	−	+	+	+	+	+
23. *C. histolyticum*[c]	+	−	−	−	−	−	−	−	−
24. *C. aurantibutyricum*	−	−	−	+	+	+	+	+	+
25. *C. novyi*									
Type A	−	+	−	+	+	−	d	−	−
Type B	V	+	−	−	+	d	+	−	−
Type C	−	−	d	−	+	−	−	−	−
26. *C. perfringens*	−	+	−	−	+	+	+	+	V
27. *C. haemolyticum*	−	+	+	−	+	V	−	−	−
28. *C. felsineum*[e]	−	−	−	−	+	+	V	+	+
29. *C. chauvoei*	−	−	−	−	+	+	+	+	−
30. *C. septicum*[e]	−	−	−	−	+	+	+	+	V
31. *C. difficile*[f]	−	−	−	−	+	+	−	−	d

[a] *C. sordellii* produces urease; *C. bifermentans* does not.
[b] *C. limosum* produces collagenase; *C. subterminale* does not.
[c] *C. histolyticum* is aerotolerant; *C. subterminale* is not.
[d] Toxin neutralization tests are necessary for identification.
[e] *C. felsineum* ferments sucrose; *C. septicum* does not.
[f] *C. difficile* hydrolyzes gelatin very slowly.
[g] V = variable.

TABLE 15.13

Carbohydrates fermented by species of genus **Clostridium** *Group II*

	12. *C. ghoni*	13. *C. bifermentans*	14. *C. sordellii*	15. *C. lituseburense*	16. *C. limosum*	17. *C. subterminale*	18. *C. mangenolii*	19. *C. sporogenes*	20. *C. botulinum* A, B, C, D, F	B, C, D, E, F	G	21. *C. plagarum*	22. *C. acetobutylicum*	23. *C. histolyticum*	24. *C. aurantibutyricum*	25. *C. novyi*, Type A	Type B	Type C	26. *C. perfringens*	27. *C. haemolyticum*	28. *C. felsineum*	29. *C. chauvoei*	30. *C. septicum*	31. *C. difficile*
Amygdalin	−	−	−	−	−	−	−	−	−	d	−	−	d	−	−	−	−	−	−	−	d	−	−	−
Arabinose	−	−	d	−	−	−	−	−	−	d	−	−	+	−	+	d	−	−	−	−	+	−	−	d
Cellobiose	−	−	−	−	−	−	−	−	−	−	−	+	+	−	+	−	−	−	d	−	+	d	+	d
Dulcitol	−	−	−	−	−	−	−	−	−	−	−	−	−	−	−	−	−	−	−	−	+	d	−	−
Esculin	−	−	−	−	−	−	−	−	−	−	−	−	d	−	−	−	−	−	−	−	d	−	d	−
Fructose	−	d	d	+	−	−	−	+	+	+	−	+	+	−	+	d	−	−	+	+	+	+	+	+
Galactose	−	−	−	−	−	−	−	−	−	d	−	+	+	−	+	−	−	−	+	−	+	+	d	−
Glucose	w	+	+	+	−	−	−	+	+	+	−	+	+	−	+	+	+	w	+	+	+	+	+	+
Glycerol	−	d	d	−	−	−	−	d	d	d	−	d	−	−	−	d	−	−	d	d	−	−	−	−
Glycogen	−	−	−	−	−	−	−	−	−	−	−	+	+	−	+	−	−	−	d	−	d	−	−	−
Inositol	−	−	−	−	−	−	−	−	−	d	−	−	d	−	−	d	+	w	+	d	−	−	−	d
Inulin	−	−	−	−	−	−	−	−	−	−	−	−	d	−	−	−	−	−	d	−	d	−	−	−
Lactose	−	−	−	−	−	−	−	−	−	−	−	+	+	−	+	−	−	−	+	−	+	+	+	−
Maltose	w	+	+	+	−	−	−	d	d	d	−	+	+	−	+	d	+	−	+	−	d	+	+	−
Mannitol	−	−	−	−	−	−	−	−	−	−	−	−	d	−	−	−	−	−	−	−	−	−	−	+
Mannose	−	d	−	d	−	−	−	−	−	+	−	+	+	−	+	−	d	−	+	−	+	+	+	+
Melezitose	−	−	−	−	−	−	−	−	−	−	−	−	−	−	−	−	−	−	−	d	−	−	−	d
Melibiose	−	−	−	−	−	−	−	−	−	d	−	−	−	−	−	−	−	−	−	−	−	−	−	−
Raffinose	−	−	d	−	−	−	−	−	−	−	−	−	d	−	+	d	−	−	d	−	d	−	−	−
Rhamnose	−	−	−	−	−	−	−	d	−	−	−	−	d	−	−	−	−	−	−	−	+	−	−	−
Ribose	−	d	d	d	−	−	−	−	−	d	−	−	−	−	−	d	d	w	d	−	−	+	d	d
Salicin	−	−	−	−	−	−	−	d	d	−	−	−	+	−	+	−	−	−	d	−	+	−	d	d
Sorbitol	−	d	−	−	−	−	−	d	−	d	−	−	d	−	−	−	−	−	−	−	−	−	−	d
Sorbose	−	−	−	−	−	−	−	−	−	−	−	−	d	−	+	−	−	−	−	−	−	−	−	−
Starch	−	−	−	−	−	−	−	−	−	d	−	d	+	−	+	−	−	−	+	−	d	−	−	−
Sucrose	−	−	−	+	−	−	−	d	d	d	−	+	+	−	+	d	−	−	+	−	+	+	−	−
Trehalose	−	−	−	−	−	−	−	d	−	d	−	−	+	−	−	−	−	−	d	−	−	−	d	d
Xylose	−	d	d	−	−	−	−	−	−	−	−	−	+	−	+	−	−	d	−	−	+	−	−	+

TABLE 15.14

Biochemical reactions of species of genus **Clostridium** *Group II*

	12. *C. ghoni*	13. *C. bifermentans*	14. *C. sordellii*	15. *C. lituseburense*	16. *C. limosum*	17. *C. subterminale*	18. *C. mangenoli*	19. *C. sporogenes*	20. *C. botulinum* A, B, C, D, F	B, C, D, E, F	G	21. *C. plagarum*	22. *C. acetobutylicum*	23. *C. histolyticum*	24. *C. aurantibutyricum*	25. *C. novyi*, type A	Type B	Type C	26. *C. perfringens*	27. *C. hamolyticum*	28. *C. felsineum*	29. *C. chauvoei*	30. *C. septicum*	31. *C. difficile*
H_2S produced	+	d	d	−	+	+	+	+	+	−	+	−	+	d	−	−	−	−	−	−	−	−	−	−
Urease	−	−	+	−	−	−	−	−	−	−	−	−	−	−	−	−	−	−	d	−	−	−	−	−
Gelatin liquefied	+	+	+	+	+	+	+	+	+	+	+	+	+	+	+	+	+	d	+	+	+	+	+	+
Casein hydrolyzed	+	+	+	+	+	+	+	+	+	−	+	+	d	+	−	−	d	−	d	−	−	−	−	−
Nitrates reduced	−	d	d	−	+	−	−	−	−	−	−	+	−	−	+	d	d	−	d	−	−	+	d	−
Acetylmethylcarbinol produced	−	−	−	−	−	−	−	−	−	−	−	−	+	−	−	−	−	−	d	−	+	−	−	−
Toxin produced	−	−	+	−	+	−	−	−	+	+	+	−	−	+	−	+	+	−	+	+	−	w	+	w
Pathogenic for laboratory animals	−	−	+	−	v	−	−	−	+	+	+	−	−	+	−	+	+	−	+	+	−	+	+	+
Blood hemolyzed	−	v	v	+	+	+	−	+	+	+	−	−	−	+	+	+	+	d	d	+	+	+	+	−

TABLE 15.15

Distinguishing characteristics of species of genus **Clostridium** *Group III*
Spores terminal, gelatin not liquefied

	Motility	Indole	Glucose	Mannitol	Lactose	Ribose	Trehalose	Maltose	Mannose	Aero-tolerance
32. *C. sphenoides*	+	+	+	+	V[a]	V	V	+	+	−
33. *C. indolis*	+	+	+	d	+	−	−	+	d	−
34. *C. scatologenes*	+	+	+	−	−	+	−	−	+	−
35. *C. malenominatum*	+	+	−	−	−	−	−	−	−	−
36. *C. tertium*	+	−	+	+	+	+	+	+	+	+
37. *C. sartagoformum*	+	−	+	+	+	V	+	+	+	−
38. *C. cellobioparum*	+	−	+	w	w	+	−	+	+	−
39. *C. thermosaccharolyticum*	+	−	+	−	+	d	d	+	+	−
40. *C. pseudotetanicum*	+	−	+	−	+	+	+	+	+	−
41. *C. carnis*	+	−	+	−	d	d	−	+	+	+
42. *C. paraputrificum*	+	−	+	−	+	V	−	+	+	−
43. *C. aminovalericum*	+	−	+	−	−	−	−	−	−	−
44. *C. glycolicum*	+	−	+	−	−	−	−	−	−	−
45. *C. sporosphaeroides*	−	−	−	−	−	−	−	−	−	−
46. *C. cochlearium*	+	−	−	−	−	−	−	−	−	−
47. *C. ramosum*	−	−	+	V	+	V	+	+	+	−
48. *C. innocuum*	−	−	+	+	−	d	+	−	+	−
49. *C. barkeri*	−	−	+	+	−	+	−	−	−	−
50. *C. perenne*	−	−	+	−	+	+	−	+	+	−

[a] V = variable; w = weak.

TABLE 15.16

Carbohydrates fermented by species of genus **Clostridium** *Group III*

	32. *C. sphenoides*	33. *C. indolis*	34. *C. scatologenes*	35. *C. malenominatum*	36. *C. tertium*	37. *C. sartagoformum*	38. *C. cellobioparum*	39. *C. thermosaccharolyticum*	40. *C. pseudotetanicum*	41. *C. carnis*	42. *C. paraputrificum*	43. *C. aminovalericum*	44. *C. glycolicum*	45. *C. sporosphaeroides*	46. *C. cochlearium*	47. *C. ramosum*	48. *C. innocuum*	49. *C. barkeri*	50. *C. perenne*
Adonitol	−	−	−	−	−	−	−	−	−	−	−	−	−	−	−	−	−	−	−
Amygdalin	−	d	−	−	+	−	−	+	+	d	d	−	−	−	−	+	−	−	−
Arabinose	d	d	+	−	d	−	+	+	+	−	d	−	−	−	−	d	d	−	−
Cellobiose	+	+	−	−	+	+	+	+	+	+	d	w	−	−	−	+	+	−	+
Cellulose	−	−	−	−	−	−	+	−	−	−	−	−	−	−	−	−	−	−	−
Dulcitol	−	−	−	−	d	−	+	−	−	−	−	−	d	−	−	d	−	−	−
Esculin	d	+	−	−	d	−	−	d	−	−	d	−	−	−	−	+	−	−	−
Erythritol	−	−	−	−	−	−	−	−	+	−	−	−	−	−	−	−	−	−	−
Fructose	+	+	+	−	+	+	+	+	+	d	+	−	+	−	−	+	+	+	+
Galactose	d	+	d	−	+	+	+	+	+	+	+	−	−	−	−	+	+	−	+
Glucose	+	+	+	−	+	+	+	+	+	+	+	+	+	−	−	+	+	+	+
Glycerol	d	−	+	−	−	−	w	−	+	−	−	−	−	−	−	−	−	−	−
Glycogen	d	−	−	−	+	+	−	+	+	−	+	−	−	−	−	−	−	−	−
Inositol	d	−	−	−	−	−	−	−	−	−	−	−	−	−	−	−	−	−	−
Inulin	−	−	−	−	d	−	−	−	+	−	d	−	−	−	−	−	+	−	−
Lactose	d	+	−	−	+	+	w	+	+	d	+	−	−	−	−	+	−	−	+
Maltose	+	+	−	−	+	+	+	+	+	+	+	−	−	−	−	+	−	−	+
Mannitol	+	d	−	−	+	+	w	−	−	−	−	−	−	−	−	d	+	+	−
Mannose	+	d	+	−	+	+	+	+	+	+	+	−	−	−	−	+	+	−	+
Melezitose	d	d	−	−	d	d	−	−	−	−	−	−	−	−	−	−	−	−	−
Melibiose	−	−	−	−	+	+	+	−	+	−	−	−	−	−	−	+	+	−	−
Raffinose	+	+	−	−	d	−	+	d	+	d	−	−	−	−	−	+	−	−	−
Rhamnose	+	d	+	−	d	−	+	−	−	−	d	−	−	−	−	d	−	−	−
Ribose	d	d	+	−	+	d	+	d	+	d	d	−	−	−	−	d	d	+	+
Salicin	+	+	−	−	+	+	+	+	+	+	+	−	−	−	−	+	+	−	+
Sorbitol	−	−	−	−	d	−	+	−	−	−	d	−	+	−	−	−	−	+	−
Sorbose	d	−	−	−	d	−	−	d	−	−	d	−	−	−	−	−	d	−	−
Starch	−	−	−	−	+	+	+	+	+	−	+	−	−	−	−	−	−	−	−
Sucrose	+	+	−	−	+	+	−	+	+	+	+	−	−	−	−	+	d	−	+
Trehalose	d	+	−	−	+	+	−	d	+	−	d	−	−	−	−	+	+	−	−
Xylose	+	+	+	−	+	+	+	+	+	−	d	−	+	−	−	d	d	+	−

TABLE 15.17

Biochemical reactions of species of genus **Clostridium** *Group III*

Gelatin not liquefied, casein not hydrolyzed. Urease and lipase not produced.

	32. *C. sphenoides*	33. *C. indolis*	34. *C. scatologenes*	35. *C. malenominatum*	36. *C. tertium*	37. *C. sartagoformum*	38. *C. cellobioparum*	39. *C. thermosaccharolyticum*	40. *C. pseudotetanicum*	41. *C. carnis*	42. *C. paraputrificum*	43. *C. aminovalericum*	44. *C. glycolicum*	45. *C. sporosphaeroides*	46. *C. cochlearium*	47. *C. ramosum*	48. *C. innocuum*	49. *C. barkeri*	50. *C. perenne*
Lecithinase produced	−	−	−	−	−	−	−	−	−	−	−	−	−	−	−	−	−	−	d
H_2S produced	+	+	+	+	−	−	−	−	−	−	−	−	+	+	d	−	−	−	w
Nitrates reduced	d	+	−	+	d	−	−	−	−	−	d	d	−	−	−	−	−	−	+
Acetylmethylcarbinol produced	−	+	−	−	−	−	d	−	−	−	−	−	−	−	−	d	−	+	−
Toxin produced	−	−	−	−	−	−	−	−	−	+	−	−	−	−	−	−	−	−	−
Pathogenic for laboratory animals	−	−	−	−	−	−	−	−	−	+	−	−	−	−	−	−	−	−	−
Blood hemolyzed	−	−	w	−	−	−	−	−	−	+	−	w	−	−	−	−	+	−	−

TABLE 15.18

Distinguishing and other characteristics of species of genus **Clostridium** *Group IV*

Spores terminal, gelatin liquefied

Lipases, urease and acetylmethylcarbinol not produced; nitrates not reduced.

	51. *C. cadaveris*	52. *C. lentoputrescens*	53. *C. putrificum*	54. *C. oceanicum*	55. *C. tetani*	56. *C. putrefaciens*
Milk digested	+	d	+	+	d	−
Casein hydrolyzed	+	+	+	+	−	−
Indole produced	+	+	−	−	d	−
Glucose fermented[a]	+	−	+	+	−	+
Maltose fermented	−	−	−	+	−	−
Spore shape[b]	O	S	O	O	S	OS
Grows at 0 C	−	−	−	−	−	+
H_2S produced	+	+	+	+	d	−
Toxin formed	−	−	−	−	+	−
Pathogenic to laboratory animals	−	−	−	−	+	−
Blood hemolyzed	w	−	+	+	+	+
Lecithinase produced	−	−	−	+	−	−

[a] Other carbohydrates listed in Table 15.15 are not fermented unless mentioned in descriptions.
[b] O = oval; S = spherical.

Genus IV. Desulfotomaculum *Campbell and Postgate 1965, 361*

Straight or curved rods, 0.3–1.5 by 3–6 μm, with rounded ends, usually single but sometimes in chains. **Motile with peritrichous flagella. Spores oval to round, terminal to subterminal, causing slight swelling of the cells. Gram-negative.** Produce black colonies in lactate-sulfate agar containing ferrous salts.

Chemoorganotrophs; **metabolism respiratory. Sulfates, sulfites and reducible sulfur compounds act as electron acceptors and are reduced to H_2S.** Lactate and pyruvate are universal electron donors; limited range of substrates utilized, rarely includes carbohydrates; acetate not oxidized. **Oxidation of organic substrates incomplete leading to formation of acetate or homologue and CO_2. Cells contain a cytochrome of the protoheme class.** Specialized media containing a reducible sulfur compound and organic growth factors are required for growth (Adams and Postgate, 1959; Campbell *et al.*, 1957; Postgate and Campbell, 1963).

Strict anaerobes.

Temperature optimum 35–55 C; maximum 70 C. Some strains grow at 30 C.

Common inhibitants of soil, fresh water, geothermal regions, certain spoiled foods, intestines of insects, and in rumen content. None has been reported to be pathogenic to man, guinea pig, mouse, rat or rabbit.

The G + C content of the DNA ranges from **41.7–45.5 moles %** (buoyant density).

Type species: *Desulfotomaculum nigrificans* (Werkman and Weaver) Campbell and Postgate 1965, 360.

Key to the species of genus Desulfotomaculum

I. Thermophilic, optimum temperature 55 C.
   1. *D. nigrificans*

II. Mesophilic, optimum temperature 30–37 C
   A. Pyruvate supports growth with or without sulfate.
      2. *D. ruminis*
   B. Pyruvate supports growth only with sulfate.
      3. *D. orientis*

Genus V. **Sporosarcina** *Kluyver and van Niel 1936, 401*

Cells spherical, occurring in regular tetrads or packets. May be motile or non-motile. **Endospores formed.** Gram-positive.

Chemoorganotrophs; energy-yielding metabolism strictly respiratory, oxygen being the terminal electron acceptor. Growth reported only in complex media; minimal nutritional requirements unknown.

Strict aerobes.

Type species: *Sporosarcina ureae* (Beijerinck) Kluyver and van Niel 1936, 401.

GENUS OF UNCERTAIN AFFILIATION

Genus **Oscillospira** *Chatton and Pérard 1913, 1159*

Large rods or filaments 3–6 μm in diameter, divided by closely spaced cross walls into numerous disc-shaped cells. Reproduction by transverse fission. Motile by means of numerous lateral flagella. Endospores may be formed. Gram-negative.

Growth in pure culture has not been reported. Exposure to air abolishes motility, thus suggesting the organisms are anaerobic.

Occur in the alimentary tract of herbivorous animals.

Type species (monotype): *Oscillospira guilliermondi* Chatton and Pérard 1913, 1159.

PART 16

GRAM-POSITIVE, ASPOROGENOUS, ROD-SHAPED BACTERIA*

Key to the Genera of Part 16

I. Rods or filaments which do not divide by visible cross-walls into disc-shaped cells.
   A. Catalase-negative, non-motile (with rare exceptions)
      1. Not pathogenic to mice.
         Family I. *Lactobacillaceae,* Genus I. *Lactobacillus,* p. 576†
      2. Pathogenic to mice.
         Genus *Erysipelothrix,* p. 597
   B. Catalase-positive, motile at 20–25 C.
         Genus *Listeria,* p. 593

II. Rods or filaments divided by visible cross-walls into disc-shaped cells.
         Genus *Caryophanon,* 598

FAMILY I. LACTOBACILLACEAE WINSLOW, BROADHURST, BUCHANAN, KRUMWIEDE, ROGERS AND SMITH 1917, 561

Straight or curved rods usually occurring singly or in chains. Non-motile, rare strains motile. Gram-positive. Anaerobic or facultative. Complex organic nutritional requirements. Highly saccharoclastic. At least half the end product carbon from carbohydrate metabolism is lactate. Lactate is not attacked anaerobically. Catalase negative (no porphyrin in apoprotein). Benzidine reaction negative. Pathogenicity unusual. Found in fermenting animal and plant products where carbohydrates are available; also found in the mouth, vagina and intestinal tracts of various warm-blooded animals, including man.

Genus I. Lactobacillus *Beijerinck 1901, 212. Nom. cons.* Opin. 38, Jud. Comm. 1971, 104

Rods, varying from long and slender to short coccobacilli. Chain formation common, particularly in later logarithmic phase of growth. Motility unusual; when present, by peritrichous flagella. Non-sporing. Gram-positive becoming Gram-negative with increasing age and acidity. Some strains exhibit bipolar bodies, internal granulations or a barred appearance with the Gram reaction or methylene blue stain.

Metabolism fermentative even though growth generally occurs in air; some are strict anaerobes on isolation.

Characteristically saccharoclastic; glucose fermented decreasing pH 1 or more units. At least

---

* **Editorial Note.** The division of genera between Part 16 and the first section of Part 17 is arbitrary and readers should consult both. Some genera have traditionally been associated with the actinomycetes and are so treated in this edition.

* All page numbers cited refer to *Bergey's Manual of Determinative Bacteriology,* 8th edition, The Williams & Wilkins Company, Baltimore, 1974.

half of end product carbon is lactate. Lactate is not fermented. Additional products may be acetate, formate, succinate, CO_2 or ethanol. Volatile acids with more than 2 carbon atoms are not produced.

Nitrate reduction highly unusual and then only where terminal pH is poised above 6.0. Gelatin not liquefied. Casein not digested but minute amounts of soluble nitrogen produced by some strains of some species. Indole and H_2S not produced.

Catalase and cytochrome negative (porphyrins absent); however, rare strains decompose peroxide by a pseudocatalase (porphyrin not present); benzidine reaction negative.

Pigment production rare; if present, yellow or orange to rust or brick red.

Complex nutritional requirements for amino acids, peptides, nucleic acid derivatives, vitamins, salts, fatty acids or fatty acid esters and fermentable carbohydrates. Nutritional requirements are generally characteristic for each species.

Surface growth on solid media often enhanced by anaerobiosis and 5–10% CO_2.

Temperature range 5–53 C; optimum generally 30–40 C.

Aciduric, optimal pH usually 5.5–5.8 or less, and generally growing at 5.0 or less; at neutral or initial alkaline reactions lag phase may be lengthened or total growth yield reduced.

Found in dairy products and effluents, grain and meat products, water, sewage, beer, wine, fruits and fruit juices, pickled vegetables, sourdough and mash; also parasitic in the mouth, intestinal tract and vagina of many homothermic animals including man. Pathogenicity is highly unusual.

The G + C content of the DNA ranges from 34.7 ± 1.4–53.4 ± 0.5 moles % (buoyant density).

Type species: *Lactobacillus delbrueckii* (Leichmann) Beijerinck 1901, 229. Opin. 38, Jud. Comm. 1971, 104.

Key to the species of genus Lactobacillus

Essential key characteristics

Type of lactic acid fermentation
Gas from glucose or gluconate
Fermentation of ribose
Growth at 45 and 15 C
Fermentation of malate
Requirement for hiochic (D-mevalonic) acid
The optical activity of the lactic acid produced is used as a primary key characteristic for some groups of species. It is an expensive and time-consuming test and can be substituted by the sugar fermentation patterns found in Table 16.1.

I. Homofermentative. Lactic acid is the major product from glucose (generally 85% or more).
   A. No gas from glucose or gluconate; ribose not fermented, thiamine not required; aldolase activity; D- or L- or DL-lactic acid produced; G + C = 34.7–50.8%; generally grows at 45 C or higher, generally not at 20 C and not at 15 C. Colonies normally rough becoming smooth and compact in the presence of Tween 80 or sodium oleate (Rogosa and Mitchell, 1950).
      1. Produce D(−)-lactic acid.*
         1. *L. delbrueckii*
         2. *L. leichmannii*
         3. *L. jensenii*
         4. *L. lactis*
         5. *L. bulgaricus*
      2. Produce DL-lactic acid.
         6. *L. helveticus*
         7. *L. acidophilus*
      3. Produce mainly L(+)- and very small amounts of D(−)-lactic acid.†

* See also No. 27, *L. vitulinus*.
† See also No. 26, *L. ruminus*.

8. *L. salivarius*

B. No gas from glucose. Gas from gluconate. Ribose, when fermented, yields lactic and acetic acids without gas. Thiamine not required. Aldolase activity. Glucose 6-phosphate dehydrogenase and inducible 6-phosphogluconate dehydrogenase activity. Variable growth at 45 C. Growth at 15 C. G + C = 45–46.4%.

1. Produce chiefly L(+)-lactic acid. Ribose fermented.

9a–9c. *L. casei* subspp. *casei*, *rhamnosus*, *alactosus*
9d. *L. casei* subsp. *tolerans*
10. *L. xylosus*

2. Produce DL-lactic acid. Ribose fermented.

9e. *L. casei* subsp. *pseudoplantarum*
11. *L. plantarum*
12. *L. curvatus*

3. Produce DL-lactic acid. Ribose fermentation equivocal.

13a. *L. coryniformis* subsp. *coryniformis*

4. Produce D(−)-lactic acid. Ribose not fermented. Favorable pH about 5.

13b. *L. coryniformis* subsp. *torquens*
14. *L. homohiochii*

II. Heterofermentative, producing about 50% of end products from glucose as lactic acid, with considerable amounts of CO_2, acetic acid and ethanol; mannitol from fructose.

A. Gas from glucose and gluconate. Ribose fermented to lactic and acetic acids without gas. Thiamine required. DL-Lactic acid produced. Aldolase is not demonstrable. Glucose 6-phosphate dehydrogenase activity and equal or very much less 6-phosphogluconate dehydrogenase activity are demonstrable.

1. Growth at 45 C; no growth at 15 C.

15. *L. fermentum*

2. Variable growth at 45 and 15 C; no growth at 48 C.

16. *L. cellobiosus*

3. Generally growth at 15 C; no growth at 45 C.

17. *L. brevis*
18. *L. buchneri*
19. *L. viridescens*
20. *L. coprophilus*

III. Heterofermentative. DL-Lactic acid, CO_2, acetate produced. Less well known organisms. Slower growing or "indifferent" toward most carbohydrates. Nutrition, enzymology, G + C, cell wall chemistry, gluconate and ribose fermentations generally not studied.

Malate, fructose, gluconate, ribose and xylose fermented; variable or less fermentation of glucose, galactose, sucrose and maltose; ethanol tolerance 15–18%; optimal initial pH range 4.5–5.5; optimal temperature 30–35 C.

21. *L. hilgardii*

Glucose and fructose fermented; malate, and generally no other carbohydrate fermented; ethanol tolerance >20%; optimal initial pH range 4.5–5.5; optimal temperature 25–30 C.

22. *L. trichodes*

Fructose fermented, weak malate activity and weak or negative glucose fermentation, gluconate fermented, ribose not fermented, no other carbohydrates fermented; ethanol tolerance 15%; optimal initial pH range 4.5–5.5; optimal temperature 25–30 C.

23. *L. fructivorans*

Arabinose fermented. Glucose, fructose and galactose may be fermented. Malate and other carbohydrates not fermented. Ethanol tolerance 15%; optimal initial pH range 5–7; optimal temperature <30 C and good growth at 10 C.

24. *L. desidiosus*

Hiochic acid (D-mevalonic acid) essential for growth. Favorable pH about 5.

25. *L. heterohiochii*

TABLE 16.1—*Carbohydrate reactions of the species of genus* **Lactobacillus**

	1. *L. delbrueckii*	2. *L. leichmannii*	3. *L. jensenii*	4. *L. lactis*	5. *L. bulgaricus*	6. *L. helveticus*	7. *L. acidophilus*	8. *L. salivarius*	9a–c. *L. casei*	9d. *L. casei* subsp. *tolerans*	9e. *L. casei* subsp. *pseudoplantarum*	10. *L. xylosus*	11. *L. plantarum*	12. *L. curvatus*	13a. *L. coryniformis* subsp. *coryniformis*	13b. *L. coryniformis* subsp. *torquens*	14. *L. homohiochii*	15. *L. fermentum*	16. *L. cellobiosus*	17. *L. brevis*	18. *L. buchneri*	19. *L. viridescens*	20. *L. coprophilus*	21. *L. hilgardii*	22. *L. trichodes*	23. *L. fructivorans*	24. *L. desidiosus*	25. *L. heterohiochii*
Amygdalin	−[a]	+	+	−	−	−	+	−	+	−	+	+	+	−	−	−	−	−	+	−	−	−	0	−	0	−	0	−
Arabinose	−	−	−	−	−	−	−	−	−	−	−	−	d*	−	−	−	−	d	+	+	+	−	+	−	−	−	+	−
Cellobiose	−	+	+	−	−	−	+	−	+	+	+	+	+	+	−	−	−	−	+	−	−	−	+	−	0	−	−	−
Fructose	+	+	+	+	+	⋕	+	+	+	+	+	+	+	+	+	+	+	+	+	+	+	+	0	+	+	+	⋕	+
Galactose	⋕	−w	+	+	+	+	+	+	+	+	+	+	+	+	+	+	−	+	+	⋕	⋕	−	0	d	−	−	⋕	−
Glucose (acid)	+	+	+	+	+	+	+	+	+	+	+	+	+	+	+	+	+	+	+	+	+	+	+	+	+	⋕	⋕	+
Glucose (gas)	−	−	−	−	−	−	−	−	−	−	−	−	−	−	−	−	−	+	+	+	+	+	+	+	+	+	+	+
Gluconate	−	−	−	−	−	−	−	−	+	+	+	+	+	+	+	+	+	+	+	+	+	+	+	+	0	+	0	+
Lactose	−	+	−	+	+	+	+	+	*	+	+	*	+	d	−	−	−	+	⋕	⋕	⋕	−	−	d	−	−	−	−
Maltose	d	+	+	+	−	+	+	+	(d)	−	+	+	+	+	+	+	+	+	+	+	+	+	+	+	d	⋕	−	−
Mannitol	−	−	−	−	−	−	−	+	+	−	+	+	+	−	+	+	+	−	−	(⋕)	(⋕)	−	−	−	−	−	−	−
Mannose	+	+	+	+	−	⋕	+	+	+	+	+	+	+	+	+	d	+	w	⋕	−	−	+	0	−	−	−	−	−
Melezitose	−	−	−	−	−	−	−	−	+	−	+	−	d	−	−	−	0	−	−	−	+	−	−	−	0	−	0	−
Melibiose	−	−	−	−	−	−	d	+	−	−	−	−	+	−	d	−	0	+	+	+	+	−	−	−	0	−	0	−
Raffinose	−	−	−	−	−	−	d	+	−	−	−	−	+	−	d	−	−	+	+	⋕	⋕	−	−	d	−	−	−	−
Rhamnose	−	−	−	−	−	−	−	*	*	−	−	−	−	−	+	−	−	−	−	−	−	−	0	−	0	−	−	−
Ribose	−	−	−	−	−	−	−	−	+	+	+	+	+	+	−	−	−	+	+	+	+	+	+	+	0	−	0	+
Salicin	−	+	+	+	−	−	+	*	+	+	+	+	+	+	d	−	+	−	⋕	−	−	−	d	−	0	−	−	−
Sorbitol	−	−	−	−	−	−	−	+	+	+	+	−	+	−	d	−	−	−	−	−	−	−	−	−	0	−	−	−
Sucrose	+	+	+	+	−	−	+	+	(d)	−	+	+	+	−	+	−	−	+	+	d	d	d	+	d	d	⋕	−	−
Trehalose	−	+	+	+	−	*	+	+	+	−	+	+	+	−	−	−	−	d	+	−	−	d	0	−	0	−	−	−
Xylose	−	−	−	−	−	−	−	−	−	−	−	+	d*	−	−	−	−	d	d	d	d	−	+	+	−	−	0	−
Esculin	−	+	+	⋕	−	−	+	*	+	−	+	−	+	+	d	−	0	−	+	d	d	−	+	−	0	−	0	0

[a] The symbols used are: + = positive reaction by 90% or more strains; d = some strains +, others − (about 89–11% positive); − = negative reaction by most strains (90% or more); ( ) = delayed reaction; w = weak reaction; ⋕ = weak, slow or negative; * = see text; 0 = not tested. Combinations of symbols, e.g. (−) = rarely positive, and then slowly; −w = negative or weak reaction.

Negative reactions by all strains from I-erythritol, D-erythrose, D-fucose, D-glucoheptose, α-methyl-D-xyloside and perseitol. All negative from α-methyl-D-glucoside except that Nos. 12, 17, 19 and 20 are variable; negative from α-methyl-D-mannoside except No. 12 variable. Except for some strains of *L. casei*, adonitol and sorbose are not fermented; glycerol, inositol, inulin and starch, dextrin, dulcitol are very rarely fermented. Tagatose (100%), turanose (100%) and D-lyxose (95%) fermented by *L. casei;* turanose (100%) by *L. plantarum* and variably by *L. curvatus;* these three substrates are not fermented by any other species. D-Arabinitol is fermented by *L. salivarius* subsp. *salivarius* and some strains of *L. plantarum.* L-Arabitol is fermented by *L. salivarius* subsp. *salicinius*, xylitol is fermented by *L. salivarius* subsp. *salivarius;* these pentitols are not fermented by any other species.

TABLE 16.1A

Differentiating characteristics of sub-species of **L. casei**

Subspecies	Growth at 45 C.	Lactose fermented	Rhamnose fermented	Group antigen	Type or Neotype, ATCC
a. *casei*	−	+	−	B or C	393 (Hansen and Lessel, 1971)
b. *alactosus*	−	−	−	B or C	27216 (Mills and Lessel, 1973)
c. *rhamnosus*	+	+	+	C	7469 (Hansen, 1968)

GENERA OF UNCERTAIN AFFILIATION

Genus **Listeria** *Pirie 1940, 383 Nom. cons.* Opin. 12. Jud. Comm. 1954, 151

Small, coccoid, Gram-positive rods with a **tendency to produce chains** of three to five or more cells, and to produce, in the rough state, elongated to filamentous forms. Smears from 18–24 hr old colonies show typical diphtheroid palisade arrangement with a few V or Y forms. Do not produce spores or capsules, not acid-fast. Gram-positive but in older cultures may be Gram-negative. **Motile** by peritrichous flagella when grown at 20–25 C. At 37 C only a few flagella found, usually one polar flagellum, occasionally two to four, sometimes none.

Acid, but no gas, from glucose and several other carbohydrates (Tables 16.2, 16.3). **Esculin is hydrolyzed.** Indole not produced, urea not hydrolyzed. Gelatin, casein and milk are not hydrolyzed. **Usually catalase positive.**

Aerobic to microaerophilic (growth enhanced under reduced oxygen and 5–10% CO_2). Growth occurs between 4 and 38 C, particularly when the medium contains small amounts of glucose.

Several antigenic types can be distinguished by means of O- and H-antigens. These share some partial O-antigenic factors with other Gram-positive bacteria but not with *Erysipelothrix*. Sensitive to many antibiotics, but not to polymyxin B.

Found in the feces of animals and man, on vegetation and in silage. Parasitic on poikilothermic and warm-blooded animals, including man. The G + C content of the DNA is 38 moles %, except *L. denitrificans* which is 56 moles %.

Type species: *Listeria monocytogenes* (Murray, Webb and Swann) Pirie 1940, 383.

Note: Stuart and Welshimer (*International Journal of Systemic Bacteriolology 24:* 177–185, 1974) suggest a new monospecific genus *Murraya* to include *Murraya grayi* subsp. *grayi* (here *Listeria grayi*) and *M. grayi* subsp. *murrayi* (here *L. murrayi*).

TABLE 16.2

Characteristics differentiating the species of genus **Listeria** *and its differentiation from genus* **Erysipelothrix**

	1. *L. monocytogenes*	2. *L. denitrificans*	3. *L. grayi*	4. *L. murrayi*	*Erysipelothrix*
Mannitol acid	−	−	+	+	−
$NO_3 \rightarrow NO_2$	−	+	−	+	−
Motile	+	+	+	+	−
Esculin hydrolyzed	+	+	+	+	−
Catalase	+	+[a]	+	+	−
Beta-Hemolysis	+	−	−	−	−
Glucose fermented	+	+	+	+	−
Glucose oxidized	+	+	+	+	+
Pathogenic to mice	+	+?	−	−	+
G + C ratio	38	56 ± 1	38 ± 1	38 ± 1	36[b]

[a] Based on ATCC strain, a subculture of what is presumably the only isolate.

[b] Reported by Stuart and Welshimer (1974); Flossmann and Erler (1972) reported a G + C ratio of 38-40 moles %.

TABLE 16.3

Acid production from carbohydrates by species of genus **Listeria**[a]

	1. *L. monocytogenes*	2. *L. denitrificans*	3. *L. grayi*	4. *L. murrayi*
L-Arabinose	–[b]	+	–	–
D-Galactose	d	+	+	(+)
Glycogen	–	+	–	–
Lactose	(d)	+	+	+
Mannitol	–	–	+	+
Melezitose	d	–	–	–
Melibiose	–	(+)	–	–
Rhamnose	d	–	–	d
Sucrose	(d)	+	–	–
Xylose	–	+	–	–

[a] All species produce acid but no gas in 24–48 hrs from amygdalin, esculin, cellobiose, dextrin, fructose, glucose, maltose, mannose, salicin, starch and trehalose. No acid in 21 days from adonitol, dulcitol, erythritol, inositol, inulin or raffinose.

[b] Key: + = acid produced 24–48 hrs (90% or more strains); – = no acid produced 21 days (90% or more strains); d = some strains positive, some negative; (+) = acid produced slowly (3–7 days); (d) = some strains produce acid slowly (3–7 days), other strains negative.

Genus Erysipelothrix *Rosenbach 1909, 367*

Rod-shaped organisms with a **tendency to form long filaments.** The filaments may also thicken and show characteristic granules. **Nonmotile.** Neither capsules nor spores produced. **Gram-positive,** older cultures having a tendency to become Gram-negative.

Acid but no gas from glucose and from certain other carbohydrates. **Catalase negative. Esculin not hydrolyzed.** Produce alpha-hemolysis but **no beta-hemolysis** on blood agar.

Aerobic; grows better in an atmosphere with reduced oxygen and containing 5–10% CO_2.

Temperature range 16–41 C. Most rapid growth at 37 C, maximum cell crop at 33 C.

Parasitic on mammals, birds and fish.

Type species: *Erysipelothrix rhusiopathiae* (Migula) Buchanan 1918, 55.

Genus Caryophanon *Peshkoff 1939, 241*

Large rods or filaments up to 3 μm in diameter, divided by closely spaced cross walls into numerous disc-shaped cells which are less than 1 μm long when growth is active. In living organisms cross walls show as dark lines, partly complete, partly developing by ingrowth from the external wall. Motile by lateral flagella. Spores not produced. Gram-positive.

Chemoorganotrophs; energy-yielding metabolism strictly respiratory, using oxygen as the terminal electron acceptor.

Strict aerobes.

Type species: *Caryophanon latum* Peshkoff 1939, 241 (Peshkoff 1948, 1004).

PART 17

ACTINOMYCETES AND RELATED ORGANISMS*

Note: This Part is divided into three groups of different rank. The Coryneform Group of Bacteria, an informal collection of genera at the family level, contains organisms which produce irregular rod-shaped cells and grow best in the presence of oxygen. The family *Propionibacteriaceae* consists of irregular rods or filaments which are anaerobic or aerotolerant. Lastly, the order *Actinomycetales* contains organisms which are aerobic (except most members of the family *Actinomycetaceae*) and form filaments with a tendency to branch. This latter characteristic is not always readily apparent and the reader is advised to observe the organisms during all stages of cultural development or in host tissue.

CORYNEFORM GROUP OF BACTERIA

The family Corynebacteriaceae Lehmann and Neumann 1907, 500 was recognized in the 6th edition of the *Manual* and included the genera *Corynebacterium*, *Listeria* and *Erysipelothrix;* in the 7th Edition *Microbacterium*, *Cellulomonas* and *Arthrobacter* were added.

The family Brevibacteriaceae Breed 1953, 13, with the genera *Brevibacterium* and *Kurthia*, was first included in the 7th edition of the *Manual*; since the type genus, *Brevibacterium*, is here considered a *genus incertae sedis*, the family has no standing in nomenclature and leaves *Kurthia* unplaced.

The coryneform bacteria present a number of unresolved problems in taxonomy and classification.

The rather widespread use of the terms "coryneform" or "diphtheroid" to describe any non-sporing Gram-positive rod of irregular outline has given rise to considerable confusion as to which organism or groups of organisms should properly constitute the genus *Corynebacterium*. Davis and Newton (1969) examined 70 strains of coryneform bacteria bearing the generic names *Corynebacterium, Arthrobacter, Brevibacterium, Microbacterium, Cellulomonas, Kurthia, Mycobacterium, Nocardia, Jensenia, Listeria* and *Erysipelothrix*. If one considers organisms which, at least at some stage in their growth cycle on artificial media, produce "straight to slightly curved rods with irregularly staining segments, and sometimes granules" (see generic description of *Corynebacterium*), one may add to the list *Bifidobacterium*, *Actinomyces*, *Propionibacterium* and the more recently described genera such as *Rothia*, *Arachnia* and *Bacterionema*. Other evidence suggests *Mycobacteriaceae* and *Nocardiaceae* might be grouped with Corynebacteriaceae and *Propionibacteriaceae;* however, experts are unable to reconcile certain differences of opinion.

Some of the above mentioned families are recognized in this part of the *Manual*. However, it seems unwise to recognize Corynebacteriaceae and instead a working concept is presented of a "Coryneform Group of Bacteria" to include *Corynebacterium*, *Arthrobacter* (with the related genera *Brevibacterium* and *Microbacterium* as *genera incertae sedis*), *Cellulomonas* and tentatively *Kurthia*.

The taxonomic position of the plant pathogenic corynebacteria poses a considerable problem. In purely morphological terms these organisms fit the generic description and, in at least one instance (*C. fascians*), are serologically related to

* **Editorial Note.** The division of genera between Part 16 and the first section of Part 17 is arbitrary and readers should consult both. Some genera have traditionally been associated with the actinomycetes and are so treated in this edition.

the human and animal pathogens. However, some are motile and all of those so far examined (except *C. fascians*) show differences in cell wall composition of a kind usually found between genera in other groups (see Table 17.1). Moreover, the DNA base ratios of the plant pathogens appear to be distinctly higher than those of the animal and human pathogens. To meet these difficulties the species have been described in three separate sections:

I. Human and Animal Parasites and Pathogens.

II. Plant Pathogenic Corynebacteria.

III. Non-pathogenic Corynebacteria.

Characteristics common to the species in each section are listed at the beginning of each section.

The anaerobic diphtheroids, represented by strains previously described as *C. acnes*, *C. avidum* and *C. granulosum*, are excluded from the genus because they are anaerobic organisms producing propionic acid as the major end-product of fermentation and because their cell wall composition differs markedly from that of the type species (Table 17.1, group 4). These strains are now included in the genus *Propionibacterium*.

Further Comments

The position of *C. pyogenes* and *C. haemolyticum* is discussed in Addendum I of Section I. These strains represent a well defined group and are a common and well recognized cause of pyogenic infections in domestic animals. However, they bear little similarity to other animal pathogenic corynebacteria; they are generally catalase negative, produce a soluble hemolysin, have a very different cell wall composition (Table 17.1, group 3) and share a cell wall polysaccharide antigen with streptococci of Lancefield group G. Their final classification must await further investigation.

A number of species of human and animal origin have been listed in Addendum II—*Species incertae sedis*. Descriptions of these are generally inadequate.

The reported G + C contents of the DNA of corynebacteria cover a wide range—48 to about 70%. However, except for a figure of 67.5% for

TABLE 17.1

Distinctive cell wall components of corynebacteria

Groups of strains	Sugars	Cell wall components (amino acids in peptidoglycan)	References
1. *C. diphtheriae* and most other human and animal pathogenic corynebacteria	Arabinose, galactose (mannose, glucose may also be present)	ala, glu, *meso*-DAP[a]	Cummins and Harris, 1958; Cummins, 1962
2. Plant pathogenic corynebacteria:			
(i) *C. fascians*	Arabinose, galactose	ala, glu, *meso*-DAP	Cummins, 1962
(ii) *C. poinsettiae*[b]	Rhamnose, mannose, galactose glucuronic acid	gly, glu, homoser, ala, orn	Perkins, 1967 Perkins, 1970
C. betae	Rhamnose, mannose, fucose glucuronic acid		
(iii) *C. tritici*	Xylose, glucose, mannose	gly, glu, DAB[c], ala	Perkins, 1968
C. insidiosum[b]	Rhamnose, mannose, galactose and fucose		Perkins, 1971[d]
C. sepedonicum[b]	Rhamnose, mannose, galactose		
3. *C. pyogenes*	Rhamnose, glucose	ala, glu, lys	Barksdale *et al.*, 1957
C. haemolyticum	Rhamnose		
4. *Propionibacterium acnes* (*C. acnes*) and "anaerobic diphtheroids"	Galactose and/or glucose and/or mannose (not all strains have all three sugars)	ala, glu, gly, L-DAP (occasional strain has *meso*-DAP)	Johnson and Cummins, 1972

[a] DAP = α,ϵ, diaminopimelic acid.

[b] Results of cell wall analysis of other strains do not always agree, see, e.g. Yamada and Komagata 1970.

[c] DAB = diaminobutyric acid.

[d] Personal communication.

C. pseudotuberculosis (Sukapure *et al.*, 1970), determinations by Tm and buoyant density would give values for the human and animal pathogens in a much narrower range—57–60%. Since the DNA from both *C. pyogenes* and *P. acnes* have G + C contents of 58–59%, it would appear that determinations of DNA base ratios are of limited value in the taxonomy of coryneform organisms.

No diagnostic key has been attempted, but a table (Table 17.2) of some differential features has been included. Descriptions of the species include information on the cell wall composition for the principal sugars identified in cell wall hydrolysates, and for the diamino acid of the peptidoglycan, since these are the features most readily used in taxonomy. The detailed structure of the peptidoglycan has been determined for a number of coryneform organisms (Fiedler *et al.*, 1970) and it appears to be of the same type for all human and animal pathogens described in Section I.

Similarities between *Corynebacterium* (Section I), *Mycobacterium* and *Nocardia*. The human and animal parasites in the genus *Corynebacterium* as here defined show a number of features which indicate a rather close relationship to the mycobacteria and nocardias. Prominent among these features are the possession of arabinose and galactose as common cell wall sugars, and the presence of a common cell wall antigen (Cummins, 1962). More recent work on the lipids of mycobacteria, corynebacteria and nocardias has reemphasized this relationship, and it now appears probable that these three groups of organisms are characterized by the presence of α-branched, β-hydroxy acids (mycolic acids) of the general type

$$\begin{array}{c} \quad\ \ \beta \qquad\ \ \alpha \\ R_1\text{—CH—CH COOH.} \\ \quad\ \ | \qquad\ \ | \\ \quad\ \ \text{OH} \qquad R_2 \end{array}$$

In mycobacteria the principal mycolic acid found in human strains of *M. tuberculosis*, for example, has the crude formula $C_{78}H_{152}O_3$, while others have similar structures, with numbers of carbon atoms between C_{79} and C_{85} (Etemadi and Lederer, 1965). In strains of *Nocardia*, similar acids are found, except that the number of carbon atoms varies from C_{48}–C_{58} (Bordet *et al.*, 1965).

In corynebacteria, corynemycolic acids occur, with a still smaller number of carbon atoms, C_{32}–C_{36}; these acids have been isolated from *C. diphtheriae*, *C. pseudotuberculosis* (*C. ovis*) and *C.*

TABLE 17.2

Some distinctive features of species of genus **Corynebacterium** *parasitic on man and animals*

Species	Hemolysis	Fermentation of sucrose	Reduction of nitrate	Gelatinase	Urease	Other features
1. *C. diphtheriae*	+	−	+	−	−	Pathogenic: specific exotoxin: specific bacteriophages
2. *C. pseudotuberculosis*	+	d[a]	d[a]	d[a]	d[a]	Pathogenic: specific exotoxin
3. *C. xerosis*	−	+	+	−	−	Not pathogenic: barred morphology often predominant
4. *C. renale*	−	−	−	−	+	Pathogenic: causes pyelitis and cystitis in exptl. animals. Zone of clearing on milk agar: alkaline reaction and burgundy-color in litmus milk
5. *C. kutscheri*	−	+	−	−	+	Specific parasite of rats and mice
6. *C. pseudodiphtheriticum*	−	−	+	−	+	Short, regular rods: strongly Gram-positive; no acid produced from any carbohydrate
7. *C. equi*	−	−	+	−	−	Oval or coccoid forms on solid media: capsulated: salmon-pink pigment: no acid produced from any carbohydrate
8. *C. bovis*	−	−	−	−	+	Lipolytic for butterfat: will grow in media containing 9% NaCl: requires long chain fatty acids

[a] Variable results reported in the literature.

pseudodiphtheriticum (*C. hofmannii*) (Senn *et al.*, 1967; Lacave, Asselineau and Toubiana, 1967; Welby-Guisse, Lanéelle and Asselineau, 1970). Furthermore *C. diphtheriae* contains a toxic glycolipid which is a 6-6′ diester of trehalose, containing corynemycolic acid ($C_{32}H_{64}O_3$) and corynemycolenic acid ($C_{32}H_{62}O_3$) in equimolar ratios. This glycolipid is therefore a lower mycolic acid analogue of the cord factor of *M. tuberculosis*, which is trehalose-6-6′ dimycolate (see Senn *et al.*, 1967; Kato, 1970).

Mycolic acids have not been reported from any other coryneforms. In contrast, for example, in strains of *Propionibacterium* (including *P. acnes*) the principal fatty acids are C_{15} branched chain fatty acids (Moss and Cherry, 1968; Moss *et al.*, 1969). Etemadi (1963) specifically noted that C_{32} acids were not found in strain 936-B of *C. parvum*, here considered a synonym of *P. acnes*.

Synopsis of recognized genera of the **Coryneform** *Group*

I. Rods with irregularly stained segments and granules; frequently club-shaped; do not break up into coccal forms; aerobic or facultatively anaerobic; do not attack cellulose.
Genus I. *Corynebacterium*

II. Irregular rods which change to coccal forms during growth in complex media; strict aerobes; do not attack cellulose.
Genus II. *Arthrobacter*

III. Irregular rods which occasionally change into coccal forms with age; attack cellulose.
Genus III. *Cellulomonas*

IV. Regular, unbranched rods occurring in chains; rods fragment into shorter forms with age; strict aerobes; do not attack cellulose.
Genus IV. *Kurthia*

Genus I. **Corynebacterium** *Lehmann and Neumann 1896, 350*

Section I. Human and Animal Parasites and Pathogens

Straight to slightly curved rods with irregularly stained segments, and sometimes granules. Frequently show club-shaped swellings. Snapping division produces angular and palisade (picket fence) arrangements of cells. Generally non-motile. Gram-positive, although some species (e.g. *C. diphtheriae*) lose the stain easily, especially in old cultures, while others are more tenacious; granules, however, are strongly Gram-positive. Not acid-fast.

Cell wall composition is characterized by the presence of *meso*-diaminopimelic acid as the diamino acid of the peptidoglycan, and by a polysaccharide containing arabinose, galactose and often mannose (Table 17.1).

Chemoorganotrophs: Carbohydrate metabolism mixed fermentative and respiratory.

Do not produce soluble hemolysins but on solid media containing blood lysis may occur of the red cells in contact with the colonies. Some pathogenic species produce exotoxins.

Aerobic and facultatively anaerobic; grow best aerobically often with a surface pellicle.

Catalase positive.

Reported G + C content of the DNA varies from 52–68%, but, if determinations by Tm and buoyant density only are considered, most species are probably in the range 57–60%.

Probably widely distributed in nature.

Type species: *Corynebacterium diphtheriae* (Kruse) Lehmann and Neumann 1896, 350.

TABLE 17.3

Distinguishing features of **Gravis, Intermedius** *and* **Mitis** *strains of* **C. diphtheriae**

Character	Gravis	Intermedius	Mitis
Morphology	Short, irregular rods, few metachromatic granules	Long, barred forms, few granules	Long, curved pleomorphic rods, many granules
Colonies on blood-tellurite medium	Gray or black center, pale periphery; crenated edge with radial striations; 2–3 mm at 24 hr; colony brittle to touch	Small, grayish black, rough or smooth, less than 1 mm at 24 hr	Dark gray to black, smooth shiny, entire edge. 1–2 mm at 24 hr. Colony soft and smears easily.
Type of growth in broth	Surface pellicle, clear fluid, coarse deposit	Fine granular turbidity which settles later	Diffuse turbidity with deposit
Reversal of pH in broth	Early	Late	pH does not revert
Fermentation of starch	+	−	−

TABLE 17.4

Distinguishing features of **C. pyogenes** *and* **C. haemolyticum**[a]

Character	*C. pyogenes S*	*C. pyogenes L*	*C. haemolyticum*
Cell size (approx.), μm	0.2 by 2.0	0.2 by 2.0–10.0	0.2 by 2.0–10.0
Colony size (24 hr blood agar), mm	<1	1–2	1–2
Fermentation of:			
Xylose	+5/5	−8/8	−2/2
Inositol	−5/5	+7/8	−2/2
Starch	+4/5	−8/8	−2/2
Gelatin liquefaction	+5/5	−8/8	−2/2
Soluble hemolysin	+	−	−
Catalase production[b]	−	w	w
Cell wall sugars	Rhamnose,[c] glucose	Rhamnose	Rhamnose
Reaction of formamide extracts with antisera to:			
C. pyogenes S	++	−	−
C. haemolyticum	−	++	++

[a] Data from Barksdale *et al.*, 1957.

[b] For comments on catalase production in these organisms, see text: w = weak positive.

[c] Small amounts of mannose have also been found in the cell walls of some strains of *C. pyogenes* S.

Section II. Plant Pathogenic Corynebacteria

Morphologically similar to the human and animal pathogens but less pleomorphic and there is generally less evidence of snapping division. Stain unevenly (granular) and on glycerol media have fatty inclusions which stain with Sudan Black B. Intracellular starch is not accumulated. Some species motile.

Starch hydrolysis is weak or negative.

Nitrate reductase negative; coagulase negative; indole, M.R. and V.P. negative; H_2S or urease rarely produced. Strains examined are sensitive to triphenyltetrazolium chloride (Lovrekovitch and Klement, 1960).

Require organic growth factors.

Strict aerobes.

The G + C content of the DNA is probably in the range 65–75 moles %.

TABLE 17.5

Differential characters for plant pathogenic species of genus **Corynebacterium**[a]

	Group 1	Group 2				Group 3	Group 4		Group 5			
	1. *C. fascians*	2. *C. rathayi*	3. *C. agropyri*	4. *C. tritici*	5. *C. iranicum*	6. *C. sepedonicum*	7. *C. insidiosum*	8. *C. michiganense*	9. *C. flaccumfaciens*	10. *C. poinsettiae*	11. *C. betae*	12. *C. oortii*
Motility	−	−	−	+	·	−	−	−[b]	+	+	+	+
Growth slow[c]	+	+	+	+	+	+	+	+	−	−	−	−
Growth opaque	+	+	+	+	+	+	+	+	±	±	−	+
Growth at 37 C	+	−	−	−	−	−	−	(+)	+	+	+	+
Urease	+	−	·	−	·	−	−	−	−	−	−	−
Esculin hydrolyzed[d]	−	+	·	+	·	+	+	+	+	+	+	·
Medium yellow pigment	+	+	+	+	+	−	−	(+)	d	−	−	−
Pink pigment	−	−	−	−	−	−	−	(−)	(−)	+	−	−
Indigoidine produced	−	−	−	−	−	−	+	−	−	−	−	−
Causes hypertrophies of stems	+	−	−	−	−	−	−	−	−	−	−	−
Causes gumming of inflorescences	−	+	+	+	+	−	−	−	−	−	−	−
Causes wilts and/or leaf spots	−	−	−	−	−	+	+	+	+	+	+	+

[a] + All of the few strains examined reported to be positive; ± translucent at 3 days, becoming more opaque later; − all of the few strains examined reported to be negative; (+) most strains positive; (−) most strains negative; d strains differ.

[b] Some authors claim to have detected motility.

[c] Colonies on nutrient dextrose agar <1 mm diameter after 3 days.

[d] Carrier, 1963.

Section III. Non-pathogenic Corynebacteria

Organisms of Section III have been most commonly isolated from soils, water, air and contaminated bacteriological media or blood *etc.* They are often poorly described and some have been at first considered *Brevibacterium* or other species and later renamed as *Corynebacterium* species. Because of the heavy or sole reliance on morphological criteria, and because at some stage of growth there may be morphological similarities, the organisms of this group could be confused with members of at least 10 other genera (see *Brevibacterium* (page 625) and the introduction to the Coryneform Group of Bacteria (page 599). In most cases there are insufficient data to permit differentiation of species.

Genus II. Arthrobacter *Conn and Dimmick 1947, 300*

Cells which in complex media undergo a marked change in form during the growth cycle. Older cultures (generally 2–7 days) are composed entirely or largely of coccoid cells. In some strains the coccoid cells are uniform in size and spherical, and resemble micrococci; in others they are spherical to ovoid or slightly elongate. In some cultures larger coccoid cells some 2–4 times the size of the remainder may occur; these may predominate under some cultural conditions. On transfer to fresh complex medium growth occurs by enlargement (swelling) of the coccoid cells followed by elongation from one or occasionally two parts of the cell to give rods which usually have a diameter less than that of the enlarged coccoid cell (Plate 17.1, Figs. 1–4). In the larger coccoid cells outgrowth may occur at two, three or rarely four parts of the cell (Plate

17.1, Figs. 5–6). In both cases subsequent growth and division give rise to irregular rods which vary considerably in size and shape and include straight, bent and curved, wedge-shaped and club-shaped forms (Plate 17.1, Fig. 7). A proportion of the rods are arranged at an angle to each other giving V formations but more complex angular arrangements often occur. Post-fission outgrowths, usually from the proximal ends of one or both cells of a pair of rods (Plate 17.1, Fig. 8), and bud-like outgrowths from segments of septate rods (Fig. 9) especially in richer media, may give the appearance of rudimentary branching but true mycelia are not formed. As the exponential phase proceeds the rods become shorter and are eventually replaced by the coccoid cells characteristic of stationary phase cultures (Plate 17.1, Fig. 10). The coccoid cells are formed either by a gradual shortening of the rods at each successive division or, especially in richer media, by multiple fragmentation of larger rods (Plate 17.1, Fig. 11). The rods are non-motile or motile by one subpolar or a few lateral flagella. Do not form endospores.

Gram-positive. However, **the rods may be readily decolorized and may show only Gram-positive granules in otherwise Gram-negative cells.** Coccoid cells are Gram-positive but may be weakly so. **Not acid fast.**

The cell walls do not contain both meso-diaminopimelic acid and arabinose.

Chemoorganotrophs: **Metabolism respiratory, never fermentative.** Molecular oxygen is the terminal electron acceptor.

Little or no acid is produced from glucose in peptone medium. **Do not attack cellulose.** Catalase positive.

All species grow in a medium containing soil extract and yeast extract. Nutrient agar (peptone + meat extract) is suitable for laboratory strains which do not require *terregens* factor or vitamin B_{12} but may be inhibitory to newly isolated strains (Topping, 1937).

Strict aerobes.

Temperature optimum 20–30 C; most strains grow at 10 C but usually not at 37 C. **Do not survive heating at 63 C for 30 min in skim milk.** Grow best at a neutral to slightly alkaline pH.

The **G + C content** of the DNA of the species examined ranges from *ca.* **60–72 moles %** (T_m).

Type species: *Arthrobacter globiformis* (Conn) Conn and Dimmick 1947, 301.

TABLE 17.6

Some characteristics of species of genus **Arthrobacter**[a]

	A. globiformis	*A. simplex*	*A. tumescens*	*A. citreus*	*A. terregens*	*A. flavescens*	*A. duodocadis*
Cell walls contain[b]							
LL-DAP	·	+	+	·	·	·	·
Lysine	+	·	·	+	+	+	+
Glycine	·	+	+	·	+	+	+
Aspartic acid	·	·	·	·	+	+	+
Serine	·	·	·	·	·	·	+
Growth factor requirements[c]							
Biotin	d	−	−	+	+	+	−
Thiamine	−	−	+	+	+	+	+
Pantothenic acid	−	−	−	−	+	−	−
Nicotinic acid	−	−	−	+	−	−	−
B_{12}	−	−	−	−	−	−	+
Terregens factor[d]	−	−	−	+	+	+	−
Gelatin hydrolyzed	+	+	+	+	−	+	+
Starch hydrolyzed	d	−	d	−	−	+	−
$NO_3 \rightarrow NO_2$	d	+	d	+	+	+	+

[a] Characteristics of *A. globiformis* based on some 22 strains, of *A. tumescens* some 6 strains, others usually 1 strain, the type.

[b] +, present as a major component; '·', absent or only small amounts detected; d, some strains positive, some negative. Alanine and glutamic acid are major components of all species.

[c] Some species require amino acids in addition.

[d] Or other sideramines, or certain chelators.

TABLE 17.7

Characteristics of ideal phenotype of **Arthrobacter globiformis**[a]

These characteristics, when used with cell wall composition (Table 17.6), are of greatest value in recognition of strains of *A. globiformis*.

1. Vitamin requirement: None or biotin only	22
2. Inorganic nitrogen utilized as sole nitrogen source[b]	22
Utilizes as carbon + energy source:	
3. Glycine	22
4. Tyramine	22
5. Uric acid	22
6. p-Hydroxybenzoate and/or 3:4-dihydroxybenzoate	22
7. m-Hydroxybenzoate	20
8. D-Xylose	20
9. Melezitose	20
10. D-Glucuronate	19
11. α-D-Glucosamine and/or *N*-Acetylglucosamine	19
12. Meso-inositol	18
13. 2-Oxoglutarate	17

[a] The ideal phenotype is positive for all of these characteristics. The figures indicate the number of strains positive for each characteristic out of 22 tested (Owens and Keddie, unpublished).

Of the 22 strains tested, 3 deviated from the ideal phenotype in three characters, the remainder conformed exactly or at most deviated in two characters. Of more than 90 coryneform strains, other than *A. globiformis*, tested by Owens and Keddie more than 90% deviated in seven or more characters and none deviated in less than five.

[b] See Owens and Keddie, 1969, for a suitable method.

TABLE 17.8

Characteristics of ideal phenotype of **Arthrobacter tumescens**[a]

These characteristics, when used with cell wall composition (Table 17.6), are of greatest value in the recognition of strains of *A. tumescens*.

1. Vitamin requirement: Thiamine only	6
2. Inorganic nitrogen utilized as sole nitrogen source[b]	6
Utilizes as carbon + energy source:	
3. D-Mannose	6
4. Raffinose	6
5. α-D-Glucosamine	6
6. Acetate	6
7. Crotonate	6
8. Citrate	6
9. Meso-inositol	6
10. D-Alanine	6
11. Thymine and/or uracil	6
12. Suberate and/or azelate	5
13. D-Glucuronate	0
14. p-Hydroxybenzoate and/or 3:4-dihydroxybenzoate	0
15. L-Ornithine	0
16. D-Phenylalanine	0
17. Uric acid	0

[a] The ideal phenotype is positive for characteristics 1–12 and negative for 13–17. The figures indicate the number of strains positive for each characteristic out of six tested (Owens and Keddie, unpublished).

Of six strains tested, five showed perfect conformity with the ideal phenotype and one strain deviated in one character. Of strains other than *A. tumescens* tested, all deviated in six or more characters.

[b] See Owens and Keddie (1969) for method.

Genera incertae sedis

Genus A. **Brevibacterium** *Breed 1953, 13*

In the 7th edition of the *Manual* this genus was used as a repository for a number of little known or ill defined bacteria formerly in *Bacterium*, a genus recognized in the 6th edition but not in the 7th.

Brevibacterium and its type species *B. linens* were described as typically short, unbranched rods, reproducing by simple cell division, and placed in the family Brevibacteriaceae proposed by Breed in 1953. There was no mention of coryneform morphology.

Breed's description of *B. linens* must have been erroneous because present strains have the morphological diversity and certain other properties of the genus *Arthrobacter*.

Genus B. **Microbacterium** *Orla-Jensen 1919, 179*

Small diphtheroid rods with rounded ends; angular and palisade arrangements of cell masses are typical. Non-motile. Gram-positive. Good growth on surface of solid media in air. Weak acid production (chiefly L(+)-lactic acid) from fermented carbohydrates. Catalase positive. Optimal growth temperature about 30 C; limited or variable growth at 15 and 35 C; no growth at 10 and

40 C; survives 72 C for 15 min or more in skim milk. Found chiefly in dairy products and on dairy utensils.

Note: The species *Microbacterium thermosphactum*, one of those included in this genus in the *Manual*, has been transferred to a new genus *Brochothrix* as the type and only species (Sneath and Jones, *International Journal of Systematic Bacteriology* *26:* 102–104, 1976). The authors suggest that the genus be placed in the family *Lactobacillaceae* (Part 16).

Genus III. **Cellulomonas** *Bergey et al. 1923, 154, emend. mut. char. Clark 1952, 50*

In young cultures, irregular rods *ca.* 0.5 μm in diameter by 0.7–2.0 μm or more in length, which may be straight, angular or slightly curved and occasionally club-shaped or beaded; a proportion of the rods are arranged at an angle to each other giving V formations (Plate 17.1, Fig. 12). Occasionally cells may show rudimentary branching but true mycelia are not formed. As cultures age the rods become shorter but only a small proportion of coccoid cells is seen in cultures a week or more old. Motile by one (usually polar or subpolar) or a few lateral flagella, or non-motile. Do not form endospores. Gram-positive: the cells are readily decolorized and often a mixture of Gram-positive and Gram-negative rods is seen, sometimes with the latter predominating; often one or more Gram-positive granules are seen in otherwise Gram-negative cells. Not acid-fast.

Chemoorganotrophs: Metabolism primarily respiratory using molecular oxygen as terminal electron acceptor; most strains produce acid from glucose under both aerobic and anaerobic conditions (see *Further Comments*).

Catalase positive. **Cellulose is attacked by all strains recognized at present.**

The cell walls contain neither meso-diaminopimelic acid nor arabinose. The walls are typical of Gram-positive organisms; they contain alanine, glutamic acid and lysine (Keddie et al., 1966; Yamada and Komagata, 1970), one or more sugars may be present but galactose is absent (Keddie *et al.*, 1966). Recent work (Lamey and Keddie, unpublished) has shown that the diamino acid in several strains tested is ornithine and not lysine. The methods used by Keddie *et al.* (1966) and Yamada and Komagata (1970) did not allow these two amino acids to be distinguished.

Biotin and thiamine are the only exogenous organic growth factors required by strains now recognized (Keddie *et al.*, 1966) and when provided with these vitamins growth occurs in suitable mineral media with glucose as carbon + energy source and an ammonium salt (or nitrate for most strains) as nitrogen source (Owens and Keddie, 1969).

Aerobic; most strains also capable of anaerobic growth (see *Further Comments*).

Temperature optimum ca. 30 C. Do not survive heating at 63 C for 30 min in skim milk. Grow well at neutral pH.

The G + C content of the DNA ranges from 71.7–72.7 moles % (T_m) (Yamada and Komagata, 1970).

Type species: *Cellulomonas flavigena* (Kellerman and McBeth) Bergey *et al.* 1923, 165.

TABLE 17.9

Characters of greatest value in the recognition of strains of **Cellulomonas flavigena**[a]

	Ideal phenotype[b]
1. Biotin and thiamine required	+
2. Inorganic nitrogen utilized as sole nitrogen source[c]	+
3. Cellulose attacked	+
Utilization of:[d]	
4. D-Ribose	−
5. Acetate	+
6. Propionate and/or butyrate and/or pentoate	−
7. One or more of: malonate, succinate, glutarate, adipate, pimelate, suberate, azelate	−
8. One or more of: ethanol, propanol, butanol, pentanol, hexanol, heptanol, octanol, nonanol	−
9. *p*-Hydroxybenzoate and/or *m*-hydroxybenzoate and/or 3:4-dihydroxybenzoate	−
10. L-Asparagine	−
11. 1:4-Butanediamine	−

[a] Of more than 100 strains of coryneform bacteria tested by Owens and Keddie other than *C. flavigena*, including named strains of *Arthrobacter* and soil and herbage isolates, more than 95% deviated in six or more characters, and none deviated in less than three characters.

[b] + Indicates 7/7 strains positive; − indicates 7/7 negative (Owens and Keddie, unpublished).

[c] See Owens and Keddie (1969) for method.

[d] See page 620 of the *Manual* for method.

Genus IV. **Kurthia** *Trevisan 1885, 92. Nom cons.* Opin. 13, Jud. Comm. 1954, 152

In young cultures (18–24 hr): **regular, unbranched rods** with rounded ends **occurring in chains** which are often parallel (Plate 17.1, Figs. 13–14); the rods are *ca.* 0.8 μm in diameter and vary in length according to the stage of growth but are generally *ca.* 2–8 μm long. **Older cultures** (3–7 days) are **usually composed of coccoid cells** formed by fragmentation of the rods; in some strains short rods are the dominant forms in such cultures. The rods are motile by peritrichous flagella. Do not form endospores.

Gram-positive. Not acid fast.

Chemoorganotrophs: **Metabolism respiratory,** never fermentative.

Molecular oxygen is the terminal electron acceptor.

Catalase positive. Oxidase negative. Do not reduce nitrate.

Do not produce acid from carbohydrates or polyols; media become alkaline presumably from peptones.

Strict aerobes.

Temperature optimum 25–30 C; range *ca.* 5–37 C. Some strains grow at 45 C, none at 50 C. Grow well in presence of 4–6% NaCl.

Type species: *Kurthia zopfii* (Kurth) Trevisan 1885, 92.

FAMILY I. **PROPIONIBACTERIACEAE** DELWICHE 1957, 569

Gram-positive, non-spore-forming, anaerobic to aerotolerant, pleomorphic, branching or regular rods or filaments. Where saccharolytic, **major products include CO_2, propionic and acetic acids or mixtures of organic acids often including butyric, formic, lactic or other monocarboxylic acids. Some species are not saccharolytic.** Lactic acid alone or lactic with an equal or greater amount of acetic acid does not accumulate as the only major products. Growth often enhanced by CO_2. Inhabitants of the skin, respiratory and intestinal tracts of most animals, and of animal products. Several species are found in soft tissue infections.

Key to the genera of family **Propionibacteriaceae**

I. Not motile.
   Anaerobic to aerotolerant.
   Produce propionic acid and acetic acids and lesser amounts of isovaleric, formic, succinic and lactic acids.
   Genus I. *Propionibacterium.*

II. Motile or non-motile.
   Obligately anaerobic. Fermentative or non-fermentative.
   Fermentative species produce mixtures of organic acids, often including large amounts of butyric, acetic, formic and lactic or other mono-carboxylic acids.
   Genus II. *Eubacterium.*

Genus I. **Propionibacterium** *Orla-Jensen 1909, 337*

Gram-positive, non-spore-forming, nonmotile rods. Usually pleomorphic, diphtheroid or club-shaped with one end rounded and the other end tapered or pointed and stained less intensely. Cells of some cultures may be coccoid, elongate, bifid or even branched. Cells usually arranged in singles, pairs or V and Y configurations, short chains or clumps in "Chinese character" arrangement.

Chemoorganotrophs: Metabolize carbohydrates, peptone, pyruvate or lactate. **Fermentation products include combinations of propionic and acetic acids and frequently lesser amounts of isovaleric, formic, succinic or lactic acids and carbon dioxide. All species produce acid from glucose.**

Anaerobic to aerotolerant. Although most strains in this genus grow most rapidly under strictly anaerobic conditions, many strains grow well in deep tubes of peptone-yeast extract-glucose broth exposed to air when a large inoculum is used.

Tween 80 stimulates growth of most strains.

Growth most rapid at 30–37 C and pH near 7. Some strains grow at 25 and 45 C. Most strains grow in glucose broth with 20% bile or 6.5% NaCl.

Ammonia produced from proteinaceous material by most strains. Hippurate usually not hydrolyzed; neutral red usually not reduced.

Growth may be white, gray, pink, red, yellow or orange.

The G + C content of the DNA of the described species ranges from 59–66 moles % (T_m).

Found in dairy products or from the skin of man and intestinal tract of man and animals. Some species may be pathogenic.

Characteristics by which species in the genus can be differentiated are given in the key and in Tables 17.10, and 17.11. Other characteristics of the species in the genus are given in the text concerning each species. The drawings (Fig. 17.A) are composites from several cultures. Individual cultures usually show somewhat less variation in cellular morphology than is depicted. (Scale for drawings: 9/16 inch = 10 μm.)

Type species: *Propionibacterium freudenreichii* van Niel 1928, 162.

TABLE 17.10
Fermentation reactions of the species of genus **Propionibacterium**[a]

	1a. *P. freudenreichii* ss. *freudenreichii*	1b. *P. freudenreichii* ss. *globosum*	1c. *P. freudenreichii* ss. *shermanii*	2. *P. thoenii*	3. *P. acidipropionici*	4. *P. jensenii*	5. *P. avidum*	6. *P. acnes*	7. *P. lymphophilum*	8. *P. granulosum*
Products from peptone-yeast extract-glucose broth	PASfl	PAsl	PAS (fl)	PA(sl *fiv*)	PAs (l*fiv*)	PA(s*fiv*)	PA(s*fiv*)	PA (Ls*fiv*)	PASlf	PA(s*iv*)
Products from lactate	P^{-}	PA	PA	PA^{-}	PA	PA	PA^{s}	P^{-}	PA^{-}	PA^{s}
Acid produced from:										
Adonitol	v	·	$+^{w}$	$+^{w}$	w^{+}	$+^{w}$	v	$-^{+}$	+	−
Amygdalin	$-^{w}$	−	−	$+^{w}$	w^{-}	w^{-}	$-^{w}$	−	w^{-}	$-^{w}$
Arabinose	$+^{w}$	+	$+^{w}$	$-^{+}$	+	v	v	$-^{w}$	−	$-^{w}$
Cellobiose	w^{-}	w	$-^{+}$	w^{-}	$+^{w}$	d	$-^{w}$	$-^{+}$	−	−
Dulcitol	−	−	−	−	−	−	−	−	−	−
Erythritol	$+^{-}$	+	$+^{w}$	+	$+^{w}$	$+^{w}$	$+^{w}$	$-^{+}$	+	−
Esculin	−	−	−	+	w^{-}	$-^{w}$	−	−	−	−
Esculin hydrol.	+	+	+	+	+	+	+	−	−	−
Fructose	+	w	$+^{w}$	+	+	+	+	$+^{-}$	$+^{-}$	+
Galactose	$+^{w}$	+	+	+	+	+	+	$+^{-}$	$-^{w}$	w^{-}
Glycerol	$+^{w}$	+	+	+	+	+	+	$+^{-}$	−	+
Glycogen	w^{-}	−	−	$-^{+}$	v	$-^{w}$	−	−	−	−
Inositol	v	w	$+^{-}$	$+^{-}$	+	$+^{-}$	$-^{+}$	$-^{w}$	v	−
Inulin	−	−	−	−	$-^{w}$	$-^{w}$	−	−	−	−
Lactose	−	+	+	d	+	v	v	−	−	−
Maltose	w^{-}	w	w	$+^{w}$	+	$+^{-}$	+	$-^{w}$	+	$+^{-}$
Mannitol	−	−	−	−	+	+	$-^{+}$	d	−	$-^{+}$
Mannose	$+^{w}$	+	+	+	+	+	+	$+^{-}$	−	$+^{w}$
Melezitose	−	w	w^{-}	v	$+^{w}$	$+^{w}$	$+^{-}$	−	−	v
Melibiose	−	w	d	$+^{-}$	v	$+^{w}$	$-^{+}$	−	−	v
Raffinose	−	−	−	$-^{+}$	d	$+^{w}$	$-^{+}$	−	−	d
Rhamnose	−	−	−	−	$+^{w}$	−	−	−	−	−
Ribose	$+^{w}$	+	v	+	$+^{w}$	$+^{w}$	$+^{w}$	$-^{+}$	+	v
Salicin	$+^{w}$	−	$-^{w}$	$+^{w}$	+	$+^{-}$	v	−	−	$-^{w}$
Sorbitol	−	−	$-^{+}$	v	+	$-^{+}$	$-^{+}$	$-^{+}$	−	−
Sorbose	−	w	w^{-}	−	$+^{-}$	$-^{w}$	$-^{w}$	$-^{w}$	−	$-^{w}$
Starch	−	w	w^{-}	$+^{w}$	$+^{w}$	v	$-^{+}$	$-^{w}$	w^{-}	$-^{w}$
Starch hydrol.	−	·	−	$-^{+}$	−	−	−	−	d	−
Sucrose	w^{-}	−	$-^{+}$	$+^{w}$	+	$+^{w}$	+	$-^{+}$	d	$+^{w}$
Trehalose	−	−	−	+	$+^{w}$	$+^{w}$	$+^{w}$	$-^{+}$	−	w^{-}
Xylose	$-^{w}$	w	w^{-}	$-^{w}$	$+^{w}$	v	$-^{+}$	$-^{w}$	−	$-^{w}$

[a] Products from PYG broth: Capital letters indicate >1 meq acid/100 ml broth; small letters <1 meq/100 ml. A, acetic; F, formic; *iv*, isovaleric; L, lactic; P, propionic; S, succinic; ( ) products of occasional strains.

+, strong acid, i.e. pH 5.5 or below in 90–100% of strains; w, weak acid, i.e. pH 5.5–6.0 in 90–100% of strains. (Uninoculated xylose and arabinose, under CO_2 (prereduced media), are often pH 5.9. For cultures in these media, we consider pH 5.4–5.7 weak acid and pH below 5.4 strong acid); −, acid not produced, i.e. pH 6.0 in 90–100% of strains; d, reaction positive 40–60% strains; v, reaction variable within a strain. When any of the above used as superscripts, reaction of 10–40% of strains; ·, not tested.

Acid products determined by procedures given in Holdeman and Moore, 1972.

Key for species of genus **Propionibacterium**

Key characteristics	
Hydrolysis of esculin	adonitol
Hydrolysis of gelatin	erythritol
Acid from:	maltose
mannitol	ribose
lactose	Nitrate reduction
trehalose	Production of indole

I. Esculin hydrolyzed.
  A. Gelatin not completely hydrolyzed.
    1. Mannitol not acid.
      a. Nitrate reduced.
        i. Lactose not acid.
          1a. *P. freudenreichii* subsp. *freudenreichii*
        ii. Lactose acid.
          1b. *P. freudenreichii* subsp. *globosum*
      b. Nitrate not reduced.
        i. Trehalose not acid.
          1c. *P. freudenreichii* subsp. *shermanii*
        ii. Trehalose acid.
          2. *P. thoenii*
    2. Mannitol acid.
      a. Nitrate reduced.
          3. *P. acidi-propionici*
      b. Nitrate not reduced.
          4. *P. jensenii*
  B. Gelatin completely hydrolyzed.
          5. *P. avidum*

II. Esculin not hydrolyzed.
  A. Indole produced.
          6. *P. acnes*
  B. Indole not produced.
    1. Adonitol, erythritol, maltose and ribose acid.
          7. *P. lymphophilum*
    2. Adonitol, erythritol, maltose or ribose not acid.
      a. Nitrate reduced.
          6. *P. acnes**
      b. Nitrate not reduced.
          8. *P. granulosum*

* For strains that are catalase negative and produce acid in sucrose also see *Arachnia propionica.*

TABLE 17.11

Biochemical reactions of the species of genus **Propionibacterium**[a]

	1a. *P. freudenreichii* subsp. *freudenreichii*	1b. *P. freudenreichii* subsp. *globosum*	1c. *P. freudenreichii* subsp. *shermanii*	2. *P. thoenii*	3. *P. acidipropionici*	4. *P. jensenii*	5. *P. avidum*	6. *P. acnes*	7. *P. lymphophilum*	8. *P. granulosum*
Gelatin	$-$	$-$	$-^{w}$	$-$	$-^{w}$	$-$	d	d^{-}	w	w^{-}
Milk	$-$	c	c^{-}	$-^{c}$	c^{-}	$-^{c}$	d^{c-}	d^{c-}	$-$	c^{-}
Indole produced	$-$	$-$	$-$	$-$	$-$	$-$	$-$	$+^{-}$	$-$	$-$
Nitrate reduced	$+$	$+$	$-$	$-$	$+$	$-$	$-$	$+^{-}$	v	$-$
Catalase	w	$+$	$+$	$+$	v	$+^{-}$	$+$	$+^{-}$	v	$+$
Gas	2^{+}	$-$	$+^{-}$	$+^{-}$	$+^{2}$	$-^{+}$	$-^{4}$	$-^{2}$	$+^{-}$	v
Acetoin	$-$	·	$-$	$-^{+}$	$-^{w}$	$-^{w}$	$-$	$-$	$-$	$-$
Growth in 20% bile	$-$	·	$+^{2}$	$+^{-}$	4^{-}	v	4^{+}	$-^{2}$	$+$	v
Lecithinase (EYA)	$-$	·	$-$	$-$	$-$	$-$	$-$	$-$	$-$	$-$
Lipase (EYA)	$-$	·	$-$	$-$	$-$	$-$	$-^{+}$	$-$	$-$	$-$
Hemolysis	b^{w}	·	$-^{b}$	b^{w}	$-^{w}$	$-^{b}$	b^{-}	$-^{b}$	$-$	$-^{b}$
Propionate ← threonine	$-$	$-$	$-$	$-$	$-$	$-$	$-$	$-^{+}$	$-$	$-$

[a] +, positive reaction—90–100% of strains; −, negative reaction—90–100% of strains; v, reaction variable within strains; hemolysis—b = beta; w = weak; gas and growth—relative amounts indicated by +, 2, 3, 4; s = stimulated. Milk—c, curd, d, digested; gelatin—w, liquefaction of chilled cultures in less than half the time of the uninoculated controls.

Any of the above used as superscripts indicates reaction of 10–40% of strains; · not tested.

Reactions determined by procedures given in Holdeman and Moore (1972).

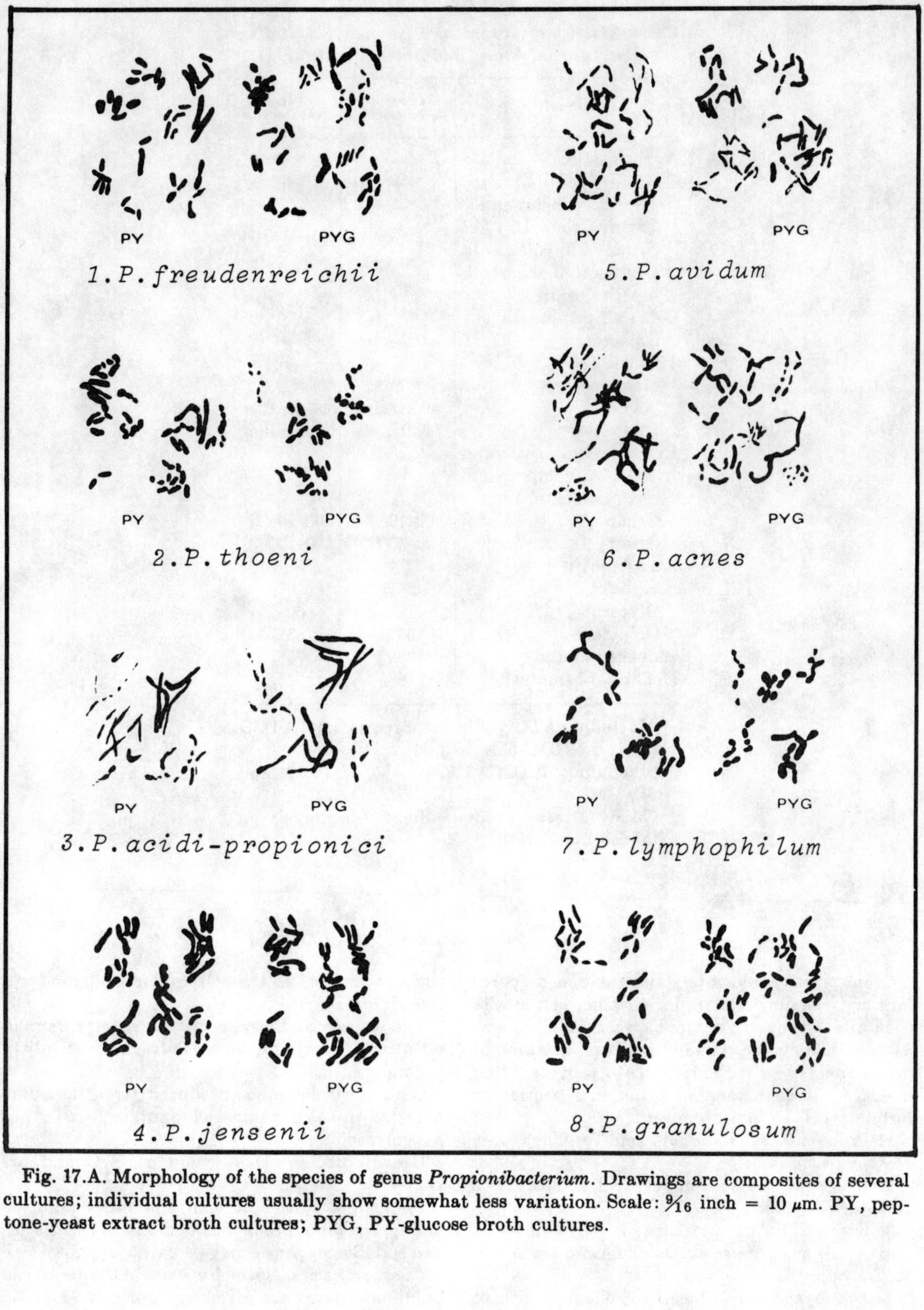

Fig. 17.A. Morphology of the species of genus *Propionibacterium*. Drawings are composites of several cultures; individual cultures usually show somewhat less variation. Scale: $\frac{9}{16}$ inch = 10 μm. PY, peptone-yeast extract broth cultures; PYG, PY-glucose broth cultures.

TABLE 17.12

Characteristics of groups of **P. acnes** *(adapted in part from Johnson and Cummins, 1972)*

Characteristic	Group I (49 strains[a])	Group II (19 strains[b])
Agglutination		
Group I cell wall antiserum absorbed with Group II	+	−
Group II cell wall antiserum absorbed with Group I	−	+
Cell wall composition		
Galactose	+	−
Glucose	+	+
Mannose	±	±
DAP-LL	49 strains	14 strains
DAP-meso	0 strains	2 strains
Relative similarity value (DNA competition experiments)		
Group I	90–100%	90–100%
Group II	90–100%	90–100%
Acid from:		
Sorbitol	86%	0%
Glycerol	72%	60%
Galactose	65%	53%
Indole produced	98%	75%
Definite β-hemolysis[c]	18%	0%

[a] Including ATCC 6919 (neotype strain), ATCC 11827 and NCDC 554.

[b] Including ATCC 11828, ATCC 11829 and NCDC 605.

[c] Many strains produce slight β-hemolysis on extended incubation.

Genus II. **Eubacterium** *Prévot 1938, 294*

Gram-positive, obligately anaerobic, non-spore-forming rods, uniform or pleomorphic, non-motile or motile. Chemoorganotrophs, saccharoclastic or non-saccharoclastic. Usually produce mixtures of organic acids from carbohydrates or peptone, often including large amounts of butyric, acetic, or formic acids. Do not produce:

1. Propionic as a major acid product (see *Propionibacterium*).

2. Lactic as the sole major acid product (see *Lactobacillus*).

3. Succinic (in the presence of CO_2) and lactic with small amounts of acetic or formic acids (see *Actinomyces*).

4. Acetic and lactic (acetic ≧ lactic), with or without formic, as the sole major acid products (see *Bifidobacterium*).

The species and strains within species vary in sensitivity to oxygen; some can be cultured only in pre-reduced media.

Catalase usually not produced (trace amounts detected in some strains); hippurate usually not hydrolyzed.

Growth usually most rapid at 37 C and pH near 7.

Found in cavities of man and other animals, animal and plant products, infections of soft tissue, and soil. Some species may be pathogenic.

Characteristics by which species in the genus can be differentiated are given in Tables 17.13–17.17.

The information on the volatile acids produced by many of the species in Table 17.17 was obtained by Duclaux distillation, a method not as reliable as the chromatographic method used for the species in Tables 17.13 and 17.15. Other characteristics are given in the text. Cellular morphology is illustrated in Figures 17.B and 17.C. Methods presented in Holdeman and Moore (1972) were used in the cultural, biochemical and chromatographic characterization of all strains listed by us.

All descriptions are based on strains in which no spores have been detected. Heat-resistant spores, often difficult to detect, have been found in organisms similar to some described species, especially motile or filamentous species; these strains are members of the genus *Clostridium*.

Type species: *Eubacterium foedans* (Klein) Prévot 1938, 294. No strains conforming to the original description have been isolated since Klein's report in 1908.

Key to species 1–20 of the genus **Eubacterium**

Key characteristics (see also Table 17.17)

Growth at 20 and 37 C
Production of butyric and caproic acids
Acid from:
- mannitol
- raffinose
- erythritol
- fructose
- mannose
- maltose
- glycogen
- lactose
- starch
- sucrose
- inositol
- amygdalin
- rhamnose

Production of indole
Hydrolysis of esculin and starch
Reduction of nitrate

I. No growth at 37 C; growth at 20 C
  1. *E. foedans*

II. Growth at 37 C.
  A. Butyric acid produced.
    1. Caproic acid produced.
      2. *E. alactolyticum*
    2. Caproic acid not produced.
      a. Mannitol acid.
        b. Raffinose acid.
          3. *E. rectale*
        bb. Raffinose not acid.
          c. Erythritol acid.
            4. *E. limosum*
          cc. Erythritol not acid.
            5. *E. ruminantium*
      aa. Mannitol not acid.
        b. Indole produced.
          6. *E. saburreum*

bb. Indole not produced.
c. Fructose acid.
d. Mannose acid.
e. Esculin hydrolyzed.
f. Maltose acid.
g. Nitrate reduced.
h. Glycogen acid.
7. *E. budayi*
hh. Glycogen not acid.
8. *E. nitritogenes*
gg. Nitrate not reduced.
h. Starch hydrolyzed.
3. *E. rectale*
hh. Starch not hydrolyzed.
9. *E. ventriosum*
ff. Maltose not acid.
g. Lactose acid.
10. *E. multiforme*
gg. Lactose not acid.
11. *E. cylindroides*
ee. Esculin not hydrolyzed.
12. *E. moniliforme*
dd. Mannose not acid.
e. Starch acid.
f. Starch hydrolyzed.
3. *E. rectale*
ff. Starch not hydrolyzed.
5. *E. ruminantium*
ee. Starch not acid.
13. *E. tortuosum*
cc. Fructose not acid.
d. Sucrose acid.
14. *E. cellulosolvens*
dd. Sucrose not acid.
15. *E. combesii*
B. Butyric acid not produced.
1. Indole produced.
16. *E. tenue*
2. Indole not produced.
a. Fructose acid.
b. Maltose acid.
c. Inositol acid.
17. *E. fissicatena*
cc. Inositol not acid.
d. Amygdalin acid.
See *Clostridium ramosum*
dd. Amygdalin not acid.
e. Rhamnose acid.
18. *E. contortum*
ee. Rhamnose not acid.
19. *E. aerofaciens*
bb. Maltose not acid.
See *Lachnospira multiparis*

aa. Fructose not acid.

20. *E. lentum*

See Table 17.17 for key-table of other species in the genus.

Clostridium ramosum includes organisms formerly recognized as *Eubacterium filamentosum* (*Catenabacterium filamentosum*); most of these strains do not sporulate readily and are not readily recognized as clostridia.

TABLE 17.13

Fermentation reactions of species of genus **Eubacterium**, *species 1–10*[a]

	1. *E. foedans*	2. *E. alactolyticum*	3. *E. rectale*	4. *E. limosum*	5. *E. ruminantium*	6. *E. saburreum*	7. *E. budayi*	8. *E. nitritogenes*	9. *E. ventriosum*	10. *E. multiforme*
Products from peptone-yeast extract-glucose	·	Abc(F c′sl)	LB(sfa)	LAb(sp)	LFb(ap)	albs24	Lba	LBAs (fp)	Fbla(s)	LAB (fsp)
Products from lactate	·	B^{c-}	−	AB^{-}	−	−	−	$-^{B}$	−	APB
Products from pyruvate	·	$AFBC^{-}$	AB^{-}	ABiv^{-}	−	ABL	AF^{-}	AB^{F}	AF^{-}	AB
Acid produced from:										
Adonitol	·	d	$-^{+}$	$+^{-}$	−	−	−	−	−	−
Amygdalin	·	$-^{+}$	d	−	−	−	$-^{w}$	$-^{w}$	$-^{+}$	−
Arabinose	·	$-^{w}$	$+^{w}$	w^{-}	+	w	w	−	w^{-}	−
Cellobiose	·	−	$+^{w}$	$-^{w}$	d	−	+	$+^{-}$	d	$-^{w}$
Dulcitol	·	−	−	−	−	−	−	−	−	−
Erythritol	·	$-^{w}$	$-^{w}$	$+^{w}$	−	w	−	−	$-^{w}$	−
Esculin	·	$-^{w}$	$-^{+}$	$-^{w}$	−	−	$-^{w}$	$-^{w}$	d	−
Esculin hydrol.	·	−	+	+	+	+	+	+	+	+
Fructose	·	+	+	+	w	w	+	$+^{w}$	$+^{w}$	$+^{w}$
Galactose	·	$-^{w}$	+	$-^{+}$	−	w	w	d	+	$-^{+}$
Glucose	−	$+^{w}$	+	+	+	+	+	+	+	+
Glycerol	·	−	−	−	−	−	−	d	$-^{w}$	−
Glycogen	·	−	$+^{-}$	−	−	−	w	−	$-^{w}$	−
Inositol	·	−	−	−	−	−	−	−	$-^{w}$	−
Inulin	·	−	$+^{-}$	−	−	−	−	−	−	−
Lactose	·	−	$+^{w}$	$-^{w}$	d	w	w^{+}	$-^{w}$	$+^{-}$	w^{+}
Maltose	·	−	$+^{w}$	$-^{+}$	d	w	w	$+^{w}$	+	−
Mannitol	·	$+^{-}$	$-^{+}$	$+^{w}$	$-^{w}$	−	−	−	−	−
Mannose	·	$-^{w}$	d	$-^{+}$	−	w	+	+	$+^{w}$	+
Melezitose	·	$-^{w}$	$+^{-}$	−	−	−	−	−	$-^{w}$	−
Melibiose	·	−	$+^{-}$	−	−	−	$-^{w}$	−	d	−
Raffinose	·	−	$+^{w}$	−	−	w	−	−	d	−
Rhamnose	·	−	$-^{+}$	$-^{w}$	−	−	−	−	d	−
Ribose	·	$-^{+}$	d	$+^{w}$	·	−	w^{+}	$-^{+}$	d	−
Salicin	·	−	$+^{-}$	$-^{w}$	d	−	w^{-}	d	$-^{w}$	-
Sorbitol	·	$-^{+}$	$+^{-}$	$-^{w}$	−	−	−	−	−	−
Sorbose	·	−	$-^{+}$	$-^{w}$	−	−	$-^{+}$	w^{-}	−	−
Starch	·	−	$+^{w}$	$-^{w}$	+	w	w^{+}	$-^{w}$	w^{+}	−
Starch hydrol.	·	−	+	$-^{+}$	−	−	$+^{-}$	−	−	−
Sucrose	·	−	$+^{w}$	$-^{w}$	d	+	−	$-^{w}$	d	−
Trehalose	·	−	$-^{+}$	−	−	w	−	$-^{+}$	−	$-^{w}$
Xylose	·	$-^{w}$	+	d	d	w	$-^{w}$	−	w^{+}	−

[a] Products from PYG broth, lactate and pyruvate: Capital letters indicate ≧ 1 meq acid/100 ml broth; small letters < 1 meq/100 ml. A, acetic; B, butyric; C, caproic; C′, caprylic; F, formic; iB, isobutyric; *iv*, isovaleric; L, lactic; P, propionic; S, succinic; 2, ethanol; 4, butanol; ( ) products of occasional strains. +, strong acid—pH 5.5 or below—for 90–100% of strains; −, negative reaction for 90–100% of strains; w, weak acid—pH 5.5–6.0—for 90–100% of strains. (Uninoculated xylose and arabinose, under CO_2, are often pH 5.9; for cultures in these media pH 5.4–5.7 are considered weak acid and pH below 5.4 strong acid.); d, reaction positive for 40–60% of strains.

When any of the above used as superscripts, reaction of 10–40% of strains.

TABLE 17.14

Biochemical reactions of species of genus **Eubacterium**, *species 1–10*[a]

	1. *E. foedans*	2. *E. alactolyticum*	3. *E. rectale*	4. *E. limosum*	5. *E. ruminantium*	6. *E. saburreum*	7. *E. budayi*	8. *E. nitritogenes*	9. *E. ventriosum*	10. *E. multiforme*
Gelatin digested	+	−	$-^{w}$	$-^{+}$	w	−	−	$-^{w}$	−	+
Milk	−	$-^{c}$	c^{-}	−	a	c	−	−	c^{-}	c^{d}
Meat	·	−	−	−	−	−	−	−	−	−
Indole produced	·	−	−	−	−	+	−	−	−	−
Nitrate reduced	·	−	−	−	d	−	+	+	−	+
Neut. red reduced	·	+	+	+	−	+	+	d	+	+
Gas	4	3^{-}	$+^{3}$	$+^{4}$	$+^{-}$	3	4	4^{2}	2^{-}	4
NH_3	·	$-^{+}$	$-^{w}$	$+^{-}$	−	+	+	$+^{w}$	$-^{w}$	+
Acetoin	·	−	$-^{+}$	$+^{-}$	w^{-}	−	−	d	d	$-^{w}$
Growth in 20% bile	·	$+^{-}$	$+^{3}$	4^{+}	$+^{2}$	−	s^{4}	2^{+}	2^{4}	2
Growth in 6.5% NaCl	·	d	−	$-^{w}$	−	−	−	−	$-^{+}$	−
Growth at pH 8.5	·	d	d	$+^{-}$	w	·	+	$+^{-}$	$-^{+}$	+
Lecithinase (EYA)	·	−	−	−	·	−	+	$-^{w}$	−	−
Lipase (EYA)	·	−	−	−	·	−	−	−	−	−
Hemolysis	·	−	−	$-^{b}$	·	a	$-^{b}$	$-^{b}$	−	b^{-}
Motility	−	−	d	−	−	−	−	−	−	+
Threonine ← propionate	·	$-^{w}$	−	$-^{+}$	−	−	$+^{-}$	+	$-^{+}$	+

[a] +, positive reaction—90–100% of strains; −, negative reaction—90–100% of strains; d, reaction + for 40–60% of strains; hemolysis—a = alpha; b = beta; w = weak; gas and growth—relative amounts indicated, by +, 2, 3, 4; s = stimulated; EYA—McClung-Toabe egg yolk agar. Milk—c, curd, d, digested; gelatin—w, liquefaction of chilled cultures in less than half the time of the uninoculated controls; meat—d, digested.

Any of the above used as superscripts indicates reaction of 10–40% of strains.

Reactions determined by procedures given in Holdeman and Moore (1972).

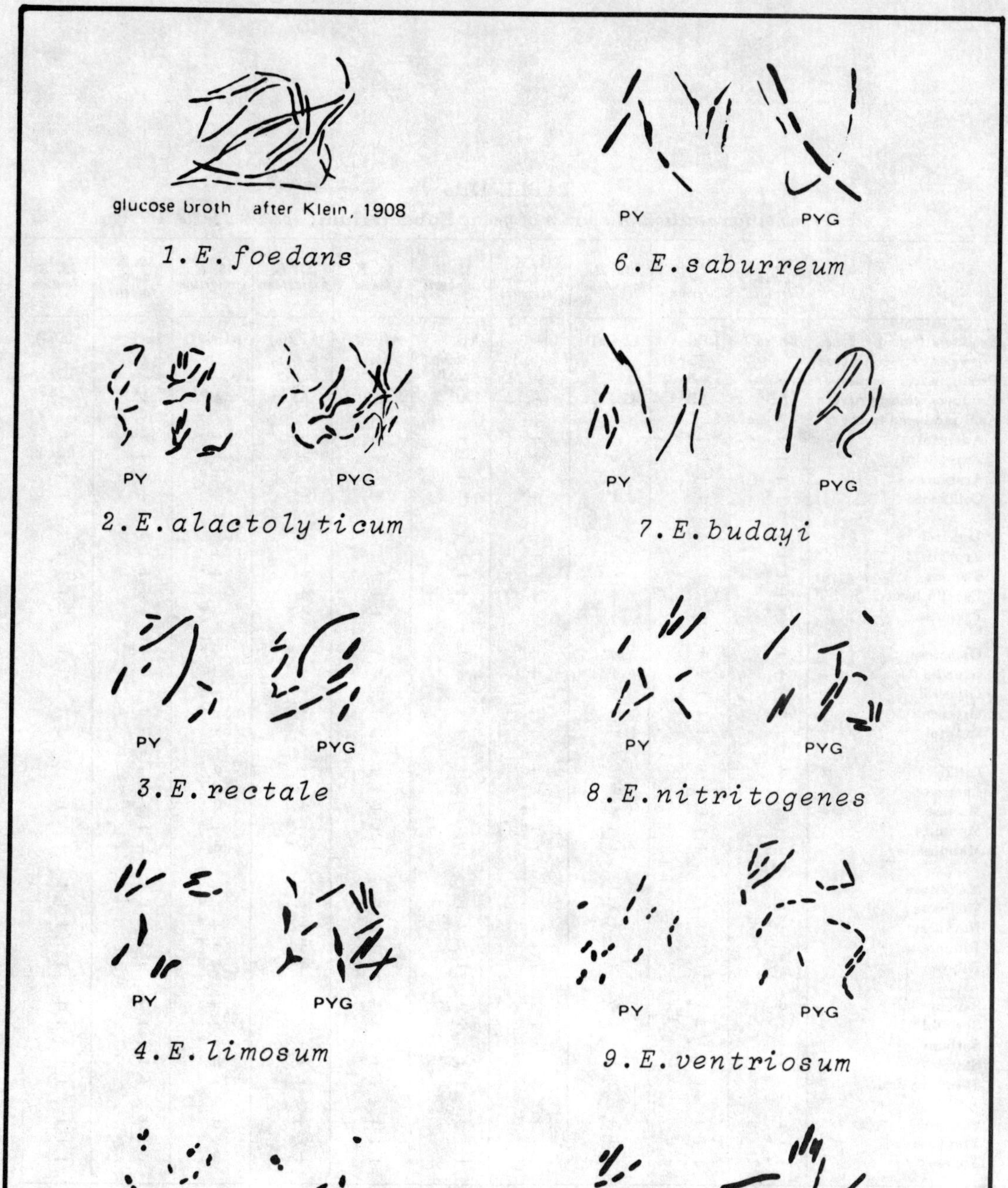

Fig. 17.B. Morphology of species 1 to 10 of the genus *Eubacterium*. Drawings are composites of several cultures; individual cultures usually show less variation. Scale: $\frac{9}{16}$ inch = 10 μm. PY, peptone-yeast extract broth cultures; PYG, PY-glucose broth cultures.

TABLE 17.15

Fermentation reaction of species of genus **Eubacterium**, *species 11–20*[a]

	11. *E. cylindroides*	12. *E. moniliforme*	13. *E. tortuosum*	14. *E. cellulosolvens*	15. *E. combesii*	16. *E. tenue*	17. *E. fissicatena*	18. *E. contortum*	19. *E. aerofaciens*	20. *E. lentum*
Products from peptone-yeast extract-glucose	Lb(asf)	LBa (fsp2)	Lbs(af)	Lfsa (bp)	AB *ibivp*	Af(lsp *ibiv*)	AF2(s)	Af(sl2)	LAF (s2)	(afsl)
Products from lactate	–	$-^{A}$	–	–	$-^{AB}$	–	–	–	–	–
Products from pyruvate	AB^{F-}	AB^{F1}	$ABFL^{-}$	–	A^{B}	AF^{-}	AF^{L2}	AF^{2}	A^{FL-}	$-^{AF}$
Acid produced from:										
Adonitol	–	$-^{w}$	–	–	–	–	–	–	–	–
Amygdalin	–	–	w^{-}	w	–	–	–	–	–	–
Arabinose	–	–	$-^{w}$	–	–	–	$-^{+}$	$+^{w}$	$-^{+}$	–
Cellobiose	$-^{w}$	–	d	$+^{w}$	–	–	–	$+^{-}$	$+^{-}$	–
Dulcitol	–	–	–	–	–	–	–	w^{-}	–	–
Erythritol	–	–	–	–	–	–	–	–	–	–
Esculin	$-^{w}$	–	–	–	–	–	–	$+^{-}$	–	–
Esculin hydrol.	+	–	+	+	+	–	+	+	$+^{-}$	–
Fructose	$+^{w}$	+	w^{+}	–	–	$-^{w}$	$+^{w}$	$+^{w}$	+	–
Galactose	–	d	w^{+}	$+^{w}$	–	–	$+^{w}$	$+^{w}$	$+^{w}$	–
Glucose	+	$+^{w}$	+	+	$+^{-}$	w	+	+	+	–
Glycerol	–	–	–	–	–	–	–	$-^{w}$	–	–
Glycogen	–	–	–	–	–	–	–	$-^{+}$	–	–
Inositol	–	–	–	–	–	–	+	–	–	–
Inulin	d	–	$-^{+}$	$+^{-}$	–	–	–	d	–	–
Lactose	–	–	d	w^{-}	–	–	$-^{+}$	d	$+^{-}$	–
Maltose	–	d	w^{+}	$+^{w}$	$-^{w}$	$-^{w}$	w^{+}	+	$+^{w}$	–
Mannitol	–	–	–	–	–	–	$-^{w}$	–	–	–
Mannose	$+^{w}$	+	–	–	–	–	w^{-}	d	+	–
Melezitose	–	–	–	–	–	–	–	–	–	–
Melibiose	–	–	w^{-}	–	–	–	–	$+^{w}$	$-^{w}$	–
Raffinose	$-^{w}$	–	w^{-}	w^{-}	–	–	$-^{w}$	$+^{w}$	$-^{+}$	–
Rhamnose	–	–	–	–	–	–	+	$+^{w}$	–	–
Ribose	$-^{w}$	$-^{w}$	–	–	–	–	d	$+^{w}$	–	–
Salicin	d	$-^{w}$	w^{-}	w^{-}	–	–	d	+	$+^{w}$	–
Sorbitol	–	–	–	–	–	–	–	–	–	–
Sorbose	–	$-^{w}$	$-^{w}$	.	–	–	$-^{w}$	$-^{w}$	–	–
Starch	–	$-^{w}$	–	–	–	$-^{w}$	–	d	–	–
Starch hydrol.	$-^{+}$	$-^{+}$	–	–	–	$-^{+}$	–	–	–	–
Sucrose	d	–	$+^{w}$	$+^{w}$	–	–	+	+	$+^{-}$	–
Trehalose	$-^{+}$	–	–	–	–	–	w^{+}	$-^{w}$	$-^{+}$	–
Xylose	$-^{+}$	–	$-^{w}$	–	–	–	$+^{-}$	d	–	–

[a] For symbols see Table 17.13.

TABLE 17.16

Biochemical reactions of species of genus **Eubacterium,** *species 11–20*

	11. *E. cylindroides*	12. *E. moniliforme*	13. *E. tortuosum*	14. *E. cellulosolvens*	15. *E. combesii*	16. *E. tenue*	17. *E. fissicatena*	18. *E. contortum*	19. *E. aerofaciens*	20. *E. lentum*
Gelatin digested	$-^{w}$	−	−	$-^{w}$	+	+	−	−	$-^{w}$	$-^{w}$
Milk	−	−	a^{-}	c	c^{d}	c^{d}	a^{-}	c	a^{c-}	−
Meat	−	−	−	−	$-^{d}$	$-^{d}$	−	−	−	−
Indole produced	−	−	−	−	−	+	−	−	−	−
Nitrate reduced	−	d	d	−	−	−	$-^{+}$	−	−	$+^{-}$
Neutral red reduced	+	+	+	+	d	+	+	+	+	$+^{-}$
Gas	$-^{3}$	4	$-^{2}$	$-^{3}$	2^{w}	2^{4}	4^{3}	2^{4}	4^{-}	$-^{+}$
NH_3	w^{-}	+	w^{-}	−	+	$+^{w}$	$+^{-}$	$-^{+}$	$+^{-}$	+
Acetoin	$-^{w}$	$-^{+}$	$-^{w}$	−	−	d	−	d	$-^{w}$	−
Growth in 20% bile	4^{-}	4^{+}	$-^{4}$	+	4^{2}	$+^{2}$	4	2^{4}	4^{-}	$+^{3}$
Growth in 6.5% NaCl	d	−	−	−	−	−	−	−	$+^{-}$	d
Growth at pH 8.5	$+^{-}$	$+^{-}$	$+^{-}$	+	−	$-^{+}$	+	d	$+^{-}$	$+^{-}$
Lecithinase (EYA)	−	−	−	.	−	+	−	−	−	−
Lipase (EYA)	−	−	−	.	$-^{+}$	−	−	−	−	−
Hemolysis	$-^{w}$	$-^{w}$	$-^{a}$	.	b^{w}	w^{-}	b^{-}	w	$-^{+}$	−
Motility	−	$+^{-}$	−	+	$+^{-}$	d	$-^{+}$	−	−	−
Threonine ← propionate	−	$-^{+}$	−	−	$-^{+}$	+	−	−	−	−

[a] For symbols see Table 17.14.

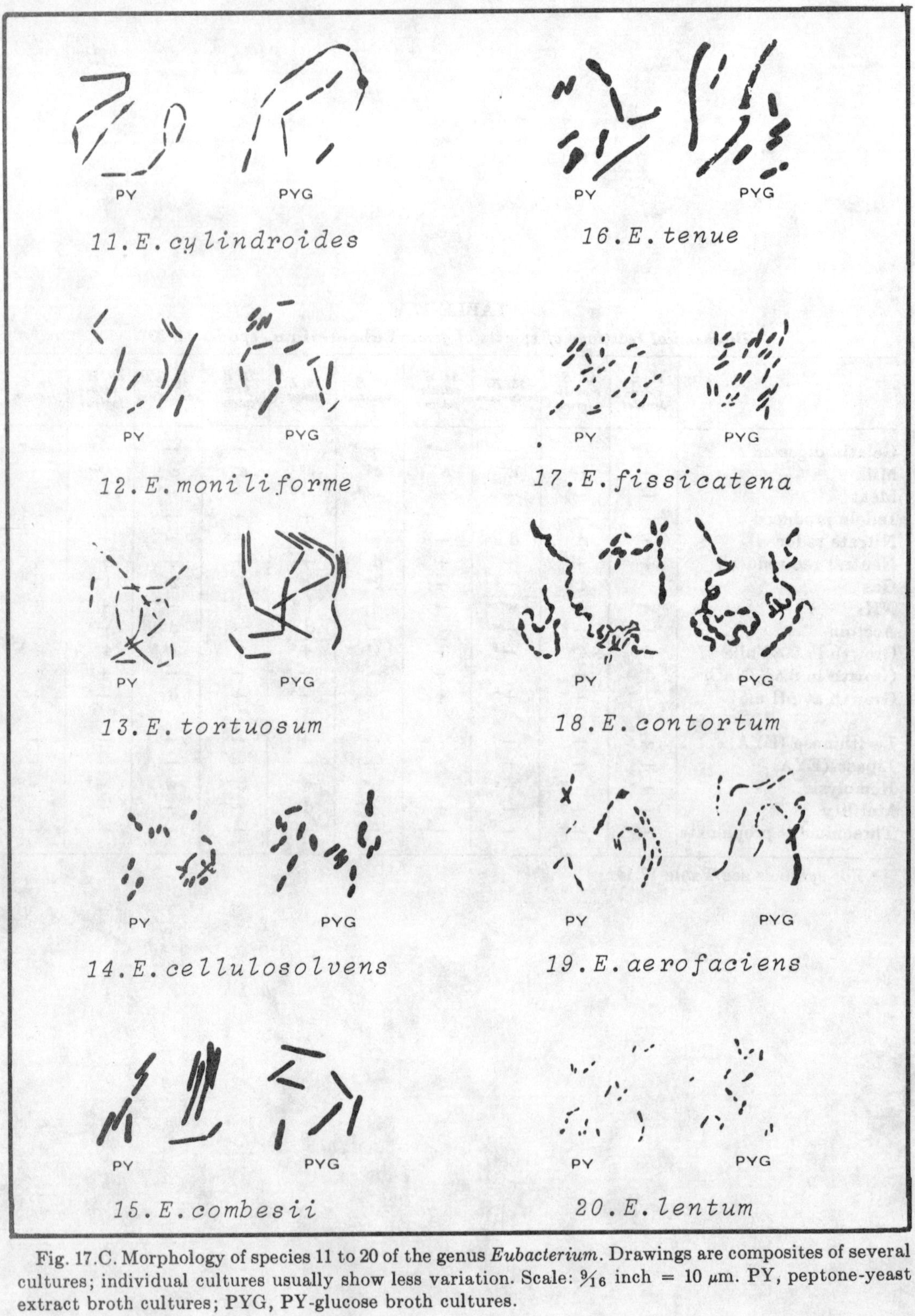

Fig. 17.C. Morphology of species 11 to 20 of the genus *Eubacterium*. Drawings are composites of several cultures; individual cultures usually show less variation. Scale: 9/16 inch = 10 μm. PY, peptone-yeast extract broth cultures; PYG, PY-glucose broth cultures.

TABLE 17.17

Characteristics of species 21–28 of genus **Eubacterium**[a]

Species	Motile	Optimum temp.	Gelatin digested	Milk	NO_3 reduction	Indole produced	Gas[b]	Fructose	Glucose	Lactose	Maltose	Mannitol	Starch	Sucrose	Additional characteristics
21. *E. endocarditidis*	**+**	37	+	−	−		4	−	+		+		−	+	Acid: arabinose; no acid: dextrin, inulin, salicin
22. *E. helminthoides*	−	37	**+**	−°	**−**		4		+		**+**	+		+	Acid: glycerol, sorbose
23. *E. pseudotortuosum*	−	37	**+**	−	**+**		4	+	+	+	+		+	+	Acid: galactose, sorbitol
24. *E. obsti*	−	37	**+**	−	−	−	4		d		**−**				"May ferment glucose; other carbohydrates are not fermented"
25. *E. ethylicum*	−	37	**−**	C	**−**	−	4	+	+	+		+	+	+	Acid: glycerol; no acid: amygdalin; hydrolyzes starch
26. *E. helwigiae*	−	55	−	a	+	−	+	+	+	+	+			+	Acid: mannose, galactose; hydrolyzes starch.
27. *E. ureolyticum*	−	†	−	C	+[Ψ]		**+**	+	+	+	+				Acid: galactose; **hydrolyzes urea**
28. *E. parvum*	−	37	−	C	−		**−**	+	+	+	+				Acid: galactose

[a] + = positive reaction or acid produced; − = negative reaction or no acid produced; d = reaction may be positive or negative; ° = see text; C = coagulated; a = acid (milk). Ψ nitrate reduced in the presence of galactose and lactose. Useful distinguishing characteristics are given in **boldface.**

[b] 4 = abundant gas; + = gas produced. † optimum temperature not given, presumably is 37 C.

ORDER I. **ACTINOMYCETALES** BUCHANAN 1917, 162

Bacteria that tend to form branching filaments which in some families develop into a mycelium. The filaments may be extremely short as in the *Mycobacteriaceae* and *Actinomycetaceae* or well developed as in *Streptomycetaceae*. The diameter varies from 0.5–2.0 μm, generally less than 1.0 μm. Filaments are not always observed because in certain families the filaments tend to fragment and can be seen only in some cultural stages of development or in host tissue. They are sometimes rare as in *Mycobacterium*. Fragmentation of filaments leads to the formation of coccoid, elongate or diphtheroid elements. In some families true spores are formed on aerial and/or substrate hyphae. Spores may be produced singly on the hypha, as a pair or as chains of various numbers of spores. If sufficient numbers of spores are present the chains can be straight, looped or spiral. Such chains arise singly from the hyphae or in a verticillate manner. In the *Actinoplanaceae*, spores are borne in a sporangium and are motile or non-motile, depending on the genus.

Gram-positive, though this reaction may vary with the age of the culture. The mycobacteria are acid-alcohol-fast and some members of the *Nocardiaceae* are weakly acid-fast. All members of the order are aerobic except for genera in the *Actinomycetaceae*, which may be anaerobic, facultatively anaerobic or aerobic.

The *Actinomycetales* are commonly isolated from soil and less commonly from fresh water. Forms pathogenic to man and animals exist in at least four families. Spores can also be allergenic to man. Some species are pathogenic to plants; others are obligate symbionts in plant root nodules and fix nitrogen.

Further Comments

New criteria for the identification of taxa in the *Actinomycetales* are now being generally adopted. In *Streptomyces* and *Streptoverticillium*, standard procedures of culture and observation are now used to characterize species (Shirling and Gottlieb, 1966).

Cell wall composition also has become widely accepted as an aid in the identification of genera. Four cell wall types are now widely accepted (Becker *et al.*, 1965; Lechevalier *et al.*, 1966; Lechevalier and Lechevalier, 1970) (Table 17.18). Other cell wall patterns have been proposed (Lechevalier and Lechevalier, 1967; Sukapure *et al.*, 1970) but more species and strains must be studied to ascertain whether these types are indeed characteristic of the genera in which they

TABLE 17.18

Major constituents of cell wall types of actinomycetes[a]

Cell wall type	DAP		Glycine	Arabinose	Galactose
	meso-	LL-			
I	−	+	+	−	−
II	+	−	+	−	−
III	+	−	−	−	−
IV	+	−	−	+	+

[a] References: Becker *et al.*, 1965; Yamaguchi, 1965; Lechevalier and Lechevalier, 1970.

All preparations also contain major amounts of glucosamine, muramic acid, alanine and glutamic acid.

The chromatographic method used does not differentiate between meso- and D-forms of 2–6 diaminopimelic acid (DAP); since D-DAP is relatively rare it is assumed it is a minor constituent, if present.

TABLE 17.19

Whole cell sugar patterns of aerobic actinomycetes

Pattern	Arabinose	Galactose	Xylose	Madurose
A	+	+	−	−
B	−	−	−	+
C	−	−	−	−
D	+	−	+	−

TABLE 17.20

Distribution of cell wall types and whole cell sugar patterns in genera of **Actinomycetales**

Cell wall type	Sugar pattern	Genera
I	NC[a]	*Streptomyces, Streptoverticillium, Microellobosporia, Sporichthya*
II	D	*Micromonospora, Actinoplanes Ampullariella, Amorphosporangium, Dactylosporangium*
III	B	*Microbispora, Streptosporangium Spirillospora, Planomonospora, Dermatophilus, Nocardia-madurae* type (*Actinomadura*)
III	C	*Actinobifida, Thermoactinomyces Geodermatophilus*
IV	A	*Mycobacterium, Nocardia, Pseudonocardia Thermomonospora, Micropolyspora*

[a] NC, no characteristic sugar pattern.

have been found. Sugar patterns of whole cells have also been suggested and combinations of cell wall and sugar patterns seem promising, although at present there are some anomalies with present classifications (Tables 17.19 and 17.20).

Base ratios in DNA may be helpful in placing some atypical organisms into the correct genera. Similarly, the homology of DNA from an unidentified species can be determined with the DNA from a known species, thus aiding in the identification of the actinomycete (Tewfik and Bradley, 1967).

Members of the *Actinomycetales*, despite the filamentous growth of some members, are now generally accepted to be bacteria and not fungi. The properties relating them to bacteria are: the absence of a nuclear membrane, small hyphal diameters, sensitivity to lysozyme, chemical nature of the cell wall, sensitivity to antibacterial agents, and flagella, when produced, of the bacterial type.

Key to the families of order **Actinomycetales**

I. Mycelium not formed; branching filaments may be produced; cells may be rod, diphtheroid or coccoid; no spores formed.
  A. Not acid-alcohol-fast; usually facultatively anaerobic; some anaerobic or aerobic; most do not contain 2,6-diaminopimelic acid in the cell walls.
    Family I. *Actinomycetaceae*, p. 659*
  B. Acid-alcohol-fast, at least in some stages of growth; cell wall type IV.
    Family II. *Mycobacteriaceae*, p. 681
II. True mycelium produced.
  A. Symbionts in plant nodules with a free stage in soil.
    Family III. *Frankiaceae*, p. 701
  B. Saprophytes or facultative parasites.
    1. Spores borne inside sporangia.
      Family IV. *Actinoplanaceae*, p. 706 and Genus *Microellobosporia*, p. 843
    2. Spores not borne inside sporangia.
      a. Mycelial filaments divide transversely and in at least two longitudinal planes to form masses of motile, coccoid elements. Aerial mycelium usually absent; cell wall type III.
        Family V. *Dermatophilaceae*, p. 723
      b. Mycelial filaments commonly fragment to give coccoid or elongate elements that are usually non-motile, though a few species are reported to be motile. Aerial spores are occasionally produced but usually are absent; cell wall type IV. Sometimes acid-fast.
        Family VI. *Nocardiaceae*, p. 726
      c. Mycelial filaments tend to remain intact and not fragment. Usually abundant aerial mycelium and long spore chains (5–50 or more); cell wall type I.
        Family VII. *Streptomycetaceae*, p. 747
      d. Mycelial filaments remain intact; spores formed singly, in pairs or short chains on either or both aerial or substrate mycelium. Cell wall type (II, III, or IV) varies with the genus.
        Family VIII. *Micromonosporaceae*, p. 846

FAMILY I. ACTINOMYCETACEAE BUCHANAN 1918, 403

Gram-positive bacteria, predominantly diphtheroid in shape, which tend to form **branched filaments** in tissue or in some stages of cultural development. **Fragmentation of the filaments readily occurs** producing diphtheroid or coccoid forms. **Aerial mycelium and spores are not formed. Non-acid-fast and non-motile.** Usually grow as **facultative anaerobes** but **some are anaerobic** and **some are aerobic;** carbon dioxide usually stimulates growth. **Carbohydrates are fermented. Catalase may or may not be formed. Chemoorganotrophs.**

* All page numbers cited refer to *Bergey's Manual of Determinative Bacteriology*, 8th edition, The Williams & Wilkins Company, Baltimore, 1974.

Key to the genera of family **Actinomycetaceae**

I. Anaerobic to facultatively anaerobic.
  A. Catalase negative or positive.
    1. Most species form a filamentous microcolony; filaments transitory, diphtheroid cells are predominant; fermentative, glucose fermentation products include acetic, formic, lactic and succinic acids but not propionic acid; cell wall contains neither diaminopimelic acid nor arabinose.
Genus I. **Actinomyces,** p. 660
  B. Catalase negative.
    1. Filamentous microcolony; filaments transitory, diphtheroid cells and spheroplasts common, fermentative, glucose fermentation products are primarily propionic and acetic acids; cell wall contains diaminopimelic acid but not arabinose.
Genus II. **Arachnia,** p. 668
    2. Smooth microcolony; filaments usually not formed, diphtheroid cells and bifid forms are common; fermentative, glucose fermentation products are primarily acetic and lactic acids; cell wall contains neither diaminopimelic acid nor arabinose.
Genus III. **Bifidobacterium,** p. 669

II. Aerobic to facultatively anaerobic.
  A. Catalase positive or negative.
    1. Filamentous microcolony; cell types include rods, filaments and characteristically filaments with a bacillus-body at one end; some strict anaerobes; fermentative, products from glucose grown cultures include CO_2, formic, acetic, propionic and lactic acids; cell wall contains both diaminopimelic acid and arabinose.
Genus IV. **Bacterionema,** p. 676
  B. Catalase positive.
    1. Smooth microcolony; growth at any given time may yield exclusively coccoid, diphtheroid or filamentous forms or a mixture of any of these; grows best aerobically; fermentative, glucose fermentation product is primarily lactic acid, propionic acid is not produced; cell wall contains neither diaminopimelic acid nor arabinose.
Genus V. **Rothia,** p. 679

Genus I. Actinomyces *Harz 1877, 485*

Gram-positive, irregularly staining bacteria; non-acid-fast, non-spore-forming and non-motile. Filaments with true branching may predominate and are particularly evident in 18–48 hr microcolonies. **Diphtheroid cells or branched rods are common;** V, Y and T forms occur (Plate 17.2, Figs. 1, 2). **Filaments,** 1 μm or less in diameter, **varying in length and degree of branching** occur in most strains.

Chemoorganotrophs. Carbohydrates are fermented with the production of **acid but no gas. End-products from glucose fermentation** may include **acetic, formic, lactic** and **succinic acids** but **not propionic acid.**

Proteolytic activity rare, and, if present, is weak. **Indole not produced; urease negative.** Some species may show greening or complete lysis of rabbit red blood cells.

Cell wall peptidoglycan contains alanine, glutamic acid and lysine with either (1) **aspartic acid or** (2) **ornithine;** diaminopimelic acid does not occur in any currently recognized species. Cell wall carbohydrates may include glucose, galactose, rhamnose and other deoxy sugars; **arabinose does not occur.**

Organic nitrogen is required for growth.

Facultative anaerobes; most are preferentially anaerobic, **one species grows well aerobically.** CO_2 is required for maximum growth. **Catalase negative or positive.**

Certain species are pathogenic for man and/or animals.

Type species: *Actinomyces bovis* Harz 1877, 485.

Further Comments

Isolation. Clinical material should be inoculated into freshly boiled thioglycollate medium (preferably with addition of serum or other enrichment) and also streaked onto a brain heart infusion agar (BHIA) and two heart infusion blood agar plates. Incubate the BHIA and one blood plate anaerobically in an atmosphere of N_2:H_2:CO_2(80:10:10) with a catalyst, otherwise in N_2:CO_2(95:5). Incubate the second blood plate aerobically with addition of 5% CO_2. Observe for microcolonies in 18–24 hr and mature colonies in 7–14 days.

Genera with similar morphology. Cell and colony morphology provides the initial basis for considering an organism to be an *Actinomyces* but is not sufficient to differentiate *Actinomyces* from certain species of *Corynebacterium*, *Bifidobacterium*, *Propionibacterium*, *Bacterionema*, *Rothia* or *Arachnia*. Confirmation of an organism as an *Actinomyces* and species identification is dependent upon physiological, cell wall, and antigenic characteristics.

Microcolony morphology, particularly on BHIA (Plate 17.2, Figs. 3–4; Plate 17.3, Fig. 2) is frequently helpful in identification because a "spider" colony with multiple branching filaments or a granular-centered colony with peripherally extended branched filaments is highly suggestive of an *Actinomyces* species. However, *A. bovis*, *A. odontolyticus* and occasional strains of *A. israelii*, *A. naeslundii* and *A. viscosus* form smooth microcolonies without branched filaments. Thus, colonial morphology is not always of differential value.

Physiological characteristics. *Actinomyces* species vary in tolerance to oxygen and in CO_2 requirements (Gerencser and Slack, 1967).

In order to obtain reproducible results in various biochemical tests it is to be emphasized that an adequate basal medium, sufficient inoculum and proper oxygen concentrations are essential. In addition, best results are obtained by the use of standardized techniques (Howell *et al*., 1959; Slack, 1968; Georg, 1970). Each species produces a fermentation pattern which is distinctive (Table 17.22) and when considered with other characteristics (Tables 17.21) is of diagnostic value.

Cell wall. In the two possible peptidoglycan patterns the amount of aspartic acid may be minimal in pattern 1 and as an additional component in pattern 2. No species currently recognized contains diaminopimelic acid or arabinose. These results are based on the study of 80 strains including *A. bovis*, *A. israelii*, *A. naeslundii* and *A. odontolyticus* (Cummins, 1962; Pine and Boone, 1967; DeWeese *et al*., 1968).

TABLE 17.21

Characteristics differentiating the species of genus **Actinomyces**[a]

	Species				
	1. *Actinomyces bovis*	2. *Actinomyces odontolyticus*	3. *Actinomyces israelii*	4. *Actinomyces naeslundii*	5. *Actinomyces viscosus*
Catalase	−	−	−	−	+
Microcolony, smooth or filamentous	S	S	F	F	F
Red mature colony on blood agar	−	+	−	−	−
Nitrate → Nitrite	d	+	d	+	+
Starch hydrolysis—wide zone	+	−	−	−	−
Fermentation (acid only)					
Arabinose	−	d	d	−	−
Ribose	−	−	+	−	−
Xylose	d	d	+	−	−
Soluble starch	+	+[s]	d	d	d
Serological group (fluorescent antibody)	B	E	D	A	F
Serotype	1, 2	1, 2	1, 2	1, 2	1, 2

[a] + = 90% of strains positive; − = 90% of strains negative; d = different reactions, positive or negative; +[s] = positive, slow.

Antigenic characteristics. Any species can be identified through the combined use of morphology and serology. Fluorescent antibody techniques provide rapid and specific results (Slack and Gerencser, 1966; Lambert *et al*., 1967). Gel diffusion techniques using culture filtrates as antigens are useful but are presently less specific for species identification (King and Meyer, 1963; Georg *et al*., 1964).

TABLE 17.22

Acid production[a] from carbon compounds by species of genus **Actinomyces**

	1. *A. bovis*	2. *A. odontolyticus*[b]	3. *A. israelii*	4. *A. naeslundii*	5. *A. viscosus*	a. *A. eriksonii*	b. *A. humiferus*	c. *A. suis*
Acetate	0/1[c]	0/0	0/4	0/1	0/2	0/12	0/27	0/0
Adonitol	0/1	6/30	0/4	0/12	0/2	0/12	0/27	7/10
Arabinose	2/26	15/30	57/223	2/54	0/56	12/12	25/27	9/10
Cellobiose	0/1	0/3	53/65	1/6	4/22	12/12	18/27	0/0
Dextrin	1/1	0/0	2/4	1/1	0/2	1/1	27/27	0/0
Dulcitol	0/1	0/30	0/4	0/12	0/2	0/12	9/27	1/10
Erythritol	0/1	0/0	0/4	0/1	0/2	0/12	0/0	0/0
Fructose	12/12	0/0	4/4	1/1	29/34	5/5	27/27	10/10
Galactose	2/2	0/0	4/4	1/1	24/34	5/5	25/27	10/10
Glucose	37/37	30/30	223/223	66/66	56/56	12/12	27/27	10/10
α-Methyl-D-glucoside	0/1	0/0	0/68	0/35	0/56	8/12	0/0	0/0
Glycerol	8/26	26/30	0/223	11/12[d]	20/23[d]	0/12	20/27	7/10
Glycogen	1/1	0/3	0/65	0/6	2/22	5/5	0/0	0/0
Inositol	1/1	14/30	137/223	40/52	48/56	0/12	2/27	0/0
Inulin	0/12	0/30	61/158	25/53	22/56	4/12	1/27	9/10
Lactate	0/1	0/0	0/4	0/1	0/2	0/12	0/27	0/0
Lactose	33/33	23/30	197/223	33/42	48/56	12/12	15/27	10/10
Maltose	29/33	29/30	158/158	32/42	32/34	12/12	25/27	10/10
Mannitol	4/37	0/30	149/223	0/56	0/56	12/12	24/27	5/10
Mannose	8/12	0/0	44/65	3/3	31/34	1/1	25/27	0/0
α-Methyl-D-mannoside	0/1	0/0	0/68	0/35	0/0	0/1	0/0	0/0
Melezitose	0/1	0/30	3/68	0/35	0/0	12/12	25/27	0/0
Melibiose	0/1	0/3	4/6	6/6	16/22	12/12	24/27	0/0
Pyruvate	0/1	0/0	0/4	0/1	0/2	4/12	27/27	0/0
Raffinose	0/16	15/30	118/223	50/53	53/56	12/12	24/27	10/10
Rhamnose	2/12	13/30	4/56	2/12	0/56	2/12	25/27	5/10
Ribose	0/12	0/0	123/158	8/42	0/34	1/1	3/27	0/0
Salicin	7/33	21/30	134/223	24/54	45/56	12/12	8/27	0/0
Sorbitol	0/1	1/30	2/61	0/1	0/34	12/12	1/27	2/10
Potato starch	13/14	0/0	0/4	0/1	0/2	1/1	0/0	0/0
Soluble starch	13/14	30/30	21/64	9/12	21/22	10/10	27/27	10/10
Sucrose	28/33	27/30	158/158	37/42	54/56	12/12	27/27	9/10
Trehalose	0/1	20/30	4/6	8/11	15/22	12/12	19/27	10/10
Xylose	11/32	16/30	206/223	1/56	0/56	12/12	24/27	10/10

[a] Acid but no gas produced.

[b] Medium thioglycollate fermentation base + 2% rabbit serum.

[c] Number of strains positive/number of strains tested.

[d] See species description.

TABLE 17.23

Biochemical reactions of species of genus **Actinomyces**

	1. *A. bovis*	2. *A. odontolyticus*	3. *A. israelii*	4. *A. naeslundii*	5. *A. viscosus*	a. *A. eriksonii*	b. *A. humiferus*
Catalase	0/30[a]	0/30	0/192	0/54	56/56[b]	0/12	0/27
Indole	0/14	0/230	0/223	0/49	0/47	0/12	0/27
$NO_3 \rightarrow NO_2$	3/12	230/230	109/220	49/55	21/22	0/12	0/27
Methyl red	12/12	202/203	57/63	12/12	21/22	12/12	27/27
Voges-Proskauer	0/8	0/203	0/192	0/37	0/48	0/12	4/27
H_2S (lead acetate paper)							
Triple sugar iron	10/10	26/30	26/26	6/14	22/22	12/12	0/0
BHIA	6/10	0/30	2/89	4/12	0/22	9/12	18/27
Urease production	0/10	0/30	0/26	0/12	0/22	0/12	0/27
Esculin hydrolysis	13/13	19/30	24/26	9/12	21/22	7/12	24/27
Starch hydrolysis							
Wide zone	30/30	0/203	0/63	0/23	0/48	0/12	0/27
Narrow zone	0/30	0/203	26/63	9/23	41/48	12/12	27/27
Gelatin hydrolysis							
Tube	1/24	0/230	0/89	0/59	0/48	0/12	12/27
Plate	0/0	0/0	34/89	0/0	0/0	0/0	0/0
Litmus milk							
Acid only	4/15	14/30	18/26	23/52	13/23	5/12	22/27
Acid and clot	8/15	10/30	8/26	24/52	10/23	7/12	0/0
No change	3/15	6/30	0/26	5/52	0/23	0/12	5/27

[a] Number of strains positive/number of strains tested.

[b] Aerobic + CO_2 cultures used, if anaerobic cultures are used expose to air 30 min before testing.

Genus II. **Arachnia** *Pine and Georg 1969, 269*

Branched diphtheroid rods 0.2–0.3 μm by 3–5 μm, and **branched filaments** 5–20 μm or longer. Cells occasionally of uneven diameter, frequently with swollen or clubbed ends. **Swollen coccoid cells** and/or spheroplasts, of diameter to 5 μm, formed during stationary growth phase. **Gram-positive, non-acid fast, non-spore-forming** and **non-motile.**

Microcolonies definitely **filamentous,** composed of branched non-septate and septate filamentous elements often originating from a common center. **No aerial filaments.**

Chemoorganotrophs: Metabolism fermentative, carbohydrates being the fermentable substrates; **fermentation of glucose anaerobically yields carbon dioxide, acetic acid, propionic acid** and traces of DL-lactic and succinic acids; glucose fermented in air to carbon dioxide and acetic acid, but **fermentative products are not oxidized beyond acetate. Does not ferment lactate or pyruvate. Does not contain catalase.**

Facultative anaerobe, CO_2 is not required for aerobic or anaerobic growth.

Cell wall contains LL-diaminopimelic acid.

Pathogenic for man and laboratory mice.

Type species: *Arachnia propionica* Pine and Georg 1969, 269.

Genus III. **Bifidobacterium** *Orla-Jensen 1924, 472*

Rods highly variable in appearance; uniform or branched, bifurcated **Y** and **V** forms and **club** or **spatulate** forms are generally characteristic of freshly isolated strains. **Morphology** may be **influenced by nutritional conditions** and on subculture straight or curved rods, of varying width within a rod, may possess breaks resembling branching. Gram-positive, non-acid-fast, non-spore-forming, non-motile bacteria. **Often stain irregularly,** occasionally two or more granules may stain with methylene blue while the remainder of the cell may be unstained.

Saccharoclastic (Table 17.26); **gas is not produced. Glucose is fermented primarily to acetic and L(+) lactic acid,** generally in a molar ratio of 3:2. Small amounts of formic and succinic acids produced. **CO_2, butyric and propionic acids are not produced.** Fructose-1,6-diphosphate aldolase is absent; glucose-6-phosphate dehydrogenase activity is generally absent but is demonstrable in strains from bees and in *B. adolescentis*. Glucose is degraded by the fructose-6-phosphate shunt in which phosphoketolase cleaves fructose-6-phosphate into acetylphosphate and erythrose-4-phosphate; pentosephosphates are then formed by transaldolase and transketolase; xylulose-5-phosphate phosphoketolase splits substrate into acetylphosphate and glyceraldehyde-3-phosphate; through a reaction sequence also occurring in glycolysis, lactate is formed from glyceraldehyde-3-phosphate.

Catalase negative (*B. indicum* positive when grown in presence of hemin); benzidine reaction negative; indole negative; nitrates not reduced. Anaerobic, although there may be infrequent oxygen tolerance in the presence of CO_2.

Optimal temperature 36–38 C.

The G + C content (cesium chloride density or T_m) varies from 57.2–64.5 moles % (211 strains).

Type species: *Bifidobacterium bifidum* (Tissier) Orla-Jensen 1924, 472.

TABLE—17.24

Fermentation reactions of named species of genus **Bifidobacterium**[a]

Previously named species	Arabinose	Xylose	Ribose	Gluconate	Cellobiose	Lactose	Mannitol	Melezitose	Salicin	Starch	Trehalose
1. *B. bifidum* Orla-Jensen	−	−	−	−	+	+	−	−	−	−	−
2. *B. adolescentis* Reuter	+	+	+	+	+	+	v	v	+	v	v
3. *B. infantis* Reuter	−	−	+	−	v	+	−	−	v	v	v
B. liberorum Reuter	−	+	+	−	v	+	−	v	v	v	v
B. lactentis Reuter	−	+	+	−	v	+	+	−	−	v	v
4. *B. breve* Reuter	−	−	+	−	+	+	+	v	v	v	v
B. parvulorum Reuter	−	−	+	−	v	+	+	−	−	+	−
5. *B. longum* Reuter	+	+	+	−	−	+	−	+	−	v	v
B. longum subsp. *animalis* Mitsuoka	+	+	+	−	−	+	−	−	v	v	v
6. *B. pseudolongum* Mitsuoka	+	+	+	−	v	v	−	v	v	+	−
7. *B. thermophilum* Mitsuoka	−	−	−	−	v	v	−	v	v	+	v
8. *B. suis* Matteuzzi *et al.*	+	+	−	−	−	+	−	−	−	−	−
9. *B. asteroides* Scardovi *et al.*	+	+	+	+	+	−	−	−	+	−	−
10. *B. indicum* Scardovi *et al.*	−	−	+	+	+	−	−	−	+	−	−
11. *B. coryneforme* Scardovi *et al.*	+	+	+	+	+	−	−	−	+	−	−

[a] + = positive; − = negative; v = variable, slight, delayed or erratic reactions by the same strain or related strains (so-called biotypes).

All bifidobacteria ferment glucose, galactose and fructose (sometimes slowly). *B. bifidum* biotype a (from adults) slowly ferments sucrose and melibiose; *B. bifidum* biotype b (from infants) does not ferment sucrose but vigorously ferments melibiose; neither ferments maltose or raffinose. Other bifidobacteria (nearly 100%) ferment maltose, melibiose, raffinose and sucrose. Inositol is fermented only by some strains of *B. liberorum* and *B. lactentis*. Inulin is fermented only by some strains of *B. adolescentis*, *B. infantis* and *B. liberorum*. Esculin and amygdalin are fermented by *B. adolescentis* and *B. breve*; *B. infantis*, *B. liberorum*, *B. longum* are negative and *B. thermophilum* is equivocal; remaining have not been tested. Sorbitol is fermented only by certain strains of *B. breve* and *B. adolescentis*. Mannose is fermented only by certain strains of *B. breve*, *B. longum*, *B. suis* and generally slowly, variably, or not at all by the others. Bifidobacteria do not ferment adonitol (ribitol), dulcitol, erythritol, glycerol, rhamnose and α-methyl-D-mannoside.

Genus IV. **Bacterionema** *Gilmour, Howell and Bibby 1961, 139*

Cells are **pleomorphic**, comprising **non-septate** and **septate filaments** 1–1.5 μm by 20–200 μm and **bacilli** 1.5–2.5 μm by 3–10 μm. **Characteristic morphology is a bacillus attached to a filament** ("whip-handle," Plate 17.4, Fig. 1). **Reproduction by filament septation and fragmentation** to form rods which germinate to yield one to four filaments. Branching is frequent with aerobic and/or acid conditions. **Gram-positive, non-acidfast, no endospores** and **non-motile.** Metachromatic granules formed.

Microcolonies are spider-like (Plate 17.4, Fig. 2).

Chemoorganotroph; carbohydrates are usually **fermented to yield acid and some gas.** Products from glucose-grown cultures include acetylmethylcarbinol and: with aerobiosis, CO_2 and propionic acid are major, formic, acetic and lactic acids are minor and/or variable; with reduced oxygen, CO_2 and lactic acid are major, formic, acetic, propionic and succinic acids are minor.

Facultative anaerobe; depending upon cultivation conditions, some isolates may appear to be strict aerobes or anaerobes, but later adapt to facultative growth.

Type species: *Bacterionema matruchotii* (Mendel) Gilmour, Howell and Bibby 1961, 139.

Genus V. **Rothia** *Georg and Brown 1967, 86*

Coccoid, diphtheroid or filamentous cells (Plate 17.4, Fig. 3). **Filamentous forms branched**, usually 1 μm in **diameter**; however, diameter may be considerably greater, up to 5 μm, due to irregular swellings and clubbed ends. Growth at any given time may consist exclusively of coccoid, diphtheroid or filamentous forms or may be a mixture of any of these. **Gram-positive, non-acid fast, non-spore-forming, non-motile.**

Mature colonies are creamy white. They may be **smooth or rough** surfaced (Plate 17.4, Fig. 4). Most are of **soft texture**; extremely rough colonies may be dry and crumbly. **No aerial hyphae** produced.

Catalase positive.

Chemoorganotroph. Ferments carbohydrates with production of acid but no gas. Major product of **glucose fermentation is lactic acid**, other products may include acetic, formic and succinic acids **but not propionic acid.** May be weakly proteolytic.

Cell wall constituents include alanine, glutamic acid, lysine and large amounts of galactose, but **no arabinose or diaminopimelic acid** (Davis and Baird-Parker, 1959; Cummins, personal communication).

Aerobic; some strains able to grow slowly at reduced oxygen pressures; CO_2 does not stimulate growth.

Optimal temperature 35–37 C.

Type species: *Rothia dentocariosa* (Onishi) Georg and Brown 1967, 86.

Further Comments

The genus *Rothia* was created to accommodate organisms which are common inhabitants of the normal mouth and throat and resemble *Actinomyces* species morphologically, but grow better aerobically, and differ significantly from members of the genera *Actinomyces* and *Nocardia* in their physiology and cell wall constituents.

For primary isolation and maintenance, enriched media such as brain heart infusion or trypticase soy, are recommended with aerobic incubation at 37 C. Cultures in deep agar butts of these media remain viable 3–4 months at room temperature. For long term storage, lyophilization of suspensions in milk is satisfactory. Little or no growth is obtained on media with inorganic nitrogen or simple peptones, such as Czapek or Sabouraud dextrose agar. A satisfactory fermentation medium is meat extract peptone base with Andrade's indicator at pH 7.4 (meat extract 3.0 g, peptone 10.0 g, NaCl 5.0 g, Andrade's indicator 10.0 ml, distilled water to 1000 ml).

Catalase test is best performed by flooding H_2O_2 over aerobic growth on slants and observing stream of bubbles. Anaerobic cultures must be exposed to air 15–20 min before testing. The Hugh and Leifson test (1953) is useful in distinguishing *Rothia* from *Nocardia*; *Rothia* ferments glucose, whereas *Nocardia* oxidizes or does not attack glucose in this medium.

FAMILY II. **MYCOBACTERIACEAE** CHESTER 1897, 63

Description as for genus *Mycobacterium.*

Genus I. **Mycobacterium** *Lehmann and Neumann 1896, 363*

Slightly curved or straight **rods,** 0.2–0.6 by 1.0–10 μm, sometimes branching; **filamentous or mycelium-like growth** may occur but **on slight disturbance usually becomes fragmented into rods or coccoid elements. Acid-alcohol-fast** at some stage of growth. Not readily stainable by Gram's method but usually considered Gram-positive. Non-motile. No endospores, conidia or capsules, no grossly visible aerial hyphae.

Genus includes obligate parasites, saprophytes and intermediate forms differing in nutritional requirements. Saprophytic strains grow on very simple substrates; others require more complex media or supplements for growth (mycobactin); and others have not been cultivated outside living cells. Species differ in amidase, catalase and other enzyme activity; all are **aerobic** although from dispersed seeding in tubed agar medium, growth of some species occurs only in the depths of the medium.

Lipid content of cells and especially cell walls high; included are waxes having component chloroform-soluble mycolic (fatty) acids with long, branched, about 80-carbon-atom chains. **Diffusible pigment rare.** Colonies of some species are regularly or variably yellow or orange, usually due to carotenoid pigments, the formation of which may or may not require exposure to light. The cell wall peptidoglycolipid contains *meso*-diaminopimelic acid, alanine, glutamic acid, glucosamine, muramic acid, arabinose and galactose (wall type IV). Mycobacteria hydrolyze phenolphthalein disulfate although some species do so only slowly.

Growth slow or very slow; easily visible colonies are produced from dilute inoculum after 2 days to 8 weeks incubation at optimum temperature, which may be near 40 C for some species but closer to 30 C for others. Although no taxonomic division of the genus on growth rate is recognized, the terms "rapid" and "slow" growers are used for convenience in identification.

Diseases produced include tuberculosis, leprosy and other usually chronic more or less necrotizing, limited or extensive granulomas.

Found in soil, water, warm-blooded and cold-blooded animals.

The G + C content of the DNA ranges from 62–70 moles % (T_m; buoyant density) (Wayne and Gross, 1968; Tewfik and Bradley, 1967).

Type species: *Mycobacterium tuberculosis* (Zopf) Lehmann and Neumann 1896, 363.

Further Comments

For convenience the genus may be divided into three groups:

1. *Slow growers*. Grossly visible colonies apparent only after 7 or more days on media seeded with fresh inocula, diluted sufficiently to yield well isolated colonies. Microcolonial test positive.

Species 1–18. See Tables 17.25 and 17.27.

2. *Rapid growers*. Grossly visible colonies in less than 7 days under similar conditions and at 25 and 37 C. Microcolonial test negative.

Species 19–28. See Tables 17.26 and 17.27.

3. Organisms which have special growth requirements or have not been cultivated *in vitro*.

Species 29–31.

In many cases it is unnecessary to carry an identification to the level of an individual species. For example, in clinical diagnosis, it may suffice to identify an organism as belonging to a complex of similar organisms, with similar significance. In these situations, the complexities of the tests required for speciation may outweigh the value received from the identification. For this reason Table 17.27 is offered as a guide to a minimal test protocol which will permit identification of some organisms at a level usually considered to be sufficient for routine practical purposes. Identification beyond these levels may require application of a larger series of tests presented in the master lists (Tables 17.25 and 17.26), as well as consideration of properties described in the narrative portions of the text. It should be recognized that intermediate growth rates are frequently seen, especially in old laboratory strains, and the groupings used are relative.

Like other organisms, a given strain of a *Mycobacterium* species may lose its capacity to exhibit a key property (e.g. pigment or a particular enzyme) and still be considered to belong to that species, on the basis of a large number of other properties. Mycobacteria are not well characterized by bacteriologic testing techniques useful for identifying members of other genera. For this reason, it is essential that tests be performed precisely as described in the literature cited (see Table 17.28).

Acid-alcohol-fastness is partially or completely lost at some stage of growth by a variable proportion of the cells of some species. Cells of rapid growers may be less than 10% acid-fast. Beading with metachromatic granules and banding with non-stained areas are common.

Most strains of mycobacteria form more than one kind of colony, but colonies of some species, as *M. tuberculosis*, are regularly rough; some, as of *M. intracellulare* on primary culture from clinical specimens, are more commonly smooth. Primary culture smooth colonies may often be very thin, *in vitro* adapted colonies in contrast are high domed. Cells of rough strains are usually compacted in curving strands; cells of smooth strains are not visibly oriented in any pattern. Some colonies, as of *M. fortuitum* and *M. xenopi*, in early growth may be mycelial, older ones exhibiting branching filamentous extensions on and into some media such as cornmeal glycerol agar; fragmentation to bacilli usually occurs in smear preparation. Aerial filamentous extensions rare, never visible without magnification (× 30–100).

According to currently available data, the mycolic acids of mycobacteria are distinctive. Corresponding acids of *Corynebacterium* and *Nocardia* species tested have smaller molecules, insoluble in chloroform.

May be isolated from sputum, soil or other contaminated sources by primary digestion and decontamination with alkali, acid or hypochlorite solution, followed by seeding on egg medium containing malachite green. Other media may be employed. Alternatively the digest, neutralized, may be injected into mice, and cultures obtained about 2 weeks later from internal organs. Isolation of mycobacteria from soil also has been attained by use of nutrient medium in which paraffin is the only added carbon source.

TABLE 17.25

Properties of slowly growing species of genus **Mycobacterium**[a]

No.	Property	1. *M. tuberculosis*	2. *M. microti**	3. *M. bovis*	4. *M. africanum**	5. *M. kansasii*	6. *M. marinum*	7. *M. simiae*	8. *M. gastri*	9. *M. nonchromogenicum**	10. *M. terrae**	11. *M. triviale*	12. *M. gordonae*	13. *M. scrofulaceum*	14. *M. paraffinicum**	15. *M. intracellulare*	16. *M. avium*	17. *M. xenopi*	18. *M. ulcerans*
3	Urease	+	+	+	+	+	+	see text	+	−	−	−	∓	+	−	−	−	−	
5	Nicotinamidase	+	+	−	+	+	+		+	+	−	−	−	+	−	+	+	+	
6	Pyrazinamidase	+	−	∓	+	∓	+		−	+	−	−	−	+	−	+	+	+	
31	Arylsulfatase (2 week)	−	−	∓		∓	+		+	+	−	+	+	±	−	+	∓	+	−
32	Growth on 5% NaCl	−	−	−		−	−		−	−	−	+	−	−	−	−	−	−	
33	Tellurite red'n																		
	(3 day)	−		−		−	−		−		−	−	−	−		±	±	−	
	(9 day)	±		±		−	+		∓		+	−	∓	±		+	+	±	
34	Nitrate reduction	+	−	−	−	+	−		−	+	+	+	−	−	−	−	−	−	−
35	Catalase 68 C	−	−	−	−	+	−		−	+	+	+	+	+	+	+	+	+	+
36	Tween agar opac. (4 week)	−		−		−	−		−	+	+	+	∓	±	−	+	+	+	
37	Nitrite red'n (7 day)	−		−		∓	−		−	−	−	+	∓	−	−	+	∓	+	
39	Niacin	+	+	−	−	−	−		−	−	−	−	−	−	−	−	−	−	∓
40	Tween 80 hydrol. (5 day)	∓	−	−	−	+	+		+	+	+	+	+	−	−	−	−	−	−
41	Catalase >45 mm foam	−	−	−	−	+	−		−	+	+	+	+	+	+	−	−	−	+
42	Pigment	N	N	N	N	P	P		N	N	N	N	S	S	S	N	N	N[b]	N
43	Resists 10 μg/ml T2H	+	−	−	−	+	+		+	+	+	+	+	+	+	+	+	+	+
44	Resists 1 μg/ml isoniazid	−		−	−	±	+		∓	+	+	+	+	+	+	+	+	∓	
45	Microcolonial test	+		+		+	+		+	+	+	+	+	+	+	+	+	+	+
46	Growth at 25 C	−	−	−	−	+	+		+	+	+	+	+	+	+	±	∓	−	±
	37 C	+	+	+	+	+	±		+	+	+	+	+	+	+	+	+	+	−
	40 C	+		+		+	−		±	+	+	∓	∓	±	−	+	+	+	−
	45 C	−	−	−	−	−	−		−	−	−	−	−	−	−	∓	±	+	−
	52 C	−	−	−	−	−	−		−	−	−	−	−	−	−	−	−	−	−

[a] Data were assembled from cited references as well as from unpublished data compiled by members of the Advisory Committee for the Family. Except for species marked with an asterisk results are expressed as follows: + = more than 84% of strains positive; ± = between 50–84% of strains positive; ∓ = between 16–49% of strains positive; − = less than 16% of strains positive. Data of strains marked with an asterisk are based on a single strain. P = photochromogenic, i.e. pigment is produced only on exposure to light; S = scotochromogenic, i.e. pigment is produced even when cultures are grown or maintained in the dark; N = no pigment, at least in young cultures. Details of techniques and interpretation of the properties cited may be found in references cited in Table 17.28. All species tested are negative for acetamidase (1), benzamidase (2) and iron uptake (38).

[b] Characteristic yellow color may be slow to develop.

TABLE 17.26

Properties of rapidly growing species of genus **Mycobacterium**[a]

No.	Property	19. *M. phlei*	20. *M. vaccae*	21. *M. diernhoferi*	22. *M. smegmatis*	23. *M. thamnopheos*	24. *M. flavescens*	25. *M. fortuitum*	26. *M. peregrinum*	27a. *M. chelonei* subsp. *chelonei*	27b. *M. chelonei* subsp. *abscessus*
	Amidases										
1*	Acetamidase	−	+	+	+	+	−	+	+	+	+
2*	Benzamidase	−	+	−	V	−	−	−	−	−	−
4*	Isonicotinamidase	−	+	−	+	−	−	−	−	−	−
5	Nicotinamidase	+	+	+	+	+	+	−	−	+	+
6	Pyrazinamidase	+	+	+	+	+	+	−	−	+	+
7	Allantoinase	−	+	−	−	+	−	+	+	−	−
8	Succinamidase	−	−	−	+	−	−	−	−	−	−
9	Decompose phenylalanine	+	+	−	−	−	−	−	−	−	−
	Carbon source										
10	Benzoate	−	V	−	+	−	−	−	−	−	−
11	Citrate	+	+	−	+	+	−	+	+	+	−
12	Malate	+	+	V	+	+	−	+	+	+	+
13	Oxalate	−	−	−	+	−	−	−	−	−	−
14	Succinate	+	+	+	+	+	V	+	+	+	+
	Acid production from carbohydrates										
15	Arabinose	+	±	+	+	−	−	−	−	−	−
16	Dulcitol	−	−	−	+	−	−	−	−	−	−
17	Fructose	+	+	+	+	V	±	+	+	−	−
18	Galactose	+	−	−	+	±	−	−	−	−	−
19	Inositol	−	+	+	+	−	−	−	−	−	−
20	Mannitol	+	+	+	+	+	V	−	+	−	−
21	Mannose	+	+	+	+	+	+	+	+	+	+
22	Rhamnose	−	+	−	+	−	−	−	−	−	−
23	Sorbitol	+	V	−	+	+	V	−	−	−	−
24	Trehalose	+	+	+	+	+	V	+	+	+	+
25	Xylose	+	+	+	+	−	−	−	−	−	−
26	Survives 60 C/4 hr	+	−	−	−	−	−	−	−	−	−
27	MacConkey agar	−	−	−	−	−	−	+	+	d	+
28	Malachite green, 0.01%	V	−	−	V	−	−	+	+	+	+
29	Methyl violet, 0.01%	V	−	−	V	−	−	+	+	+	+
30	Pyronin B, 0.01%	V	−	−	+	−	−	+	+	+	+
31	Arylsulfatase (3-day)	−	−	−	−	−	−	+	+	+	+
	(4 week)	+	+	+	+	+	+	+	+	+	+
32	Growth on 5% NaCl	+	+	−	+	+	+	+	+	−	+
33	Tellurite reduction (3 day)	−	+	+	+	−	∓	±	±	−	±
34	Nitrate reduction	+	V	∓	+	−	+	+	+	−	−
35	Catalase, 68 C	+	+	−	∓	−	+	+	+	+	+
38	Iron uptake	+	+	+	+	+	−	+	+	−	−
39	Niacin	−	−	−	−	−	−	−	−	−[b]	−
40	Tween hydrolysis (5 day)	+	+	−	+	+	+	±	±	−	−
41	Catalase >45 mm foam	+	+	−	+	−	+	+	+	+	+
42	Pigment	S	P/S	N/P	N	N	S	N	N	N	N
46	Growth at 25–32 C	+	+	+	+	+	+	+	+	+	+
	37 C	+	+	+	+	−	+	+	+	+	+
	40 C	+	+	+	+	−	+	+	−	V	+
	45 C	+	−	−	+	−	−	−	−	−	−
	52 C	+	−	−	−	−	−	−	−	−	−

[a] Data assembled and results expressed as for Table 17.25, except the notation V which indicates character is inconstant, i.e. the same strain may sometimes test positive, sometimes negative.

All species tested give a negative microcolonial test and are negative for salicylamidase and malonamidase, positive for urease (4 hr test). All species positive on glucose, negative on lactose and raffinose.

[b] Some strains of *M. chelonei* may be niacin positive (the former *M. borstelense* var. *niacinogenes*).

TABLE 17.27

Selected properties for routine identification of species and/or complexes of mycobacterial species[a]

No.	Property	Species or complex																	
		1. *M. tuberculosis*	3. *M. bovis*	5. *M. kansasii*	6. *M. marinum*	8. *M. gastri*	9-11. *M. nonchromogenicum-terrae-triviale complex*	12. *M. gordonae*	13. *M. scrofulaceum*	15–17. *M. avium-intracellulare-xenopi* complex[b]	18. *M. ulcerans*	19. *M. phlei*	20. *M. vaccae*	21. *M. diernhoferi*	22. *M. smegmatis*	23. *M. thamnopheos*	24. *M. flavescens*	25–26. *M. fortuitum-peregrinum* complex	27. *M. chelonei* complex
3	Urease	+	+	+	+	+	−	−	+	−		+	+	+	+	+	+	+	+
27	MacConkey agar											−	−	−	−	−	−	+	d
31	Arylsulfatase (3 day)											−	−	−	−	−	−	+	+
32	Growth on 5% NaCl	−	−	−	−	−	d	−	−	−		+	+	−	+	+	+	+	d
33	Tellurite red'n (3 day)	−	−	−	−	−	−	−	−	d		−	+	+	+	−	d	d	d
34	Nitrate reduction	+	−	+	−	−	+	−	−	−	−	+	d	d	+	−	+	+	−
35	Catalase 68 C	−	−	+	−	−	+	+	+	+	+	+	+	−	d	−	+	+	+
38	Iron uptake	−	−	−	−	−	−	−	−	−	−	+	+	+	+	+	−	+	−
39	Niacin	+	−	−	−	−	−	−	−	−	∓	−	−	−	−	−	−	−	−
40	Tween 80 hydrol. (5 day)	d	−	+	+	+	+	+	−	−	−	+	+	−	+	+	+	d	−
41	Catalase: >45 mm foam	−	−	+	−	−	+	+	+	−	+	+	+	−	+	−	+	+	+
42	Pigment	N	N	P	P	N	N	S	S	N	N	S	P/S	N/P	N	N	S	N	N
43	Resists 10 μg/ml T2H	+	−	+	+	+	+	+	+	+	+	+	+		+		+	+	+
46	Growth at 37 C	+	+	+	d	+	+	+	+	+	−	+	+	+	+	−	+	+	+
	45 C	−	−	−	−	−	−	−	−	d	−	+	−	−	+	−	−	−	−
	52 C	−	−	−	−	−	−	−	−	−	−	+	−	−	−	−	−	−	−

[a] Identifications based on this table will serve most practical needs; more definitive speciation may be obtained by use of additional properties given in Tables 17.25 and 17.26. See footnotes to Table 17.25 for meaning of symbols +, −, N, S and P; d = strain-to-strain variation, some strains positive, others negative.

[b] *M. xenopi* is distinguished on cellular and colonial morphology, drug susceptibility and late pigment production.

TABLE 17.28

References to specific techniques and interpretation of properties cited in Tables 17.25, 17.26 and 17.27

Property numbers	Reference
1, 2, 3, 4, 5, 6, 7, 8	Bönicke, 1961, 7
9	Gordon, 1966, 329
10, 11, 12, 13, 14, 15, 16, 17, 18, 19, 20, 21, 22, 23, 24, 25, 26	Gordon and Mihm, 1959, 736
27, 28, 29, 30	Jones and Kubica, 1963, 355
31	Jones *et al.*, 1966, 790
32	Kestle *et al.*, 1967, 1041
33	Kilburn *et al.*, 1969, 94
34[a] (rapid growers)	Kubica and Dye, 1967, 44, 70
35	Kubica and Pool, 1960, 387
36	Wayne, 1966, 919
37	Wayne and Doubek, 1965, 738
34[a] (slow growers), 38, 39, 40, 41, 42, 43, 44	Wayne and Doubek, 1968, 925
45	Wayne *et al.*, 1957, 451
46	Wayne *et al.*, 1967, 88

[a] Note that different techniques are cited for nitrate reduction (property 34) among slowly and rapidly growing mycobacteria.

FAMILY III. FRANKIACEAE BECKING 1970, 201

Symbiotic, filamentous, mycelium-forming bacteria which induce and live in root nodules on a wide variety of non-leguminous dicotyledonous plants belonging to 6 orders, 7 families and 14 genera (Table 17.29). The nodules are capable of fixing molecular nitrogen. These bacteria have also a free stage in the soil.

TABLE 17.29

Non-leguminous nodule-bearing plants with **Frankia** *symbiosis*

Plant order	Plant family	Plant genus	*Frankia* species
Fagales	Betulaceae	*Alnus*	1. *F. alni*
Rhamnales	Elaeagnaceae	*Elaeagnus*	
		Hippophaë	2. *F. elaeagni*
		Shepherdia	
	Rhamnaceae	*Discaria*	3. *F. discariae*
		Ceanothus	4. *F. ceanothi*
Coriariales	Coriariaceae	*Coriaria*	5. *F. coriariae*
Rosales	Rosaceae	*Dryas*	6. *F. dryadis*
		Purshia	7. *F. purshiae*
		Cercocarpus	8. *F. cercocarpi*
Myricales	Myricaceae	*Myrica*	
		Gale	9. *F. brunchorstii*
		Comptonia	
Casuarinales	Casuarinaceae	*Casuarina*	10. *F. casuarinae*

Genus I. **Frankia** *Brunchorst 1886, 174*

A true mycelium is produced; the mycelium is septate and branched, but branching is not necessarily correlated with cross-wall formation. The hyphae of most species are 0.3–0.5 μm in diameter, but may be larger in some species (Table 17.30). In most cases the slender hyphae in the host tissue lack the refractivity necessary for study by light microscopy. Electron micrographs show that the hyphae contain fine strands of DNA not enclosed by a nuclear membrane (Plate 17.5); they are thus procaryotic. Distinct membranous bodies, most probably identical with the so-called plasmalemmosomes (Edwards and Stevens, 1963) or mesomes (Fitz-James, 1960; Chen, 1964) of some actinomycetes and Gram-positive bacteria can be observed.

In an active, nitrogen-fixing nodule, the center of the host cell is filled by a hyphal mass; **near the periphery, spherical or club-shaped terminal swellings** (Plate 17.6, Fig. 1.) **are formed on radially arranged hyphae close to the plant cell wall. These spherical bodies are often called "vesicles."** Both vesicles and club-shaped structures have been considered as a kind of sporangium but there is little evidence for this view; the vesicles lack complete cell walls and in older stages are disintegrated by the living host-cell cytoplasm surrounding them. **The vesicles and club-shaped structures are probably associated with nitrogen fixation** as they are abundant in tissues actively fixing nitrogen. In acetylene reduction tests, tissue slices of alder root nodules containing the endophyte predominantly in the vesicular form showed much more nitrogen fixation that similar slices containing the endophyte but poor in vesicles.

In host cells which have not survived invasion by *Frankia* species, vesicles or club-shaped structures are not found; instead **the mycelium fragments into small particles or bacteria-like cells** (Plate 17.6, Fig. 2) sometimes termed "bacteroids," which fill the host cell completely (Table 17.30). When the host cell ruptures these particles are released into adjacent host cells or into the soil. They may represent a resting stage, capable of surviving in soil and of propagating the endophyte. Species in which bacteria-like cells are commonly found nodulate easily in water cultures when crushed nodules are applied to the root system. In species where these cells have not been observed, nodulation has been found only when planted in habitat soil.

Hyphae, vesicles and club-shaped structures are Gram-variable. The rod-shaped structures or bacteria-like cells are Gram-positive.

Molybdenum is required for the nitrogen-fixation process; non-leguminous nodulated plants deficient in this element show restricted nitrogen fixation and when grown under nitrogen-deficient conditions in soil or water culture produce only poor growth (Hewitt and Bond, 1961; Bond and Hewitt, 1961; Becking, 1961). **Cobalt is also essential** for nitrogen fixation (Bond and Hewitt, 1962; Hewitt and Bond, 1966). Vitamin B_{12} analogues, containing cobalt, have been isolated from non-leguminous root nodules (Kliewer and Evans, 1962, Bond *et al.*, 1965).

No species has as yet been grown on artificial media; in previous claims of isolation Koch's postulates were not fulfilled. The organisms have been grown in root nodule tissue culture, but are often lost in subsequent transplants of the tissue because of the difficulty of infecting newly formed callus tissue (Becking, 1965, 1966, 1970). So far the endophyte-containing callus tissue does not show nitrogen fixation, probably because the callus tissue can grow only on a complex organic medium containing much combined nitrogen (Becking, unpublished).

Probably micro-aerophilic. In *Alnus* and *Hippophaë* optimum nitrogen fixation took place in the presence of 12% oxygen, in *Myrica* and *Casuarina* in about 20% (Bond, 1959, 1961, 1964).

The organisms inhabit cortical parenchyma of root nodule tissues as symbionts, but are also present in a free state in soil, presumably as a viable, resting stage, since sterile non-leguminous seedlings nodulate when planted in suitable soil.

The endophyte enters the plant through a root hair, which has undergone strong curvation as a result of auxin formation by the endophyte (Becking, 1966, 1968, 1970). It reaches the cortex of the root by means of an infection thread and by proliferation of plant cells of a side root primordium a nodule is formed. A root nodule is thus a deformed side root of inhibited or arrested growth. The infected cells are enlarged and have a large and misshapen nucleus, which is probably polyploid (Becking, 1966, 1968, 1970). Two types of nodules are produced; in most plants a coralloid nodule is produced by continuous branching of the deformed side root with reduced apical meristematic activity; in *Casuarina* and representatives of the *Myricaceae*, the apex of each nodule lobe gives rise to a normal but negatively geotropic root and the nodules in these species become covered with upward growing rootlets. Such a root nodule may be called a rhizothamnion. Upward growing rootlets are also formed on the nodular lobes in some ineffective cross-inoculation associations of alder species, such as the root nodule endophyte of *Alnus glutinosa* on *A. rubra* and on *A. jorullensis* (pseudo-rhizothamnia).

Moreover, it is not certain that *Comptonia peregrina* (*Myricaceae*) produces rhizothamnia (see below).

The endophyte is specific for a certain plant genus or a group of related genera. In contrast to the *Rhizobiaceae* no exceptions are known; the barriers of incompatibility are strict. It therefore seems appropriate to give species status to these specific cross-inoculations groups.

Type species: *Frankia alni* (Woronin) Von Tubeuf 1895, 118.

TABLE 17.30

Characteristics differentiating the species of genus **Frankia**[a]

Species	Hyphae				Bacteria-like cells or "bacteroids"
	Diameter	Terminal swellings			
		Spherical (S) or club-shaped (C)	Diameter[b]	Length	
	μm		μm	μm	μm
A. Nodules coralloid					
1. *F. alni*	0.3–0.5	S	3.0–5.0 6.0–8.0[c]		0.5–1.5
2. *F. elaeagni*	0.3–0.5	S	2.5–4.0 1.8–2.2[c]		0.3–0.9
3. *F. discariae*	0.3–0.4	S	*ca.* 4.0		Not known
4. *F. ceanothi*	0.3–0.4	SC	1.5–3.0[d] aver. 2 0–2.5		Not known
5. *F. coriariae*	0.4–0.7	C	1.2–1.3	9.0–12.0	Not known
6. *F. dryadis*	0.5–0.8	C	1.5–2.0	1.5–5.0	1.0–2.0
7. *F. purshiae*	0.3–0.5	SC	2.2–4.4	2.2–5.5	Not known
8. *F. cercocarpi*	0.3–0.5	C	3.0	4.0	Not known
B. Rhizothamnia produced					
9. *F. brunchorstii*	1.2–2.8	C	1.6–2.4	7.5–12.5	1.5–2.5
10. *F. casuarinae*	0.3–0.5	C	0.6–1.5	3.0–4.0	0.4–1.0

[a] Species 2, 3, 7, 8 have been studied by light microscopy only; all others by both light and electron microscopy. Data on species 4, 5, 6, 10 are unpublished results of author.

[b] Diameter of spherical cells or vesicles; diameter of club-shaped structures at widest point.

[c] Rarely these dimensions.

[d] Club-shaped in younger stages with gradually tapering base.

FAMILY IV. **ACTINOPLANACEAE** COUCH 1955, 269

Bacteria that form a distinct mycelium, intramatrical but in some forms also aerial. The diameter of the filaments 0.2–2.6 μm. Sporangia characteristic of the family, spherical, sub-spherical, lobed, cylindrical, digitate, club-shaped, to very irregular in shape; formed always in the air above the surface of the culture media on branched or unbranched, septate, palisade hyphae or on undifferentiated hyphae (Plate 17.7). Palisade hyphae up to 2 μm in diameter. Spores arranged in the sporangia in one or more spirals, or in parallel rows, or irregularly, few to many in a sporangium or in single file with one to several spores in one sporangium. Sporangial dehiscence variable. Spores spherical, sub-spherical, oval, rod-shaped or spiral, with flagella and motile or without flagella and non-motile; contain one or several small spherical lipoid globules. Conidia formed in some species on penicillate conidiophores or in moniliform series. Gram-positive; not acid fast.

Cell wall composition of some genera and species of the family is given in Table 17.31.

Growth on certain agars brilliantly colored to white; flocculent, pasty to compact and roughish.

Chemoorganotrophs: Metabolism respiratory. Molecular oxygen is the universal electron acceptor.

Aerobic.

Temperature range: 15–45 C.

Abundant in all soils which contain humus and less common in fresh water habitats, on dead plant parts as pollen, leaves, etc. and animal material as hair and other keratin and chitin substrates.

TABLE 17.31

Cell wall composition[a] *of some species in the family* **Actinoplanaceae**

Organism (morphologic designation)	Strain no.	DAP	HDAP	Galactose	Glucose	Mannose	Arabinose	Xylose	Deoxyhexose
Actinoplanes philippinensis	2	+++		+++	+++	++	++	++	++
A. utahensis	260		+++	+++		++		TR	TR
A. missouriensis	431		+++	+++	++	TR	TR	TR	
Amorphosporangium auranticola	253	++	+	+++	+++	++	++	++	++
Ampullariella regularis	79		+++	+++	+++	++	++	++	++
A. regularis	168		+++	+++	+++	++	++	++	++
A. regularis	28		+++					+++	++
A. digitata	33		+++	+++			++		
A. lobata	72		+++	+++		++		TR	
A. campanulata	65	++	+	+++		++			
Pilimelia anulata	1	+++		+	+++	TR	++	++	
P. terevasa	1	+++		+	+++	+++	++	+	
Spirillospora albida (y)	761	+++	+	+					
Streptosporangium album (y)	S-16-1	++	TR	+++	+	++			
S. amethystogenes (y)	S-5	++	TR						
S. roseum (y)	276-1	++	TR	TR	TR	TR			
S. viridialbum (y)	S-20	++	TR	TR	TR	+			
S. vulgare (y)	S-1	++		TR	TR	+			

[a] With the exception of *Spirillospora* and *Streptosporangium* which lack glycine as a major cell wall component, all members of the Actinoplanaceae studied thus far have major amounts of glucosamine, muramic acid, glutamic acid, glycine and alanine as cell wall components. *Ampullariella lobata* and *A. campanulata* also contain heptose and traces of galactosamine. Determinations by Szaniszlo and Gooder (1967) except those marked (y) which are by Yamaguchi (1967).

Key to the genera of family **Actinoplanaceae**

I. Sporangia spherical, cylindrical or highly irregular and containing a few to several thousand spores per sporangium. Spores in coiled or in parallel chains within the sporangia.

A. Sporangia spherical, subspherical to very irregular; spores arranged in coiled chains within the sporangia.

1. Spores motile.

a. Zoospores spherical to subspherical with a tuft of polar flagella; agar colonies usually orange or reddish orange.

Genus I. *Actinoplanes*, p. 708

b. Zoospores rod-shaped, curved or spiral with one to three sub-polar flagella; agar colonies white, pale yellow, light gray or bright blue.

Genus II. *Spirillospora*, p. 711

2. Spores non-motile, spherical or slightly elongated.

Genus III. *Streptosporangium*, p. 711

B. Sporangia exceedingly irregular, spores short rods.

Genus IV. *Amorphosporangium*, p. 715

C. Sporangia cylindrical, ovoid or irregular; spores rod-shaped, arranged in parallel chains within the sporangia.

1. Zoospores rod-shaped, 0.5–1.0 by 2.0–4.0 μm, with a polar tuft of flagella; agar colonies orange, brown, green-brown or black.

Genus V. *Ampullariella*, p. 716

2. Zoospores rod-shaped, 0.3–0.7 by 0.8–1.5 μm, with a single polar flagellum or one to four

lateral flagella; agar colonies yellow-brown, yellow-gray, yellow or bright lemon yellow; organism usually occurring on hair and perhaps other keratin and chitin material.

Genus VI. *Pilimelia*, p. 718

II. Sporangia club-shaped, finger-like or pyriform containing one to six spores each occurring singly or arranged in double parallel rows on the aerial hyphae.

A. Sporangia containing one spore each, arranged in double parallel rows along the aerial hyphae.

Genus VII. *Planomonospora*, p. 719

B. Sporangia finger-like or club-shaped, two to several spores arranged linearly within each sporangium.

1. Sporangia finger-like or linear; zoospores exhibiting lophotrichous or peritrichous flagella.

a. Zoospores rod-shaped, formed in longitudinal *pairs* on the aerial mycelium.

Genus VIII. *Planobispora*, p. 720

b. Zoospores oval to pyriform, formed in a single row of three to four in sporangia grown directly from the vegetative mycelium.

Genus IX. *Dactylosporangium*, p. 721

2. Sporangia club-shaped; zoospores uniflagellate.

Genus X. *Kitasatoa*, p. 722

Genus I. Actinoplanes *Couch 1950, 89 emend. mut. char. 1955, 153*

Sporangia, 3–20 by 6–30 μm, spherical, subspherical, cylindrical with rounded ends, to very irregular. **Spores, globose to subglobose,** 1–1.5 μm in diameter, in coils, nearly straight chains, or irregularly arranged in sporangia; **motile** by a tuft of polar flagella 2–6 μm in length (see Plate 17.7).

Hyphae, 0.2–2.6 μm in diameter, branched, irregularly coiled, twisted or straight with few septa. **Vertical pallisade hyphae** formed on certain agars; **aerial mycelium scanty,** except in *A. armeniacus.*

Most species brilliantly colored on peptone Czapek and certain other agars; colors: orange, red, yellow, violet and purple. Some form diffusible pigments which color the agar blue, red, yellow, brownish or greenish. No organic growth factors are required.

H_2S is produced by some species. Cell wall composition is given in Table 17.31.

Strict aerobes.

Temperature range 18–35 C.

Occur on a wide variety of plant material, less often on parts of dead animals such as hair, hoofs, snake skin.

The G + C content of the DNA (of two species studied) ranges from 72.1–72.6 moles %.

Type species: *Actinoplanes philippinensis* Couch 1950, 89.

Key to the species of genus Actinoplanes

I. Sporangia globose to subglobose.

1. Sporangia usually more than 15 μm in diameter.

a. Sporangia 8–25 μm in diameter.

1. *A. philippinensis*

b. Sporangia up to 50 μm in diameter.

2. *A. armeniacus*

2. Sporangia 6–14 μm in diameter.

3. *A. missouriensis*

II. Sporangia irregular in shape, 5–18 μm in diameter.

4. *A. utahensis*

Genus II. Spirillospora *Couch 1963, 61*

Hyphae 0.2–1 μm in diameter. Sporangia 5–24 μm in diameter, spherical to vermiform. **Spores** developed from one or more coils in the sporangium, **short to long rods to spiral** in shape, motile by one to three sub-polar flagella (Plate 17.7).

May be grown on a wide variety of culture media.

For cell wall composition see Table 17.31.

Temperature range 18–35 C optimum 25 C.

The G + C content of the DNA of the single species examined is 72.9 moles %.

Type species: *Spirillospora albida* Couch 1963, 65.

Genus III. **Streptosporangium** *Couch 1955, 148*

Sporangia spherical to ovoid; 7–48 μm (usually 7–20 μm). **Sporangiospores arranged in a single coil within the sporangium,** spherical to ovoid, 1.0–1.3 μm by 1.5–3.5 μm, **non-motile** (Plate 17.7). Mycelium as in *Streptomyces;* hyphae much branched, sparingly septate; 0.5–1.2 μm in diameter, palisade hyphae absent. Growth less vigorous than *Actinoplanes* on various agars but usually with greater development of aerial hyphae.

Organic growth factors required by some species. For cell wall composition see Table 17.31.

Strict aerobes.

Temperature range 18–35 C (55 C in one species).

The G + C content of the DNA of the two species studied is 69.5–70.6 moles %.

Type species: *Streptosporangium roseum* Couch 1955, 151.

Key to the species of genus **Streptosporangium**

Key characteristics

Color of aerial mycelium on oatmeal agar
Size of spores and sporangia
Growth at 42 C
Production of violet crystals
Hydrolysis of starch
Reduction of nitrate

I. Aerial mycelium on oatmeal agar pale pink.
  A. Spores 1.0–1.3 by 1.5–1.9 μm.
    1. No growth at 42 C.
      a. Violet crystals not produced.
        1. *Streptosporangium roseum*
        2. *Streptosporangium vulgare*
      b. Violet crystals produced.
        3. *Streptosporangium amethystogenes*
    2. Growth at 42 C; violet crystals not produced.
      a. Starch hydrolyzed.
        4. *Streptosporangium pseudovulgare*
      b. Starch not hydrolyzed.
        5. *Streptosporangium nondiastaticum*
  B. Spores 0.6–0.8 by 1.5–3.5 μm, three times longer than wide.
        6. *Streptosporangium longisporum*
II. Aerial mycelium on oatmeal agar other than pink.
  A. Aerial mycelium white.
    1. Spores spherical 0.2 by 0.4 μm.
        7. *Streptosporangium viridogriseum*
      a. Nitrate reduced to nitrite.
        7a. *Streptosporangium viridogriseum* subsp. *viridogriseum*
      b. Nitrate not reduced.
        7b. *Streptosporangium viridogriseum* subsp. *kofuense*
    2. Spores spherical to ovoid 0.8–2 by 1.5–1.9 μm.
      a. Sporangia 6–8 μm.
        8. *Streptosporangium album*
      b. Sporangia 10–30 μm.
        9. *Streptosporangium albidum*
  B. Aerial mycelium gray, yellowish or greenish.
    1. Sporangia 6–14 μm; aerial mycelium light gray to yellowish.
        10. *Streptosporangium viridialbum*

a. Nitrates not reduced.

10a. *Streptosporangium viridialbum* subsp. *viridialbum*

b. Nitrates reduced to nitrites.

10b. *Streptosporangium viridialbum* subsp. *reducens*

2. Sporangia 8–11 μm; aerial mycelium dark gray.

11. *Streptosporangium rubrum*

Genus IV. Amorphosporangium *Couch 1963, 65*

Sporangia very irregular in shape, much lobed, 6–25 μm wide by 8–15 μm high, width usually greater than height. Spores rod-shaped, 0.5–0.7 by 1.0–1.5 μm; motile by two to three polar flagella (Hanton, 1968) (see Plate 17.7).

Mycelium brilliant shades of orange and in this and cell wall composition (see Table 17.31) resembles *Actinoplanes*.

No organic growth factors required.

Type species: *Amorphosporangium auranticolor* Couch 1963, 65.

Genus V. Ampullariella *Couch 1964, 29*

Sporangia bottle- or flask-shaped, digitate or lobate; 5–20 by 8–30 μm. **Spores arranged in parallel chains within the sporangium, rod-shaped,** 0.5–1 by 2–4 μm, **motile** by polytrichous, polar flagella of 3.5–6 μm in length (see Plate 17.7). Hyphae much as in *Actinoplanes*.

For cell wall composition of various species see Table 17.31.

Temperature range 18–35 C; optimum 25 C.

The G + C content of the DNA, of the single species examined is 72.3 moles %.

Type species: *Ampullariella regularis* (Couch) Couch 1964, 29.

Key to the species of genus Ampullariella

I. Sporangia bottle-shaped, cylindrical, 5–14 by 8–30 μm.

1. *Ampullariella regularis*

II. Sporangia irregular, rarely cylindrical.

A. Sporangia mostly bell-shaped, 5–15 by 6–12 μm.

2. *Ampullariella campanulata*

B. Sporangia mostly irregular, lobed or fan-shaped in vertical section, 4–20 by 12–26 μm.

3. *Ampullariella lobata*

C. Sporangia mostly digitate, 3–9 by 6–15 μm.

4. *Ampullariella digitata*

Genus VI. **Pilimelia** *Kane 1966, 225*

Sporangia large, globose or cylindrical. **Rod-shaped spores produced end to end in parallel chains, approximately 1,000 per sporangium;** motile by means of flagella (Plate 17.7). Loses ability to produce sporangia on repeated sub-culturing.

Hyphae up to 2.2 μm wide and up to 300 μm long, irregularly septate. **Requires complex organic substrate** such as peptone for growth. Growth on peptone-containing agar **usually yellow** yellow-brown or gray; colonies compact 1–2 mm, soft pasty, **slight or no penetration of mycelium into agar.**

For cell wall composition see Table 17.31.

Obligate aerobes requiring Petri dish cultivation for maintenance; poor growth on primary test tube culture and no growth upon subculture.

World-wide distribution in soil.

Type species: *Pilimelia terevasa* Kane 1966, 225.

Note: There are two species recognized in the genus. *P. terevasa* produces globose sporangia and those of *P. anulata* are cylindrical.

Genus VII. **Planomonospora** *Thiemann, Pagani and Beretta 1967, 28*

Substrate and aerial mycelia formed, the former growing profusely into the medium and forming a compact layer on surface of agar.

Sporangia formed on aerial mycelia only, each containing a single large spore motile by peritrichous flagella. Not acid-fast. Gram-positive.

Grows on a variety of substrates; best growth on complex media. Pigments may give characteristic colors to the aerial and/or vegetative mycelium. Cell wall contains *meso*-diaminopimelic acid.

Starch hydrolyzed. Nitrates reduced to nitrites.

Aerobic.

Temperature optimum 28–37 C; no growth at 45 C.

Found in soils, relatively rare occurring in 10 of 453 samples of temperate and tropical soils.

Type species: *Planomonospora parontospora* Thiemann, Pagani and Beretta 1967, 29.

Note: There are two species recognized in the genus. *P. parantospora* produces sporangia arranged in a double parallel row whereas *P. venezuelensis* produces sporangia along the aerial hyphae, giving a palm leaf pattern.

Genus VIII. **Planobispora** *Thiemann and Beretta 1968, 157*

Substrate mycelium 1 μm in diameter, well developed, highly branched. Gram-positive. Not acid fast.

Aerial mycelium 1.0 μm, long and sparsely branched; hyphae usually parallel to agar surface.

Sporangia containing a longitudinal pair of large spores formed only on aerial mycelia; 1.0–1.2 by 6–8.0 μm (Plate 17.8, Fig. 4). Spores motile by peritrichous flagella.

Cell wall contains *meso*-diaminopimelic acid.

Arabinose, glucose and xylose support good growth; raffinose, sucrose and cellulose not metabolized. Starch hydrolyzed. Nitrates reduced to nitrites.

Aerobic.

Optimum temperature 28–37 C; no growth at 45 C.

Isolated from soil collected from river bank in Venezuela. Rare, found in only 2 of 454 samples.

Type species: *Planobispora longispora* Thiemann and Beretta 1968, 157.

Note: There are two species recognized in the genus. *P. longispora* produces a white aerial mycelium and *P. rosea* produces an aerial mycelium with a light rose tinge.

Genus IX. **Dactylosporangium** *Thiemann, Pagani and Beretta 1967, 43*

Substrate mycelium 0.5–1.0 μm in diameter; hyphae long, irregularly branched, twisted and occasionally coiled. Septation occurs but is not frequent. Do not fragment. Produce tough leathery growth on surface of agar. Gram-positive. Not acid-fast.

Aerial mycelium not formed or is in a very rudimentary form as short and usually unbranched hyphae.

Sporangia, 1.0–1.2 by 4.0–6.0 μm, **formed in clusters on surface of solid media,** emerging from vegetative mycelium on short sporangiophores (Plate 17.8, Figs. 5–9). Usually straight but in some strains long, wavy and branched sporangia can be found. Each sporangium contains a **single row of three to five spores. Large globose spores** or "sporangioles" **formed on the mycelium embedded in the agar.**

Spores motile usually by polytrichous polar flagella, in some strains peritrichous.

Cell wall contains *meso*-diaminopimelic acid, pentoses and hexoses.

Utilizes arabinose, dextrin, fructose, galactose, glucose, inulin, lactose, maltose, mannitol, mannose, raffinose, rhamnose, starch, sucrose and xylose. Does not utilize dulcitol, glycerol, inositol or sorbitol.

Casein hydrolyzed. H_2S produced. Litmus milk not coagulated but completely peptonized.

Grows on a variety of substrates.

Aerobic.

Optimum temperature 37 C; no growth at 45 C. pH optimum 6–7; very slight growth at pH 5.0.

Isolated from cultivated soil; widely distributed.

Type species: *Dactylosporangium aurantiacum* Thiemann, Pagani and Beretta 1967, 43.

Note: There are two species recognized in the genus. *D. aurantiacum* does not utilize ribose, does not hydrolyze tyrosine and gelatin and reduces nitrates to nitrites. *D. thailandense* utilizes ribose, hydrolyzes tyrosine and gelatin and does not reduce nitrates.

Genus X. **Kitasatoa** *Matsumae and Hata in Matsumae, Ohtani, Takeshima and Hata, 1968, 616*

Sporangia club-shaped, 2–2.5 by 5 μm. Spores diplococcus-like, spherical, ellipsoidal or cylindrical, arranged in a single chain within the sporangium and motile by a single polar flagellum. Aerial hyphae abundant and of two types: one type, 1.0–1.2 μm wide, produces long chains of cylindrical conidiospores; the other 1.2–1.5 μm wide produces either terminal sporangia or rounded vesicular bodies. Vegetative mycelium on liquid or solid media produces club-shaped sporangia at the hyphal tips; vegetative hyphae 0.8–1.2 μm wide, highly branched with few septa.

No organic growth factors required. Produce the antibiotic chloramphenicol. Temperature range 10–37 C; pH range 5.0–9.0.

Type species: *Kitasatoa purpurea* Matsumae and Hata in Matsumae, Ohtani, Takeshima and Hata 1968, 617.

TABLE 17.32

Characteristics of species of genus **Kitasatoa**

		1. *K. purpurea* Matsumae and Hata in Matsumae *et al.*, 1968, 617	2. *K. diplospora* Matsumae, Ohtani and Hata in Matsumae *et al.*, 1968, 621	3. *K. kauaiensis* Matsumae, Ohtani and Hata in Matsumae *et al.*, 1968, 621	4. *K. nagasakiensis* Matsumae and Hata in Matsumae *et al.*, 1968, 622
Zoospore		Globose single or in pairs 2.0–2.8 by 1.6–2.3 μm	Ellipsoidal single or in pairs 2.5–3.0 by 1.5 μm	Cylindrical single or in pairs 2.4–5.0 by 1.1–1.35 μm	Rod single or in parallel 2.9 by 0.8 μm
Conidia		Ellipsoidal, smooth 1.5–1.8 by 0.8–1.0 μm	Cylindrical, smooth 1.5–1.3 by 0.7 μm	Cylindrical, warty 1.2–1.5 by 0.8 μm	Ellipsoidal, smooth 1.2–1.6 by 0.8–1.0 μm
Glycerol-Czapek's agar	G.[a]	Dark brownish purple	Pale yellowish brown	Reddish brown	Grayish brown
	A.M.	White	White	White	White
	S.P.	Dark brownish purple	Light brown	Reddish brown	Pale brown
Sucrose-Czapek's agar	G.	Grayish yellow brown	Pale yellow	Colorless	Brownish white
	A.M.	Light pinkish gray	Pinkish orange	Brownish white	Dark brownish gray
	S.P.	None	None	None	None
Oatmeal agar	G.	Yellowish brown	Yellowish gray	Pale olive	Light brownish gray
	A.M.	Light brownish gray	Brownish white	White	Gray white
	S.P.	Yellowish brown	Grayish yellow brown	Pale olive gray	Brownish black
Milk coagulation		−	−	±	+
Milk peptonization		−	−	−	+
Hemolysis		−	+	±	−
Serum liquefaction		−	+	−	−
Collection site		Kauai Island, Hawaii	Kauai Island, Hawaii	Kauai Island, Hawaii	Nagasaki Prefecture, Japan
Type strain		KA-279	KA-280	KA-281	KA-282

[a] G., growth; A.M., aerial mycelium; S.P., soluble pigment.

FAMILY V. **DERMATOPHILACEAE** AUSTWICK 1958, 42, *emend. mut. char.* Gordon 1964, 521

Mycelial filaments or muriform thalli which divide transversely and in at least two longitudinal planes to form masses of coccoid or cuboid cells which characteristically become motile. Aerial mycelium ordinarily absent but sometimes inducible by 10% CO_2. Gram-positive. Not acid fast.

Cell walls of type III (Becker *et al.*, 1965) (see Table 17.18). Pigment production common.

Aerobic.

Includes pathogenic organisms causing skin lesions in mammals, including man.

Type genus: *Dermatophilus* Van Saceghem 1915, 357.

Key to the genera of family **Dermatophilaceae**

I. Mycelium of narrow tapering filaments with lateral branching at right angles; septa formed in transverse and in horizontal and vertical longitudinal planes. Weakly fermentative. Madurose present.

Genus I. *Dermatophilus*

II. Mycelium rudimentary. Produce a muriform, tuber-shaped, non-capsulated, multilocular thallus containing masses of cuboid cells 0.5–2.0 μm in diameter. Madurose absent.

Genus II. *Geodermatophilus*

Genus I. **Dermatophilus** *Van Saceghem 1915, 357, emend. mut. char. Gordon 1964, 521*

Aerial mycelium ordinarily absent. Substrate mycelium consists of **long tapering filaments, branching laterally** at **right angles; septa formed in transverse and in horizontal and vertical longitudinal planes, giving rise to up to eight parallel rows of coccoid cells (spores), each of which becomes motile by a tuft of flagella.** Gram-positive. Cell walls of type III (Becker *et al.*, 1965) (see Table 17.18). Whole cell hydrolysates contain madurose (Lechevalier *et al.*, 1971).

Chemoorganotrophs. **Non-fermentative** but **acid is produced** from certain carbohydrates. **Catalase positive.** Not acid fast.

Growth reported only on complex media; minimum nutritional requirements unknown.

Aerobic and facultatively anaerobic.

Temperature optimum *ca.* 37 C.

Pathogenic for mammals, invading only the uncornified epidermis.

Type species: *Dermatophilus congolensis* Van Saceghem 1915, 357.

Genus II. **Geodermatophilus** *Luedemann 1968, 1857*

Organisms producing a **muriform, tuber-shaped, non-capsulated, holocarpic multilocular thallus** (see Cross, 1970 and glossary) containing masses of cuboid cells 0.5–2.0 μm in diameter. The thallus under favorable environmental conditions breaks up releasing cuboid and coccoid non-motile cells. Some of these cells may develop into **elliptical to lanceolate zoospores** which appear to be **propelled by a terminal tuft of long flagella.** Germinating spores or resting zoospores may divide directly to produce a new thallus or may produce a germ tube and an irregularly constricted and branched filament. The contents of these tubes first divide transversely by septa that do not appear to involve the outer layer of the cell wall, giving rise to a tube of longitudinally compressed cells. Septa are formed later in horizontal and vertical longitudinal planes and give rise to rows of cuboidal cells. **Mycelium rudimentary, aerial mycelium not produced.** Gram-positive.

Cell wall type III (see Table 17.18). Whole cell hydrolysates do not contain madurose (Lechevalier *ct al.*, 1971).

Aerobic.

Type species: *Geodermatophilus obscurus* Luedemann 1968, 1857.

Further Comments

The organisms in this genus appear related to *Dermatophilus* through an obscure filamentous phase. This filamentous condition may be difficult to discover and sometimes may be induced by using a water culture or lean broth medium. The filamentous phase does not appear as a true mycelium but rather like the pseudomycelium seen in certain yeasts as loosely united filaments representing elongations of buds, characteristically pinched and constricted at irregular intervals. These filaments or tubes resemble the holocarpic sporangia of fungi, producing within the tubes spores, which, when liberated, may germinate directly or become zoospores. Occasionally cells of the thallus may germinate *in situ*, emerging as lanceolate zoospores or may produce a germ tube and small filament. The zoospores appear unique in that additional zoospores may develop as terminal buds often forming sluggishly motile chains of lanceolate or elliptical cells. The filamentous phase and zoospore stage may be difficult to observe in some isolates. In other isolates the thallus is reduced to a few cells and may appear reminiscent of miniature fungal dictyospores or sarcinaform packets. The muriform, multilocular thallus and the large number of morphological forms capable of being produced under different conditions of culture and by various strains are the most distinctive and constant features of members of this genus.

FAMILY VI. NOCARDIACEAE CASTELLANI AND CHALMERS 1919, 1040

Aerobic actinomycetes having cell walls of type IV *sensu* Lechevalier and Lechevalier (1965) consisting of *meso*-diaminopimelic acid, arabinose and galactose as distinguishing components. Gram-positive. Mycelium production may be rudimentary or extensive. It can be limited to the substrate or both aerial and substrate hyphae may be common. Spore production varies with the genus.

Key to the genera of family **Nocardiaceae**

I. Spores not produced on differentiated hyphae. Reproductive bodies are mycelial fragments, formed irregularly in substrate or aerial hyphae.

Genus I. *Nocardia*, p. 726

II. Spores produced on differentiated hyphae. Spores borne in chains. Cylindrical spores formed acropetally.

Genus II. *Pseudonocardia*, p. 746

Genus I. **Nocardia** *Trevisan 1889, 9; Nom. cons.* Opin. 13, Jud. Comm. 1954, 153: see Int. Code of Nom. 1958, 166 and note in Index Bergeyana p. 753

Microscopic morphology varies from forms producing sparse mycelium due to early division into coccoid and/or bacillary fragments, to those in which fragmentation is delayed, allowing abundant mycelial production (Plate 17.11, Figs. 1–3). Branched filaments may be found in most cultures but are not common in older cultures of some species. Aerial hyphae are formed by some organisms which may fragment irregularly into coccoid and bacillary cells similar to those produced by substrate mycelia. Members of the genus may be divided into **three morphological groups according to the degree of mycelial development.**

Group I. Members of this group have a limited amount of mycelial development due to the onset of fragmentation in the center of the colony after 12–14 hr incubation (Plate 17.11, Fig. 1). The colonial texture is soft, butyrous or mucoid.

Group II. In members of this group, fragmentation is delayed for 18–20 hr so that an extensively branched mycelium develops (Plate 17.11, Fig. 2). Aerial hyphae are not produced. Colonial texture tends to be flaky, pasty or doughy with a matt surface.

Group III. In members of this group, fragmentation is often delayed for several days allowing abundant mycelial development (Plate 17.11, Fig. 3). Aerial hyphae are commonly produced by members of this group, which may fragment into unequal coccoid and bacillary elements.

The wall of the hypha remains as the wall of the fragments in the manner of arthrospore production in fungi; hence they are not considered spores *sensu stricto*, but mycelial fragments. Colonial texture of members of this group may be crusty, cartilaginous or leathery.

Characteristically, a colony starts by elongation of a coccoid or bacillary mycelial fragment. A branch begins to form after several hours incubation with secondary and tertiary branches following, though fragmentation usually precludes branching in Group I nocardias.

Pre- and post-fission bending can be characterized and related to the morphological groups (McClung, 1949). Fragmentation begins in the center of the young colony producing bacillary elements which continue to elongate and again fragment. The outermost bacillary elements in the center of the microcolony are surrounded by branching peripheral mycelium.

Coccoid cells which have been called **microcysts or chlamydospores** are produced by some members of Groups I and II. These structures are more heat resistant than vegetative cells, surviving 80 C for several hours.

Gram-positive. Some species **acid-fast to partially acid-fast.**

Obligate aerobes. Non-motile.

Pigments are produced by several species of *Nocardia;* carotenoid pigments are the most common and best known.

Generally resistant to lysozyme. Virulent phages are known for a few species; evidence for a lysogenic system has been reported for at least one species.

The G + C content of nocardial DNA varies from *ca* 60–72 moles %. Distribution appears discontinuous and can be roughly correlated with the morphological groups: Group I, 61–63% G + C; Group II, 66–68% G + C and Group III, 68–72% G + C.

Type species: *Nocardia farcinica* Trevisan 1889, 9.

Key to the species of genus **Nocardia**

Key characteristics

Morphological group (see above)
Colony texture and form
Acid fastness
Color of substrate mycelium
Pathogenicity
Reduction of nitrates to nitrites
Hydrolysis of:
- xanthine
- gelatin
- urea
- DNA
- casein

Production of brown exopigment on organic media
Grows on mineral salts medium without added carbon source (the key uses the term "chemoautotrophic," but Hirsch (1958, 1961) states that the organisms probably use CO_2 as a carbon source and volatile hydrocarbons as an energy source)
Resistance to lysozyme
Resistance to penicillin
Acid from arabinose and lactose
Production of blue crystals of indigotin on indole agar
Utilization of dextrin or phenol as carbon and energy source

I. Extensive mycelium produced because fragmentation does not begin until after 5 days incubation. Colonial texture, leathery, dry, crusty or flaky. Thin aerial mycelium commonly present. Long, branched filaments persist in old cultures. Morphological Group III.
  A. Acid-fast or partially acid-fast.
    1. Substrate mycelium white, gray or buff.
      1. *Nocardia farcinica*
    2. Substrate mycelium some shade of yellow, orange or red.
      a. Pathogenic.
        b. Nitrates reduced to nitrites.
        c. Xanthine hydrolyzed.
          2. *Nocardia otitidis-caviarum*

cc. Xanthine not hydrolyzed.

d. Gelatin hydrolyzed.

3. *Nocardia brasiliensis*

dd. Gelatin not hydrolyzed.

4. *Nocardia asteroides*

bb. Nitrates not reduced to nitrites.

5. *Nocardia transvalensis*

aa. Non-pathogenic.

b. Gelatin hydrolyzed.

c. Nitrates reduced to nitrites.

6. *Nocardia formicae*

cc. Nitrates not reduced to nitrites.

7. *Nocardia coeliaca*

bb. Gelatin not hydrolyzed.

c. Brownish exopigment on organic media.

8. *Nocardia polychromogenes*

cc. Exopigment not produced.

9. *Nocardia paraffinae*

AA. Not acid-fast.

1. Chemoautrophic.

a. Substrate mycelium yellow, citron or ochre.

10. *Nocardia petroleophila*

aa. Substrate mycelium not as above.

b. Colonies with raised center surrounded by a wide margin, suggesting Saturn and her rings.

11. *Nocardia saturnea*

2. Heterotrophic.

12. *Nocardia kuroishii*

II. Mycelial development limited due to fragmentation beginning after 20 hr of incubation. Colonial texture soft, pasty or butyrous. Hyphal fragments up to 15–20 μm long persist in mature cultures. Microcysts 0.8–1.5 μm diameter common in older cultures. Morphological Group II.

A. Colonies white, cream or buff.

13. *Nocardia rugosa*

AA. Colonies some shade of yellow, pink or red.

1. Gelatin hydrolyzed.

14. *Nocardia rhodnii*

2. Gelatin not hydrolyzed.

a. Lysozyme resistant.

b. Penicillin sensitive.

c. Acid from L-arabinose.

15. *Nocardia vaccinii*

cc. No acid from L-arabinose.

16. *Nocardia minima*

bb. Penicillin resistant.

c. Urea hydrolyzed.

17. *Nocardia blackwellii*

cc. Urea not hydrolyzed.

18. *Nocardia convoluta*

aa. Lysozyme sensitive.

b. DNA hydrolyzed.

c. Casein hydrolyzed.

19. *Nocardia cellulans*

cc. Casein not hydrolyzed.

d. Nitrates reduced to nitrites.

20. *Nocardia lutea*

dd. Nitrates not reduced to nitrites.

e. Blue crystal of indigotin from indole agar.

21. *Nocardia globerula*

ee. Not as above.

22. *Nocardia rubropertincta*

bb. DNA not hydrolyzed.

c. Dextrin utilized as carbon and energy source.

d. Nitrates reduced to nitrites.

e. Phenol utilized as carbon and energy source.

23. *Nocardia corallina*

ee. Phenol not utilized as carbon and energy source.

24. *Nocardia salmonicolor*

dd. Nitrates not reduced to nitrites.

25. *Nocardia rubra*

cc. Dextrin not utilized as carbon and energy source.

26. *Nocardia opaca*

III. Mycelium production extremely limited due to the early (12–14 hr) initiation of fragmentation. Branching is also sparse causing an ephemeral mycelium which soon fragments into unbranched hyphal elements. Aerial mycelium lacking. Colonial texture soft, pasty and sometimes mucoid. Short bacillary and coccoid cells predominate in older cultures. Morphological Group I.

A. Partially acid-fast.

1. Casein hydrolyzed.

27. *Nocardia calcarea*

2. Casein not hydrolyzed.

a. Nitrates reduced to nitrites.

28. *Nocardia restricta*

aa. Nitrates not reduced to nitrites.

29. *Nocardia erythropolis*

AA. Not acid-fast.

1. Isolated from marine environments.

a. Produce acid from lactose. Lemon-yellow colonies.

30. *Nocardia marina*

aa. Do not produce acid from lactose. Orange-yellow colonies.

31. *Nocardia atlantica*

Genus II. **Pseudonocardia** *Henssen 1957, 408*

Substrate hyphae 0.4–2.6 μm in diameter; aerial hyphae 0.4–1.8 μm in diameter. A characteristic feature is the growth of the hyphae by budding; a constriction is produced just behind the tip of the terminal segment, the tip then elongates to form a new segment, another constriction is formed near the tip and the process repeated (see Figs. 13 and 15, IJSB 1971, *21:* 32).

Spores may be formed on aerial or substrate mycelium in three ways:

1. The usual way is by successive acropetal formation, designated as Pseudonocardia-type (Henssen and Schnepf, 1967), the spore chains arise terminally or laterally.

2. Fragmentation spores produced irregularly along the hyphae.

3. Two or three blastospores develop per segment in a basipetal direction (see Figs. 4 to 7, IJSB 1971, *21:* 30). Spores are smooth walled or spiny and vary greatly in size, usually 0.5–1.0 μm wide by 1.5–3.0 μm long (length may vary from 1.0–4.5 μm).

Gram-positive, not acid-fast. Cell wall type IV (Table 17.18).

Colonies colorless or yellow to orange. Soluble pigment rarely formed. Aerial mycelium white, powdery or forming a thick cover.

Starch not hydrolyzed, milk unchanged.

Grows well on a variety of organic and synthetic media. Minimum growth requirements not known.

Aerobic.

Type species: *Pseudonocardia thermophila* Henssen 1957, 408.

Note: There are two species recognized in the genus. *P. thermophila* has an optimum growth temperature of 40–50 C, whereas *P. spinosa* will not grow at 37 C.

FAMILY VII. **STREPTOMYCETACEAE** WAKSMAN AND HENRICI 1943, 339

Vegetative hyphae (0.5–2.0 μm in diameter) produce a well developed branched mycelium that does not fragment readily. Reproduction by germination of the aerial spores; sometimes by growth of fragments of the vegetative mycelium.

Gram-positive. Aerobic. *Sporichthya* reported as Gram-variable and facultatively anaerobic.

Cell walls contain L-diaminopimelic acid, glycine and no major amounts of arabinose (cell wall Type I of Lechevalier *et al.* 1966, 153).

Primarily soil forms, sensitive to anti-bacterial agents including antibiotics.

G + C content of the DNA, in two genera examined (*Streptomyces* and *Streptoverticillium*) ranged from 69–73 moles %.

Key to the genera of family **Streptomycetaceae**

I. Sporangia-like vesicles not formed.
  A. Non-motile aerial spores produced.
    1. Spores not borne on verticillate sporophores.
      Genus I. *Streptomyces*, p. 748
    2. Spores borne on verticillate sporophores.
      Genus II. *Streptoverticillium*, p. 829
  B. Motile aerial spores produced.
      Genus III. *Sporichthya*, p. 842
II. Sporangia-like vesicles formed.
      Genus IV. *Microellobosporia*, p. 843

Genus I. **Streptomyces** *Waksman and Henrici 1943, 339*

Slender, **coenocytic** hyphae, 0.5–2.0 μm diameter. **The aerial mycelium at maturity forms chains of three to many spores 0.5–2.0 μm in diameter.** The cell wall contains L-diaminopimelic acid and no readily detectable arabinose.

Gram-positive.

On isolation, **colonies are small** (1–10 mm diameter) discrete and lichenoid, leathery or butyrous; **initially relatively smooth surfaced** but **later develop a weft of aerial mycelium** that may appear granular, powdery, velvety or floccose. **Produce a wide variety of pigments** responsible for colors of vegetative mycelium, aerial mycelium and substrate. Many strains produce one or more anti-bacterial, anti-fungal, anti-algal, anti-viral, anti-protozoal or anti-tumor antibiotics. Many strains are sensitive to anti-bacterial agents (antibiotics and synthetic compounds) and to actinophages.

Heterotrophs: **Highly oxidative,** acids not generally detectable in fermentation tests.

Glucose is utilized for growth by all species tested.

Generally hydrolyze one or more of the following: gelatin, casein and starch. Generally reduce nitrates to nitrites.

Aerobes.

Temperature optimum 25–35 C; some species grow at temperatures within the thermophilic range.

Optimal pH range for growth 6.5–8.0.

Species of *Streptomyces* may form synemmata-like, sclerotial or pycnidial structures. There are isolated reports on the occurrence of a definite sexual cycle comprising formation and fusion of motile gametes. Hybridization of wild-type strains has not been reported, but certain auxotrophic mutant strains may be induced to hybridize. Heterokaryosis occurs.

Type species: *Streptomyces albus* (Rossi Doria) **Waksman and Henrici 1943, 339.**

Further Comments

Beyond these general characteristics, groups within the genus *Streptomyces* may be recognized, based on the following:

A. Color of mature sporulated aerial mycelium (spores *en masse*): White; Gray (gray to brownish); Yellow (yellowish to greenish yellow); Red (tan, pink and rose shades); Blue (bluish to grayish blue shades); Green (greenish to grayish green shades); Violet.
B. Spore chain morphology: (See Plate 17.12).
   1. *Rectus Flexibilis* (RF)—spores in straight or flexuous chains.
   2. *Spira* (S):
      a. Spore chains in the form of hooks, open loops or greatly extended coils of large diameter (5–10 μm) (*Retinaculum-Apertum*).
      b. Spore chains in the form of short, gnarled or compact coils or extended, long and open coils (*Spira*).
C. Melanoid pigments: Ability to produce brown, dark brown or black melanin-like pigments (commonly but usually incorrectly referred to as "diffusible or soluble pigments") which color the media in or on which the organisms are grown.
D. Spore wall ornamentation: (See Plate 17.13). Glabrous (smooth), Verrucose (warty), Echinulate (spiny) and Hirsute (hairy).
E. Ability to utilize, under standardized conditions, particular carbon-containing compounds for growth. Those commonly used are: D-glucose, D-xylose, L-arabinose, L-rhamnose, D-fructose, D-galactose, raffinose, D-mannitol, *i*-inositol and sucrose.
F. Ability or inability to produce particular antibiotic factors: groupings depend upon precise identification of particular antibiotics or families of antibiotics; cross-antagonism spectra; antibiotic spectra and the like.
G. Color of vegetative mycelium: colorless, yellow, brown to black, orange, pink, red and red-blue violet.

TABLE 17.33

Primary key to tables of species[a]

Melanoid pigment	Spore surface	White (W)		Gray (GY)		Yellow (Y)		Red (R)		Blue (B)		Green (G)		Violet (V)	
		RF	S	RF	S	RF	S	RF	S	RF	S	RF	S	RF	S
C+	SM	17.41a (751)	17.41c (751)	17.42a (758)	17.42c (764)	17.43a (793)		17.44a (804)	17.44c (808)	17.45a (820)	14.45b (820)				
	WTY								17.44d (812)		17.45c (820)				
	SPY		17.41d (751)		17.42d (768)		17.43c (802)		17.44e (812)		17.45d (820)				17.47b (826)
	H		17.41e (751)		17.42e (768)						17.45e (820)		17.46a (825)		
C−	SM	17.41b (751)	17.41f (753)	17.42b (762)	17.42f (771)	17.43b (795)	17.43d (802)	17.44b (808)	17.44f (814)					17.47a (826)	
	WTY				17.42g (782)				17.44h (818)						
	SPY				17.42h (782)				17.44g (818)		17.45f (820)		17.46b (825)		
	H				17.42i (786)								17.46c (825)		
?	?	17.41g (756)		17.42j (788)		17.43e (802)		17.44i (818)				17.46d (825)			

[a] Melanoid pigments produced C+, not produced C−. Spore wall ornamentation: SM = smooth, WTY = warty, SPY = spiny, H = hairy (see Plate 17.13). Spore color *en masse* indicated as White (W), Gray (GY) *etc.* RF = Rectus flexibilis; spore chains straight or flexuous: S = Spira; Spore chains in form of hooks, open loops and coils (see Plate 17.12).

Figures in parentheses indicate page on which tables on utilization of carbon compounds may be found for each particular group or groups; descriptions of species covered immediately follow the table or begin on a nearby page.

TABLE 17.34a–e

White series[a]

Names of species and subspecies	Strain designation	Utilization of carbon compounds (P & G basal)											
		No carbon control	D-Glucose	D-Xylose	L-Arabinose	L-Rhamnose	D-Fructose	D-Galactose	Raffinose	D-Mannitol	*i*-Inositol	Salicin	Sucrose
17.34a W; RF; C+; SM													
1. Streptomyces albolongus**[b]	T		+	+	+	−	−	+	+	−	−	+	−
2. S. viridaris	T			+		+			+	−	−		+
17.34b W; RF; C−; SM													
3. S. alboniger	T	−	+	+	+	−	+	+	−	+	+	+	−
4. S. albosporeus subsp. labilomyceticus (see no. 392)***[c]	T		+	+	−	−	−	+	+		−	+	+
5. S. albovinaceus	T	−	+	+	+	+	+	+	−	+	−	+	−
6. S. aureocirculatus	T		+		−	−	+	+	−	+	+		−
7. S. baarnensis	T	−	+	+	+	+	+	+	−	+	+		
8. S. clavifer	T	−	+	+	−	+	+	+	−	+	−	−	
9. S. galtieri	T	−	+	−	−	−	+	+	−	−	−	+	
17.34c W; S; C+; SM													
10. S. bobili	T	−	+	+	+	+	+	+	+	−	+	−	+
11. S. longispororuber	R												
17.34d W; S; C+; SPY													
12. S. longisporus	R		+	+	+	+	+		+	+	+		+
17.34e W; S; C+; H													
13. S. herbeus	T		+	−	−	+		+	−	+	−		−

[a] In all tables the reactions for utilization of carbon compounds are for the type (T), neotype (N) or reference (R) strain listed and were obtained on P & G basal medium (Pridham and Gottlieb, 1948) unless noted otherwise. + indicates carbon compound utilized; − not utilized; ± very slight utilization.

[b] ** (Double asterisk) indicates carbon utilization basal medium not specified.

[c] *** (Triple asterisk) indicates carbon utilization basal medium, Czapek's solution.

TABLE 17.34f
White series[a]

Names of species and subspecies	Strain designation	Utilization of carbon compounds (P & G basal)											
		No carbon control	D-Glucose	D-Xylose	L-Arabinose	L-Rhamnose	D-Fructose	D-Galactose	Raffinose	D-Mannitol	*i*-Inositol	Salicin	Sucrose
17.34f W; S; C−; SM													
14. S. albofaciens	T		+	±	+	−	+		+	+	+		±
15. S. albus (see nos. 16, 17 and 35 and *incertae sedis* nos. D 2 and 3).	N	−	+	+	−	−	±	+	−	+	−	+	
16. S. albus subsp. bruneomycini (see nos. 15, 17 and 35 and *incertae sedis* nos. D 2 and 3).	T			−	−	−	+		−	+			
17. S. albus subsp. pathocidicus (see nos. 15, 16 and 35 and *incertae sedis* nos. D 2 and 3).	T	−	+	+	+	+	−	+	−	−	+	−	
18. S. almquisti	T	−	+	+	−	−	+		−	+	−	+	−
19. S. aminophilus	T	−	+	+	+	−	+	+	−	+	−	+	
20. S. cacaoi	T	−	+	+	+	−	+		±	+	−		±
21. S. chrestomyceticus	T	−	+	−	−	−	+	+	−	+	−	−	
22. S. flocculus	T	−	+	+	+	−	+	+	+	+	+	+	
23. S. gibsonii	T	−	+	+	+	−	−		−	+	−	+	
24. S. herbescens	T		+	+	+	+		+	+	+	+		+
25. S. iodoformicus	T		+	−	+	−	+	+	−		−		+
26. S. ochraceiscleroticus	T	−	+	+	+	+	+	+	+	+	+	+	
27. S. rangoon	T		+	+	±	−	+		−	+	−		−
28. S. rimosus (see nos. 29 and 30).	T	−	+	−	+	−	+	+	+	+	+	−	
29. S. rimosus subsp. paromomycinus (see nos. 28 and 30).	T		+	−	−	−	+	+	+	+	+	−	−
30. S. rimosus subsp. pseudoverticillatus (see nos. 28 and 29).	T	−	+	−	+	−	+	+	+	+	+	+	
31. S. spiroverticillatus	T	−	+	+	+	−	+		−	−	−		
32. S. subflavus	T		+		+	−		+	−	+	+		−
33. S. varsoviensis	T	−	+	−	−	−	+	+	−	+	+	+	−
34. S. xantholiticus	T		+	−	−	−		+	−				−

[a] For explanation of meanings of symbols see footnotes to Tables 17.33 and 17.34a–e.

TABLE 17.34g

White series[a]

Names of species and subspecies	Strain designation	Utilization of carbon compounds (P & G basal)											
		No carbon control	D-Glucose	D-Xylose	L-Arabinose	L-Rhamnose	D-Fructose	D-Galactose	Raffinose	D-Mannitol	*i*-Inositol	Salicin	Sucrose
17.34g W; ?; ?; ?													
35. S. albus subsp. fungatus	T												
36. S. hydrogenans	T												
37. S. vendargus	T		+	−	−	−	+		+	+	+		−

[a] For explanation of meanings of symbols see footnotes to Tables 17.33 and 17.34a–e.

TABLE 17.35a

Gray series[a]

Names of species and subspecies	Strain designation	Utilization of carbon compounds (P & G basal)											
		No carbon control	D-Glucose	D-Xylose	L-Arabinose	L-Rhamnose	D-Fructose	D-Galactose	Raffinose	D-Mannitol	*i*-Inositol	Salicin	Sucrose
17.35a GY; RF; C+; SM													
38. S. achromogenes (see no. 80).	T	−	+	+	+	+	+	+	−	+	+	+	−
39. S. antibioticus	T	−	+	+	+	+	+	+	−	+	+	−	−
40. S. bikiniensis	T	−	+	+	−	−	−	+	−	−	−	±	±
41. S. cacaoi subsp. asoensis (see no. 20).	T		+	+	+		+	+	+		+		+
42. S. cinereoruber (see no. 43).	T		+	+	+	−	−	+	−	−	−	+	−
43. S. cinereoruber subsp. fructofermentans (see no. 42).	T	−	+	+	+	+	+	+	−	−	−	+	
44. S. cylindrosporus subsp. piceus (see *incertae sedis* no. A3).	T												
45. S. ederensis	T												
46. S. fulvoviolaceus	T	−	+	+	+	+	+	+	+	+	+	+	+
47. S. fulvoviridis	T	−	+	+	+	+	+	+	−	+	−	−	−
48. S. gardneri	T	−	+	+	+	+	+	+	+	−	−	+	
49. S. globosus	R	−	+	+	+	−							
50. S. griseorubiginosus	R		+	+	+	+	+		+	+	+		+
51. S. herbaricolor	T	−	+	+	+	−	+	+	+	−	−	−	+
52. S. indigoferus	T	−	+	+	+	−	−	+	−	−	−	−	
53. S. litmocidini	T		+	±	+	−	±		−	−	−		−
54. S. narbonensis (see no. 73).	T	−	+	+	+	+	+		+	−	−		+
55. S. nashvillensis	T	−	+	+	+	−	−	+	−	−	−	+	±
56. S. noboritoensis	T		+	+	+	−	+		+	+	+	+	±
57. S. phaeopurpureus	T	−	+	+	+	+	+	+	+	+	+	−	
58. S. purpeofuscus	T	−	+	+	+	−	−	+	−	−	−	−	
59. S. showdoensis	T		+	+	±	−	+	+	−	−	−	+	±
60. S. tanashiensis (see no. 265).	T	−	+	+	+	−	−	+	−	−	−	+	−
61. S. violaceorectus	T		+	+	+	−	+		−	−	−		+
62. S. zaomyceticus	T	−	+	+	+	−	−	+	−	−	−	+	+

[a] For explanation of meanings of symbols see footnotes to Tables 17.33 and 17.34a–e.

TABLE 17.35b

Gray series[a]

Names of species and subspecies	Strain designation	Utilization of carbon compounds (P & G basal)											
		No carbon control	D-Glucose	D-Xylose	L-Arabinose	L-Rhamnose	D-Fructose	D-Galactose	Raffinose	D-Mannitol	*i*-Inositol	Salicin	Sucrose
	17.35b	GY; RF; C—; SM											
63. S. aburaviensis	T	−	+	±	−	−	±	−	−	−	−	−	−
64. S. caeruleus	T	−	+	−	−	−							
65. S. catenulae	T	−	+	−	−	−	+	+	−	+	−	−	
66. S. chrysomallus subsp. fumigatus (see no. 288).	T	−	+	+	+	−	−	+	−	−	−	−	
67. S. flavogriseus	T	−	+	+	+	+							
68. S. flavovirens	T	−	+	+	+	+	+	+	+	+	−	−	−
69. S. gelaticus	T	−	+	+	−	+	−	+	+	−	−	+	+
70. S. griseolus	T	−	+	+	+	−	+	+	+	−	−	−	
71. S. halstedii	T	−	+	+	+	−	+	+	+	−	−	+	−
72. S. misakiensis	T		+	−	−	−	+		+	−	−		+
73. S. narbonensis subsp. josamyceticus (see no. 54).	T		+	+	+	−	+	+	+	−	−		+
74. S. nigrifaciens	T	−	+	+	+	+	+	+	+	+	+	−	
75. S. nitrosporeus	T	−	+	+	+	+	−	+	−	−	−	−	−
76. S. olivarius	T												
77. S. omiyaensis	T	−	+	+	−	+	−	+	−	−	−	−	
78. S. ramulosus	T	−	+	−	−	−	±	+	+	+	−	−	−
79. S. xanthocidicus	T		+	+	+	−	+	+	−	−	−	−	+

[a] For explanation of meanings of symbols see footnotes to Tables 17.33 and 17.34a–e.

TABLE 17.35c

Gray series[a]

Names of species and subspecies	Strain designation	Utilization of carbon compounds (P & G basal)											
		No carbon control	D-Glucose	D-Xylose	L-Arabinose	L-Rhamnose	D-Fructose	D-Galactose	Raffinose	D-Mannitol	i-Inositol	Salicin	Sucrose
17.35c GY; S; C+; SM													
80. S. achromogenes subsp. rubradiris (see also no. 38).	T	±	+	+	+	+	+	+	+	+	±	±	+
81. S. anandii	T		+	+	+	−	+	+	+	+	+	−	
82. S. aurantiogriseus	R		+	+	+	+	+		+	+	+		+
83. S. bobili subsp. sporificans (see also no. 10).	T												
84. S. cinerochromogenes	T			−	+	+			−	−	−	+	+
85. S. cirratus	T	−	+	+	+	−	+	+	−	−	−	−	+
86. S. collinus	T		+	+	+	+	+		+	+	+		+
87. S. eurythermus	T	−	+	+	+	−	+	+	+	+	−	−	+
88. S. galbus	T	−	+	+	+	−	+		−	+	+		−
89. S. galilaeus	T	−	+	+	+	+							
90. S. griseoruber	T	−	+	+	+	+	+	+	−	−	+	+	−
91. S. griseosporeus	T		+	+	+	+	+	+	+	+	+	+	+
92. S. hygroscopicus subsp. ossamyceticus (see also nos. 152, 153, 154, 155 and *Incertae sedis* no. B17).	T	−	+	+	+	+	+	+	+	+	+	−	+
93. S. kurssanovii	T	−	+	+	+	−	+	+	+	−	−	−	+
94. S. luteogriseus	T	−	+	+	+	+	+	+	+	+	+	+	+
95. S. massasporeus	T	−	+	+	+	+	+	+	+	+	+	+	+
96. S. mirabilis	R	−	+	+	+	+							
97. S. multispiralis	T		+	+	+	−	+	+	+	+	+	+	+
98. S. naganishii	T	−	+	+	+	+	+	+	+	+	+	+	−
99. S. neyagawaensis	T	−	+	+	+	+	+	+	+	+	+	+	+
100. S. nojiriensis	T		+	−	−	−	−	−	−	−	−	+	−
101. S. olivochromogenes	T	−	+	+	+	+	+	+	+	+	+	+	
102. S. phaeofaciens	T	−	+	+	+	+							
103. S. pulveraceus	T	−		+	−	+	+	+	+	−	−	+	−
104. S. rameus	T	−		+	+	−	+	+	+	+	−	+	+
105. S. resistomycificus	T		+	+	+	+	+		+	+	+		+
106. S. rishiriensis	T	−	+	+	+	+	+	+	+	−	+	+	+
107. S. thermoviolaceus (see also no. 267).	T												
108. S. violaceochromogenes	R		+	+	+	+	+		+	+	+		+

[a] For explanation of meanings of symbols see footnotes to Tables 17.33 and 17.34a–e.

TABLE 17.35d, e

Gray series[a]

Names of species and subspecies	Strain designation	Utilization of carbon compounds (P & G basal): No carbon control	D-Glucose	D-Xylose	L-Arabinose	L-Rhamnose	D-Fructose	D-Galactose	Raffinose	D-Mannitol	i-Inositol	Salicin	Sucrose
17.35d GY; S; C+; SPY													
109. S. afghaniensis* **	T		+	+	+	+	+	+	+	±	±		+
110. S. arenae	T	−	+	+	+	+	+	+	+	+	+		+
111. S. atrocyaneus	T												
112. S. chromofuscus	R		+	+	+	+	+		+	+	+		−
113. S. durhamensis	T	−	+	+	+	−	+	+	+	+	+	−	
114. S. echinatus	T	−	+	+	+	+	+	+	+	+	+	−	−
115. S. filipinensis	T	−	+	+	+	−	+	+	+	+	+	−	+
116. S. fimbriatus	R	−	+	+	+	+	+	+	+	+	+	−	
117. S. griseochromogenes	T	−	+	+	+	−	+	+	+	+	+	−	+
118. S. iakyrus	T	−	+	+	+	+	+	+	+	+	+	+	
119. S. lucensis	T		+	+	+	−	+		−	+	−		+
120. S. malachitofuscus	T		+	+	+	+	+		−	+	+		+
121. S. malachitorectus	T		+	+	+	+	+		−	+	+		+
17.35e GY; S; C+; H													
122. S. pilosus	T		+	+	+	+	+		−	+	+		−

[a] For explanation of meanings of symbols and asterisks, see footnotes to Tables 17.33 and 17.34a–e.

TABLE 17.35f

Gray series[a]

Names of species and subspecies	Strain designation	Utilization of carbon compounds (P & G basal)											
		No carbon control	D-Glucose	D-Xylose	L-Arabinose	L-Rhamnose	D-Fructose	D-Galactose	Raffinose	D-Mannitol	*i*-Inositol	Salicin	Sucrose
17.35f GY; S; C−; SM													
123. S. albidofuscus	T	−	+	−	−	−	±	±	−	−	−	−	−
124. S. albogriseolus	T	−	+	+	+	+	+	+	+	+	+	+	+
125. S. ambofaciens	T	−	+	+	+	+	+		−	+	+		+
126. S. anthocyanicus	T		+		+	+	+	+	−	−	+		−
127. S. antimycoticus	T	−	+	+	+	+	+		−	+	+	+	+
128. S. argenteolus	T	−	+	+	+	+	+	+	−	+	−	+	−
129. S. atratus	T	−	+	+	−	+	+	+	+	−	−	+	
130. S. aureofaciens	T	−	+	±	+	−	+	+	−	−	−	−	+
131. S. avellaneus	T	−	+	±	−	−	+		−		−		+
132. S. caesius	T		+		−	+	+	+	−	+	+		−
133. S. carnosus	T	−	+	+	+	+							
134. S. chibaensis	T	−	+	+	+	+	+	+	+	+	+	−	+
135. S. coelescens	T		+		+					−			
136. S. coelicolor subsp. achrous (see also nos. 137, 138, 139 and 290).	R	−	+	+	+	+	+	+	+	+	+	+	
137. S. coelicolor subsp. coelicoferus (see also nos. 136, 138, 139 and 290).	T		+		+	+	+	+	−	+	+		+
138. S. coelicolor subsp. coelicolatus (see also nos. 136, 137, 139 and 290).	T		+		+	+	+	+	+	−	+		−
139. S. coelicolor subsp. coelicovarians (see nos. 136, 137, 138 and 290).	T		+		+	+	+	+	−	−	+		−
140. S. corchorusii	T	−	+	+	+	+	+		+	+	+		+
141. S. cyanogenus	T		+	+	+	+	+		+	+	+		+
142. S. diastaticus subsp. ardesiacus (see also *incertae sedis* no. A4).	T	−	+	+	+	+							
143. S. diastatochromogenes subsp. bracus (see also *incertae sedis* no. A5).	T		+	−	−	+	+	+	+	+	+	+	+
144. S. endus	T	−	+	+	+	+	+	+	−	+	−	+	−
145. S. erumpens	T	−	+	−	+	−	+	+	+	+	+	−	−
146. S. griseoaurantiacus	T		+		+	+	+	+	+	+	−		−
147. S. griseofuscus	T		+	+	+	−	+		−	+	−		−
148. S. griseolosuffuscus	T	−	+	+	+	−	+	+	−		−	−	−
149. S. griseoluteus	T	−	+	+	−	−	+	+	−	+	−	−	
150. S. griseus subsp. difficilis (see also nos. 302, 303, 304, 305 and *incertae sedis* nos. B15 and B16).	T												
151. S. humidus	T		+	+	+	+	+	+	−	+	+	+	−
152. S. hygroscopicus (see also nos. 92, 153, 154, 155 and *incertae sedis* no. B17).	R	−	+	+	+	+	+		−		−	+	
153. S. hygroscopicus subsp. angustmyceticus (see also nos. 92, 152, 154, 155 and *incertae sedis* no. B17).	T	−	+	−	−	−		±	±	+	−	−	+

TABLE 17.35f—*Continued*

Names of species and subspecies	Strain designation	Utilization of carbon compounds (P & G basal)											
		No carbon control	D-Glucose	D-Xylose	L-Arabinose	L-Rhamnose	D-Fructose	D-Galactose	Raffinose	D-Mannitol	*i*-Inositol	Salicin	Sucrose
154. S. hygroscopicus subsp. decoyicus (see also nos. 92, 152, 153, 155 and *incertae sedis* no. B17).	T	−	+	+	−	−	+	+	−	+	+	−	
155. S. hygroscopicus subsp. globosus (see also nos. 92, 152, 153, 154 and *incertae sedis* no. B17).	T	−	+	+	−	−	+	+	+	+	+	−	+
156. S. libani (see also no. 157).	T		+	+	−	−	+		+		+		+
157. S. libani subsp. rufus (see also no. 156).	T		+	+	+	−	+		+		+		+
158. S. lividans	T				−		+	+	−	+	+		−
159. S. lusitanus	T	−	+	−	±	−	+		−	−	±	−	+
160. S. lydicus	T	−	+	+	+	−	+	+	+	+	−		+
161. S. melanosporofaciens	T		+	+	+	+	+		+	+	+		−
162. S. misionensis	T	−	+	+	+	−	+	+	+	+	+	−	−
163. S. murinus	T	−	+	+	−	−	+		−	+			−
164. S. mutabilis	T	−	+	+	+	+	+	+	−	+	+	−	±
165. S. nigrescens	T		+	+	−	−	+		+	+	+		+
166. S. nodosus	T	−	+	+	−	+	+		−	+	+		−
167. S. nogalater	T	−	+	+	+	+	+	+	+	+	+	−	−
168. S. olivaceiscleroticus	T	−	+	+	+	+							
169. S. olivaceoviridis	T		+	+	+	+	+		+	+	+		+
170. S. olivaceus	T	−	+	+	+	+	+	+	+	+	+	−	−
171. S. parvullus	T	−	+	+	+	+	+		−	+	+	+	+
172. S. platensis	T	−	+	−	−	−	+		+	+	+		
173. S. plicatus	T	−	+	+	+	+	+	+	−	+	+		−
174. S. poonensis	T	−	+	+	+	+	+	+	+	+	+	+	
175. S. psammoticus	T	−	+	−	−	−	+		−	−	−	−	+
176. S. purpurogeniscleroticus	T	−	+	+	+	+	+	+	+	+	+	+	
177. S. recifensis	T		+	+	+	−	+	+	+	+	−	+	+
178. S. rochei	T	−	+	+	+	+	+	+	−	+	+	+	−
179. S. rokugoensis	T	−	+	+	+	+	+		−	+	+	+	
180. S. roseodiastaticus	T	−	+	+	+	+							
181. S. rutgersensis subsp. castelarense (see also no. 319).	T		+	+	+	+	+	+	+	+	−	+	−
182. S. sayamaensis	T			−	−	−	+	+	−	−	−	−	+
183. S. sendaiensis	T			−	−	+	−	+	−	+	+	+	−
184. S. sioyaensis	T		+	+	−	−	+	+	+	+	+	−	+
185. S. tendae	T	−	+	+	+	+	+		−	+	+		+
186. S. thermovulgaris	T		+	+	+	+	+		+	+	+		+
187. S. tricolor	T												
188. S. tubercidicus	T	−	+	−	−	−	+		+	+	+		+
189. S. tumemacerans	T	−	+	−	−	+	+	+	−		+		−
190. S. vastus	T		+	+	+	+	+	+	+	+	+	+	+
191. S. violaceolatus	T		+	+	+	+	+		+	+	+		+
192. S. violaceus-niger (see also *incertae sedis* no. B33).	R	−	+	+	+	+	+		+	+	+	+	
193. S. violaceus-ruber	T	−	+	+	+	+	+	+	−	+	+	+	−
194. S. viridifaciens	T	−	+	−	−	−	+	+	−	−	−	−	+

[a] For explanation of meanings of symbols see footnotes to Tables 17.33 and 17.34a–e.

TABLE 17.35g–h

Gray series[a]

Names of species and subspecies	Strain designation	Utilization of carbon compounds (P & G basal)											
		No carbon control	D-Glucose	D-Xylose	L-Arabinose	L-Rhamnose	D-Fructose	D-Galactose	Raffinose	D-Mannitol	*i*-Inositol	Salicin	Sucrose
17.35g GY; S; C−; WTY													
195. S. atroolivaceus (see also no. 204).	T		+	+	+	+	+						
196. S. cyanocolor	T		+		−	+	+	+	−		+		−
197. S. graminofaciens	T	−	+	+	+	+							
198. S. griseoplanus	T	−	+	+	+	−	+	+	+	−	−	−	−
17.35h GY; S; C−; SPY													
199. S. albaduncus	T		+	+	+	+	+	+	+	+	+	+	±
200. S. albospinus	T		+	±	−	−	+	+	+	+	+	+	−
201. S. albulus	T	−	+	−	−	−	+	+	−	+	+	+	
202. S. althioticus	T		+	+	+	+	+	+	−	+	+		±
203. S. arabicus	T	−	+	+	+	+	+		−	+	+		
204. S. atroolivaceus subsp. mutomycini (see also no. 195).	T												
205. S. canus	T	−	+	+	+	+	+	+	±	+	+		+
206. S. chattanoogensis	T	−	+	−	−	−	+		+	+	+		+
207. S. chlorobiens	T		+	+	+	+		+	+	+	+		+
208. S. cuspidosporus	T	±	+	+	+	+	+	+	+	+	+	+	+
209. S. gancidicus	T	−	+	+	+	+	+	+	−	+	+	−	−
210. S. griseoflavus	R		+	+	+	+	+		−	+	+		−
211. S. griseoincarnatus	T		+	+	+	+	+		−	+	±		+
212. S. griseorubens	T		+	+	±	+	+		−	+	±		−
213. S. macrosporeus	T	−	+	+	+	+	+		−	+	+		−
214. S. malachiticus	T	−	+	+	+	+	+	+	+	+	+	−	+
215. S. matensis	T	−	+	+	+	+	+	+	−	+	+		−
216. S. noursei	T	−	+	−	−	−	+	+	−	+	+	−	+
217. S. olivoviridis	T	−	+	+	+	+	+	+	−	+	−	−	−
218. S. pseudogriseolus	T	−	+	+	+	+	+	+	−	+	+	+	
219. S. rubiginosus	T		+	+	±	+	+		−	+	+		+
220. S. sparsogenes	T	−	+	+	+	+	+	±	+	+	±	−	+
221. S. viridiviolaceus	T												
222. S. virido-diastaticus	T		+	+	+	+	+			+	+		

[a] For explanation of meanings of symbols see footnotes to Tables 17.33 and 17.34a–e.

TABLE 17.35i

Gray series[a]

Names of species and subspecies	Strain designation	Utilization of carbon compounds (P & G basal)											
		No carbon control	D-Glucose	D-Xylose	L-Arabinose	L-Rhamnose	D-Fructose	D-Galactose	Raffinose	D-Mannitol	i-Inositol	Salicin	Sucrose
17.35i GY; S; C−; H.													
223. S. calvus	T	−	+	+	+	+	+	+	+	+	+	+	+
224. S. cyanoalbus	T	−	+	+	+	+	+	+	+	+	−	−	+
225. S. finlayi	T		+	+	+	+	−		−	−	−		±
226. S. flaveolus	T	−	+	+	+	+	+	+	+	+	+	+	+
227. S. geysiriensis	T												
228. S. herbiferis	T		+	+	+	+		+	+	+	−		+
229. S. pactum	T	−	+	−	−	−	−	+	−	−	−	−	−

[a] For explanation of meanings of symbols see footnotes to Tables 17.33 and 17.34a–e.

TABLE 17.35j
Gray series[a]

Inadequately described species in the gray series; names of species and subspecies	Strain designation	Morphological section	Melanoid pigment formation	Spore surface configuration	Utilization of carbon compounds (P & G basal)											
					No carbon control	D-Glucose	D-Xylose	L-Arabinose	L-Rhamnose	D-Fructose	D-Galactose	Raffinose	D-Mannitol	*i*-Inositol	Salicin	Sucrose
17.35j GY; ?; ?; ?																
230. S. akitaensis	T	S	−?			+	−	−	+	−	+	±	+	−	±	−
231. S. akiyoshiensis**	T	S	+		−		+	+	+	+	+	+	+	+	−	
232. S. alanosinicus	T	S	+			+	+	+	−	+	+	+	+	+	+	
233. S. albidus subsp. invertens	R	S	+?													
234. S. albochromogenes**	T	S	+?			+	+	+	+	+	+	+	+	+	−	+
235. S. ansochromogenes (see also no. 236).	T	S	+		−	+	+	−	−	+	+	−	+	−	−	−
236. S. ansochromogenes subsp. pallens (see also no. 235)	T	S	+		−	+	+	−	−	+	+	−	+	−	−	−
237. S. avidinii	T	S	+				−	−				−				
238. S. carcinomycicus	T	S	−				−		−			+			+	
239. S. castaneoglobisporus	T	RF	+		−	+	+	−	+	+	+	+	+	+	−	−
240. S. castaneus	T		+		−	+	+	−	−	+	−	−	−	−	−	+
241. S. cyanoflavus	T	RF	+		−	+	−	−	−	+		+	−	+	−	+
242. S. djakartensis	T	S	+													
243. S. erythrochromogenes subsp. narutoensis (see also *incertae sedis* no. A6).	T															
244. S. glomerochromogenes	T	S			−	+	−	−	−	+	−	−	−	+	−	−
245. S. grisinus	T	S	−	Hairy?												
246. S. haranomachiensis	T	RF	−			+	+	−	−	+	+	+	+	+	+	+
247. S. hygrostaticus	R	S	−				+	+	+			+	+	+	+	+
248. S. insulatus	T	S		Smooth												
249. S. inversochromogenes	T	S		Smooth	−	+	+	−	−	+	+	−	+	+	−	−
250. S. kitazawaensis	T	RF	+		−	+	+	+	−	±	+	+	+	+	+	−
251. S. mariensis**	T	S	−		−		−	+		+	+	−	+	+	+	+
252. S. minutiscleroticus	T	S	−			+		+	+	+	+	−	+		+	−
253. S. mitakaensis	T	S	−			+	+	+	+	+	+	+	+	+	−	+
254. S. nigrogriseolus	T	RF		Smooth												
255. S. ogaensis	T	S	+			+	−	−		−	+	+		+		−
256. S. piedadensis	T		−		−	+	+	+	+	+	+	+	−	−	−	
257. S. regensis	T	S	+			+	+	+	−	+	−		+	+		+
258. S. robefuscus	T	S	+			+		+	+	+	+	+		+		+
259. S. robeus	T	S	+			+		+	+	+	+	+		+		+
260. S. robustrus	T	S	+			+		+	+	+	+	+		+		+
261. S. roseogriseolus	T	RF		Smooth												
262. S. roseogriseus	T	S		Smooth												
263. S. sahachiroi	T	S	−			+	+	+	−	+	+	−	+	−	+	+
264. S. senoensis	T	RF	+			+	+	+	−	+	−	−	−	−	+	
265. S. tanashiensis subsp. cephalomyceticus*** (see also no. 60).	T	RF	+?			+	+	+	−	+	+	−	−	−	+	+
266. S. thermonitrificans	T	RF	+			+	−	−	−	−	+	−	+	+	−	−
267. S. thermoviolaceus subsp. apingens (see also no. 107).	T	S		Warty?												
268. S. viridoniger	T	S	−				+	+	−	+	+	+	+	+	+	+
269. S. werraensis	T	S	−													

[a] For explanation of meanings of symbols and asterisks, see footnotes to Tables 17.33 and 17.34a-e.

TABLE 17.36a

Yellow series[a]

Names of species and subspecies	Strain designation	Utilization of carbon compounds (P & G basal)											
		No carbon control	D-Glucose	D-Xylose	L-Arabinose	L-Rhamnose	D-Fructose	D-Galactose	Raffinose	D-Mannitol	*i*-Inositol	Salicin	Sucrose
17.36a Y; RF; C+; SM													
270. S. alboflavus	T	−	+	+	+	−	+	+	+	+	+	−	+
271. S. bacillaris	T	−	+	+	−	−	+	+	+	+	+	+	
272. S. cavourensis	T	−	+	+	−	−	+		−	+	−		
273. S. cyaneofuscatus	T	−	+	+	−	+	+	+	−	+	−	+	+
274. S. fulvissimus	R	−	+	+	+	−	+		−	+	+	+	
275. S. griseobrunneus	T	−	+	+	−	−	+	+	+	+	−	+	+
276. S. michiganensis	T	−	+	+	−	−	+		±	+	+		−
277. S. tsusimaensis	T		+	+	+	−	+	+	−	+	−	+	+
278. S. xanthochromogenus (*sic*)	T		+	+	±	±	+		±	+	±		±

[a] For explanation of meanings of symbols see footnotes to Tables 17.33 and 17.34a–e.

TABLE 17.36b

Yellow series[a]

Names of species and subspecies	Strain designation	Utilization of carbon compounds (P & G basal)											
		No control carbon	D-Glucose	D-Xylose	L-Arabinose	L-Rhamnose	D-Fructose	D-Galactose	Raffinose	D-Mannitol	*i*-Inositol	Salicin	Sucrose
	17.36b	Y; RF; C=; SM											
279. S. albidoflavus	R	−	+	+	+	−	+		−	+	−		−
280. S. alboviridis	R	−	+	+	−	+	+		−	+	−		−
281. S. anulatus	T	−	+	+	+	+	+		−	+	−		−
282. S. badius	T	−	+	+	+	−	+		−	+	−		−
283. S. californicus	T	−	+	+	−	−	+	+	−	+	−	−	−
284. S. canescens	T	−	+	−	+	−	+		−	−	−	−	−
285. S. celluloflavus	T	−	+	−	−	−							
286. S. cellulosae	T	−	+	+	+	+	+		−	+	+		
287. S. champavatii	T	−	+	+	+	−	+	+	−	+			−
288. S. chrysomallus (see no. 66).	T	−	+	+	+	+	+	+	−	+	−	+	−
289. S. citreofluorescens	T	−	+	+	+	+	+	+	−	+	−	−	−
290. S. coelicolor (see no. 136, 137, 138 and 139).	T	−	+	+	+	−	+		−	+	−		
291. S. felleus	R	−	+	+	+	+			−	+	−	+	
292. S. fimicarius	T	−	+	+	+	+	+	+	−	+	−	−	
293. S. floridae	T		+	+	−	−	+	+	−	+	−	−	
294. S. fluorescens	T	−	+	+	+	−	+	+	−	+	−	+	−
295. S. globisporus (see nos. 296, 297, 298 and *incertae sedis* no. B14).	R	−	+	+	+	+	+		−	+	−		−
296. S. globisporus subsp. caucasicus (see nos. 295, 297, 298 and *incertae sedis* no. B14).	T	−	+	+	+	−							
297. S. globisporus subsp. flavofuscus (see nos. 295, 296, 298 and *incertae sedis* no. B14).	T	−	+	+	+	+							
298. S. globisporus subsp. vulgaris (see nos. 295, 296, 297 and *incertae sedis* no. B14).	R		+	+	+	+	+	+	−	+	−		−
299. S. gougeroti	T	−	+	−	+	−							
300. S. griseinus	T		+	+	+	+	+	+	−	+	−	−	−
301. S. griseoloalbus	T		+	+	+	+	+			+	+		+
302. S. griseus (see nos. 150, 303, 304, 305 and *incertae sedis* nos. B15 and B16).	R	−	+	+	−	−	+	+	−	+	−	+	−
303. S. griseus subsp. alpha (see nos. 150, 302, 304, 305 and *incertae sedis* nos. B15 and B16).	T	−	+	+	−	+							
304. S. griseus subsp. cretosus (see nos. 150, 302, 303, 305 and *incertae sedis* nos. B15 and B16).	R	−	+	+	−	+	+		−		−	+	
305. S. griseus subsp. solvifaciens (see nos. 150, 302, 303, 304 and *incertae sedis* nos. B15 and B16).	T	−	+	+	+	−	+		−	+	−	+	
306. S. intermedius	R	−	+	+	+	−	+	+	+	+	−	+	+
307. S. kanamyceticus	T	−	+	+	+	−	+	+	+	+	−	+	
308. S. levoris	T	−	+	+	+	−	+	+	−	+	−	+	−

TABLE 17.36b—*Continued*

Names of species and subspecies	Strain designation	Utilization of carbon compounds (P & G basal): No carbon control	D-Glucose	D-Xylose	L-Arabinose	L-Rhamnose	D-Fructose	D-Galactose	Raffinose	D-Mannitol	*i*-Inositol	Salicin	Sucrose
17.36b Y; RF; C−; SM—(*Continued*)													
309. S. limosus	T		+	+	+	−	+		−	+	−		−
310. S. lipmanii	T	−	+	+	−	+	+	+	+	+	−	+	±
311. S. microflavus	T	−	+	+	−	+	+	+	−	+	−	+	−
312. S. odorifer	R	−	+	+	+	−	+		−	+	+	+	
313. S. parvus	R	−	+	+	+	+	+	+	−	+	−	−	
314. S. pluricolorescens	T	−	+	+	−	+	+		−	+	−		−
315. S. pneumonicus	T		+	+	−	−		+	−	+			−
316. S. praecox	T	−	+	+	+	+	+	+	+	+	−	+	
317. S. puniceus	T	−	+	+	+	−	+	+	−	+	−	+	±
318. S. raffinosus	T	−	+	+	−	+	+	+	+	+	−	+	+
319. S. rutgersensis	T	−	+	+	+	−	+	+	+	+	−	+	−
320. S. sampsonii	T		+	+	+	−	+		−	+	−		−
321. S. setonii	T	−	+	+	+	+	+		−	+	−		−
322. S. sindenensis	T	−	+	+	+	−	+	+	−	+	−	+	
323. S. sulphureus	R	−	+	+	+	−	+		+		−	+	
324. S. willmorei	T	−	+	+	−	+							

[a] For explanation of meanings of symbols see footnotes to Tables 17.33 and 17.34a–e.

TABLE 17.36c–e

Yellow series[a]

Names of species and subspecies	Strain designation	Utilization of carbon compounds (P & G basal): No carbon control	D-Glucose	D-Xylose	L-Arabinose	L-Rhamnose	D-Fructose	D-Galactose	Raffinose	D-Mannitol	*i*-Inositol	Salicin	Sucrose
17.36c Y; S; C+; SPY													
325. S. hawaiiensis	T	−	+	+	+	+	+	+	+	+	+		+
17.36d Y; S; C−; SM													
326. S. albohelvatus	T		+		+	−		+	−	−	−		−
327. S. aurigineus	T		+	−	+	−	−	+	−	−	−		−
328. S. canarius	T	−	+	+	+	+	+	+	+	+	+	−	+
329. S. chryseus	T		+		+	−		+	−	−	−		−
330. S. flavidovirens	T	−	+	+	+	+			−	−			+
331. S. helvaticus	T		+		+	−		+	−	−	−		−
332. S. longisporoflavus	R	−	+	+	+	+	+	+	−	+	−	+	−
333. S. niveus	T	−	+	+	+	+	+	+	−	+	−	−	
334. S. paucidiastaticus	T		+	−	+	−	+	+	+		+	−	−
335. S. spheroides	T	−	+	+	−	+	+		−	+	−		
17.36e Y:?:C−:?													
336. S. pimprina	R		+		+	+	+	+	+	+		+	+

[a] For explanation of meanings of symbols see footnotes to Tables 17.33 and 17.34a–e .

TABLE 17.37a, b

Red series[a]

Names of species and subspecies	Strain designation	Utilization of carbon compounds (P & G basal)											
		No carbon control	D-Glucose	D-Xylose	L-Arabinose	L-Rhamnose	D-Fructose	D-Galactose	Raffinose	D-Mannitol	*i*-Inositol	Salicin	Sucrose
17.37a R; RF; C+; SM													
337. S. capoamus	T		+	+	+	−	+	+		+	+	+	+
338. S. cinnabarinus	T		+	+	+	+	+		+	+	+		+
339. S. crystallinus	R		+										−
340. S. flavotricini	T		+	−	−	−	±		−	−	−		−
341. S. gobitricini	T	−	+	+	+	+	+	+	−	−	+	−	
342. S. lincolnensis	T	−	+	+	+	+	+	+	+	+	+	+	+
343. S. melanogenes	T	−	+	+	+	−	+		+	+	+		±
344. S. phaeochromogenes (see no. 345).	R	−	+	+	+	+	+	+	+	+	+	+	
345. S. phaeochromogenes subsp. chloromyceticus (see no. 344).	T	−	+	+	+	−	+	+	−	−	−	+	
346. S. pseudovenezuelae	T		+	+	−	+	−	+	−	−	+		−
347. S. roseoviridis	T	−	+	+	+	−	−	+	−	−	−	−	
348. S. spectabilis	T	−	+	+	−	−	+	+	+	+	+	−	−
349. S. subrutilus	T		+	−	−	−	+	+	−	−	−	−	+
350. S. umbrinus	T	−	+	+	+	+	+	+	+	+	+	−	
351. S. venezuelae	T	−	+	+	+	+	+	+	−	−	−	+	
352. S. xanthophaeus	T		+	−	−	−	−		−	−	−		−
17.37b R; RF; C−; SM													
353. S. aureomonopodiales	T	−	+	+	+	−	+	+	−	+	+	+	
354. S. exfoliatus	T	−	+	+	+	+	+	+	+	−	−	+	+
355. S. filamentosus	T	−	+	+	+	−	−	+	−	−	−	−	+
356. S. prunicolor	R		+	+	+	+	+		+	+	+		
357. S. roseofulvus	T	−	+	+	+	+	+	+	+	−	−	+	+
358. S. roseolus	T		+	+	+	+	±		−	−	−		−
359. S. roseosporus	T		+	+	+	+	−		−	−	−	+	−
360. S. rubiginosohelvolus	T		+	+	+	+	+		−	+	−		−
361. S. termitum	T		+	+	−		−		−	−	−		

[a] For explanation of meanings of symbols see footnotes to Tables 17.33 and 17.34a–e.

TABLE 17.37c, d, e

Red series[a]

Names of species and subspecies	Strain designation	Utilization of carbon compounds (P & G basal)											
		No carbon control	D-Glucose	D-Xylose	L-Arabinose	L-Rhamnose	D-Fructose	D-Galactose	Raffinose	D-Mannitol	i-Inositol	Salicin	Sucrose
	17.37c R; S; C+; SM												
362. S. cinnamonensis (see no. 420)	T	−	+	−	−	−	+		−	−	−	+	
363. S. colombiensis	T	−	+	−	−	−							
364. S. goshikiensis	T		+	−	−	−	+		−	−	−		−
365. S. katrae	T	−	+	−	−	−	+	+	+	−	−	−	
366. S. lavendofoliae	T	−	+	+	+	−	−	+	−	−	+	−	−
367. S. lavendulae (see nos. 368, 369 and 370).	R	−	+	−	−	−	−	+	+	−	−	+	−
368. S. lavendulae subsp. avireus (see nos. 367, 369 and 370).	T	−	+	−	−	−	−	+	−	−	−	+	−
369. S. lavendulae subsp. brasilicus (see nos. 367, 368 and 370).	T	−	+	−	−	−	−	+	−	−	−	−	
370. S. lavendulae subsp. grasserius (see nos. 367, 368 and 369).	T	−	+	−	−	−	−	+	−	−	−	+	−
371. S. lavendulocolor	T	−	+	+	+	−	−	+	−	−	+	−	−
372. S. luridus	T		+	+	+	−	±		−	−	+		−
373. S. orchidaceus	T												
374. S. racemochromogenes	T		+	−	+	−	−		−	−	−		+
375. S. syringae	T		+	−	−	−	+	−	−	−	−		−
376. S. toxytricini	T		+	−	−	−	±		−	±	−		−
377. S. tuirus	T		+	+	+	+	+	+	+	+	+	−	+
378. S. vinaceus	T	−	+	−	−	−							
379. S. virginiae	T	−	+	−	−	−	+		−	−	+	+	
	17.37d R; S; C+; WTY												
380. S. lateritius	T		+	+	+	+	+		−	−	±		−
	17.37e R; S; C+; SPY												
381. S. flavovariabilis	T		+	+	+	+	+	+	+	+			+
382. S. janthinus	T	−	+	+	+	+	+	+	+	+	+	+	+
383. S. purpurascens	T		+	+	+	+	+		+	+	+	+	+
384. S. roseospinus	T			+	+	+	−	−	+	−	+		
385. S. roseoviolaceus	T		+	+	+	+	+		+	+	+		+
386. S. violaceus (see Nos. 387 and 388)	R	−	+	+	+	+	+	+	+		+		+
387. S. violaceus subsp. confinus (see Nos. 386 and 388)	T	−	+		+	+	+	+	+		+		+
388. S. violaceus subsp. vicinus (see Nos. 386 and 387)	T	−	+		+	+	+	+	+		+		+
389. S. violarus	T	−	+		+	+	+	+	+		+		+
390. S. violatus	T	−	+	+	+	+	+	+	+		+		+
391. S. yokosukanensis	T	−	+	+	+	+	+	+	+	+	+	+	+

[a] For explanation of meanings of symbols see footnotes to Tables 17.33 and 17.34a–e.

TABLE 17.37f, g, h, i

Red series[a]

Names of species and subspecies	Strain designation	Utilization of carbon compounds (P & G basal)											
		No carbon control	D-Glucose	D-Xylose	L-Arabinose	L-Rhamnose	D-Fructose	D-Galactose	Raffinose	D-Mannitol	i-Inositol	Salicin	Sucrose
17.37f R; S; C−; SM													
392. S. albosporeus	R	−	+	+	+	+	+	+	+	+	+	−	
393. S. aurantiacus	R		+	±	+	+	+	+	±	+	+		±
394. S. aureoverticillatus	T	−	+	+	+	−	+	+	+	+	+	+	
395. S. aurini	R												
396. S. cremeus	T	−	+	+	+	−	+	+	−	−	−	−	−
397. S. daghestanicus	T		+	+	+	+	+		−	+	−		−
398. S. fradiae	T	−	+	+	+	−	−	+	+	−	−	−	−
399. S. fragilis	T	−	+	+	+	−	−	+	−	−	−	−	±
400. S. fumanus	T		+	+	+	+	+		+	+	−		−
401. S. glomeroaurantiacus	T	−	+	+	−	−	+	+	−	+	+	−	−
402. S. griseoviridis	T	−	+	+	+	+	+	+	−	+	−		−
403. S. niveoruber	T	−	+	+	+	+							
404. S. peucetius	T**		+	+	−	−	+		+	+			+
405. S. phaeoviridis	T**	−	+	+	+	+	+		+	+			+
406. S. roseiscleroticus	T	−	+	+	+	+	+	+	−	+	−	−	
407. S. roseoflavus	T	−	+	+	+	−	−	−	−	−	−		−
408. S. roseolilacinus	T	−	+	−	+	−	±		−	−	−		−
409. S. rubro-cyaneus	T		+	+	+	+	+	+	+	+	+		+
410. S. tauricus	T			+	+	+	+		+	−			
411. S. vinaceus-drappus	T	−	+	+	+	+	+		+	+	+	+	
412. S. virocidus	T		+	+	+	+	+	+	−	+	+		+
17.37g R; S; C−; SPY													
413. S. erythraeus	R	−	+	+	+	+	+	+	+	+	+	+	
17.37h R; S; C−; WTY													
414. S. luteofluorescens	T		+	+	+	+	+		−	+	−		+
17.37i R; ?; ?; ?													
415. S. erythrogriseus	T			+		+			−	+		+	+
416. S. garyphalus	T												
417. S. lavendularectus	T	−	+	−	−	−	−	+	−	−	−	−	−
418. S. nagasakiensis	T		+	+	+	+	+	+	+	−			+
419. S. rubrolavendulae	T	−	+	+	+	+	+	+	−	+	+	−	−
420. S. cinnamonensis subsp. proteolyticus (see no. 362).	T												

TABLE 17.38a–f
Blue series

Names of species and subspecies	Strain designation	Utilization of carbon compounds (P & G basal): No carbon control	D-Glucose	D-Xylose	L-Arabinose	L-Rhamnose	D-Fructose	D-Galactose	Raffinose	D-Mannitol	i-Inositol	Salicin	Sucrose
	17.38a B; RF; C+; SM												
421. S. ashchabadicus	T												
422. S. polychromogenes	T	−	+	+	+	−	+	+	−	−	−	+	
	17.38b B; S; C+; SM												
423. S. amakusaensis	T	−	+	−	±	−	−	−	−	−	−	−	±
424. S. caelestis	T	−	+	+	+	+	+	+	+	−	+	−	+
	17.38c B; S; C+; WTY												
425. S. azureus	T	−	+	+	+	+	+	+	+	+	+		+
	17.38d B; S; C+; SPY												
426. S. bellus	T	−	+	+	+	+	+		+	+	+		+
427. S. chartreusis	T	−	+	+	+	+	+	+	+	+	+	+	+
428. S. coeliatus	T		+		−	+	+	+	+	+	+		+
429. S. coerulatus (see Nos. 430 and 444).	T		+		−	+	+	+	−	+	+		+
430. S. coerulatus subsp. amylolyticus (see nos. 429 and 444).	T		+		−	+	+	+	+	+	+		+
431. S. coeruleofuscus (see *incertae sedis* no. B10).	R		+	+	+	+	+		+	+	+		+
432. S. coeruleorubidus	T	−	+	+	+	+	+	+	+	+	+	+	+
433. S. coerulescens	T	−	+	+	+	+	+		+	+	+		+
434. S. curacoi	T	−	+	+	+	+	+	+	+	+	+	+	+
435. S. cyaneus	R	−	+	+	+	+							
436. S. cyanoglomerus	T		+		+	+	+	+	+	+	+		−
437. S. indigocolor	T		+		−	+	+	+	+	+	+		−
438. S. lanatus	T	−	+	+	+	+	+		+	+	+		+
439. S. lazureus	T		+		−	+	+	+	−	+	+		−
440. S. valynus	T												
441. S. viridochromogenes	T	−	+	+	+	+	+	+	+	+	+	+	
	17.38e B; S; C+; H												
442. S. glaucescens	T												
	17.38f B; S; C−; SPY												
443. S. bluensis	T		+	+	+	+	+	+	+	+	+	+	+
444. S. coerulatus subsp. anaseuli (see nos. 429 and 430).	T		+			+	+	+		+	+		
445. S. coeruleoroseus	T												
446. S. ipomoeae	R		+	+	+	+	+		+	+	+		+
447. S. spinosus	T												

[a] For explanation of meanings of symbols see footnotes to Tables 17.33 and 17.34a–e.

TABLE 17.39a–d

Green series[a]

Names of species and subspecies	Strain designation	Utilization of carbon compounds (P & G basal)											
		No carbon control	D-Glucose	D-Xylose	L-Arabinose	L-Rhamnose	D-Fructose	D-Galactose	Raffinose	D-Mannitol	i-Inositol	Salicin	Sucrose
17.39a GN; S; C+; H.													
448. S. griseomycini	T		+	+	+	+	+		−	+	+		−
449. S. griseostramineus	T		+	+	+	+	+			+	+		−
450. S. prasinosporus	T		+	+	+	+	+			+	+	+	−
17.39b GN; S; C−; SPY													
451. S. ghanaensis	T												
452. S. hirsutus	T	−	+	+	+	+	+		+	+	+		+
453. S. prasinus	T	−	+	+	+	+	+	+	−	+	+	−	+
454. S. viridosporus	T	−	+	+	+	+	+	+	−	+	+	−	±
17.39c GN; S; C−; H													
455. S. acrimycini	T	−	+	+	−	+	+		−	+	+		−
456. S. bambergiensis	T												
457. S. prasinopilosus	T	−	+	+	+	+	+	+	−	+	+	−	
17.39d GN; ?; ?; ?													
458. S. horton	T	−	+	+	+	−	+	+	−	+	−	−	

[a] For explanation of meanings of symbols see footnotes to Tables 17.33 and 17.34a–e.

TABLE 17.40a, b
Violet series[a]

Names of species and subspecies	Strain designation	Utilization of carbon compounds (P & G basal)											
		No carbon control	D-Glucose	D-Xylose	L-Arabinose	L-Rhamnose	D-Fructose	D-Galactose	Raffinose	D-Mannitol	*i*-Inositol	Salicin	Sucrose
17.40a V; RF; C−; SM													
459. S. rectiviolaceus	T	−	+	+	+	+	+	+	+	+	+		+
17.40b V; S; C+; SPY													
460. S. lilacinofulvus	T												
461. S. mauvecolor***	T		+	−	+	−	−	+	+	−	−	+	−
462. S. violans	T	−	+		+	+	+	+	+		+		+
463. S. violascens	T	−	+	+	+	−	+		+	−	±		±

[a] For explanation of meanings of symbols and asterisks, see footnotes to Tables 17.33 and 17.34a–e.

Genus II. Streptoverticillium *Baldacci 1958, 15, emend. mut. char. Baldacci, Farina and Locci 1966, 168*

Substrate mycelium, 1–2 μm in diameter, branching. **The aerial mycelium is characteristic; three to six short branches** (1–10 μm long, average 3–5 μm) **are produced in a whorl (verticil) at more or less regular intervals along the aerial mycelium giving a "barbed wire" appearance** at *ca.* 100 × magnification. **These branch in turn to produce 2–12 or more secondary branches in an umbel-like manner; chains of rounded to oblong spores form at the ends of the secondary branches.** A chain may be straight, flexuous or terminate in hooks.

Reproduction occurs either from particles of substrate or aerial mycelium or from germination of spores.

Gram-positive. Cell wall Type I (Table 17.18).

On primary isolation **colonies are small and discrete, initially smooth, later developing a weft of aerial mycelium** that may appear velvety or floccose. **Substrates containing starch or other polysaccharides are especially good for production of spores;** on less suitable media spores are not formed. **Produce a wide variety of pigments** and substrate and aerial mycelium usually colored (Table 17.41).

Hydrogen sulfide *per se* seldom produced on peptone-yeast extract-iron agar (peptone-iron agar), or in traces only, when lead acetate paper is used as a detecting agent; cultivation of streptoverticillia on this medium, however, often results in the formation of bluish black to black diffusible pigment which correlates with melanin production and the production of brown, deep brown or black diffusible pigments in other organic media. Starch and gelatin usually hydrolyzed. Tyrosine is utilized by all species; xanthine is utilized by only two species. Produce a variety of compounds (antibiotics) which exhibit anti-bacterial, anti-fungal, anti-protozoal and anti-tumor activity. Sensitive to anti-bacterial agents and to actinophages.

Aerobic. Optimum growth range about 25 C to about 35 C at pH 6.5–8.0. Saprophytic soil forms.

The G + C content of the DNA of the species examined ranges from 69–73 moles %.

Type species: *Streptoverticillium baldaccii* Farina and Locci 1966, 48.

TABLE 17.41

Color characteristics of series of genus **Streptoverticillium**

Series	Color of substrate mycelium	Color of aerial mycelium	Species
I. Baldaccii	pink-red to orange-red	pink, gray-pink and violet-pink	1. *Stv. baldaccii* 2. *Stv. fervens* 2b. *Stv. fervens* subsp. *melrosporus* 3. *Stv. rubrochlorinum*
II. Biverticillatum	colorless, reddish and orange, yellow to brick red	pinkish white	4. *Stv. biverticillatum* 5. *Stv. aureoversales* 6b. *Stv. pentaticum* subsp. *jenense* 7. *Stv. roseoverticillatum* 7b. *Stv. roseoverticillatum* subsp. *albosporum* 8. *Stv. rubroverticillatum*
III. Hiroshimense	brick red	beige to pink-beige	9. *Stv. hiroshimense*
IV. Salmonis	brick-red to orange	white with pink and yellow shades	10. *Stv. salmonis*
V. Luteoverticillatum	yellow, yellowish to brown	light yellow, yellowish to beige	11. *Stv. luteoverticillatum* 12. *Stv. olivoreticuli* 13. *Stv. waksmanii*
VI. Griseocarneum	brownish yellow	pinkish beige with lilac shades	14. *Stv. griseocarneum*
VII. Cinnamoneum	yellow to greenish yellow and brown-yellow	pinkish with beige and lilac shades	15. *Stv. cinnamoneum* 15b. *Stv. cinnamoneum forma azacoluta* 15c. *Stv. cinnamoneum* subsp. *albosporum* 15d. *Stv. cinnamoneum* subsp. *lanosum* 15e. *Stv. cinnamoneum* subsp. *sparsum* 16. *Stv. hachijoense*
VIII. Ardum	light yellow to yellowish to pinkish yellow	basically white with yellow, pink and gray	17. *Stv. ardum* 18. *Stv. abikoense* 19. *Stv. albireticuli* 20. *Stv. eurocidicum* 21. *Stv. kishiwadense* 22. *Stv. mashuense* 23. *Stv. olivoverticillatum* 24. *Stv. orinoci* 25. *Stv. parvisporogenes*

TABLE 17.41—*Continued*

Series	Color of substrate mycelium	Color of aerial mycelium	Species
IX. **Kentuckense**	yellow, yellowish to hazel-nut yellow	beige with shades toward yellow, pink and cinnamon	26. *Stv. kentuckense* 27. *Stv. album* 28. *Stv. distallicum* 29. *Stv. ehimense* 30. *Stv. flavopersicum* 31. *Stv. griseoverticillatum* 32. *Stv. netropsis* 33. *Stv. rectiverticillatum* 34. *Stv. septatum*
X. **Mobaraense**	yellow to greenish yellow	grayish green	35. *Stv. mobaraense* 36. *Stv. blastmyceticum* 37. *Stv. lavenduligriseum*
XI. **Lilacinum**	brown	pinkish white	38. *Stv. lilacinum* 39. *Stv. kashmirense*
XII. **Thioluteum**	brown-yellow to greenish	light yellow	40. *Stv. thioluteum*

TABLE 17.42

Biochemical properties of species and subspecies of genus **Streptoverticillium**[a]

	Melanin production	H_2S production	Starch hydrolysis	Casein hydrolysis	Gelatin liquefaction	Growth on carbohydrates, ISP medium 9										
						Control (no sugar)	Arabinose	Cellulose	Fructose	Glucose	Inositol	Mannitol	Raffinose	Rhamnose	Sucrose	Xylose
Series I. Baldaccii																
1. *Stv. baldaccii*	+	tr	+	−	+	−	−	−	−	+	(+)	tr	tr	−	+	−
2. *Stv. fervens*	+	+	+	+	+	−	−	−	tr	+	tr	tr	tr	tr	+	tr
2b. *Stv. fervens* subsp. *melrosporus*	+	−	+	+	+	tr	−	−	(+)	+	+	(+)	(+)	(+)	+	tr
3. *Stv. rubrochlorinum*	+	tr	+	−	+	−	−	−	−	+	tr	−	−	−	(+)	−
Series II. Biverticillatum																
4. *Stv. biverticillatum*	+	tr	+	(+)	(+)	−	−	−	(+)	+	tr	−	(+)	tr	(+)	(+)
5. *Stv. aureoversales*	+	−	+	−	+	tr	−	−	−	+	+	−	tr	tr	+	−
6b. *Stv. pentaticum* subsp. *jenense*	+	−	+	−	+	−	−	−	+	+	tr	−	−	−	(+)	−
7. *Stv. roseoverticillatum*	+	−	+	−	+	−	−	−	−	+	−	−	−	−	+	−
7b. *Stv. roseoverticillatum* subsp. *albosporeum*	+	−	+	(+)	+	tr	−	−	tr	+	+	−	−	tr	−	−
8. *Stv. rubroverticillatum*	+	−	+	−	+	tr	−	−	tr	+	tr	−	tr	tr	tr	−
Series III. Hiroshimense																
9. *Stv. hiroshimense*	+	tr	+	−	+	−	−	−	−	+	+	−	−	−	tr	−
Series IV. Salmonis																
10. *Stv. salmonis*	+	−	+	+	+	−	−	−	−	+	(+)	−	−	−	tr	−
Series V. Luteoverticillatum																
11. *Stv. luteoverticillatum*	+	tr	+	(+)	+	−	−	−	+	+	−	+	tr	−	+	−
12. *Stv. olivoreticuli*	+	−	+	+	+	−	−	−	−	+	tr	−	tr	−	+	−
13. *Stv. waksmanii*	+	−	+	+	+	tr	−	tr	+	+	(+)	+	(+)	tr	(+)	tr
Series VI. Griseocarneum																
14. *Stv. griseocarneum*	+	tr	+	−	+	−	−	−	tr	+	+	−	−	−	tr	−
Series VII. Cinnamoneum																
15. *Stv. cinnamoneum*	−	tr	+	−	+	tr	−	tr	(+)	+	+	tr	tr	tr	+	tr
15b. *Stv. cinnamoneum* *forma azacoluta*	−	(+)	+	−	+	tr	tr	−	tr	+	−	tr	tr	tr	(+)	tr
15c. *Stv. cinnamoneum* subsp. *albosporum*	−	tr	+	(+)	+	−	−	−	−	−	(+)	−	−	−	−	−
15d. *Stv. cinnamoneum* subsp. *lanosum*	−	−	+	+	+	tr	+	(+)	(+)	+	+	(+)	(+)	tr	(+)	tr
15e. *Stv. cinnamoneum* subsp. *sparsum*	−	tr	+	−	+	−	−	−	tr	+	+	−	−	−	−	−

16. *Stv. hachijoense*	−	−	+	(+)	+	tr	−	tr	(+)	+	+	tr	tr	tr	(+)	tr
Series VIII. Ardum																
17. *Stv. ardum*	+	(+)	−	−	+	tr	−	−	+	+	+	−	(+)	−	+	−
18. *Stv. abikoense*	+	tr	+	+	+	−	−	−	−	−	−	−	−	−	−	−
19. *Stv. albireticuli*	+	tr	+	+	+	tr	tr	−	tr	+	+	tr	tr	tr	+	tr
20. *Stv. eurocidicum*	+	+	+	+	+	−	−	−	tr	+	(+)	−	−	−	+	−
21. *Stv. kishiwadense*	+	tr	+	+	+	−	(+)	−	+	+	+	−	tr	−	+	−
22. *Stv. mashuense*	+	−	+	+	+	−	−	−	+	+	+	−	(+)	−	+	−
23. *Stv. olivoverticillatum*	+	(+)	+	+	+	−	tr	−	−	+	(+)	−	tr	−	+	−
24. *Stv. orinoci*	−	tr	−	−	+	−	−	−	tr	+	−	−	tr	−	tr	−
25. *Stv. parvisporogenes*	+	tr	+	−	+	−	−	−	tr	+	(+)	−	−	−	(+)	−
Series IX. Kentuckense																
26. *Stv. kentuckense*	−	−	+	+	+	−	−	−	(+)	+	+	tr	(+)	−	+	tr
27. *Stv. album*	−	−	+	−	+	−	−	−	−	+	tr	−	−	−	−	−
28. *Stv. distallicum*	+	tr	+	−	+	−	−	−	(+)	+	+	−	tr	tr	(+)	tr
29. *Stv. ehimense*	+	−	+	+	+	−	−	−	(+)	+	(+)	tr	−	tr	tr	tr
30. *Stv. flavopersicum*	+	−	+	−	+	−	−	−	tr	+	+	−	−	−	tr	−
31. *Stv. griseoverticillatum*	−	−	+	−	−	−	−	−	−	+	+	−	tr	−	+	−
32. *Stv. netropsis*	−	tr	+	−	+	−	−	−	−	+	+	−	−	tr	−	tr
33. *Stv. rectiverticillatum*	+	+	+	+	+	−	−	−	+	+	+	+	tr	tr	(+)	(+)
34. *Stv. septatum*	+	−	+	−	+	tr	tr	tr	tr	+	+	−	tr	tr	tr	tr
Series X. Mobaraense																
35. *Stv. mobaraense*	−	(+)	(+)	(+)	−	(+)	(+)	tr	+	+	(+)	tr	−	−	+	tr
36. *Stv. blastmyceticum*	+	(+)	+	+	+	−	−	−	−	+	(+)	−	tr	tr	+	−
37. *Stv. lavenduligriseum*	−	tr	+	−	+	tr	tr	tr	+	+	+	(+)	+	tr	+	(+)
Series XI. Lilacinum																
38. *Stv. lilacinum*	+	−	+	+	+	−	−	−	−	+	tr	−	−	−	tr	−
39. *Stv. kashmirense*	+	−	+	+	+	−	−	−	−	−	−	−	−	−	−	−
Series XII. Thioluteum																
40. *Stv. thioluteum*	−	−	−	−	(+)	−	−	−	−	+	(+)	−	−	−	tr	−

[a] Degree of activity is indicated in decreasing order +, (+), tr, −. Tyrosine crystals disappear from the medium with all species and varieties. Xanthine crystals disappear from the medium only with *Stv. kentuckense* (+) and *Stv. flavopersicum* +.

Genus III. **Sporichthya** *Lechevalier, Lechevalier and Holberl 1968, 279*

Hyphae, 0.5–1.2 μm in diameter, short (10–25 μm) branched, **form only aerial mycelium that is attached to the surface of solid media by means of holdfasts** which originate from the wall of the hyphal base. **The aerial hyphae divide into smooth-walled spores** which may be spherical (0.5–1.2 μm), rod-shaped (0.5–1.2 by 0.75–1.5 μm), balloon-shaped or pisciform (up to 6 μm long). Spores exhibit a collar-like scar and a flagellum originates at this end of the spore; up to three flagella arranged in a tuft are sometimes seen. **Spores become motile** when exposed to water. The **cell wall contains L-diaminopimelic acid. Gram-variable.**

On poor media, growth is mold-like; on rich media, bacteria-like.

Heterotrophic. Acids not generally detectable. Starch is hydrolyzed.

Casein is hydrolyzed; gelatin is not liquefied; hypoxanthine and tyrosine are not attacked.

Facultatively anaerobic.

Temperature optimum 28–37 C; growth range 22–42 C, no growth at 55 C.

Optimum pH for growth *ca.* 7.0.

Type species (monotype): *Sporichthya polymorpha* Lechevalier *et al.* 1968, 279.

Genus IV. **Microellobosporia** *Cross, Lechevalier and Lechevalier 1963, 422*

Slender hyphae, about 1 μm in diameter, developing a substrate mycelium which grows into and forms a compact layer on the top of the medium and **an aerial mycelium. Both substrate and aerial mycelia bear sporangia** on short sporangiophores; the **sporangia contain a single longitudinal row of non-motile sporangiospores,** usually one to five. Spores are formed by simultaneous division of an intrasporangial hypha. **Arthrospores are not produced. Cell wall Type I.**

Heterotrophs. The type species hydrolyzes gelatin and starch. Carbon utilization of the various species is given in Table 17.43.

Aerobic.

Temperature optimum 30 C; growth occurs at 25–30 C. Optimal pH range for growth at 30 C 6.0–8.5.

Antibiotic production by some strains.

Type species: *Microellobosporia cinerea* Cross, Lechevalier and Lechevalier 1963, 422.

TABLE 17.43

Carbon utilization patterns of the species of genus **Microellobosporia**

Carbon source	Utilization[a]			
	1. *M. cinerea*	2. *M. violaceus*	3. *M. flavea*	4. *M. grisea*
No carbon control	−	−	−	0
D-(+) Glucose control	+++	+++	++	+++
D-(+) Xylose	++	‡	++	+++
L-(+) Arabinose	++	+	++	+++
L-(+) Rhamnose	+++	++	++	+++
Fructose	0	0	0	+++
D-(+) Galactose	++	++	++	+++
Maltose	++	−	+	++
Lactose	++	−	+	++
DL-Sucrose	++	0	+++	++
Raffinose	+++	+++	++	++
Starch	+++	++	++	++
Inulin	0	0	0	+++
Cellulose	0	0	0	‡
D-(−) Mannitol	+++	++	+++	++
D-(+) Sorbitol	+++	‡	++	‡
Sodium acetate	++	−	++	0
Sodium citrate	++	‡	+	++
Sodium malate	++	0	+	++

[a] Basal medium: Pridham and Gottlieb (1948, 107); − = growth absent; ‡ = very weak, nil, or doubtful growth; + = weak growth; ++ = satisfactory or moderate growth; +++ = good growth or grows well; 0 = no data.

Key to the species of genus **Microellobosporia**

I. Surfaces of sporangia free of crystals or spines.
  A. Vegetative mycelium colored in tints and shades of violet.
    1. In Pridham and Gottlieb basal medium utilizes sodium acetate, lactose, maltose, D(+)-sorbitol and D-xylose.
      1. *Microellobosporia cinerea*
    2. Does not utilize sodium acetate, lactose or maltose and utilizes D(+)-sorbitol and D-xylose poorly.
      2. *Microellobosporia violacea*
  B. Vegetative mycelium not colored violet.
      3. *Microellobosporia flavea*
II. Surfaces of sporangia covered with crystals forming serrate spines.
      4. *Microellobosporia grisea*

FAMILY VIII. MICROMONOSPORACEAE KRASIL'NIKOV 1938, 272

Aerial mycelium present, except in *Micromonospora*. Spores are formed singly, in pairs or short chains on either or both the aerial and substrate mycelium. Sporophores are lacking or very short, in some cases showing dichotomous branching. Generally aerobic, a few anaerobic species are known. Mostly mesophilic, some thermophilic. Primarily saprophytic soil forms.

Key to the genera of family **Micromonosporaceae**

I. Aerial mycelium absent.
  A. Cell wall Type II.
    1. Single spores on substrate mycelium only.
      Genus I. *Micromonospora*, p. 846
II. Aerial mycelium present.
  A. Cell wall usually Type III. (One species in *Actinobifida* is Type II and one in *Thermomonospora* Type IV.)
    1. Single spores on both the aerial and substrate mycelium.
      a. Sporophores lacking or very short.
        Genus II. *Thermoactinomyces*, p. 855
      aa. Sporophores dichotomously branched.
        Genus III. *Actinobifida*, p. 856
    2. Single spores on aerial mycelium only.
      Genus IV. *Thermomonospora*, p. 858
    3. Longitudinal pairs of spores on aerial mycelium only.
      Genus V. *Microbispora*, p. 859
  B. Cell wall Type IV.
    1. Short chains of spores on both the aerial and substrate mycelium.
      Genus VI. *Micropolyspora*, p. 861

Genus I. **Micromonospora** *Ørskov 1923, 147*

Well developed, branched, septate mycelium averaging 0.5 μm in diameter. **Spores borne singly, sessile or on short or long sporophores** which often occur in branched clusters. Sporophore development monopodial or in some cases sympodial.

Aerial mycelium absent or in some cultures appearing irregularly as a restricted white or grayish bloom. Gram-positive. Not acid-fast. **Cell wall Type II.**

Colonies on agar limited in peripheral growth, rising steeply from the surface of the medium giving a convolute walnut-kernel-like appearance.

Most isolates are strongly proteolytic, cellulolytic and diastatic.

Most species aerobic, two anaerobic. Sensitive to pH environs below 6.0. Good growth occurs normally between 20–40 C but not above 50 C. Spores generally resistant to water temperatures of 60 C for 90 min but killed at 90 C in 15 min.

Usually saprophytic occurring as normal populations in soils and aquatic environments.

Type species: *M. chalcea* (Foulerton) Ørskov 1923, 156.

Further Comments

Immature or non-sporulating colonies of *Micromonospora* may often resemble the glabrous colonies of orange-colored species of *Nocardia*, *Actinoplanes* or *Amorphosporangium*. Sporulating colonies of micromonosporae may be distinguished from these genera by the development of a characteristic dark brown to black, waxy or moist layer composed of dark pigmented spores. Mature cultures of *Micromonospora* may often be composed of coccoid spores and mycelial fragments which resemble in microscopic appearance species of *Nocardia*. Despite the superficial resemblance to other genera, the manner by which micromonosporae produce their spores is distinctive and if the origin of the spore is sought, misidentifications can be avoided. Micromonosporae produce their spores as single, well defined, terminal or lateral structures which occur on the surface and throughout the colony but not in a powdery aerial mycelium characteristic of the streptomycetes, the genus *Actinobifida*, or *Thermoactinomyces*.

Key to the species of genus **Micromonospora**

Key characteristics

Relationship to oxygen
Utilization of:
- α-melibiose
- raffinose
- mannitol
- rhamnose

Arrangement of sporophores
Growth on Czapek's sucrose agar
Spore shape
Pigment production
Hydrolysis of cellulose
Growth on protein-containing media
Growth on potato
Source of isolate
Production of formic or propionic acid

I. Aerobic.
- A. α-Melibiose utilized (for species in which carbohydrate data are lacking see IC or ID).
  - 1. Raffinose utilized.
    - a. Sporophores arranged monopodially.
      - i. Growth poor on Czapek's sucrose agar.
        - 1. *Micromonospora chalcea*
      - ii. Growth fair to good on Czapek's sucrose agar.
        - 2. *Micromonospora halophytica*
  - 2. Raffinose not utilized.
    - a. Sporophores arranged sympodially.
      - 3. *Micromonospora carbonacea*
    - b. Spores often produced on monopodially arranged, short lateral spikes. In broth culture mycelium often formed into cluster colonies which fragment readily.
      - i. D-Mannitol not utilized.
        - 4. *Micromonospora narashinoensis*

ii. D-Mannitol utilized.

See *Species incertae sedis* (p. 854).

c. Spores produced monopodially either on long sporophores or in flower-like clusters.

5. *Micromonospora melanosporea*

B. α-Melibiose not utilized.

1. L-Rhamnose utilized. Spores echinulate.

6. *Micromonospora echinospora*

2. L-Rhamnose not utilized. Spores aberrant.

7. *Micromonospora purpurea*

C. Produce characteristic pigments.

1. Produce extensive dark brown to black pigment diffusing into agar.

8. *Micromonospora purpureochromogenes*

2. Do not produce brown or black diffusible pigment.

a. Mycelial pigment green or blue-green.

i. Cellulose hydrolyzed.

9. *Micromonospora bicolor*

ii. Cellulose not hydrolyzed.

10. *Micromonospora coerulea*

D. Pigments non-characteristic; colonies in shades of pale yellow, orange, brown or black.

1. Growth vigorous on protein-containing media.

11. *Micromonospora globosa*

2. Growth not vigorous on protein-containing media.

a. Does not grow on potato. Cellulose hydrolyzed.

12. *Micromonospora elongata*

b. Grows on potato. Cellulose not hydrolyzed.

i. Isolated from soil.

13. *Micromonospora parva*

ii. Clinical isolate.

14. *Micromonospora gallica*

II. Anaerobic.

A. Cellulose not utilized; acetic and formic acids produced.

15. *Micromonospora acetoformici*

B. Cellulose utilized; acetic and propionic acids produced.

16. *Micromonospora propionici*

Genus II. Thermoactinomyces *Tsiklinsky 1899, 501*

The genus *Thermoactinomyces* is **characterized by the formation of single spores on both the aerial and substrate mycelium.** Occasionally longitudinal pairs of spores are formed on the aerial mycelium. Simple or branching sporophores may be so short that the spores often appear to be produced sessily or directly on the mycelium. The genus comprises, so far, **only forms capable of growing at temperatures between 45–60 C.** Gram-variable.

Cell wall may be of Type III.

Aerobic.

Type species: *Thermoactinomyces vulgaris* Tsiklinsky 1899, 501.

Note: There are two species recognized, *T. vulgaris* and *T. sacchari*, separated on their growth response on various agar media.

Genus III. Actinobifida *Krasil'nikov and Agre 1964, 939*

Fine **Gram-positive mycelium** (0.7–1.0 μm) differentiated into **substrate mycelium and aerial mycelium; hyphae show characteristic dichotomous branching particularly evident in the sporophores.** The mycelium does not break up into regular rod-shaped or coccoid cells. **Single spores on substrate and aerial mycelium,** repeated dichotomous branching of the sporophores can lead to the formation of dense spore clusters. Young apical regions of the sporophore exhibit conspicuous **heart-shaped terminal swellings** which progressively develop into two, or occasionally as a result of further division, four spores. The **spores** of two species investigated in detail have a **structure similar to bacterial endospores,** contain **dipicolinic acid** and are **resistant to heat and desiccation.**

Characteristics differentiating the species are given in Table 17.44.

Chemoorganotrophs.

Strict aerobes.

Type species: *Actinobifida dichotomica* Krasil'nikov and Agre 1964, 939.

TABLE 17.44

Differential characteristics of species of genus **Actinobifida**

	Aerial mycelium color	Reverse colony color	Optimum temperature
1. *A. dichotomica*	Pale yellow	Yellow	50–58 C
2. *A. alba*	White	Colorless	37–45 C
3. *A. chromogena*	Cinnamon	Brown	55–58 C

Genus IV. **Thermomonospora** *Henssen 1957, 398*

Single spores formed only on aerial mycelium.

Vegetative hyphae flexuous and non-septate. Aerial mycelium white; aerial hyphae simple or branched, arising from side or terminal branches of vegetative hyphae. Single spores are usually formed acropetalously at the tip of the sporophores; sporophores occasionally branched dichotomously. Gram-variable. Not acid-fast. Colonies colorless or yellow.

Cell wall may be of Type III or IV.

Facultative aerobe.

Thermophilic; optimal growth temperature 45–55 C.

Occurs in soils, manure and hay.

Type species: *Thermomonospora curvata* Henssen 1957, 401 (Henssen and Schnepf 1967, 227).

Note: There are two species recognized in the genus. *T. curvata* produces a white aerial mycelium, and does not hydrolyze gelatin or starch or coagulate milk. *T. viridis* produces a gray-green aerial mycelium and green colony reverse, hydrolyzes gelatin and starch and coagulates milk.

Genus V. **Microbispora** *Nonomura and Ohara 1957, 307*

Spores in characteristic longitudinal pairs are formed on aerial mycelium; (Plate 17.8, Fig. 12) as a rule **no spores formed on substrate mycelium** (occasional production on substrate mycelium reported for one unidentified strain, Lechevalier and Lechevalier, 1967). Spores (1.2–1.8 μm diameter) may be sessile or on short sporophores; paired spores may be closely arranged along the aerial hyphae, giving the appearance of a catkin or borne at longer intervals (Plate 17.8, Figs. 10, 11). In most species aerial mycelium is pink.

Gram-positive. **Purified cell wall Type III.**

Several species form bronze-violet needle-like crystals of iodinin on certain solid media (e.g. oatmeal agar) and release this compound in submerged culture (Gerber and Lechevalier 1964) (Table 17.45). B vitamins, particularly thiamine essential for growth.

Aerobic to facultatively anaerobic. Mesophiles and facultative thermophiles.

Normal habitat appears to be soil and lake muds; one species, *M. rosea*, has been implicated in a case of pericarditis and pleuritis in man (Louria and Gordon, 1960).

Type species: *Microbispora rosea* Nonomura and Ohara 1957, 307.

Genus VI. **Micropolyspora** *Lechevalier, Solotorovsky and McDurmont 1961, 13*

Substrate and aerial mycelium formed (about 1 μm in diameter) **both bearing short chains of 1–20 spores.** Spores (1.2–1.5 μm in diameter) are formed basipetally on lateral branches or terminally. Gram-positive.

Purified cell walls Type IV, contain *meso*-diaminopimelic acid, arabinose and galactose.

Mesophiles and facultative thermophiles.

Type species: *Micropolyspora brevicatena* Lechevalier, Solotorovsky and McDurmont 1961, 13.

TABLE 17.45

Characteristics differentiating the species of genus **Microbispora**[a]

	1. *M. rosea*	2. *M. aerata*	3. *M. amethystogenes*	4. *M. bispora*	5. *M. chromogenes*	6. *M. diastatica*	7. *M. parva*	8. *M. thermodiastatica*	9. *M. thermorosea*
Aerial mycelium									
color: white	−	−	−	+	−	−	−	−	−
pink	+	+	+	−	+	+	+	+	+
Carbon sources:									
glucose	3	3	d	3	3	3	3	3	2
glycerol	3	3	3	0	3	3	3	2	3
arabinose	3	3	3	0	3	3	3	3	3
rhamnose	3	1	0	1	0	1	1	0	0
inositol	3	0	1	2	3	0	0	0	0
Starch hydrolysis	0	3	0	0	d	3	d	3	0
$NO_3 \rightarrow NO_2$	+	+	+	−	+	−	−	−	−
Gelatin liquefaction	d	1	d	0	d	d	1	1	1
Milk peptonization	d	3	2	0	d	1	1	3	1
Melanoid pigment	0	1	0	0	d	0	0	0	0
Soluble pigment	0	0	2	0	2	d	0	0	0
Iodinin production	−	+	+	−	−	−	d	−	−
Growth at: 25 C	+	−	+	−	+	+	+	−	−
50 C	−	+	−	+	−	−	+	+	+
55 C	−	+	−	+	−	−	−	+	+

[a] Extent of growth, utilization or pigment: 0 = none; 1 = slight; 2 = good; 3 = excellent. + = most strains positive; − = most strains negative; d = some (less than 90%) strains positive, some negative.

TABLE 17.46

Characteristics differentiating the species of genus **Micropolyspora**

	1. *M. brevicatena*	2. *M. angiospora*	3. *M. caesia*	4. *M. faeni*	5. *M. rectivergula*	6. *M. rubrobrunea*	7. *M. thermovirida*	8. *M. viridinigra*
Color of aerial mycelium								
white/cream/yellow	+	+	−	+	+	−	−	−
blue/green/gray	−	−	+	−	−	+	+	+
Color of colony								
yellow/orange	+	+	−	+	+	−	−	−
green/gray	−	−	+	−	−	−	+	+
brown	−	−	−	−	−	+	−	−
Average no. spores in chain								
1–5	−	−	+	−	−	−	−	−
5–10	+	−	−	+	+	−	+	−
5–20	−	+	−	−	−	+	−	+
Spore surface								
smooth	−	−	+	+	+	−	−	−
warty	+	−	−	−	−	−	−	−
fine teeth/spines	−	+	−	−	−	+	+	+
Growth temperature C								
25	−	+	+	−	−	−	+	−
28	+	+	+	−	−	−	+	−
37	+	+	+	+	+	+	+	+
45–maximum	−	−	+	+	+	+	+	+

PART 18

THE RICKETTSIAS

Key to the orders of Part 18

I. Divide by binary fission in host cells.

Order I. *Rickettsiales*, p. 882*

II. Reproduction in host cells by a characteristic developmental cycle in which the infectious elementary body changes into a larger, thin-walled, non-infectious form. This form divides by fission and then becomes reorganized into elementary bodies to complete the cycle. (The developmental cycle is best seen in the electron microscope and if that is out of reach of the reader there is a group-specific antigen which can be easily detected by a complement fixation test).

Order II. *Chlamydiales*, p. 914

ORDER I. **RICKETTSIALES** GIESZCZKIEWICZ 1939, 25

The majority are rod-shaped, coccoid and often pleomorphic microorganisms which have typical bacterial cell walls, no flagella, are Gram-negative and multiply only inside host cells. They may be cultivated in living tissues such as embryonated chicken eggs or vertebrate cell cultures. Except for binary fission, which is common to all members of this order, there are notable exceptions to any one of the characteristics listed above. For example, microorganisms are included that appear ring-shaped in stained preparations, or have a flagellum, or are Gram-positive or multiply on bacteriological media of moderate complexity. All are regarded as parasitic or mutualistic. The parasitic forms are associated with the reticuloendothelial and vascular endothelial cells or erythrocytes of vertebrates and often with various organs of arthropods which may act as vectors or primary hosts. May cause disease in man or in other vertebrate and invertebrate hosts. The mutualistic forms in insects are regarded as essential for development and reproduction of the host.

In the 7th edition of The Manual, Class III *Microtatobiotes* included two orders, *Rickettsiales* and *Virales*. With the recognition that members of the order *Rickettsiales* are procaryotic microorganisms with no relationship to the non-organismic viruses, the class *Microtatobiotes* no longer serves any useful function. The order *Rickettsiales* is retained.

Three families are accepted in the order *Rickettsiales*, i.e. *Rickettsiaceae*, *Bartonellaceae* and *Anaplasmataceae*.

Key to the families of order **Rickettsiales**

I. Parasites, intracellular or intimately associated with tissue cells other than erythrocytes or with certain organs in arthropods; rarely extracellular in arthropods. Transmitted by arthropod vectors.

Family *Rickettsiaceae*, p. 883

II. Parasites, intracellular or facultatively extracellular; found characteristically in or on the erythrocytes of vertebrates, and in one genus in fixed tissue cells. Small, rod-shaped, bacteria-like cells.

---

* All page numbers cited refer to *Bergey's Manual of Determinative Bacteriology*, 8th edition, The Williams & Wilkins Company, Baltimore, 1974.

At least one species, when cultured, may show a single polar flagellum. Arthropod transmission established for some members of the family. Have been cultivated on media without living cells.

Family *Bartonellaceae*, p. 903

III. Very small, virus-like particles occurring in the erythrocytes of vertebrates. Transmitted by arthropods.

Family *Anaplasmataceae*, p. 906

FAMILY I. **RICKETTSIACEAE** PINKERTON 1936, 186

Small, rod-shaped, coccoid and diplococcus-shaped, often pleomorphic **organisms which are often intimately associated with arthropod tissues, usually in an intracellular position.** Gram-negative. With one exception the species pathogenic for vertebrates **have not been cultivated to date in cell-free media.** May be parasitic in man and other animals causing disease (typhus and related ills) that may be transmitted by invertebrate vectors (chiefly lice, fleas, ticks and mites). Information is still inadequate for the systematic assignment of many of the species which inhabit arthropod hosts and which were originally described in this family.

The family *Rickettsiaceae* as accepted, includes three tribes, *Rickettsieae*, *Ehrlichieae* and *Wolbachieae*.

Key to the tribes of family **Rickettsiaceae**

I. Adapted to existence in arthropods but capable of infecting suitable vertebrate hosts including man; pathogenic for man who may be the primary host, but most frequently is an incidental host. Cells rod-shaped, coccoid and diplococcoid; occasionally filamentous.

Tribe I. *Rickettsieae*, p. 883

II. Only a few species adapted to invertebrate existence; pathogenic for certain mammals but not for man; cells spherical, occasionally pleomorphic.

Tribe II. *Ehrlichieae*, p. 893

III. Adapted to existence in arthropods as symbiotes but not in vertebrates as highly pathogenic parasites; cells pleomorphic, coccoid to short or long and curved rods.

Tribe III. *Wolbachieae*, p. 897

TRIBE I. RICKETTSIEAE PHILIP 1953, 486

Small, pleomorphic, usually intracellular organisms found in arthropods and pathogenic for man and certain other vertebrate hosts.

Key to the genera of tribe **Rickettsieae**

I. Rod-shaped to coccoid, sometimes filamentous, not cultivable in absence of host cells. Growth generally in cytoplasm, not in vacuoles, sometimes in nucleus of certain vertebrate and arthropod cells. Rapidly inactivated at 56 C, unstable when separated from host compounds.

Genus I. *Rickettsia*

II. Resembles *Rickettsia,* but usually in an extracellular environment in the arthropod host, and cultivable on certain bacteriological media.

Genus II. *Rochalimaea*

III. Growth preferentially in vacuoles of host cells, highly resistant to physical and chemical agents in an extracellular environment.

Genus III. *Coxiella*

Genus I. **Rickettsia** *da Rocha-Lima 1916, 567 Nom. gen. cons.* Opin. 19, Jud. Comm. 1958, 158

Short rods, 0.3–0.6 by 0.8–2.0 μm, some species up to 4 μm long prior to cell division. No flagella or capsules, but an outer layer of amorphous material is occasionally seen in electron micrographs of cells subjected to a minimum of laboratory manipulation. **Gram-negative.** Retain basic fuchsin when stained by the method of Giménez (1964), a modification of the procedure of Macchiavello (1937).

Have not been cultivated in the absence of host cells. Growth generally occurs in the cytoplasm, but not in vacuoles and, sometimes, in the nucleus of certain vertebrate and arthropod cells. The organisms grow to high titer in the entodermal cells lining the yolk sac of the developing chick embryo (Cox, 1941). Excellent growth has also been obtained *in vitro* in chick embryo cells and in several established mammalian cell lines. Plaques have been produced on chick embryo fibroblasts. Temperature optima for growth, 32–35 C. Multiplication by transverse binary fission only. Division time in chick embryos during phase of most rapid growth is about 8 hrs. Nutritional requirements for independent rickettsial cell growth not known.

Growth inhibited by *p*-aminobenzoic acid (approximate minimal inhibitory concentration for most species: 0.2–0.4 mg/egg); inhibition reversed by *p*-hydroxybenzoic acid. Sulfonamides do not affect growth. Growth is inhibited by several antibiotics. The following amounts (per egg) have a measurable effect on multiplication, i.e. delay the deaths of infected chick embryos by about 1½ days: chloramphenicol, 0.08 mg; chlortetracycline, 0.05 mg; oxytetracycline, 0.008 mg; erythromycin, 0.08 mg (in some species, 0.008 mg) (Ormsbee *et al.*, 1955). The organisms are affected to a lesser extent by penicillin and streptomycin in eggs, but these

TABLE 18.1

Differentiation of biotypes and species of the genus **Rickettsia**

Biotype	Typhus group	Spotted fever group	Scrub typhus group
Species[a]	1. *R. prowazekii* 2. *R. typhi*	4. *R. rickettsii* 5. *R. sibirica* 6. *R. conorii* 7. *R. australis* 8. *R. akari*	10. *R. tsutsugamushi*
Intracellular location			
Cytoplasm	+	+	+
Nucleus	−	+	−
Cultivation in chick embryos			
Opt. temp. C.	35	32–34	35
Peak titer			
Just prior to death	+	−	+
24–72 hrs after death	−	+	−
Plaque formation[b]			
Days required	8–10	5–8	11–17
Size, mm	1	2–3	1
Relative susceptibility to infection of			
Guinea pig[c]	+++	+++	+ to ++
Mouse	+(1)[d] ++(2)	+(4–6) ++(7) +++(8)	+++
Specificity of CF antigen			
Soluble, extracted by ether			
Group	+	+	−
Unsatisfactory	−	−	+
Washed cells			
Species	+	+	+
Type	−	−	+

[a] *R. canada* and *R. parkeri* are not included, but are discussed in the text.

[b] On monolayers of chick embryo monolayers (from McDade, Stakebake and Gerone, 1969; McDade and Gerone, 1970).

[c] The following clinical manifestations of the disease in the guinea pig are often used as criteria for species differentiation; Fever only—*R. prowazekii;* Fever and scrotal swelling—*R. typhi*, *R. conorii*, *R. australis* and *R. akari;* Fever and scrotal necrosis—*R. rickettsii* and *R. sibirica*. These reactions are subject to considerable strain variation.

[d] Species number in parentheses.

antibiotics may have a pronounced effect in cell cultures.

Rapidly inactivated at 56 C and generally unstable when separated from host components, except in diluents containing proteins such as skim milk or plasma albumin or a solution consisting of sucrose, potassium phosphate and glutamate (SPG) (Bovarnick *et al.*, 1950).

Human pathogens. Man is the reservoir of the type species, an incidental host of the other species. Small rodents and other vertebrates serve as reservoirs or disseminate the organisms. Arthropods play primary roles in the life cycles of all species, often as main reservoirs, and almost exclusively mediate natural transmission among vertebrates.

The guanine + cytosine (G + C) content of the DNA (in the species examined) ranges from **30–32.5 moles %.**

Type species: *Rickettsia prowazekii* da Rocha-Lima 1916, 567.

Genus II. **Rochalimaea** (*Macchiavello*) *Krieg 1961, 162*

Closely resembling organisms of the genus *Rickettsia* in morphology, staining properties and in dependence on arthropod and vertebrate hosts for natural survival. **Usually reside in an extracellular environment in the arthropod host. Can be cultivated in host cell-free media.**

Type species: *Rochalimaea quintana* (Schmincke) Krieg 1961, 163.

Genus III. **Coxiella** (*Philip*) *Philip 1948, 58*

Short rods, usually 0.2–0.4 by 0.4–1.0 μm, resembling organisms of the genus *Rickettsia* in staining properties, dependence on host cells for growth and close natural association with arthropod and vertebrate hosts.

Grows preferentially in the vacuoles of the host cell (rather than in cytoplasm or nucleus as do the species of *Rickettsia*). **Grows well in the yolk sac of chick embryos. Resists drying and relatively elevated temperatures,** that generally destroy the viability of *Rickettsia*.

Type species: *Coxiella burnetii* (Derrick) Philip 1948, 58.

TRIBE II. EHRLICHIEAE PHILIP 1957, 948

Minute, **rickettsia-like organisms pathogenic for certain mammals,** but, unlike those of Tribe *Rickettsieae*, **not infectious for man.** Differential characteristics, such as cytotropisms, hosts and vectors, are given in Table 18.2. Cells grow in the cytoplasm, but not in the nucleus, usually as compact, single to multiple morula-like colonies. Less often seen as single cells (early termed initial bodies), or as cells scattered throughout the cytoplasm. Also adapted to existence in ticks or, in one case, in trematodes, but not in insects. **Cell walls present. Non-motile.** Gram-negative. No Weil-Felix (*Proteus* OX) antibodies elicited. Includes the genera *Ehrlichia*, *Cowdria* and *Neorickettsia*.

TABLE 18.2

Differential characteristics of genera of tribe **Ehrlichieae**

Genus	Mammalian hosts	Tissues parasitized	Invertebrate vectors	Maintenance in vectors	Distribution
Ehrlichia	Canidae; domestic ruminants	Circulating leukocytes	Ticks: *Rhipicephalus* spp., *Hyalomma* sp.	Transovarial and/or transstadial only	Various places in Old and New Worlds.
	?Equines	Ditto	?*Dermacentor* sp.	?	Calif.
Cowdria	Domestic and certain wild ruminants	Vascular endothelial cells	Ticks: *Amblyomma* spp.	Transstadial only	Africa and Madagascar; possibly Dalmatia
Neorickettsia	Canidae	Reticular cells of lymphoid tissues	Trematode: *Nanophyetus salmincola*	Transovarial and through all stages	Pacific Coast, U.S.A.

Genus IV. **Ehrlichia** *Moshkovski 1945, 18*

Small, often **pleomorphic, coccoid to ellipsoidal organisms occurring intracytoplasmically,** either singly or in compact colonies (morulae) **in certain circulating leukocytes** of susceptible mammalian hosts. Adapted to growth in certain species of ixodid ticks associated with such hosts. **Non-motile. Not cultivable in cell-free media or chicken embryos.** The etiological agents of tick-borne diseases chiefly of dogs, cattle, sheep, goats and probably horses.

Type species: *Ehrlichia canis* (Donatien and Lestoquard) Moshkovski 1945, 18.

Note: There are two species recognized in the genus. *E. canis* is pathogenic for canines and *E. phagocytophila* is pathogenic for ruminants.

Genus V. **Cowdria** *Moshkovski 1947, 62*

Small, pleomorphic, coccoid or ellipsoidal, occasionally **rod-shaped, organisms** occurring **intracytoplasmically** but not intranuclearly, and characteristically localized in clusters inside vacuoles in the **cytoplasm of vascular endothelial cells** of ruminants (Table 18.2). **Not passed transovarially in tick vectors. Non-motile.** Gram-negative. **Have not been cultivated in cell-free media.** The etiological agent of heartwater, a septicemic disease of domestic ruminants in Africa.

Type species: *Cowdria ruminantium* (Cowdry) Moshkovski 1947, 62.

Genus VI. **Neorickettsia** *Philip, Hadlow and Hughes 1953, 257*

Small, coccoid, often pleomorphic, intracytoplasmic (but not intranuclear) **organisms** which occur **primarily in reticular cells of lymphoid tissues** of *Canidae*. Also seen in certain tissues of mature **fluke vectors; all** other **fluke stages**, eggs, rediae-cercariae and metacercariae have been proven **infectious** by injection into susceptible vertebrate hosts, which confirms that the infectious cycle includes **transovarial transmission in the vector.** Filterable forms not reported, but suspected. Gram-negative. Non-motile. **Not cultivable in cell-free media or in chicken embryos.** Sensitive to tetracycline antibiotics. The etiological agent of a helminth-borne disease of *Canidae* on the mid-West Coast of U.S.A.

Type species: *Neorickettsia helminthoeca* Philip, Hadlow and Hughes1953, 257.

TRIBE III. WOLBACHIEAE PHILIP 1955, 271

A miscellaneous group of organisms associated with invertebrates, mostly arthropods, but not usually invading vertebrates. Most of these organisms have not been cultivated outside their hosts, and only in rare instances have they been grown on bacteriological media.

Four genera are accepted. They include many species which are rickettsia-like in morphology, staining properties and growth and which were formerly assigned to the genus *Rickettsia*. Characterization is usually not as adequate as for the preceding genera, which are pathogenic to vertebrates, and differentiation has been chiefly on the basis of presumed host specificity in arthropods, although differences in development and morphology are often noted. There is no evidence of an evolutionary relationship among the genera of this tribe.

Key to the genera of tribe **Wolbachieae**

I. Not pathogenic for larvae of insects or other invertebrate hosts.
  A. Associated with arthropods, seldom pathogenic for their hosts, seldom develop in mycetomata.
  Genus VII. *Wolbachia*
  B. Symbiotic; located in special organs (mycetomata) in the insect host, but found also in other tissues; highly pleomorphic.
  Genus VIII. *Symbiotes*
  C. Symbiotic in cockroaches; located in dispersed cells (mycetocytes) in the fat body and in ovaries; rod-shaped.
  Genus IX. *Blattabacterium*

II. Pathogenic for insect larvae and for other invertebrate hosts.
  Genus X. *Rickettsiella*

Genus VII. **Wolbachia** *Hertig 1936, 472*

Heterogeneous group of small coccoid forms, rods and spheres, which morphologically and tinctorially resemble either *Rickettsia* or *Chlamydia*. **Gram-negative.** Stain well with Giemsa's and Macchiavello's stains.

Only *Wolbachia melophagi* has been cultivated on a cell-free medium, but this work has not been confirmed since 1924. The other two species have not been cultivated and have been seen only in close association with host cells. Differential characteristics are summarized in Table 18.3.

Associated with arthropods, but unlike *Rickettsiella*, **seldom pathogenic for their hosts.** Extracellular or intracellular but, unlike *Symbiotes*, seldom developing inside mycetomes. **Not associated with diseases of vertebrates.**

Type species: *Wolbachia pipientis* Hertig 1936, 472.

TABLE 18.3

Characteristics differentiating the species of genus **Wolbachia**

	1. *W. pipientis*	2. *W. melophagi*	3. *W. persica*
Morphology	Coccoid forms to small rods, 0.25–0.5 by 0.5–1.3 μm	Small rods, 0.3 by 0.6 μm	Small spheres, 0.5 by 0.9 μm, often flattened
Cultivation	Not reported	On glucose-blood-bouillon-agar and in yolk sac of chick embryos	Profuse growth in yolk sac of chick embryos and cell culture
Host	Mosquito (gonads in *Culex pipiens*, gut epithelium in other species)	Sheep ked (*Melophagus ovinus*); possibly some species of lice	Tick (*Argas persicus* and *A. arboreus*)
Growth in host	Intracellular	Extracellular	Intracellular

Genus VIII. **Symbiotes** *Philip 1956, 267*

Rickettsia-like, pleomorphic organisms of bedbugs (*Cimex*), living chiefly **intracellularly in** specialized paired organs called **mycetomes.** The intracellular forms coccoid or diplococcoid, 0.2 by 0.4–0.5 μm, in young hosts; extracellular bacillary, lanceolate or thread forms, 0.25–0.3 by 3.0–8.0 μm, migrate to other tissues in maturing hosts. Transmitted via ovaries to embryos.

Biochemical, nutritional and physiological characters unknown. Therefore, separation of the pleomorphic complex into pathogenic and symbiotic forms, as was done by Krieg (1961), is unwarranted.

Source and habitat identical. Members of the genus *Symbiotes* characteristically occurring as intracellular symbionts in mycetomes and much of the body tissue of species of *Cimex*.

Type species: *Symbiotes lectularius* (Arkwright *et al.*) Philip 1956, 267.

Genus IX. **Blattabacterium** *Hollande and Favre 1931, 754*

Straight or slightly curved rods, 0.8–0.9 by 1.5–8 μm, occurring singly, or occasionally in pairs in tandem **in the host insect**. The longer forms occur **in mycetocytes** in the abdominal fat body, the shorter forms in the **germ tissue and oocytes.** Microorganisms in mycetocytes are **encapsulated by a host-provided membrane** which is continuous with host-cell organelles. **Non-motile. No resting stages known. Gram-positive,** becoming indeterminate (speckled, barred or negative) in malnourished hosts. **Cell wall 5–10 nm composed of several alternating electron opaque and transparent layers.** Muramic acid and glucosamine present. **Complex fibrillar mesosomes** near site of formation of cross-walls.

Biochemical, nutritional and physiological characters unknown.

Growth in the insect host inhibited by penicillin, streptomycin, chloramphenicol, chlortetracycline, oxytetracycline, sulfathiazole and egg white lysozyme.

Source and habitat identical, members of the genus *Blattabacterium* characteristically occurring as **intracellular symbionts** in mycetocytes in abdominal fat body tissue and in ovaries and eggs, **of all species of cockroaches.**

Type species: *Blattabacterium cuenoti* (Mercier) Hollande and Favre 1931, 754.

Genus X. **Rickettsiella** *Philip 1956, 267*

Gram-negative **rods,** usually smaller than those of the genus *Rickettsia*. **Pathogenic for their insect larval hosts and other invertebrate hosts,** but of very limited virulence for vertebrates. Have not been cultivated in host cell-free media.

Type species: *Rickettsiella popilliae* (Dutky and Gooden) Philip 1956, 267.

FAMILY II. **BARTONELLACEAE** GIESZCZYKIEWICZ 1939, 25

Parasites of the erythrocytes in man and other vertebrates. **Cell wall present.** Rod-shaped, coccoid, ring- or disc-shaped, often beaded or filamentous and usually less than 3 μm in greatest diameter. Erythrocytic forms stain lightly with many aniline dyes but distinctly with Giemsa's stain after methyl alcohol fixation; and then do not exhibit a eucaryotic nucleus which distinguishes them from protozoa which parasitize red blood cells. Gram-negative. Not acid-alcohol-fast. At least one species bears unipolar flagella in culture. **Cultivated *in vitro* on non-living media.** Arthropod transmission has been established for one genus. Cause bartonellosis in man and grahamellosis in other vertebrates.

Key to the genera of family **Bartonellaceae**

I. Occur in or on erythrocytes and within fixed tissue cells. Often have flagella. Found in man and *Phlebotomus* spp.

Genus I. *Bartonella*

I. Intra-erythrocytic. Not known to multiply in fixed tissue cells. Flagella not described. Not found in man.

Genus II. *Grahamella*

Genus I. **Bartonella** *Strong, Tyzzer and Sellards 1915, 808*

Microorganisms **found in fixed tissue cells** (Plate 18.2, Figs. 3 and 4) **and in or on erythrocytes** (Plate 18.2, Figs. 1 and 2). In stained blood films, the organisms appear as rounded or ellipsoidal forms or as slender, straight, curved or bent rods occurring either singly or in groups. **Characteristically occur in chains of several segmenting organisms,** sometimes swollen at one or both ends and frequently beaded (Strong *et al.*, 1913). In the tissues they are situated within the cytoplasm of endothelial cells as isolated elements or are grouped in rounded masses (Plate 18.2, Figs. 3, 6).

Has a cell wall, formation of which can be inhibited by penicillin. Reproduce by binary fission. In cultures possess a tuft of 1–10 unipolar flagella (Peters and Wigand, 1955) (Plate 18.2, Fig. 5); flagella have not been demonstrated in tissues.

Gram-negative. Not acid-fast. Stain poorly or not at all with many aniline dyes, but satisfactorily with Romanowsky's and Giemsa's stains.

May be cultivated by unlimited serial transfers on cell-free media.

Occur spontaneously **in man and in arthropod vectors** (*Phlebotomus* spp.) One species has been recognized, and it is known to be **established only on the South American continent** and perhaps in Central America. Human bartonellosis may be manifested clinically as a progressive anemia (Oroya fever) or as a cutaneous eruption (Verruga peruana), usually one after the other in succession, rarely simultaneously; asymptomatic infections also occur.

Type species: *Bartonella bacilliformis* (Strong *et al.*) Strong, Tyzzer and Sellards 1915, 808.

Genus II. **Grahamella** *Brumpt 1911, 517*

Microorganisms observed **within the erythrocytes of mammals, excluding man.** Not known to multiply in fixed tissue cells. Morphologically these organisms bear a resemblance to but are less polymorphic than *Bartonella*. **Have a cell wall.** **Motility has not been demonstrated; flagella have not been seen.**

Gram-negative. Not acid-fast.

Both species have been cultivated on non-living media; growth is favored by the addition of hemoglobin. In cultures, rods and coccoid forms with indistinct contours commonly appear together in dense masses. Infective cultures were first grown by Tyzzer (1942); Wu Lien-Teh and Jettmar may have cultivated Grahamella in 1930 but their cultures failed to infect.

Aerobic.

Not affected by arsenicals.

Parasitic. The etiological agent of grahamellosis of rodents and other mammals. May possibly occur in other vertebrates and in insects. Not known in man. Wide geographical distribution.

Type species: *Grahamella talpae* Brumpt 1911, 517.

Note: There are two species recognized in the genus. *G. talpae* is pathogenic for moles and *G. peromysci* is pathogenic for deer mice.

FAMILY III. **ANAPLASMATACEAE** PHILIP 1957, 980

Obligately parasitic organisms found within or on erythrocytes or free in the plasma of various wild and domestic vertebrates. No demonstrable multiplication in other tissues.

In blood smears treated with Giemsa's stain, appear as rod-shaped, spherical, coccoid- or ring-shaped bodies staining reddish violet and measuring 0.2–0.4 μm in diameter. May occur in short chains or irregular groups in blood plasma or within erythrocytes. **Each organism enclosed in a membrane,** with internal structure resembling that of rickettsiae. Multiply by binary fission. Gram-negative; not acid-fast. **Have not been cultivated.** Organisms transmitted by arthropods. Blood or blood-containing tissue homogenates can cause infection by any parenteral route.

Infection may or may not cause disease but usually results in long term persistence of the agent, with concomitant resistance to clinically demonstrable reinfection. Anemia most prominent feature of clinical disease. Members of the family occur throughout the world.

Influenced by tetracycline compounds, but not by penicillin or streptomycin.

Key to the genera of family **Anaplasmataceae**

I. Parasites form inclusions in erythrocytes.
  A. Inclusions in erythrocytes round, 0.3–1.0 μm in diameter. Several parasites may be found in each inclusion.
    1. No appendages to inclusions. Infect ruminants only.
       Genus I. *Anaplasma*, p. 907
    2. Appendages to inclusions. Infect cattle but not deer.
       Genus II. *Paranaplasma*, p. 908
  B. Inclusions in erythrocytes, 0.3–3.9 μm in diameter. Infect birds.
       Genus III. *Aegyptianella*, p. 909

II. Parasites within, or outside, erythrocytes.
  A. Parasites in and on erythrocytes. Stain intensely by Romanowsky methods. Ring structures rare or absent.
       Genus IV. *Haemobartonella*, p. 910
  B. Parasites on erythrocytes and in plasma. Ring forms common in preparations stained by Romanowsky methods.
       Genus V. *Eperythrozoon*, p. 912

Genus I. **Anaplasma** *Theiler 1910, 7*

With Romanowsky-type stains, **this organism appears in the erythrocytes as dense, homogeneous,** bluish purple **round structures, 0.3–1.0 μm in diameter. Electron microscopy reveals that these structures are inclusions separated from the cytoplasm of the erythrocyte by a limiting membrane. Each inclusion contains from one to eight subunits or initial bodies, which are the actual parasitic bacteria, each 0.3–0.4 μm in diameter that are dense aggregates of fine granular material embedded in an electron-lucid plasma and all enclosed in a double membrane (Ristic and Watrach, 1961).**

The organism (initial body) enters erythrocyte by causing invagination of cytoplasmic membrane and subsequent formation of a vacuole. In the vacuole initial body multiplies by binary fusion and forms an inclusion. Thus, the **formation of the inclusion body,** which **is most frequently encountered during the acute and convalescent phases of infection,** represents only a phase in the developmental cycle of the initial body. **Spores or resistant stages not formed.** Apparently aerobic (Pilcher *et al.*, 1961), produce catalase (Wallace and Dimopoullos, 1965). Do not produce pigments. Incorporate amino acids from plasma *in vitro* (Mason and Ristic, 1966). The adenosine triphosphate and glutathione concentrations of erythrocytes remain essentially unchanged, regardless of the intensity of infection and only a small quantity of methemoglobin occurs in parasitized erythrocytes (Mann, 1967). **Histochemical analysis** of parasites **reveals DNA, RNA,** protein and organic iron (Moulton and Christensen, 1955).

Obligate parasites of vertebrates; transmitted by arthropod vectors. Ticks are probable biological vectors and tabanidae and other biting arthropods are mechanical vectors. **Host range limited to ruminants. Do not infect common laboratory animals.** A review of *Anaplasma* can be found in Ristic, 1968.

Type species: *Anaplasma marginale* Theiler 1910, 7.

Further Comments

Species differentiation in this genus is based entirely on host range and location of inclusions in the erythrocytes. Morphologically the initial bodies and individual erythrocytic inclusions of the species are indistinguishable by conventional microscopy and electron microscopy. Some antigens are common to all members of the genus; others are unique to one species (Kreier and Ristic, 1963; Schindler *et al.*, 1966; Kuttler, 1966, 1967). Antigenic analysis, if developed, could be used for species differentiation.

Anaplasma marginale is the most pathogenic of the three species and is the causative agent of severe bovine anaplasmosis. The disease is widespread in tropical and subtropical areas of the world. *Anaplasma centrale* causes a mild form of bovine anaplasmosis in Africa and *A. ovis* causes anaplasmosis of sheep and goats in various regions of the world. Common laboratory animals are all refractory: rabbits, guinea pigs, rats, mice, ferrets, dogs, cats and chickens.

TABLE 18.4

Differential characteristics of species of genus **Anaplasma**

Speics	Host	Location in erythrocyte	Disease
1a. *A. marginale* subsp. *marginale*	Cow Deer	Predominantly marginal	Severe anaplasmosis
1b. *A. marginale* subsp. *centrale*	Cow	Predominantly central	Generally mild anaplasmosis
2. *A. ovis*	Sheep Deer Goats	Predominantly marginal	Mild to severe disease

Genus II. Paranaplasma *Kreier and Ristic 1963, 701*

Inclusion bodies and initial bodies morphologically indistinguishable from species of *Anaplasma* in Giemsa-stained blood films studied by light microscopy. **By** techniques using wet preparations of blood or **fluorescent antibody stain, and by electron microscopy, appendages to inclusion bodies are revealed** which may resemble tails, loops or rings or connect two inclusion bodies in a "dumbbell" form. Fluorescent antibody and cross-immunity studies show organisms to be in part antigenically distinct from *A. marginale.* Infectious for cattle but not deer; may or may not grow in sheep (Kreier and Ristic, 1963).

Type species: *Paranaplasma caudatum* Kreier and Ristic 1963, 701.

Note: There are two species recognized, *P. caudatum* and *P. discoides*, separated on the basis of the morphology of their inclusion bodies.

Genus III. Aegyptianella *Carpano 1929, 12*

Parasitic in species of wild and domestic birds and in some poikilothermal animals. **With Romanowsky-type stains, purple staining intracytoplasmic inclusions containing bacteria occur in erythrocytes.** Thin sections of inclusion bodies examined by electron microscopy revealed up to 26 bacteria or initial bodies enclosed in a double membrane (Gothe, 1967; Bird and Garnham, 1967) and with **internal morphology strikingly similar to *Anaplasma*.** Within the inclusion, **initial bodies reproduce by binary fission.** Has not been cultivated *in vitro*.

Organisms sensitive to dithiosemicarbazones and tetracyclines (Barrett *et al.*, 1965; Lämmler and Gothe, 1967).

Type species: *Aegyptianella pullorum* Carpano 1929, 12.

Genus IV. Haemobartonella *Tyzzer and Weinman 1939, 143*

Obligate parasites on or within erythrocytes of many vertebrate species. Organisms coccoid or rod-shaped as seen by light microscopy; rods seem to be chains of coccoids. Occur **singly, in pairs or in groups in shallow or deep indentations** on the erythrocyte surface, **sometimes in vacuoles within the erythrocytes, rarely in the plasma.** Stain well with Romanowsky-type stains; poorly with many other aniline dyes. **Possess single or a double limiting membrane, have neither cell wall nor distinct nuclear structures** (Tanaka, *et al.*, 1965); **Gram-negative;** not acid-fast. Have **not** been **cultivated** outside the host.

The experimental **host range** of any species is **restricted**; a species occurring naturally in a given host may be infective for closely related host species but not for the whole range of animals susceptible to other species of *Haemobartonella*. Morphology of the organism may vary in different hosts. **Growth inhibited by arsenicals and tetracyclines, not by penicillin** or **streptomycin.**

Most species are pathogenic but clinical disease, characterized by anemia, is usually not apparent unless the animal is splenectomized, except in the case of *Haemobartonella felis* in cats. The organisms are transmitted by arthropods and *H. felis* may also be transmitted by ingestion of blood (Flint *et al.*, 1958).

Type species: *Haemobartonella muris* (Mayer) Tyzzer and Weinman 1939, 143.

Note: There are three species recognized in the genus. *H. muris* is pathogenic for rodents, *H. felis* for cats and *H. canis* for dogs.

Genus V. Eperythrozoon *Schilling 1928, 1854*

Obligate parasites in the blood of various vertebrate species including some rodents, ruminants and pigs.

Stains bluish or pinkish violet with Giemsa's and Romanowsky-type stains. In such stained preparations cells show **no differentiation of nucleus and cytoplasm** and **appear as rings or coccoids, 0.4–1.5 μm in diameter, characteristically round with numerous annular or disc-shaped elements, rarely rods. Occur on erythrocytes and free in plasma with about equal frequency.**

Swarm-like clusters of ring-shaped eperythrozoa may occur on the surface of erythrocytes. Rod-shaped forms may occur partly or entirely circling an erythrocyte. In fresh preparations observed by **dark-field** and **phase microscopy, coccoids but no ring forms are seen** (Wigand, 1958; Kreier and Ristic, 1963). By electron microscopy organism appears pleomorphic, surrounded by a single limiting membrane, with no cell wall, nucleus or other organelles (Tanaka *et al.*, 1965). Has not been **cultivated in cell-free media.** Some species have been shown to be transmitted by arthropod vectors. Experimental transmission by inoculation of blood occurs readily.

Splenectomy activates latent infection with most species.

Growth inhibited by arsenicals and tetracyclines but not by penicillin or streptomycin.

Type species: *Eperythrozoon coccoides* Schilling 1928, 1854.

Further Comments

Differentiation from *Haemobartonella* is in many cases difficult, and possibly arbitrary. Differentiation is based on the fact that haemobartonellae rarely occur as ring forms while eperythrozoa commonly do and on the fact that eperythrozoa occur with about equal frequency on the erythrocytes and free in the plasma while haemobartonallae rarely occur free in the plasma.

Note: There are five species recognized, which are separated on the basis of morphology and host. *E. coccoides* is a parasite of certain rodents; *E. ovis* of sheep, goats and deer; *E. suis* of swine; *E. parvum* of swine; and *E. wenyonii* of cattle.

ORDER II. CHLAMYDIALES STORZ AND PAGE 1971, 334

Coccoid microorganisms whose obligately intracellular mode of multiplication is characterized by change of the small, rigid-walled infectious form of the organism (elementary body) into a larger, thin-walled, non-infectious form (initial body) that divides by fission. The developmental cycle is complete when daughter cells reorganize and condense to become elementary bodies which survive extracellularly to infect other host cells. Metabolically limited, Gram-negative parasites of vertebrates in which they may cause various diseases. Occasionally found in arthropods.

FAMILY I. CHLAMYDIACEAE RAKE 1957, 957

Coccoid microorganisms, 0.2–1.5 μm in diameter, which multiply only within the cytoplasm of host cells by a developmental cycle characterized by change of a small elementary body into a larger initial body that divides by fission. The cycle is complete when daughter cells reorganize and condense to become elementary bodies which survive extracellularly to infect other host cells. Elementary bodies contain compactly arranged nuclear material and ribosomes and are bounded by a rigid, trilaminar cell wall that is chemically similar to that of Gram-negative bacteria. Initial bodies contain amorphous nuclear material less electron-dense than that of elementary bodies, have thin, fragile cell walls and apparently are non-infectious. Metabolically limited, Gram-negative parasites of vertebrates in which they may cause various diseases. Occasionally found in arthropods. Cultivatable in yolk sac of chicken embryos. Sensitive to tetracycline antibiotics. Stain with aniline dyes.

Genus I. Chlamydia *Jones, Rake and Stearns 1945, 55**

Non-motile, spheroidal organisms, 0.2–1.5 μm in diameter depending upon their stage of development in a unique, obligately intracellular growth cycle. The principal developmental stages are (1) the **elementary body** which is a small, electron-dense spherule, 0.2–0.4 μm in diameter containing a nucleus and numerous ribosomes surrounded by a multilaminated wall; (2) the **initial body** which is a large, thin walled, reticulated spheroid, 0.8–1.5 μm in diameter containing nuclear fibrils and ribosomal elements; (3) an **intermediate body** representative of a transitional stage between the initial body and elementary body (Plate 18.6). The **elementary body is the infectious form** of the organism. The **initial body is the vegetative form** which divides by fission intracellularly but is apparently non-infectious when separated from the host cell.

Growth, multiplication and maturation of the organisms occur over a period of 40 hrs or less as microcolonies within a cytoplasmic vesicle whose envelope originates from invaginated host cell membrane (Plate 18.6). The elementary body is phagocytized by the host cell and within several hours reorganizes and enlarges to become the initial body which divides by fission. Daughter cells continue to divide, then develop an electron-dense nuclear mass and laminated walls (intermediate form) and decrease in size to become the mature, infectious elementary bodies. When the wall of the vesicle and host cell are disrupted, organisms are released to repeat the cycle in other host cells. **Phagocytosis of more than one elementary body per host cell may occur** resulting in the presence of several developing chlamydial microcolonies in the cytoplasm of a host cell.

Gram-negative. Cell walls are similar in chemical composition to those of Gram-negative bacteria. May be stained by Giemsa's, Macchiavello's, Gimenez's or Castenada's methods, or may be seen readily in unstained wet mounts of infected cells with a phase-contrast optical system.

When separated from host cells, chlamydiae have **few detectable metabolic activities,** and these are dependent upon an exogenous source of organic and inorganic cofactors including high energy phosphates (e.g. ATP). Nevertheless, when extracellular chlamydiae are provided with cofactors, they can **catabolize glucose, pyruvic acid or glutamic acid and produce carbon dioxide.** They also can **synthesize certain lipids.** The growth of one species is inhibited by sodium sulfadiazine thereby reflecting the ability of the organisms to synthesize folates intracellularly. Since they are unable to synthesize their

*An historical review of the classification and nomenclature of this genus has been given by Page (1966, 223).

own high energy compounds, they have been described as **"energy parasites." To date, extracellular propagation of chlamydiae has not been achieved.**

Multiplication is inhibited by tetracyclines, penicillin and 5-fluorouracil. An antigenically similar, 100 C-stable lipoglycoprotein is present in all organisms of the genus. A spectrum of specific antigens is present in the cell walls of various strains. Mosaics of five or six different antigens may be shared by strains isolated from different host species that have varying clinical syndromes.

After appropriate adaptation, all organisms of the genus **may be propagated in the yolk sac of chicken embryos or in cultures of vertebrate tissues.** Multiplication in chicken embryos occurs over a temperature range of 33–41 C but sensitivity to either end of the range or the optimum for growth varies with the species.

The organisms are **parasites of tissue cells of vertebrates.** However, they have been found in or on ectoparasitic arthropods. Many strains produce generalized infections in several host species, but some strains characteristically localize and cause pronounced inflammation in one or more tissues or organs of a certain host species. Groups of strains are found in **three major ecological niches: man,** where they cause oculourogenital and respiratory diseases; **birds,** where they cause respiratory disease and generalized infection; **mammals** (excluding primates) where they cause respiratory, placental, arthritic and enteric diseases. Some crossing between niches by certain strains occurs, especially between birds and man.

RNA and DNA have been demonstrated in both initial and elementary bodies, but the ratio of RNA to DNA in the initial body is twice that in the elementary body. G + C content of the DNA ranges between 39 and 45 moles % (T_m). The DNA weight is approximately 6.6×10^8 daltons. The number of nucleotide pairs per chlamydial organism is less than 1×10^6.

Only two species are recognized.

Type species: *Chlamydia trachomatis* (Busacca) Rake 1957, 958.

Key to the species of genus **Chlamydia**

I. Forms compact microcolonies within cytoplasmic vesicles. Produces iodine-staining compounds within vesicle. Growth in the yolk sac of chicken embryos is inhibited by sodium sulfadiazine (1 mg/embryo).

1. *Chlamydia trachomatis*

II. Forms microcolonies within cytoplasmic vesicles which tend to rupture early in microcolony development, and the organisms in various stages of growth become distributed throughout the host cell cytoplasm. Does not produce iodine-staining compounds in vesicles. Growth in the yolk sac of chicken embryos is not inhibited by sodium sulfadiazine (1 mg/embryo).

2. *Chlamydia psittaci*

PART 19

THE MYCOPLASMAS

CLASS MOLLICUTES EDWARD AND FREUNDT 1967, 267

Procaryotic organisms bounded by a single triple-layered (Plate 19.2, Fig. 4) membrane; they lack a true cell wall and are incapable of synthesizing cell wall precursors, such as muramic and diaminopimelic acids. Cells small, sometimes ultramicroscopic (about 200 nm), highly pleomorphic, coccoid to filamentous (Plate 19.2, Figs. 3 and 4), with a more or less pronounced tendency of the filaments to produce truly branched myceloid structures. The method of reproduction is controversial, but appears to take place by the development within the filaments of tiny coccoid structures ("elementary bodies") and their subsequent release by fragmentation and disintegration of the filaments, and/or by binary fission; also, reproduction through budding may occur. Usually non-motile. No resting stages known. Gram-negative.

The species recognized thus far can be grown on artificial cell-free media of diverse complexity. Colonies are minute; they have a marked tendency to grow down into solid media and usually have a characteristic "fried egg" appearance (Plate 19.1, Fig. 1). Most species completely resistant to penicillin and its analogues. Growth and metabolism specifically inhibited by antibody. Several of the above-mentioned characteristics of this Class, such as the morphological instability, the inability to retain the dye-iodine complex of the Gram stain, the tendency to penetrate into the depth of solid media, the insusceptibility to penicillin and the susceptibility to antibody, can be attributed to the lack of a cell wall. The mycoplasmas may be saprophytic, parasitic or pathogenic. The pathogens cause diseases of animals and possibly of plants.

ORDER I. MYCOPLASMATALES FREUNDT 1955, 71

Only one order, *Mycoplasmatales* is accepted in the class *Mollicutes;* the description of the order is therefore the same as for the class.

Two families are accepted in the order *Mycoplasmatales*.

Key to the families of order **Mycoplasmatales**

I. Require sterol for growth.

Family I. *Mycoplasmataceae*, p. 930*

II. Do not require sterol for growth.

Family II. *Acholeplasmataceae*, p. 949

Note: There are two genera included in an addendum to this Part. Members of the genus *Thermoplasma* are thermophilic and do not require sterols and members of the genus *Spiroplasma* require sterols and form helical filaments.

Since publication of the *Manual* a genus of strictly anaerobic mycoplasmas, *Anaeroplasma*, which contains both sterol requiring and non-requiring strains (Robinson *et al.*, *International Journal of Systematic Bacteriology 25:* 173–181, 1975; Robinson and Allison, *ibid*, 182–186) has been named and

* All page numbers cited refer to *Bergey's Manual of Determinative Bacteriology*, 8th edition, The Williams & Wilkins Company, Baltimore, 1974.

described. The genus contains two species, *A. abactoclasticum*, which does not lyse bacterial cells, and *A. bactoclasticum*, which is bacteriolytic. The reader is referred to the two papers cited above for complete descriptions.

FAMILY I. MYCOPLASMATACEAE FREUNDT 1955, 71

Sterol required for growth. Other characters as for the class and order.

Note: The *Manual* lists the genus *Mycoplasma* as the sole genus in this family. However, since publication of the *Manual* a group called the "T-mycoplasmas," described as *species incertae sedis* on p. 947 of the *Manual*, have been assigned to a new genus and species, *Ureaplasma urealyticum* (Shepard *et al.*, *International Journal of Systematic Bacteriology 24:* 160–171, 1974). This genus is differentiated from the genus *Mycoplasma* by its ability to hydrolyze urea. The reader is referred to the original paper for a complete description of the genus.

Genus I. **Mycoplasma** *Nowak 1929, 1349 Nom. cons.* Jud. Comm. Opin. 22, 1958, 166

Cells highly pleomorphic, varying in shape **from spherical or slightly ovoid,** about 125–250 nm in diameter in early exponential growth phase, **to slender branched filaments** of uniform diameter ranging in length from a few to 150 μm (Plate 19.2, Fig. 1). Chains of coccoid bodies develop at a later growth phase (Plate 19.2, Fig. 2). **Possibly endowed with alternative modes of replication:** binary transverse fission of coccoid and filamentous cells, multiple release of elementary bodies produced within the filaments and budding. The filaments also branch, giving rise to a pseudo-mycelial structure; later dense corpuscles appear in the filaments which are transformed into chains of spherical bodies as a result of constrictions between the spheres but without formation of septa; the chains fragment liberating free elementary bodies.

Cells lack a true cell wall and are bounded by a single triple-layered membrane, about 7.5–10 nm thick. Usually non-motile but gliding motility has been described in some species. No resting stages known. **Gram-negative.**

Colonies on solid medium are hemispherical, slightly flattened and very small, their diameter generally ranging from 10–600 μm, although occasionally colonies are found up to 4 mm. **The typical colony is biphasic, with a "fried egg" appearance,** i.e. consisting of an opaque, granular central area which grows down into the medium and a flat, translucent peripheral zone (Plate 19.1, Fig. 1). On blood agar, colonies of most species produce zones of α,α' or β hemolysis, under appropriate experimental conditions (Cole *et al.*, 1968; Aluotto *et al.*, 1970). Peroxide is the major hemolysin in most species. The colonies of some species are able to adsorb on their surfaces erythrocytes and/or other animal cells in suspension (Taylor-Robinson and Manchee, 1967; Sobeslavsky *et al.*, 1968; Manchee and Taylor-Robinson, 1969). Growth in fluid media is usually a barely visible opacity.

Chemoorganotrophs; metabolism is mainly fermentative, although a few species have an oxidative metabolism. **Most species utilize either glucose or arginine** as the major source of energy; rarely both are used, sometimes neither. Glucose-fermenting strains may or may not ferment mannose (Table 19.1); they usually also ferment fructose, galactose, maltose, glycogen, dextrin and starch. Reports on utilization of other carbohydrates are conflicting.

Generally not proteolytic; a few species liquefy gelatin, coagulated serum or casein (Edward, 1954; Freundt, 1958; Aluotto *et al.*, 1970). Reduction of nitrate or production of indole has not been reported; hydrogen sulfide is produced by a few species only. **Urea not hydrolyzed.** Phosphatase activity a distinctive property of some species (Table 19.1).

2,3,5-Triphenyltetrazolium chloride, methylene blue and potassium tellurite are **frequently reduced under anaerobic conditions,** less frequently under aerobic.

All species require cholesterol, or certain other sterols, **for growth.** All species so far tested also require long chain fatty acids for growth and morphological and osmotic stability. Nutritional requirements are usually met by use of a complex medium containing serum or 1–2% PPLO Serum Fraction (Difco), although some species require yeast extract or yeast extract dialysate (see Aluotto *et al.*, 1970; Hayflick and Chanock, 1965).

Most species are facultative anaerobes although growth is better aerobically. One or two species are aerobic, and several of primate origin prefer

TABLE 19.1

Characteristics useful in the differentiation of the species of genus **Mycoplasma**

All species require sterol for growth; none hydrolyzes urea or esculin. Do not grow at 45–62 C or at pH 1.0–3.0. All sensitive to 5% sodium-polyanethol-sulfonate except *M. anatis*, *M. gallinarum*, *M. iners* (strain PG30), *M. gateae* and *M. maculosum* which may be either totally resistant or show a slight partial inhibition only. The usual designation that + = 90% or more positive strains applies only to hydrolysis of arginine and glucose data. In tests marked *, generally only a few strains, including the type strain, of each species have been tested. d = variable reaction.

	Arginine hydrolyzed	Glucose acid	O-F test*	Mannose acid*	Phosphatase*	Film and spots on egg yolk*	Tetrazolium reduced* Aerobic/anaerobic	G + C ratio*
1. *M. mycoides*	−	+	F	+	−	d	+/+	26.1–26.8
1b. *M. mycoides* subsp. *capri*	−	+		+	−	d	+/+	23.6–25.8
2. *M. bovirhinis*	−	+	F	−	−	−	+/+	24.5–25.5
3. *M. dispar*	−	+	O	+	−	−	+/+	28.5
4. *M. ovipneumoniae*	−	+					+/+	
5. *M. conjunctivae*	−	+	F	+	−	−	+/	
6. *M. gallisepticum*	−	+	F	+	−	−	+/+	32.0–35.5
7. *M. anatis*	−	+	F	+	+	+	−/+	
8. *M. synoviae*		+				+		34.0
9. *M. neurolyticum*	−	+	F	+	−	−	−/+	23.0–26.5
10. *M. pulmonis*	−	+	d	+	−	+	d/+	27.5–28.3
11. *M. felis*	−	+	F	−	+	+	−/+	25.0
12. *M. feliminutum*	−	−			−	+[1]	−/+	
13. *M. canis*	−	+	F	−	−	−	−/+	28.5–29.0
14. *M. edwardii*	−	+	F	−	−	+	−/+	29.2
15. *M. cynos*	−	+			+	+	∓/+	
16. *M. hyorhinis*	−	+	O	−	+	−	d/+	27.0–28.0
17A. *M. hyopneumoniae*	−	+						
18. *M. flocculare*	−	+			−		−/	
19. *M. pneumoniae*	−	+	O	+	−	+[2]	+/+	38.6–40.8
20. *M. bovigenitalium*	−	−		−	+	+	−/+	28.0–32.0
21. *M. agalactiae*	−	−			+	d	+/+	33.5–34.0
22. *M. alkalescens*	+	−			+		−/	25.9
23. *M. gallinarum*	+	−			−	+	+/+	27.0–28.0
24. *M. iners*	+	−		−	−	+	−/−	29.0–29.5
25. *M. meleagridis*	+	−		−	+	d	−/+	28.0–28.5
26. *M. arthritidis*	+	−		−	+	−	−/−	30.0–33.7
27. *M. gateae*	+	−		−	−	−	−/±	28.5
28. *M. maculosum*	+	−		−	+	+	−/+	26.5–29.5
29. *M. spumans*	+	−		−	+	−	−/−	28.5–29.0
30. *M. hyosynoviae*	+	−			−		−/	
31. *M. arginini*	+	−			−	−	−/+	28.6
32a. *M. orale* type 1	+	−		−	−	−	−/−	24.0–28.0
32b. type 2	+	−		−	+	−	−/+	26.1
32c. type 3	+	−		−	−			
33. *M. salivarium*	+	−		−	−	d	−/d[1]	27.0–31.5
34. *M. hominis*	+	−		−	−	−	−/−	27.3–29.2
35. *M. primatum*	+	−			+		−/	28.6
36. *M. fermentans*	+	+	F	−	−	+	−/+	27.5–28.5

[1] Weak reaction.

[2] Film but no spots.

anaerobic conditions, especially on primary isolation. The addition of 5% CO_2 may or may not improve growth.

Temperature range 22–41 C; optimum about 36–37 C. Grows well at pH 7 or slightly acid or alkaline.

Relatively resistant to lysis by osmotic shock but very sensitive to lysis by surface-active agents and lipolytic agents. Lysed by taurocholate at concentrations ot 0.03% when suspended in M sucrose-0.05 M NaCl solution; by digitonin in concentrations of 10–15 μg/ml.

All species moderately to highly sensitive to 1.5% digitonin using a disc growth inhibition technique.

With a few notable exceptions (*M. anatis*, *M. gallinarum*, *M. iners*, (strain PG 30), *M. gateae* and *M. maculosum*) the growth of most species is inhibited by 5% sodium-polyanethol-sulfonate (Kunze, 1971; Andrews, Freundt, unpublished).

Resistant to 1000–10,000 μg/ml of penicillin G and to 1000–4000 μg/ml of ampicillin, chloraxillin, methicillin or nafcillin. Susceptibility to other antibiotics variable; susceptibility to erythromycin and certain other macrolides particularly variable and may be useful for differentiation (Taylor-Robinson and Path, 1967; Arai *et al.*, 1967). Usually tolerates 1:2000–1:4000 thallium acetate.

Parasites and pathogens of a wide range of mammalian and avian hosts.

The G + C content of the DNA ranges from 23–40 moles % (T_m, buoyant density and chemical analysis). The genome size in the species examined is 4.4–4.8 × 10^8 daltons as determined by the renaturation method of Wetmur and Davidson (1968).

Type species: *Mycoplasma mycoides* (Borrel *et al.*) Freundt 1955, 73.

FAMILY II. **ACHOLEPLASMATACEAE** EDWARD AND FREUNDT 1970, 1

Sterol not required for growth. Other characters as for the class and order.

Genus I. **Acholeplasma** *Edward and Freundt 1970, 1*

Cells spherical, with a minimum diameter of about 125–220 nm, **and filamentous,** usually about 2–5 μm in length. Modes of reproduction as for genus *Mycoplasma*. **Cells bounded by a triple-layered membrane,** approximately 7 nm thick. Non-motile. Gram-negative.

Colonies on solid medium usually fried egg appearance, relatively large, up to 3 mm in diameter.

Chemoorganotrophs; metabolism fermentative, carbohydrates being the fermentable substrates. **Serum or cholesterol not required for growth.** Minimal nutritional requirements only partially defined.

Arginine and urea not hydrolyzed. Phosphatase activity weak or negative. **Film or spots not produced** on horse serum agar or egg yolk medium; no clearing.

Facultative anaerobes.

Temperature range about 20–40 C.

More susceptible to lysis by osmotic shock at 37 C than genus *Mycoplasma*.

All species resistant or only very slightly sensitive to 1.5% digitonin. Highly resistant to 5% sodium-polyanethol-sulfonate (Kunze, 1971; Andrews, unpublished; Freundt, unpublished).

Absolutely resistant to penicillins, not being inhibited by penicillin G, ampicillin, cloxacillin or methicillin in concentrations of at least 4000 μg/ml.

Apparently free-living saprophytes as well as mammalian and avian parasites, and possibly pathogens, with a fairly wide host range.

The G + C content of the DNA is about 30.0–33.0 moles % (T_m and buoyant density). The genome size is 1.0 × 10^9 daltons as determined by the DNA renaturation method of Wetmur and Davidson.

Other characters as for genus *Mycoplasma*.

Type species: *Acholeplasma laidlawii* (Sabin) Edward and Freundt 1970, 1.

Note: There are five species recognized in the genus but they are not described fully enough for construction of a key or table that would aid in identification. They can be distinguished antigenically and the reader is referred to the *Manual*, pp. 950–952, for mention of the tests. The five species are: *A. laidlawii*, *A. granularum*, *A. axanthum*, *A. modicum* and *A. oculusi*.

ADDENDUM TO ORDER **MYCOPLASMATALES**

GENUS OF UNCERTAIN AFFILIATION

Genus **Thermoplasma** *Darland, Brock, Samsonoff and Conti 1970, 1418*

Cells pleomorphic, varying in shape from spherical (0.3–2 μm) to filamentous structures, believed to reproduce by budding. Cells lack a true cell wall and are surrounded by a single triple-layered membrane, approximately 10–12 nm thick. Non-motile as far as known. No resting stages known. Gram-variable.

Colonies on solid medium containing 1.2% Ion Agar attain a diameter of about 0.3 mm at pH 2; they are rather flat, coarsely granulated, dark brown in color and some of them exhibit a typical "fried egg" appearance with a translucent peripheral zone.

Biochemical and nutritional characters relatively poorly defined. Growth occurs in fluid medium containing 0.02% $(NH_4)_2SO_4$, 0.05% $MgSO_4$, 0.025% $CaCl_2 \cdot 2H_2O$, 0.3% KH_2PO_4, 0.1% yeast extract and 1.0% glucose. The cell yield is proportional to the concentration of yeast extract at concentrations less than 0.2%; no growth on the ether-extractable fraction of yeast. Growth is not stimulated by the addition of glucose, galactose, sucrose, ribose or glycerol. **Does not require sterol.**

Apparently strict aerobe.

Optimum temperature for growth about 59 C, no growth occurring at 37 or 65 C.

Grows well between pH 0.96 and 3.0; **the optimum pH is about 1–2.**

Cells rapidly lysed by sodium lauryl sulfate.

Insensitive to vancomycin, a specific inhibitor of cell wall synthesis, at concentrations of at least 5 mg/ml. Inhibited by novobiocin at a concentration of 0.1 μg/ml. Sensitivity to penicillin could not be determined because of the acid lability of this antibiotic.

Apparently a true saprophyte, having been isolated from a burning coal refuse pile and from acid hot springs.

Reports of the G + C content of the DNA are conflicting, ranging from 25 to 40 moles % (T_m and buoyant density).

Type species: *Thermoplasma acidophilum* Darland *et al.* 1970, 1418.

ADDENDUM TO CLASS **MOLLICUTES**

GENUS OF UNCERTAIN AFFILIATION

Genus **Spiroplasma** *Saglio, Lhospital, Laflèche, Dupont, Bové, Tully and Freundt 1973*, 201

Cells pleomorphic, varying in shape **from spherical or slightly ovoid,** with a minimal diameter of about 100–250 nm, **to helical and branched non-helical filaments.** The helical forms, about 120 nm in diameter and 2–4 μm in length in the logarithmic and considerably longer in the postlogarithmic phase of growth, predominate in liquid cultures. Only non-helical and round cells have been demonstrated in agar cultures. Mode of replication not known. **Cells lack a true cell wall and are bounded by a single triple layered membrane,** about 7–8 nm thick. An **additional outer layer of short projections,** which often appear periodic, is present on the surface of the bounding membrane. Helical filaments are **motile,** showing two types of motility: a rapid rotary or "screw" motion, and a slow undulation. Flagella, axial filaments, or other organelles of locomotion not demonstrated. **Gram-positive.**

Colonies on solid medium attain a diameter of about 0.2 mm and exhibit a **typical biphasic "fried-egg" appearance.** Marked turbidity produced in both cultures.

Chemoorganotroph. **Acid produced from glucose and mannose.** Does not hydrolyze esculin, arginine or urea. Reduction of tetrazolium variable. Phosphatase positive. Spots, but no film, produced on horse serum agar. No liquefaction of coagulated horse serum. Agar colonies produce zones of α type hemolysis by the overlay technique, using guinea pig red blood cells. One of two strains tested adsorb guinea pig red blood cells. Does not synthesize pigmented carotenoids.

Cholesterol, or possibly other sterols, required for growth. Growth occurs in a conventional mycoplasma medium containing 10% yeast extract and 20% horse serum, modified by the addition of 7% sorbitol, 1% tryptone, 1% sucrose, 0.1% fructose, and 0.1% glucose. On adaptation, growth may occur in less complex media.

Facultative anaerobe, maximal growth occurring at 5% CO_2 in nitrogen.

Temperature range 20–37 C, growth being very sparse below 25 C and above 35 C; **sharp optimum at 32 C.**

Resistant to 10,000 i.u./ml of penicillin G; completely inhibited by 0.04–0.08 μg/ml of erythromycin, 0.16–0.32 μg/ml of tetracyclin, 20–40 μg/ml of neomycin, 15–25 μg/ml of amphotericin B, and 1,280–2,560 μg/ml of thallium acetate. Sensitive to 1.5% of digitonin, using a disc test; sensitivity to sodium-polyanethol-sulfonate varies with growth conditions.

Isolated from the leaves of Citrus plants affected by "Stubborn" disease, the sieve tubes of which contain morphologically identical structures. The disease has not been reproduced experimentally with the organism.

The G + C content of the DNA is 25.0–26.35 moles % (T_m and buoyant density); the **genome size is 10^9 daltons.**

Type species: *Spiroplasma citri* Saglio *et al.* 1973, 202.

Further Comments

Although several of the properties described for the genus would seem to allow its classification within the Class *Mollicutes*, Order *Mycoplasmatales*, other of its characteristics are apparently inconsistent with the definition of the *Mycoplasmatales*. While helical shape and motility need not rule out inclusion in the *Mycoplasmatales*, the demonstration of a surface layer of special structure, retention of the Gram stain complex, and infection by a tailed bacteriophage of classic type B morphology suggest the presence of a modified cell wall or a structure reminiscent of a cell wall. Hence, the affiliation of the genus with the Class *Mollicutes* and its exact position must await further ultrastructural and biochemical studies, including a search for cell wall precursors.

Glossary

No attempt has been made to provide a dictionary of bacteriology. Words which in our opinion may not be too well known are given for convenience and to save the user time. Words found in the ordinary dictionary, e.g. *Webster's Seventh New Collegiate*, are not included.

Abbreviations (as used in etymologies).
Gr. = Greek, usually latinized Greek. The original word is transliterated into the Latin alphabet and, in most instances, is given in its Latin form.
L. = Latin; usually this indicates that the word is classic Latin and found in an unabridged Latin dictionary.
M.L. = Modern Latin; a word not necessarily classic Latin, but treated and used as a Latin word.
Med.L. = Medieval Latin; many words derived from languages other than Latin and Greek were latinized during the Middle Ages and used in fields such as alchemy, biology and pharmacy.
adj., adjective; dim., diminutive; n., noun; part., participle; part. adj., participial adjective; pl., plural (note: names of all taxa higher than genus are plural and feminine); sub., substantive or noun; fem., feminine gender; masc., masculine gender; neut., neuter gender; nom., nominative case; gen., genitive case.

Acrasin-like. Similar in action to acrasin, the diffusible substance which promotes cellular aggregation in a group of eukaryotic protists, the *Acrasiales*.

Acropetally. Produced successively in the direction of the apex, i.e. the apical member is the youngest.

Actinophage. A virus able to infect and eventually cause lysis of actinomycetes.

Aerial (of mycelium). Mycelial growth above the substrate.

Aleuriospore. Small, terminal spores whose position, color, form, structure and dimensions approach those of conidia. They differ from conidia in that they are not ephemeral and immediately separated from the mycelium by dehiscence but remain attached by a structure equal in diameter to the hyphal branch from which it was formed. They are normally solitary and terminal and present in great numbers, e.g. *Micromonospora, Thermomonospora*.

Alluvial Fan. An erosion deposit of a stream where it empties out from a gorge upon an open plain; delta-like.

Alpha-Reaction (or Alpha-hemolysis). The production of a green discoloration around a colony growing on a blood agar medium.

Andersen Sampler. A device for collecting, sizing and enumerating viable airborne particles (Andersen, 1958).

Apical Meristem. Growing point of a plant.

Arborescent. Branched in a treelike manner.

Arthrospore. An asexual spore (but not sporangiospore, chlamydospore or endospore) formed by the walling off a pre-existing hypha or branch to give a chain of spores, e.g. *Streptomyces*.

Articulate. Constrictions between cells in a chain so deep that cells appear connected by a thread.

Arylsulfatase. An enzyme characteristic of mycobacteria capable of splitting the bond between the sulfate group and the aromatic ring in compounds having the general formula $ROSO_3H$, wherein R represents the aromatic ring.

Asexual (of spores). Spores produced vegetatively, i.e. without involving the union of nuclei in a sexual process.

Auxotrophic (of mutants). Having a growth requirement for certain compound(s) or nutrients not required by the parental strain.

Axial Fibrils. Long, slender, threadlike processes of regular thickness which are intertwined with the protoplasmic cylinder of spirochetes. Each fibril arises from subterminal attachment discs at either end of the protoplasmic cylinder and overlaps fibrils originating at the opposite end for varying portions of their length. Probably responsible for motility.

Axial Filament. A term which has been used synonymously with axial fibril and axistyle.

Axistyle. The pair of axial fibrils observed in *Leptospira.*

Bacteroid. In *Rhizobium* a special deformed type of cell, probably associated with nitrogen fixation. In *Frankia* used for bacteria-like cells. In *Blattabacterium* bacteria-like intracellular microorganisms found in various species of insects.

Basionym. The name of the species whose specific epithet is adopted for a new combination, a senior synonym.

Basipetally. Development in the direction of the base, i.e. the apical part is oldest.

Battey Bacillus. Properly, a vernacular name for a strain of the species *Mycobacterium intracellulare*, but often used loosely for any strain of the group or related species, including *M. avium*, usually associated with human disease or resembling such bacteria.

Beta-Hemolytic Reaction (or Beta hemolysis). The elaboration of a hemolysin that lyses red blood cells with the resultant clearing around a colony growing on a blood agar medium.

Bifid. A rod with one or both ends forked. The ends may be short and clavate or short and rounded and of greater diameter than the rod.

Blastospore. Spores found as enlargements at the ends of aerial hyphae in acropetal succession.

Bloom. A surface texture characterized by the projection of very short aerial hyphae appearing minutely velvety or powdery.

Bulla. A bleb or bubble-like projection extending from the outer cell envelope of Spirochetes. May be observed with dark field or electron microscopy.

Camp Test. A blood agar plate is streak-inoculated across the center of the plate with a β-lysin producing strain of *Staphylococcus aureus* (*note:* β refers to the type of lysin produced and not the hemolytic reaction). At right angles to the *Staphylococcus* streak, but not touching it, a suspected *Streptococcus agalactiae* is streaked. After incubation (37 C for 24 hr), a synergistic clearing (complete hemolysis) in the proximity of the two organisms indicates a strong possibility that the unknown organism is *Streptococcus agalactiae* (usually used to determine the etiologic agent of bovine mastitis).

Cephalosporial. Like a rounded mass (head) of spores in a slimy matrix at tip of sporophore. Literally, "resembling the spore production in the fungal genus *Cephalosporium*."

Chemolithotroph. An organism that uses carbon dioxide as its principal source of carbon for growth and obtains its energy by the oxidation of inorganic compounds.

Chemoorganotroph. An organism that obtains its energy by the oxidation of organic compounds. These are also its principal sources of carbon.

Chalmydospores. Terminal or intercalary, thick walled, resistant mycelial cells generally with a larger diameter than the parent hyphae. These cells differ from true conidia in their irregularity of size and shape and haphazard occurrence.

Chlorobium Vesicles. Photopigment-bearing elements of the green sulfur bacteria (Chlorobiaceae). The vesicles are oblong bodies, 30 to 40 nm wide and 100 to 150 nm long; they are completely bounded by a thin (3 nm) electron-dense nonunit membrane which lies adjacent to the cell membrane but is distinguishable from it (Plate 1.4).

Chromogenicity. Ability to produce color (pigment); often referring to the dark brown to blackish pigments produced *in situ* in certain tyrosine-containing media as a result of the formation of melanin or melanin-like compounds.

Coenobium. Several cells of an organism remain together after division to form an ordered (*Thiopedia*) or disordered clump.

Conidia (substrate). Asexual spores produced in the substrate medium.

Conidiospore. Any asexual spore (but not a sporangiospore or endospore). In general use for the spores formed by several genera of the Actinomycetales, e.g. *Streptomyces* and *Micromonospora*, but see definitions for aleuriospore and arthospore.

Cords. Term applied to mycobacteria compacted side by side and end to end in more or less curving ("serpentine") strands. Cord-forming strains of tubercle bacilli are more virulent than strains lacking cord formation.

Coremium (plural coremia). A fascicle of parallel aerial hyphae on the culture surface; usually the individual hyphae bear spores.

Cortex. Tissue of a plant located between the epidermis and the vascular system (from the Latin word for bark or rind).

Coxal Organs. Secretory organs of glandular nature located close to the attachment of the first pairs of legs of *Ornithodoros*, which excrete a filtrate of ingested blood through an outlet pore during or after feeding.

Crista. The large bundle of axial fibrils observed in *Cristispira*, forming a spiral around the cell, known as the "crest."

Cyst. In the Myxobacterales, the sporangium.

Dextrorse (of spore chains). Spiralling clockwise when viewed from above.

Diagnostic Titer for Fluorescent Antibody Work. One-half of the highest dilution of a conjugated antiserum which gives 4+ fluorescence with its homologous antigen.

Dictyospores. A muriform spore; a spore having both longitudinal and transverse septations, example: spores of the fungal genus *Alternaria*.

Diffusible (of pigments). Penetrating into the medium from their source. However, pigmentation in the medium does not prove the pigment to be diffusible. Generally, the melanoid pigments in a medium are developed *in situ*.

Diphtheroid. Cells of irregular morphology. Usually an irregularly staining rod of uneven diameter, frequently enlarged at one end, which may appear granular and may have lateral short branches. Frequently arranged as palisades or having a V, Y or T shape.

Endophyte. Microorganisms living in symbiosis with higher plants or lower plants (e.g. algae).

Endospore. A resting spore produced endogenously by bacteria with a characteristic fine structure and exhibiting degrees of resistance to heat, chemicals, and adverse environmental changes. Single endospores are normally formed within a vegetative bacterial cell (mother cell), e.g. *Bacillus*, *Clostridium*. Other species may form more than one spore per cell, e.g. *Metabacterium polyspora*, and certain genera of the mycelial Actinomycetales may produce many endospores within a thallus.

Epilimnion. The upper layer of lakes, subject to disturbance by winds, lying above the thermocline.

Etymology. The origin of a word. Those who name taxa often do not indicate the source of the name or do so in very general terms; the editors have indicated the possible derivation to the best of their ability. Pronunciation of names is not attempted; names are syllabilized and the usual place of emphasis indicated.

The letter k is rare in Latin and the Greek letter kappa has been transliterated as c.

According to Stearn (*Botanical Latin*, Thomas Nelson, London, 1966) the suffix-*cola* is a substantive and not an adjective; it is therefore not declined.

Eurythermal. Capable of good growth at 28 C as well as 50 C and above.

Extant. In existance; not destroyed; applied to cultures available and viable.

Filament. In this edition, any long threadlike form which may or may not be segmented and includes the forms sometimes referred to as trichomes (q.v.).

Floccose. Appearing like tufts of wool.

Fluorescent Antibody. Immune globulin conjugated to a fluorochrome.

Fragmentation. The division of a rod or filament into smaller rods by transverse fission.

Fragmentation Spores. Spores formed by breaking up of hyphae into separate cells (arthrospore).

Fruiting Body. In the Myxobacterales, the (usually) macroscopically visible structure containing the resting cells.

Gamma Reaction. Indicates no change in the appearance of red blood cells immediately surrounding a colony on a blood agar medium.

Gliding Motility. A continuous and regular movement across solid substrates performed by certain prokaryotes (e.g. myxobacteria, many blue-green algae), which do not possess detectable locomotor organelles.

Glossy Colony. Referring to a streptococcal colony that contains cells which are deficient in type-specific M capsular polysaccharide.

Granule. A term which has been used synonymously with spirochetal sphere.

Halophile (halophilic). Have a requirement for NaCl greater than marine forms. May be divided into (a) extreme halophiles which require from about 15% to saturated NaCl plus other ions (*Halobacterium*, *Halococcus*); (b) moderate halophiles which require from about 5 to 15% NaCl (*Paracoccus halodenitrificans*, *Vibrio costicola*). Marine strains usually require undiluted sea water or NaCl at concentrations of 1.5 to 5%, i.e. they have a requirement for added salt greater than that found in the usual laboratory media or in physiological saline (0.85%).

Heloplankton. Floating vegetation of a marsh.

Heptaene (heptaenic) (of antibiotics). With seven conjugated double bonds. Also see Polyene (antibiotics).

Heterokaryosis. A condition in which genetically different nuclei are associated in the same protoplast.

Hexaene (hexaenic). With six conjugated double bonds. Also see Polyene (antibiotics).

Hibitane (or chlorohexidine acetate). Bis-*p*-chlorophenyldiguanidohexane diacetate, a general murobicide.

Holdfast. A secreted mass of material which cements a cell or organism to the substrate or a surface.

Holocarpic. Condition in an organism in which the entire vegetative body (mycelium) is converted into a reproductive structure (sporangium).

Holotype Strain. The one strain designated by the author as the nomenclatural type of a species. A synonym of type.

Hormogonium (plural **hormogonia**). Groups of cells which detach from parent filaments, exhibit motility and develop into new filaments.

Hygroscopic (of aerial mycelium). Becoming moist and usually darkened.

Hypolimnion. The portion of certain lakes below the thermocline which receives no heat from the sun and no aeration by circulation.

Hypothallus (plural **hypothalli**). A mass of hardened slime at the base of a fruiting body in species of the Myxobacterales.

Illegitimate Name. A name of a taxon that does *not* conform to the International Code of Nomenclature. Such names are to be rejected.

Immobilizin. An antibody which deprives an organism of its locomotory ability, i.e. immobilizes it.

Incurvation. The process of one end of a spirochete folding over and gliding closely around cell toward opposite end; may precede tranverse fission which then occurs at looped end of the intertwined organism.

Indicator Pigment. A pigment that changes color, depending upon the pH of the medium.

Indicator Strain. A strain used by several groups of investigators in lieu of a type of neotype strain.

Intercalary. A bud or spore or propagation cell formed *in* a filament or hypha, or chain of cells, but not terminally.

Intracytoplasmic Membrane Systems. The photopigment-bearing intracytoplasmic membrane system of purple nonsulfur and sulfur bacteria is an extension of the cytoplasmic unit membrane. 1. *Vesicular type:* The intracytoplasmic membrane system originates as vesicular invaginations at many points over the cytoplasmic membrane, and extends inward in the form of isolated and connected vesicles (Plate 1.2, Fig. 1). 2, *Lamellar type:* The intracytoplasmic membrane system originates in one or several points of the cytoplasmic membrane. Tubular intrusions extend to flattened discs which form stacks of lamellae, if several disc-shaped infoldings become appressed against each other (Plate 1.2, Figs. 2 and 3; Plate 1.3, Fig. 1). 3. *Tubular type:* The intracytoplasmic membrane system extends in the form of many tubular intrusions from the cytoplasmic membrane. There may either be many groups of parallel tubes lying in different directions within the cell (Plate 1.3, Fig. 2) or the tubes may form an irregular three-dimensional network. 4. *Thylakoids:* The term is used by some authors to designate photopigment-bearing intracytoplasmic membrane systems, preferably of the lamellar type. Also intracytoplasmic membrane systems of nitrifying bacteria have been called thylakoids.

Intrasporangial Hypha. A hyphal element within a sporangium.

Isoniazid. Isonicotinic acid hydrazide; an antituberculosis drug.

Johnin. A tuberculin-like agent for testing skin sensitivity to *Mycobacterium paratuberculosis* (*M. johnei*).

Lamellae. The flattened discs formed from infolding of the intracytoplasmic membrane systems (q.v.) of photosynthetic bacteria.

Lectotype. A specimen selected from the original material to serve as nomenclatural type when the holotype was not designated at the time of publication.

Macrolide (antibiotic). An antibiotic having a large ring lactone structure.

Meristem. Formative tissue adding new cells to the plant body.

Meromictic (lakes). Lakes partly mixed in which thermal turnover occurs only in top layers, bottom layers are stagnant and anaerobic in contrast to holomictic lakes in which mixing is complete.

Merosporangium. A tubular sporangium containing a single longitudinal row of sporangiospores, e.g. *Microellobosporia, Actinomycetes.*

Mesosomes. A membranous infolding derived from the cell membrane, found in the cells of certain bacteria. Synonym: plasmalemmosome (q.v.)

Metacyclic Forms. Small, irregular forms of *Borrelia*, often resembling granules, that appear after the "negative phase." The term is a survivor of the concept according to which *Borrelia* undergo a life cycle from fully developed forms through an invisible (by ordinary light microscope) stage.

Microarthrospores. Small spores resulting from the fragmentation of a hypha. According to Snell and Dick (1957) "an arthrospore produced in the substrate mycelium of actinomycetes."

Microconidium. A small conidium; a round, small, single celled spore produced either on a conidiophore or sessile.

Microcyst. In the Myxobacterales, a myxospore surrounded by a definite capsule of hardened slime. In the Spirochaetales, a term used synonomously with spirochetal sphere. In *Nocardia*, coccoid structures produced by shortening and condensing of bacillary elements; microcysts survive 80 C for an hour or more, vegetative cells are killed on exposure to 60 C for 30 min.

Mixotroph (mixotrophic). As used of phototrophic bacteria, the ability to grow in either the presence or the absence of light; also of organisms able to use various combinations of inorganic and organic substrates as energy and carbon sources; "a commingling of alternative modes of energy generation and/or carbon assimilation" (Rittenberg, 1969).

Monopodial. A main axis that continues its original line of growth giving off lateral branches (Jackson, 1928).

Monotype. A single strain, clearly designated by the author, upon which the species description was based.

Mother Cell. The part of the sporangium (vegetative bacterial cell) outside the developing spore (endospore) in *Bacillus* and *Clostridium*.

Mucoid Colony. A colony with a gummy consistency, resulting from the abundant synthesis of capsular material.

Muriform. Having both longitudinal and transverse septa; actually wall-like, resembling a stone or brick wall.

Mycetocytes. Specialized cells of some insects, usually in the fat body, containing symbiotic fungal or bacterial inclusions.

Mycetome. A specialized organ of some insects containing symbiotic microorganisms, an aggregation of mycetocytes.

Mycobactin. An agent, derived from some species of *Mycobacterium*, required for or stimulating *in vitro* growth of *M. paratuberculosis* and some other species; an iron-chelating growth factor.

Mycolic Acid. Long chain, branched fatty acids of high molecular weight, characteristic of species of *Mycobacterium;* similar acids of lower molecular weight occur in species of *Nocardia* and *Corynebacterium.*

Myxospore. The resting cell usually contained within the fruiting body of a myxobacter and substantially more resistant than a vegetative cell.

Necrohormones. A substance responsible for autoantagonism of Actinomycetales.

Negative Phase. When *Borrelia* is taken up by an arthropod, it disappears from the stomach after a short time, and if found again in the celomic cavity after a few days. The interim is called the "negative phase." It may be due to the *Borrelia* becoming small and slender while penetrating the gut wall and their number so small that they are missed on microscopic examination. The arthropod is not considered infective during that period.

Neotype. A strain proposed in accordance with the Bacteriological Code to replace a type strain no longer extant or a type description of an organism which was not cultivated or grown in pure culture or was represented by a nonliving specimen. The present Code accepts a neotype as soon as published in the *International Journal of Systematic Bacteriology*. Some authors have included suggested neotypes—these had not been published in *International Journal of Systematic Bacteriology* when THE MANUAL went to press and until they are properly proposed have no nomenclatural significance and are synonymous with reference strain.

Nomen Dubium. A name representing a taxon that is not identifiable from the original diagnosis or from the type material.

Nomen Nudum. A name published without any definition, description or indication of the taxon to which it is intended to apply, and hence without status in nomenclature.

Nonphotochromogen. Not requiring light for pigment formation; usually applied to mycobacteria having (1) little or no pigmentation or (2) color developing late in colony development.

0/129. A vibriostatic compound, 2,4-diamino-6,7-diisopropyl pteridine.

Optical Antipode. The D and L isomers of an optically active organic compound.

Outer Envelope. A thin flexible structure surrounding the protoplasmic cylinder and axial fibrils of spirochetes.

Palisade. Rods lined-up side by side as pales in a picket fence.

Parenchyma. Fundamental or ground tissue of a plant; a tissue composed of living cells variable in their morphology and physiology, but generally having thin walls and a polyhedral shape.

Pedicel. A small stalk; in the Myxobacterales one of the ultimate divisions of a common stalk, bearing a single sporangium.

Pentaene (pentaenic) (antibiotics). With five conjugated double bonds. Also see Polyene (antibiotics).

Periplast. From Greek *plassein*, to form. Peripheral protoplasm of a cell; matrix of a part or organ; cell wall. (*Taber's Cyclopedic Medical Dictionary*, 12th edition). The use of this term by some authors originated probably at the time when *Borrelia* was classified with protozoa.

Phalangiform (of spores). Spores which, when joined in chains, give the appearance of phalanges (fingers).

Photochromogen. Organism (mycobacterium) which exhibits pigmentation (carotenoid) only if exposed to light during growth.

Photolithotrophic. Able to grow with light in a strictly inorganic medium.

Photoorganotrophic. Able to grow with light at the expense of organic compounds.

Pileus (plural pilei). The "cap" of a mushroom.

Plasmalemma. Synonym of cell (cytoplasmic) membrane (botanical usage).

Plasmalemmosome. Infolding of the cell membrane (see Mesosome).

Polyene (polyenic) (antibiotics). Characterized by having conjugated carbon to carbon double bonds, most of which give characteristic ultraviolet spectra.

PPD. A tuberculin ("purified protein derivative") derived from autoclaved culture filtrate by ammonium sulfate precipitation or comparable product from culture filtrates of mycobacteria other than tubercle bacilli.

Prostheca. A semirigid appendage, extending from a procaryotic cell, the diameter of which is always smaller than that of the mature cell and which is bounded by the cell wall. Includes the cellular stalks of caulobacters and the hyphae of hyphomicrobia but not the secreted stalks of *Gallionella* or *Nevskia*.

Protoplasmic Cylinder. The central structure of the spirochete which includes the nuclear apparatus, ribosomes and other intracellular organelles surrounded by a three layer unit membrane.

Pseudoplasmodium. In the Myxobacterales, a swarm or vegetative colony resembling the plasmodium of a slime mold (myxomycete) but composed of bacteria and slime. Cells not amoeboid.

Pseudostalk. In *Asticcacaulis*, an appendage similar in dimensions, fine structure and development to the stalk in *Caulobacter*, but which lacks adhesive material.

Purine (antibiotics). Antibiotics containing a purine, a nitrogen heterocyclic ring system.

Pyriform. Pearlike in form.

Rapid Grower (mycobacterium). One which exhibits its complete growth to maturity within a week at 37 C and comparable growth at 25 C.

Rectus Flexibilis. A term to describe streptomycetes

having spore chains that are straight, flexuous or fascicled (see Plate 17.12).

Reference Strain. A strain which is used for reference in the production of a specific compound, antigen, or one which appears to meet the description of the type but has not been proposed as a neotype.

Retinaculum Apertum. A term to describe streptomycetes having spore chains that are hooks, open loops or greatly extended spirals (see Plate 17.12).

Rhizothamnion (plural **rhizothamnia**). Root nodules infected by two species of the genus *Frankia* become covered with negatively geotrophic rootlets growing from the apical meristem of each lobe of the nodule. A nodule covered with upward growing rootlets may be called a rhizothamnion.

Sclerotium (plural **sclerotia**). A hard resting body resistant to unfavorable conditions composed of a hardened mass of hyphae.

Scotochromogen. A mycobacterium exhibiting pigmentation in presence or absence of light.

Sedum-like. *Sedum*, a genus of fleshy green plants, a common example is stonecrop.

Segmentation Spores. Spores produced by the septation or successive contraction of the protoplasts of the tips of aerial filaments.

Sensitin. "... A non-antigenic substance, prepared from a microorganism (virus, bacterium or fungus), capable of revealing sensitivity of the delayed type evoked by the organism. Human and avian tuberculin, histoplasmin, and coccidioidin are examples of the 'sensitin'" (Magnusson, 1961, p. 57).

Serophilic. A bacterium that grows better when serum is added to the medium.

Sessile. Of cells, attached directly to the substrate or a surface without a stalk; of spores, budded directly from a hypha without an intervening sporophore.

Sinus, Apical. Crater-like concavity on top of colony of *Zoogloea ramigera*.

Sorption. The removal of antibodies from an antiserum by the addition of the antigen and then removal of the antigen-antibody complex to obtain a sorbed antiserum. Applicable to either adsorption or absorption.

Sorus (plural **sori**). In the Myxobacterales, a mass of sporangia enclosed in a common slime layer or envelope.

Spheroplast. In a general sense, used to designate a bacterial cell which has rounded up and become osmotically sensitive as a result of the partial or complete destruction of the peptidoglycan layer of the wall. In the genus *Actinomyces* it refers to the coccoid cells which occur in cultures of these normally filamentous organisms.

Spicule. A straight, pointed, rigid terminal extension of spirochetes; found on some cristispiras from fresh water bivalve mollusks.

Spider Colony. A colony composed almost entirely of radiating filaments which seem to originate from a central point.

Spira. Term to describe streptomycetes having spore chains that are either short, gnarled or compact spirals, or extended long and open spirals (see Plate 17.12).

Spirochetal Sphere. A form of spirochete in which the protoplasmic cylinder is either partially or completely coiled within a saclike outer envelope.

Sporangiole. A small sporangium generally having a small number of spores.

Sporangiospore. An asexual spore (but not an endospore) formed within a sporangium or spore vesicle, typically in the Actinoplanaceae.

Sporangium. A spore case or envelope, the sac in which spores are produced. (1) Spore vesicle containing one to many motile or nonmotile asexual spores (sporangiospores) in Actinoplanaceae, Actinomycetaceae. Differs from the sporangium formed in fungi, e.g. Mucorales; in the actinomycetes, hyphae grow inside the spore vesicle and eventually divide by annular ingrowth of the hyphal wall to give the individual sporangiospores; (2) the organ or part of the bacterium specially fashioned for the development of its spore (endospore) in *Bacillus*, *Thermoactinomyces* and *Actinobifida;* (3) in the Myxobacterales, a saclike structure delimited by a discrete slime layer or membrane and containing the myxospores.

Spore. A general term used for the reproductive or resting body produced in fungi and bacteria. See aleuriospore, arthrospore, chlamydospore, conidia, condiospore, endospore, microcyst, sporangiospore, zoospore.

Spore Vesicle. Term proposed by Cross (1970) for the envelope within which spores are produced by several of the Streptomycetaceae and which previously has been loosely referred to as a "sporangium."

Sporogenic. Producing spores.

Sporophore. Any structure which bears spores or spore chains.

Stalk. A filiform structure, either cellular or secreted, which anchors an organism to the substrate or a surface.

Streptotrichosis. A skin disease of sheep, cattle, horses and goats, especially in tropical Africa (also spelled streptothrichosis).

Subspecies. (variety). A subdivision of the species in which the name is a trinomial consisting of a generic name followed by a specific epithet and this, in turn, by a subspecific epithet. The epithets are preferably separated by the abbreviation "subsp.".

Substrate Mycelium. The portion of the mycelium of an actinomycete in contact with the substrate.

"Sulfur Granule." A small yellowish to white granule, 1 to 2 mm in diameter, usually present in pus from human or animal actinomycosis. Microscopically composed of branching filaments of *Actinomyces*; the ends of some of the filaments extend from the periphery and are encased in a protein-polysaccharide complex. The ashed granule contains CaO, P_2O_5 and acid phosphate probably occurring as poorly crystallized apatite.

Swarm (swarming). The spread of motile organisms over the surface of solid media due to the movement of individuals, groups or colonies; characteristic of the vegetative cells of myxobacteria and of a few flagellated bacteria, e.g. *Proteus*, *Bacillus*, *Clostridium*.

Sympodial. A dichotomy where at each forking one branch continues to develop and the other aborts (Jackson, 1928).

Synergistic (of antibiotics). The combined effect of two antibiotics when the total activity is greater than the sum of the two effects taken independently.

Synnema. A group of parallel hyphae joined together forming an elongated erect spore-bearing structure.

Tetraene (tetraenic) (antibiotics). With four conjugated double bonds. Also see Polyene (antibiotics).

Thallus. As used in *Dermatophilaceae*, a general term for the vegetative or assimilative phase of a non-vascular plant or organism.

Thermocline. The layer in a thermally stratified body of water within which the temperature decreases rapidly with increasing depth usually at a rate greater than 1 C per meter.

Thylakoids. See intracytoplasmic membrane systems.

Trichome. The meaning of this term in bacteriology seems to vary. It comes from the Greek n. *trichoma* a growth of hair or hair generally. In botany it usually refers to any hairlike outgrowth of the epidermis such as a hair or bristle. In the blue-green algae a trichome refers to a row of cells, i.e., a filament apart from its investing sheath. In the sheathed bacteria, trichome refers to a hairlike organism in which the septa are not necessarily a part of the fission mechanism. The septa may not be completely closed, allowing transfer of protoplasm from one compartment to the next. The term has also been used for cells, not necessarily filamentous or hairlike, in which at one time septation is found in various stages of completion. Because of this confusion the term is not used in this edition of THE MANUAL

Tuberculin. Derivative of tubercle bacilli or their culture filtrate capable of eliciting skin reactions resulting from sensitization by tubercle bacilli.

Tweens. A group of non-ionic surface active agents. A trade name of Atlas Powder Co. Tween 20—polyoxyethylene sorbitan monolaurate; Tween 80—polyoxyethylene sorbitan monooleate.

Twitching. A movement on solid surfaces of both flagellated and nonflagellated bacteria but not due to action of flagella. Cell movement intermittent and jerky, does not follow the long axis of cell and is predominately of single cells.

Tycholimnetic. Fresh water forms which are at first fixed, but later break loose and float.

Umbel. The arrangement of three or more spore chains borne at the apex of each branch of a verticil in *Streptoverticillium*.

Umbellate (of spore chains). Branched spore chains in which the branches arise at about the same point from the spore-bearing hypha..

Vegetative Mycelium. See substrate mycelium.

Verticil. The three or more side branches which arise in a whorl at the same level and at more or less regular intervals along the main aerial hyphae in *Streptoverticillium*. Each side branch may subdivide to produce umbels (q.v.).

Verticillate (or spore chains). Whorled chains of spores arising radiately at nodes on a spore-bearing hypha like the spokes of a wheel.

V Factor. An accessory growth substance required by certain species of *Haemophilus* replaceable by niacinamide adenine dinucleotide or one of its riboside precursors. It is destroyed by autoclaving and deteriorates when stored, and especially rapidly above freezing temperatures.

Whole Cell Hydrolyzates (of actinomycetes). The residue from the acid hydrolysis of whole cells with 6 N HCl at 100 C for 18 hrs.

Xenodiagnosis. Diagnosis based on feeding of uninfected vectors on a patient suspected of harboring an organism and examining the vector for development in it of the organism in question.

X Factor. An accessory growth substance required by certain species of *Haemophilus* replaceable by hemin or other iron porphyrin compounds and stable to autoclaving.

Zoogloeal Floc. Growth form of *Zoogloea ramigera* in mature cultures in high C:N media, sewage or natural waters.

Zoospore. A spore motile by means of flagella. The term is used more appropriately in botany and zoology.

INDEX OF SCIENTIFIC NAMES OF BACTERIA

This is an abbreviated index of those names mentioned in this book and all page numbers cited refer to the *Shorter Bergey's Manual of Determinative Bacteriology*. Page numbers given for taxa of genus or above refer to the page on which the description begins. Page numbers given for species or subspecies refer to the page on which the taxon is mentioned in a table, key or generic description. Names of serotypes of *Salmonella* appear only under the generic name. A more complete index of scientific names is found in the *Manual* beginning on page 1147.